Lectures on Real-valued Functions

Alexander Kharazishvili

Lectures on Real-valued Functions

Alexander Kharazishvili
Andrea Razmadze Mathematical Institute
Tbilisi State University
Tbilisi, Georgia

ISBN 978-3-031-95368-2   ISBN 978-3-031-95369-9 (eBook)
https://doi.org/10.1007/978-3-031-95369-9

© The Editor(s) (if applicable) and The Author(s), under exclusive license to Springer Nature Switzerland AG 2025

This work is subject to copyright. All rights are solely and exclusively licensed by the Publisher, whether the whole or part of the material is concerned, specifically the rights of translation, reprinting, reuse of illustrations, recitation, broadcasting, reproduction on microfilms or in any other physical way, and transmission or information storage and retrieval, electronic adaptation, computer software, or by similar or dissimilar methodology now known or hereafter developed.
The use of general descriptive names, registered names, trademarks, service marks, etc. in this publication does not imply, even in the absence of a specific statement, that such names are exempt from the relevant protective laws and regulations and therefore free for general use.
The publisher, the authors and the editors are safe to assume that the advice and information in this book are believed to be true and accurate at the date of publication. Neither the publisher nor the authors or the editors give a warranty, expressed or implied, with respect to the material contained herein or for any errors or omissions that may have been made. The publisher remains neutral with regard to jurisdictional claims in published maps and institutional affiliations.

This Springer imprint is published by the registered company Springer Nature Switzerland AG
The registered company address is: Gewerbestrasse 11, 6330 Cham, Switzerland

If disposing of this product, please recycle the paper.

Preface

Without any doubt, one of the main objects in mathematics is the real line $\mathbb{R}$ endowed with its natural structures: Dedekind complete dense linear order, Euclidean topology, canonical algebraic binary operations (i.e., addition and multiplication of real numbers).

By definition, a real-valued function is any mapping f acting from an arbitrary nonempty set X into $\mathbb{R}$ or, in other words, is any element f of the product set $\mathbb{R}^X$ (such f is also called a functional on X).

So, one can associate to a set X the set $\mathbb{R}^X$ of all those real-valued functions whose domains coincide with X. The standard structures of $\mathbb{R}$ produce corresponding structures on $\mathbb{R}^X$. In this way, $\mathbb{R}^X$ becomes a partially ordered vector space over $\mathbb{R}$, a commutative ring with unit, a conditionally complete lattice, and so on.

Let Y be another nonempty set (not necessarily distinct from X) and let $\phi : X \to Y$ be a mapping from X into Y. Then we immediately come to the induced mapping

$$u_\phi : \mathbb{R}^Y \to \mathbb{R}^X$$

defined by the formula

$$u_\phi(f) = f \circ \phi \quad (f \in \mathbb{R}^Y).$$

It is not difficult to check that u_ϕ is a homomorphism of the ring $\mathbb{R}^Y$ to the ring $\mathbb{R}^X$.

If $Y = X$ and ϕ is the identity mapping of X onto itself, then u_ϕ is the identity automorphism of the ring $\mathbb{R}^X$.

Also, for any three nonempty sets X, Y, Z and for any two mappings

$$\phi : X \to Y, \quad \psi : Y \to Z,$$

one readily obtains $u_{\psi \circ \phi} = u_\phi \circ u_\psi$. This circumstance indicates that, actually, we have a contravariant functor of the form

$$X \to \mathbb{R}^X \quad (X \text{ is a nonempty set}),$$

$$\phi \to u_\phi \quad (\phi : X \to Y,\ u_\phi : \mathbb{R}^Y \to \mathbb{R}^X)$$

acting from the category of nonempty sets to the category of commutative rings.

Notice that the analogous construction works when instead of the category of nonempty sets, denoted usually by **Set**$_0$, we are given a certain subcategory of **Set**$_0$.

In particular, we can consider the subcategory of **Set**$_0$ consisting of all nonempty compact topological spaces and of all continuous mappings between them. In this case, it is natural to associate with a nonempty compact space X the Banach space $C(X) \subset \mathbb{R}^X$ of all real-valued continuous functions on X (here we mean that $C(X)$ is equipped with the standard sup-norm).

Similarly, we may consider the subcategory of **Set**$_0$ consisting of all measurable spaces and of all measurable mappings between such spaces. In this case, it is natural to associate with a measurable space X the family $B(X, \mathbb{R}) \subset \mathbb{R}^X$ of all Borel mappings acting from X into $\mathbb{R}$.

Also, we can consider the subcategory of **Set**$_0$ consisting of all nonempty partially ordered sets and of all increasing mappings between such sets. In this case it is natural to associate with a nonempty partially ordered set $(X, \leq)$ the family $\text{Incr}(X, \mathbb{R}) \subset \mathbb{R}^X$ of all increasing mappings from X into $\mathbb{R}$.

Obviously, many other analogous examples for various structures might be given in this context.

From the above examples one can see that if a nonempty ground set X carries a certain structure, then a corresponding structure emerges on the family $\mathbb{R}^X$ (or on an appropriate subfamily of $\mathbb{R}^X$).

It makes sense to present a short list of important real-valued functions that are often considered in standard courses on mathematical analysis.

(1) The family of all real-valued semicontinuous functions on a topological space X
(2) The family of all real-valued monotone functions on a partially ordered set X
(3) The family of all real-valued convex functions on a nonempty convex subset X of a real vector space
(4) The family of all real-valued continuous linear functionals on a topological vector space X
(5) The family of all real-valued positive linear functionals on a partially ordered vector space X
(6) The family of all real-valued measures (finitely additive or countably additive) on a measurable space X
(7) The family of all real-valued Lebesgue integrable functions on a σ-finite measure space X
(8) The family of all those real-valued functions on a non-degenerate subsegment $[a, b]$ of $\mathbb{R}$ which have finite variation
(9) The family of all those real-valued functions on a non-degenerate subsegment $[a, b]$ of $\mathbb{R}$ which are absolutely continuous

(10) The family of all those real-valued functions on a non-degenerate subsegment $[a, b]$ of $\mathbb{R}$ which are derivatives (in the classical sense)

Of course, the above list of important classes of real-valued functions can be substantially expanded.

In this concise course of lectures we intended to cover several topics of mathematical analysis which are closely connected with significant properties of real-valued functions of various types.

An essential part of the course was presented by the author at Batumi State University (Georgia), during a semester of the academic year 2018–2019. It comprises 33 lectures and 5 appendixes.

The lectures are primarily oriented to university mathematics students. Alongside fairly standard themes of real analysis (such as semicontinuous functions, monotone functions, convex functions, restrictions and extensions of real-valued functions, measurable functions, Lebesgue measure and integration theory, etc.), less traditional questions are also considered in the course (for instance, discontinuous functions on resolvable topological spaces, Luzin sets on the real line $\mathbb{R}$, pointwise limits of finite sums of periodic functions, some general results about invariant and quasi-invariant measures, nonmeasurable sets and functions, the Baire property of functions, uncountable products of probability measures, etc.).

As is widely known, modern approaches to numerous questions of real analysis and measure theory systematically use the basic concepts and methods of contemporary set theory. For instance, at present the Axiom of Choice (**AC**) and its logical equivalents (e.g., the Kuratowski–Zorn Lemma), the principle of transfinite induction (transfinite recursion), and some additional axioms such as the Continuum Hypothesis (**CH**) and Martin's Axiom (**MA**) turn out to be necessary tools for solving many profound problems connected with the structure of point sets on the real line $\mathbb{R}$ and with the behavior of real-valued functions of different types. Because of this, in certain sections of this course set-theoretical aspects of the studied topics are emphasized and discussed.

In general, we have tried to present the material in a self-contained form accessible to graduate and postgraduate students of average level. If a potential reader is more or less familiar with elements of set theory (e.g., the system of axioms of Zermelo and Fraenkel, the method of transfinite induction, cardinal and ordinal numbers), with basic notions of general topology (compact spaces, completely regular spaces, normal spaces) and with elements of measure theory (algebras of sets, finitely additive measures, countably additive measures, the Lebesgue measure on $\mathbb{R}$), then they should not encounter significant difficulties in the process of reading this book.

Also, we would like to stress that the chapters (each playing the role of a separate lecture) and five appendixes of this manuscript all end with a number of exercises. These exercises provide potential readers with additional information concerning questions and themes touched upon in the book. Some are almost trivial or quite easy, while others are closely related to deep mathematical facts and need substantial effort to solve. Notice that many exercises are accompanied by hints.

The list of references presented at the end of the book is far from being complete and primarily serves to stimulate the reader's interest in various questions of classical real analysis and measure theory.

Acknowledgment The author is very grateful to Dr. Habiba Kalantarova for her substantial help in the process of preparing the final version of this manuscript.

Tbilisi, Georgia   Alexander Kharazishvili

Contents

1	Unary and Binary Relations	1
2	Partial Functions and Functions	11
3	Elementary Facts on Cardinal Numbers	23
4	Some Properties of the Continuum	37
5	The Oscillation of a Real-Valued Function at a Point	47
6	Points of Continuity and Discontinuity of Real-Valued Functions	57
7	Real-Valued Monotone Functions	65
8	Real-Valued Convex Functions	77
9	Semicontinuity of a Real-Valued Function at a Point	87
10	Semicontinuous Real-Valued Functions on Quasi-Compact Spaces	95
11	The Banach–Steinhaus Theorem	105
12	A Characterization of Oscillation Functions	113
13	Semicontinuity Versus Continuity	121
14	Outer Measures	131
15	Finitely Additive and Countably Additive Measures	141
16	Extensions of Measures	153
17	Caratheodory's and Marczewski's Extension Theorems	163
18	Positive Linear Functionals	175
19	The Nonexistence of Universal Countably Additive Measures	187
20	Radon Measures	199
21	Invariant and Quasi-Invariant Measures	209

22	Pointwise Limits of Finite Sums of Periodic Functions	223
23	Absolutely Nonmeasurable Sets in Commutative Groups	235
24	Radon Spaces	249
25	Nonmeasurable Sets with Respect to Radon Measures	261
26	The Radon–Nikodym Theorem	271
27	Decompositions of Linear Functionals	281
28	Linear Continuous Functionals and Radon Measures	289
29	Linear Continuous Functionals on a Real Hilbert Space	307
30	The Baire Property in Topological Spaces	321
31	The Stone–Weierstrass Theorem	333
32	More on the Function Space $C(X)$	345
33	Uniformization of Plane Sets by Relatively Measurable Functions	357
A	Lebesgue Integration of Real-Valued Functions	375
B	Product Measures	389
C	Comparing the Riemann and Lebesgue Integrals	405
D	The Lax–Milgram Theorem	417
E	Resolvable Topological Spaces	429
Bibliography		439
Index		447

Chapter 1
Unary and Binary Relations

We begin this lecture course by recalling several general concepts about sets, relations, and functions. At the same time, we assume that the reader is more or less familiar with elementary logic and basic notions of "naive" set theory, such as: a subset of a set, a family of sets, the union of sets, the intersection of sets, the difference of two sets, the symmetric difference of two sets, etc. (see [64]).

Recall that a successful formalization of "naive" set theory was carried out by E. Zermelo and A. Fraenkel and finally was realized in their symbolic logical system **ZF**. The abbreviation **ZFC** is commonly used for the same system enriched by the Axiom of Choice (**AC**) (for more details, see e.g. [76, 108, 111]).

Let E be a *base* (*ground*) *set*. This phrase just means that we will primarily be concerned with E, its elements, its subsets, and other objects associated with E (cf. [35, 111]).

Consider a *unary relation* $S(x)$ for elements x from E. In other words, $S(x)$ is a certain property of an arbitrary (variable) element x of E. Naturally, for some $y \in E$ this property may hold and for some other $z \in E$ this property may be false. Thus, by definition, the set

$$X_S = \{x \in E : S(x)\}$$

consists of all those elements from E for which $S(x)$ holds true.

The existence of the above-mentioned set X_S is a consequence of one of the axioms of contemporary Zermelo–Fraenkel set theory **ZF** (the so-called *axiom of separation*; see [76, 108, 111]). Conversely, every subset X of E can be treated as a unary relation on E, or as a certain property $S(x)$ of a variable element x of E. Namely, we may put

$$X = X_S = \{x \in E : S(x)\},$$

where $S(x)$ is the standard membership relation $x \in X$.

It is easy to see that if $R(x)$ and $S(x)$ are two logically equivalent relations over E, then $X_R = X_S$. In other words, the formula

$$(\forall x \in E)(R(x) \Leftrightarrow S(x))$$

implies the equality $X_R = X_S$.

Conversely, $X_R = X_S$ implies $(\forall x \in E)(R(x) \Leftrightarrow S(x))$.

Example 1.1 Obviously, the equalities

$$\emptyset = \{x \in E : x \neq x\}, \qquad E = \{x \in E : x = x\}$$

hold, where the symbol $\emptyset$ denotes the (unique) *empty set* (i.e., $\emptyset$ does not have any elements).

These are, respectively, the *empty* unary relation on E and the *total* unary relation on E.

If $R(x)$ and $S(x)$ are any two unary relations on E and $R(x) \Rightarrow S(x)$ for all $x \in E$, then $X_R \subset X_S$. Conversely, if $X_R \subset X_S$, then we have $R(x) \Rightarrow S(x)$ for all $x \in E$.

Example 1.2 Put $E = \mathbb{R}$, where $\mathbb{R}$ is the real line, and let $S(x)$ denote the relation $0 < x$ (here $<$ is the standard strict linear ordering of $\mathbb{R}$). Then the set X_S coincides with the set of all strictly positive real numbers.

Example 1.3 According to the equivalence stated before Example 1.1, the set of all unary relations on E is the set of all subsets of E (or the *power set* of E, or the *Boolean* of E) which is usually denoted by the symbol $\mathcal{P}(E)$. The existence of $\mathcal{P}(E)$ is precisely one of the axioms of **ZF** set theory (see, for instance, [76, 108, 111]). All subsets of E distinct from $\emptyset$ and E are called *proper subsets* of E. Actually, $\mathcal{P}(E)$ is a canonical representative of general Boolean algebras. Moreover, by virtue of M. Stone's fundamental theorem, every Boolean algebra is isomorphic to some subalgebra of $\mathcal{P}(E)$ for a suitable ground set E, where $\mathcal{P}(E)$ is equipped with the standard set-theoretical operations $\cap$, $\cup$, $\setminus$ (see, e.g., [19, 84, 111]).

It is not hard to prove in **ZF** theory that, for any two sets x and y, there exists a set $\{x, y\}$ whose elements are precisely x and y. In particular, if $x = y$, then $\{x, y\} = \{x\}$ is a one-element set or a *singleton*. Now, let us introduce the notation

$$(x, y) = \{\{x\}, \{x, y\}\}.$$

The object (x, y) is called the *ordered pair* of x and y (or, in short, the pair of x and y). This definition of an ordered pair was given by K. Kuratowski within the framework of naive set theory. The main property of ordered pairs is the following implication:

$$(x, y) = (x', y') \Rightarrow (x = x' \,\&\, y = y').$$

1 Unary and Binary Relations

The converse implication is trivial. If we have the pair $z = (x, y)$, then x and y are uniquely determined by z and we write $x = \mathrm{pr}_1(z)$ and $y = \mathrm{pr}_2(z)$.

Now, let E and F be any two ground sets. From the axioms of **ZF** theory it follows that there exists a unique set $E \times F$ which consists of all pairs z such that $\mathrm{pr}_1(z) \in E$ and $\mathrm{pr}_2(z) \in F$. The set $E \times F$ is called the *Cartesian product* of E and F. Thus, we have by definition

$$(x, y) \in E \times F \Leftrightarrow (x \in E \ \& \ y \in F).$$

Any unary relation $S(z)$ on $E \times F$ is called a *binary relation* between elements of E and F or, simply, a binary relation on $E \times F$. Therefore, according to the definition of unary relations, $S(z)$ can be identified with a certain subset Z of $E \times F$. Evidently, we may introduce a relation $R(x, y)$ of two variables $x \in E$ and $y \in F$ by the formula

$$R(x, y) \Leftrightarrow S((x, y))$$

and, consequently, we come to the logical equivalence

$$S(z) \Leftrightarrow R(\mathrm{pr}_1(z), \mathrm{pr}_2(z)).$$

We thus see that the notion of a binary relation is reducible to the notion of a unary relation. But in mathematical practice such a reduction is not always convenient. Moreover, binary relations have certain specific features which do not occur in the case of unary relations.

In our further considerations we will identify $S(z)$ and $R(x, y)$ with the corresponding subset of $E \times F$. Sometimes, we will omit the variables z, x, y in $S(z)$ and $R(x, y)$. So we will write S and R instead of $S(z)$ and $R(x, y)$.

In the special situation when $E = F$, we have $E \times F = E^2$ and it is usually said that $S \subset E^2$ is a binary relation on E.

Example 1.4 Every ground set E is endowed with three canonical binary relations: $\emptyset$ (the *empty* binary relation), E^2 (the *total* binary relation), and $\triangle_E = \{(x, x) : x \in E\}$ (the *diagonal* in E^2). If a binary relation R on E contains $\triangle_E$, then R is called a *reflexive* binary relation. In other words, R is reflexive if $R(x, x)$ holds true for all $x \in E$.

Example 1.5 For any binary relation $R \subset E \times F$, we have its *converse* (*reverse*) binary relation $R^{-1} = \{(y, x) \in F \times E : (x, y) \in R\}$ on the product set $F \times E$. Evidently, $(R^{-1})^{-1} = R$.

Now, we introduce several standard terms associated with an arbitrary binary relation $R \subset E \times F$.

The *first projection* of R or the *domain* of R is defined by

$$\mathrm{pr}_1(R) = \mathrm{dom}(R) = \{x : (\exists y) R(x, y)\}.$$

The *second projection* of R or the *range* of R is defined by

$$\operatorname{pr}_2(R) = \operatorname{ran}(R) = \{y : (\exists x) R(x, y)\}.$$

The *image* $R(X)$ of a set X, with respect to R, is defined by

$$R(X) = \{y : (\exists x \in X) R(x, y)\}.$$

It is easy to check that $R(X) = \operatorname{pr}_2((X \times F) \cap R)$.
The *pre-image* $R^{-1}(Y)$ of a set Y, with respect to R, is defined by

$$R^{-1}(Y) = \{x : (\exists y \in Y) R(x, y)\}.$$

It is also easy to check that $R^{-1}(Y) = \operatorname{pr}_1((E \times Y) \cap R)$.
In general, we may consider binary relations independently of base sets E and F. Indeed, we always have the inclusion

$$R \subset \operatorname{pr}_1(R) \times \operatorname{pr}_2(R),$$

so the sets $\operatorname{pr}_1(R)$ and $\operatorname{pr}_2(R)$ can play the role of E and F respectively. Briefly speaking, we may define a binary relation as any set of pairs.

Let R and S be two binary relations. The *composition* $R \circ S$ of these relations is introduced as follows:

$$(x, y) \in R \circ S \Leftrightarrow (\exists z)((x, z) \in S \ \& \ (z, y) \in R).$$

The composition operation $\circ$ is thus defined for any two binary relations R, S and yields the uniquely determined binary relation $R \circ S$.

Theorem 1.1 *The operation $\circ$ is associative, i.e.,*

$$R \circ (S \circ T) = (R \circ S) \circ T$$

for any three binary relations R, S, and T.

Proof The membership $(x, y) \in R \circ (S \circ T)$ is equivalent to

$$(\exists z)((x, z) \in S \circ T \ \& \ (z, y) \in R)$$

which is equivalent to

$$(\exists z)((\exists t)((x, t) \in T \ \& \ (t, z) \in S) \ \& \ (z, y) \in R).$$

Since $(z, y) \in R$ does not contain a variable t, the formula $(x, y) \in R \circ (S \circ T)$ is equivalent to

1 Unary and Binary Relations

$$(\exists z)(\exists t)((x, t) \in T \ \& \ (t, z) \in S \ \& \ (z, y) \in R).$$

Further, the membership $(x, y) \in (R \circ S) \circ T$ is equivalent to

$$(\exists t)((x, t) \in T \ \& \ (t, y) \in R \circ S)$$

which is equivalent to

$$(\exists t)((x, t) \in T \ \& \ (\exists z)((t, z) \in S \ \& \ (z, y) \in R)).$$

Since $(x, t) \in T$ does not contain a variable z, the formula $(x, y) \in (R \circ S) \circ T$ is equivalent to

$$(\exists t)(\exists z)((x, t) \in T \ \& \ (t, z) \in S \ \& \ (z, y) \in R).$$

We thus can conclude that

$$(x, y) \in R \circ (S \circ T) \Leftrightarrow (x, y) \in (R \circ S) \circ T,$$

which yields the required equality $R \circ (S \circ T) = (R \circ S) \circ T$. $\square$

Example 1.6 If R is any binary relation on a ground set E, then we have

$$\emptyset \circ R = R \circ \emptyset = \emptyset, \qquad R \circ \Delta_E = \Delta_E \circ R = R.$$

The above equalities show us that $\emptyset$ plays the role of zero and Δ_E plays the role of the neutral element in the set of all binary relations on E.

Theorem 1.2 *If R is a binary relation and $\{X_i : i \in I\}$ is an arbitrary family of sets, then the following formulas hold:*

(1) $R(\cup\{X_i : i \in I\}) = \cup\{R(X_i) : i \in I\}$;
(2) $R(\cap\{X_i : i \in I\}) \subset \cap\{R(X_i) : i \in I\}$;
(3) $R^{-1}(\cup\{X_i : i \in I\}) = \cup\{R^{-1}(X_i) : i \in I\}$;
(4) $R^{-1}(\cap\{X_i : i \in I\}) \subset \cap\{R^{-1}(X_i) : i \in I\}$.

Also, if X and Y are any two sets such that $X \subset Y$, then

(5) $R(X) \subset R(Y)$;
(6) $R^{-1}(X) \subset R^{-1}(Y)$.

Proof We will only prove formula (1), because the argument is similar in all other cases (notice by the way that (5) follows from (1) and (6) follows from (3)).

The membership relation $y \in R(\cup\{X_i : i \in I\})$ is equivalent to

$$(\exists x \in \cup\{X_i : i \in I\})((x, y) \in R)$$

which, in its turn, is equivalent to

$$(\exists i \in I)(\exists x)(x \in X_i \ \& \ (x, y) \in R).$$

The latter is equivalent to $y \in \cup\{R(X_i) : i \in I\}$ and, therefore, implies the validity of (1). □

Using an analogous argument, one can obtain the next theorem.

Theorem 1.3 *If R and R_i ($i \in I$) are binary relations, then the following formulas hold:*

(1) $R \circ (\cup\{R_i : i \in I\}) = \cup\{R \circ R_i : i \in I\}$;
(2) $(\cup\{R_i : i \in I\}) \circ R = \cup\{R_i \circ R : i \in I\}$;
(3) $R \circ (\cap\{R_i : i \in I\}) \subset \cap\{R \circ R_i : i \in I\}$;
(4) $(\cap\{R_i : i \in I\}) \circ R \subset \cap\{R_i \circ R : i \in I\}$.

We omit the proof of Theorem 1.3 and leave it to the reader.

Let E and F be two ground sets and let $R(x, y)$ be a binary relation on $E \times F$. For any $x_0 \in E$, we have the unary relation $R(x_0, y)$ on F, and for any $y_0 \in F$ we have the unary relation $R(x, y_0)$ on E.

These two induced unary relations are called the *sections* of $R(x, y)$ corresponding to x_0 and to y_0 respectively.

We may also say that $R(x_0, y)$ is the x_0-*vertical section* of $R(x, y)$ and $R(x, y_0)$ is the y_0-*horizontal section* of $R(x, y)$.

Let now E be a base set and let $\{S_i : i \in I\}$ be a family of unary relations on E. A binary relation $R(x, y)$ on E is called *universal* for this family if, for each index $i \in I$, there exists an element $y = y_i$ of E such that

$$(\forall x \in E)(S_i(x) \Leftrightarrow R(x, y_i)).$$

The next classical result is due to Cantor and vividly demonstrates his celebrated diagonal argument (cf. [18, 111, 167]).

Theorem 1.4 *There exists no binary relation on E universal for the family of all unary relations on E.*

Proof Suppose otherwise, i.e., there exists a binary relation $R(x, y)$ on E such that the family $\{R(x, y) : y \in E\}$ consists of all unary relations on E. Now, consider the unary relation $\neg R(x, x)$ on E. By virtue of our assumption, there is a $y_0 \in E$ such that

$$\neg R(x, x) \Leftrightarrow R(x, y_0) \qquad (x \in E).$$

Putting in this formula $x = y_0$, we come to a contradiction

$$\neg R(y_0, y_0) \Leftrightarrow R(y_0, y_0),$$

which ends the proof of Cantor's theorem. □

1 Unary and Binary Relations

As a corollary of Theorem 1.4, one can deduce Cantor's famous inequality

$$\text{card}(\mathcal{P}(E)) > \text{card}(E)$$

between the cardinalities of $\mathcal{P}(E)$ and E (we will be dealing with this inequality in the sequel).

Cantor's diagonal method plays an important role in many branches of mathematics, especially in mathematical analysis, descriptive set theory, mathematical logic, and general topology.

Exercises

1. Show the validity of the implication

$$(x, y) = (x', y') \Rightarrow (x = x' \,\&\, y = y')$$

   which reflects the main property of all ordered pairs.
   Infer from the above implication that if z is an ordered pair, i.e., if

$$(\exists x)(\exists y)(z = (x, y)),$$

   then the terms $\text{pr}_1(z)$ and $\text{pr}_2(z)$ are uniquely determined by z.

2. For a binary relation $R \subset E \times F$ and for any two sets X and Y, verify that the formulas

$$R(X) = \text{pr}_2((X \times F) \cap R), \qquad R^{-1}(Y) = \text{pr}_1((E \times Y) \cap R)$$

   hold true.

3. Give an example of a binary relation R and of two sets X and Y such that the formula $R(X \cap Y) = R(X) \cap R(Y)$ is false.

4. Let E be a ground set containing two distinct elements a and b. Put

$$S = E \times \{a\}, \qquad T = E \times \{b\}, \qquad R = E \times E.$$

   Check that $R \circ (T \cap S) \neq (R \circ S) \cap (R \circ T)$.

5. A binary relation $S(x, y)$ is called *symmetric* if $S(x, y) \Rightarrow S(y, x)$ (equivalently, if $S = S^{-1}$).
   A binary relation S is called the *symmetric closure* of a given binary relation R if S is a least (by inclusion) symmetric binary relation containing R.
   Demonstrate that the symmetric closure of R coincides with $R \cup R^{-1}$.

6. A binary relation $T(x, y)$ is called *transitive* if $T(x, y) \,\&\, T(y, z)$ implies $T(x, z)$.

A binary relation T is called the *transitive closure* of a given binary relation R if T is a least (by inclusion) transitive binary relation containing R.
Show that the transitive closure T of R can be defined by the formula

$$(x, y) \in T \Leftrightarrow (\exists n \in \mathbb{N} \setminus \{0, 1\})(\exists x_1, x_2, ..., x_n)(x_1 = x \ \& \ x_n = y \ \&$$

$$R(x_1, x_2) \ \& \ R(x_2, x_3) \ \& \ldots \& \ R(x_{n-1}, x_n)).$$

Here $\mathbb{N} = \{0, 1, 2, ..., n, ...\}$ denotes the set of all natural numbers.

7. A binary relation R on E is called an *equivalence relation* between elements of E if R is reflexive, symmetric, and transitive.
Prove that any such R produces a partition of E into sets having the form $R(x) = \{y : R(x, y)\}$, where $x \in E$ (these sets are usually called *R-equivalence classes*). In other words, the following assertions hold:

(a) $R(x) \neq \emptyset$ for each $x \in E$;
(b) if $R(x) \cap R(y) \neq \emptyset$, then $R(x) = R(y)$;
(c) $\cup \{R(x) : x \in E\} = E$.

The set of all R-equivalence classes is called the *quotient set* (or *factor set*) of E with respect to R and is usually denoted by E/R.

Conversely, prove that if one has the equality $E = \cup \{X_i : i \in I\}$, where all X_i are nonempty pairwise disjoint sets (i.e., are pairwise without common elements), then there exists a unique equivalence relation R on E which produces the partition $\{X_i : i \in I\}$ of E, and the sets X_i ($i \in I$) are precisely the R-equivalence classes.

8. Let $\mathbb{R}$ be the real line, let $\mathbb{Q}$ denote the field of all rational numbers, and let a binary relation V be defined as follows:

$$V(x, y) \Leftrightarrow (x \in \mathbb{R} \ \& \ y \in \mathbb{R} \ \& \ x - y \in \mathbb{Q}).$$

Verify that for V these two assertions hold true:

(a) V is an equivalence relation on $\mathbb{R}$;
(b) V as a subset of the plane $\mathbb{R}^2$ is a union of (countably many) straight lines, all of which are parallel to the diagonal of $\mathbb{R}^2$.

Remark 1.1 V is called *Vitali's equivalence relation* on $\mathbb{R}$ and the partition of $\mathbb{R}$ into V-equivalence classes is called *Vitali's partition of* $\mathbb{R}$ (see [26, 53, 75, 134, 141]).

9. By definition, a binary relation R on a base set E is a partial quasi-ordering of E (or a partial pre-ordering of E) if R is reflexive and transitive.
If, in addition, R is *antisymmetric* (i.e., $R(x, y) \ \& \ R(y, x)$ implies $x = y$), then R is called a *partial ordering* of E.

1 Unary and Binary Relations

According to commonly adopted mathematical practice, for a partial quasi-ordering (respectively, for a partial ordering) R, the notation $x \leq y$ or $x \preceq y$ is used instead of $R(x, y)$.

Let R be a partial quasi-ordering of E and let a binary relation S be defined as follows:

$$S(x, y) \Leftrightarrow (R(x, y) \,\&\, R(y, x)).$$

Check that S is an equivalence relation on E and there exists a unique partial ordering R' of E/S produced by R, via the equivalence S, such that R' is compatible with R. Namely, for any two sets $X \in E/S$ and $Y \in E/S$, one has

$$R'(X, Y) \Leftrightarrow (\exists x \in X)(\exists y \in Y) R(x, y).$$

10. Let E be an arbitrary ground set.
    Demonstrate that the power set $\mathcal{P}(E)$ endowed with the standard inclusion relation $\subset$ is a partially ordered set. Moreover, consider any partial order $\leq$ on E and, for each element $x \in E$, introduce the notation

    $$E_x = \{z \in E : z \leq x\}.$$

    Show that $x \leq y \Leftrightarrow E_x \subset E_y$ and conclude from this fact that $(\mathcal{P}(E), \subset)$ is a universal partially ordered set for all partial orderings of E. In other words, for every partial order $(E, \leq)$, there exists a subset $\mathcal{F}$ of $\mathcal{P}(E)$ such that $(\mathcal{F}, \subset)$ is isomorphic to $(E, \leq)$.

11. On the set $\mathbb{N} \setminus \{0\}$ consider the following binary relation:

    $$R(m, n) \Leftrightarrow m \text{ divides } n.$$

    Verify that:

    (a) $R(m, n)$ is a partial ordering of $\mathbb{N} \setminus \{0\}$;
    (b) $R(m, n) \Rightarrow m \leq n$, where $\leq$ is the standard ordering of $\mathbb{N} \setminus \{0\}$;
    (c) the implication $m \leq n \Rightarrow R(m, n)$ is false for infinitely many pairs (m, n) of nonzero natural numbers.

12. A partial ordering $R(x, y)$ of a ground set E is called a *linear ordering* if, for any two elements $x \in E$ and $y \in E$, one has the disjunction $R(x, y) \vee R(y, x)$. Show that if E is a finite set, then any partial ordering $\leq$ of E is contained in some linear ordering of E.
    For this purpose, use induction on card(E).

    Remark 1.2 Using the Axiom of Choice or its equivalent Kuratowski–Zorn lemma, it can be proved that the result of Exercise 12 remains true for an arbitrary base set E equipped with a partial ordering $\leq$. Moreover, we have the following: every partial ordering $\leq$ of E coincides with the intersection of all those linear orderings of E which extend $\leq$ (see, e.g., [18, 111]).

13. Let E, F, G be three ground sets and let $(E \times F) \times G$ denote the Cartesian product of $E \times F$ and G. By definition, the elements of $(E \times F) \times G$ are all of the form $((x, y), z)$, where $(x, y) \in E \times F$ and $z \in G$.

    Any binary relation on $(E \times F) \times G$ is called a *ternary relation* between elements of E, F, and G.

    Demonstrate that the notion of a ternary relation is reducible to the notion of a unary relation.

    Furthermore, for every natural number $n \geq 3$ introduce, by using induction on n, the notion of an *n-ary relation* and show that this notion can also be reduced to the notion of a unary relation.

Chapter 2
Partial Functions and Functions

A binary relation $S(x, y)$ is called a *functional relation* with respect to a variable y if, for every $x \in \text{dom}(S)$, there exists exactly one y such that $S(x, y)$ holds true.

In other words, $S(x, y)$ is functional with respect to y if any vertical section of $S(x, y)$ contains at most one element. According to another terminology (adopted in descriptive set theory), such $S(x, y)$ is called *uniform* with respect to y (cf. [119, 167]).

If, in addition, $\text{dom}(S) \subset E$ and $\text{ran}(S) \subset F$ for some sets E and F, then S is called a *partial function* (*partial mapping*) acting from E into F.

Moreover, if a partial function S is such that $\text{dom}(S) = E$ and also $\text{ran}(S) \subset F$, then S is called a *function* (*mapping*) acting from E into F.

Example 2.1 The relation $\{(x, y) \in \mathbb{R}^2 : y = x^2\}$ is a function from $\mathbb{R}$ into itself. The relation $\{(x, y) \in \mathbb{R}^2 : y^2 = x \ \& \ y \geq 0\}$ is a partial function from $\mathbb{R}$ into itself.

The standard symbols used for denoting various partial functions and functions are $f, g, h, \phi, \psi, \chi, \ldots$

For each x from the domain of a partial function f, the symbol $f(x)$ denotes the unique element y such that $(x, y) \in f$.

Naturally, f is identical with its graph, i.e., we have

$$f = \{(x, y) : (x, y) \in f\} = \{(x, f(x)) : x \in \text{dom}(f)\}.$$

By definition, a partial function g extends a partial function f if $f \subset g$.

More precisely, we say that g is an *extension* of f defined on the set $\text{dom}(g)$. Clearly, $\text{dom}(f) \subset \text{dom}(g)$.

We also say that a partial function f is a *restriction* of a partial function g if $f \subset g$. More precisely, we say that f is a restriction of g to the domain of f. In short, this situation is described by the formula $f = g|\text{dom}(f)$.

If a partial function g (or a function g) acts from E into F, then the notation $g : E \to F$ is commonly used. If E and F are fixed in our considerations, then the notation $x \to g(x)$ is often used, without mentioning E and F.

A partial function g acting from E into F is called *injective* (or an *injection*) if $x \neq x'$ implies $g(x) \neq g(x')$ for any elements x and x' of dom(g).

A partial function g acting from E into F is called *surjective* (or a *surjection*) if ran(g) = F.

A function g acting from E into F is called *bijective* (or a *bijection*) if g is simultaneously injective and surjective.

Example 2.2 Let E be an arbitrary base set and let f be defined by the formula $f = \{(x, x) : x \in E\}$. Then f is a bijection of E onto itself. This bijection is called the *identity transformation* of E and is denoted by Id_E. Obviously, $\mathrm{Id}_E = \triangle_E$.

A function $g : E \to E$ is usually called a *unary operation* on E. Analogously, a function $h : E \times E \to E$ is called a *binary operation* on E. More generally, for any natural number $n \geq 1$, a function acting from E^n into E is treated as a certain n-ary operation on E.

Theorem 2.1 *The following six assertions hold:*

(1) g is an injection if and only if g^{-1} is a partial function;
(2) the composition of two injections is again an injection;
(3) the composition of two surjections is again a surjection;
(4) a partial function $g : E \to F$ is surjective if and only if there exists a function $h : F \to E$ such that $g \circ h = \mathrm{Id}_F$;
(5) a function $g : E \to F$ is injective if and only if there exists a partial function $h : F \to E$ such that $h \circ g = \mathrm{Id}_E$;
(6) if g is an arbitrary partial function and $\{X_i : i \in I\}$ is an arbitrary family of sets, then $g^{-1}(\cap\{X_i : i \in I\}) = \cap\{g^{-1}(X_i) : i \in I\}$.

Proof It is left to the reader to verify assertions (1), (2), (3), and (6).

To demonstrate (4), let a partial function $g : E \to F$ be a surjection. Consider the family $\{g^{-1}(y) : y \in F\}$ of nonempty pairwise disjoint subsets of E. Choose in each set $g^{-1}(y)$, where $y \in F$, an element $x = x(y)$ and put $h(y) = x(y)$. Notice the essential use of the Axiom of Choice (**AC**) at this point of the argument. Now, it can easily be checked that the obtained function $h : F \to E$ satisfies the equality $g \circ h = \mathrm{Id}_F$.

On the other hand, if $h : F \to E$ is any function satisfying $g \circ h = \mathrm{Id}_F$, then $(g \circ h)(y) = y$ for each $y \in F$, so $y = g(h(y))$. The last equality shows us that $y \in \mathrm{ran}(g)$, and therefore g is a surjection.

To demonstrate (5), let a function $g : E \to F$ be injective. Then, according to (1), g^{-1} is a partial function. Putting $h = g^{-1}$, one can readily verify that $h \circ g = \mathrm{Id}_E$.

On the other hand, if $h : F \to E$ is any partial function satisfying $h \circ g = \mathrm{Id}_E$, then the equality $g(x) = g(x')$ for $x \in E$ and $x' \in E$ implies

$$x = h(g(x)) = h(g(x')) = x',$$

2 Partial Functions and Functions

which shows us that g is an injection. $\square$

Assertions (1) and (6) of Theorem 2.1 imply that if g is an injection, then

$$g(\cap\{X_i : i \in I\}) = \cap\{g(X_i) : i \in I\}$$

for every nonempty family of sets $\{X_i : i \in I\}$ (cf. Theorem 1.2 of Chap. 1). Also, if g is an injection, then $g(X \setminus Y) = g(X) \setminus g(Y)$ for any two sets X and Y.

Remark 2.1 Within **ZF** set theory, assertion (4) of Theorem 2.1 is equivalent to the Axiom of Choice (**AC**).

The next theorem immediately follows from Theorem 2.1.

Theorem 2.2 *If E is an arbitrary base set, then the family of all bijections of E onto itself is a group with respect to the composition operation $\circ$.*

The above-mentioned group is usually called the group of all permutations of E or the *symmetric group* of E and is denoted by $\mathrm{Sym}(E)$.

A mapping $f : E \to E$ is called an *involution* of E if $f \circ f = \mathrm{Id}_E$.

Assertions (4) and (5) of Theorem 2.1 immediately imply that any involution f of E is a bijection of E onto itself, so $f \in \mathrm{Sym}(E)$.

Example 2.3 Let E be a ground set. To any subset X of E one can associate its *characteristic function* (or *indicator*) $\chi_X : E \to \{0, 1\}$, defined as follows:

$$\chi_X(x) = \begin{cases} 1 & \text{if } x \in X, \\ 0 & \text{if } x \in E \setminus X. \end{cases}$$

This notion allows one to establish a bijective mapping (bijective correspondence) between the power set $\mathcal{P}(E)$ and the family of all $\{0, 1\}$-valued functions on E. Under the above-mentioned correspondence, the standard set-theoretic operations over subsets of E are transformed to the respective algebraic operations over the characteristic functions. For instance, one has

$$\chi_{X \cup Y} = \sup(\chi_X, \chi_Y),$$
$$\chi_{X \cap Y} = \inf(\chi_X, \chi_Y) = \chi_X \cdot \chi_Y,$$
$$\chi_{X \cup Y} + \chi_{X \cap Y} = \chi_X + \chi_Y,$$
$$\chi_{X \setminus Y} = \chi_X - \chi_{X \cap Y}.$$

In particular, if $X \cap Y = \emptyset$, then $\chi_{X \cup Y} = \chi_X + \chi_Y$. It is also clear that $\chi_\emptyset$ is identically zero and χ_E is identically 1.

Every family of sets $\{X_i : i \in I\}$ may be treated as a function

$$G : I \to \mathcal{P}(\cup\{X_i : i \in I\})$$

such that $G(i) = X_i$ for each $i \in I$. In this case, mathematicians quite often speak of the *multi-function* G (or of the *set-valued function* G) given on I.

By the standard definition, the *graph* of G is the set

$$\mathrm{Gr}(G) = \{(i, G(i)) : i \in I\},$$

where $G(i) = X_i$ for all indices $i \in I$.

However, if one treats G as a multi-function, then the second definition of the graph of G is frequently used, namely,

$$\mathrm{Gr}(G) = \{(i, y) : i \in I, y \in G(i)\}.$$

The next example shows that every partial function may be considered as a special case of a multi-function.

Example 2.4 Let $g : E \to F$ be a partial function. We can associate with g a multi-function $G : E \to \mathcal{P}(F)$ defined as follows:

$$G(x) = \begin{cases} \{g(x)\} & \text{if } x \in \mathrm{dom}(g), \\ \emptyset & \text{if } x \in E \setminus \mathrm{dom}(g). \end{cases}$$

It is not hard to show that, given G, the original partial function g can uniquely be reconstructed.

Example 2.5 For any partial function $g : E \to F$, define a multi-function $G : F \to \mathcal{P}(E)$ by putting $G(y) = g^{-1}(y)$ for each $y \in F$. Again, it is easy to see that, given G, the partial function g can uniquely be reconstructed.

If a binary relation R is given, then we may also consider it as a multi-function G defined on $\mathrm{dom}(R)$. Indeed, for any $x \in \mathrm{dom}(R)$, we can put

$$G(x) = R(x) = \{y : (x, y) \in R\}.$$

The important question arises whether there exists a uniformization (selector) of the above multi-function, i.e., whether there exists a function

$$g : \mathrm{dom}(R) \to \mathrm{ran}(R)$$

such that

$$(\forall x \in \mathrm{dom}(R))((x, g(x)) \in R).$$

This question has various aspects. From a purely set-theoretical point of view, it is closely connected with the Axiom of Choice (cf. Exercise 4 below). In topology, it corresponds to finding a continuous selector g, whenever R possesses sufficiently nice topological properties (e.g., Michael's theorems; see [130, 131, 148]). In measure-theory, it asks us to find a measurable selector, whenever R possesses sufficiently good measurability properties (e.g., the theorem of Kuratowski and Ryll-Nardzewski [112]).

In general, theorems on selectors are helpful in the study of many profound mathematical problems and questions.

In subsequent chapters, we will be dealing with real-valued functions on topological spaces (or on measurable spaces) and, specifically, with functions acting from the real line $\mathbb{R}$ into itself. Most topics discussed in further sections of this lecture course belong to the theory of real-valued functions or, briefly, to real analysis.

Let us formulate two important questions of modern real analysis.

(1) For a given function f, find a maximally large subset X of dom(f) such that the restriction $f|X$ has good descriptive properties.
(2) For a given partial function g with good descriptive properties, find an extension g^* of g to a maximally large set Y such that g^* preserves these properties on Y.

Remark 2.2 Needless to say, the term "good descriptive properties" is specified in concrete situations which are of interest from different viewpoints: purely set-theoretical, topological, measure-theoretical, analytical, algebraic, etc. In real analysis there are certain types of partial functions and functions which are important in various respects: continuous, differentiable, monotone, convex, semicontinuous, measurable, integrable, and so forth. The questions (1) and (2) formulated above are primarily concerned with partial functions and functions of the just listed types.

Exercises

1. Recall that a function $f : \mathbb{R} \to \mathbb{R}$ is *affine* if f has the form
$$f(x) = ax + b \quad (x \in \mathbb{R}),$$
where a and b are some real constants.
Let f and g be two affine functions acting from $\mathbb{R}$ into $\mathbb{R}$:
$$f(x) = ax + b, \quad g(x) = cx + d \quad (x \in \mathbb{R}).$$
Show that $f \circ g = g \circ f$ if and only if $d(a-1) = b(c-1)$.
Infer from this fact that there are infinitely many pairs (f, g) of affine functions from $\mathbb{R}$ into $\mathbb{R}$ which do not commute.

2. Check that if E contains at most two elements, then the group Sym(E) is cyclic (of order not exceeding 2) and hence is commutative.

3. Let a base set E consist of exactly three distinct elements.
Verify that there are two mappings $f \in \text{Sym}(E)$ and $g \in \text{Sym}(E)$ such that

$$(\forall x \in E)(g(f(x)) \neq f(g(x))).$$

Deduce from this fact that if there are at least three distinct elements of a set X, then the group $\mathrm{Sym}(X)$ is not commutative.

4. Working within **ZF** set theory, demonstrate that the following assertions are equivalent:

    (a) the Axiom of Choice (**AC**);
    (b) for every surjection $g : E \to F$, there exists a function $h : F \to E$ such that $g \circ h = \mathrm{Id}_F$;
    (c) any multi-function (binary relation) admits a uniformization.

5. Show the validity of the formulas presented in Example 2.3.
6. Let $\mathbb{Q}$ denote the set of all rational numbers in $\mathbb{R}$ and let $\chi_\mathbb{Q}$ be the characteristic function of $\mathbb{Q}$ (i.e., the *Dirichlet function*).
   Check that the following assertions hold true:

    (a) the function $\chi_\mathbb{Q}$ is discontinuous at all points of $\mathbb{R}$;
    (b) the restriction of $\chi_\mathbb{Q}$ to the set $\mathbb{R} \setminus \mathbb{Q}$ is identically zero, so is continuous on $\mathbb{R} \setminus \mathbb{Q}$;
    (c) $\mathbb{R} \setminus \mathbb{Q}$ is a maximal (by inclusion) subset X of $\mathbb{R}$ with the property that the restriction of $\chi_\mathbb{Q}$ to X is continuous.

7. Let $f : \mathbb{R} \to \mathbb{R}$ be a partial function and let z be a point of $\mathbb{R}$ satisfying the following condition: there are two sequences

$$\{x_n : n \in \mathbb{N}\} \subset \mathrm{dom}(f), \qquad \{y_n : n \in \mathbb{N}\} \subset \mathrm{dom}(f)$$

both converging to z such that there exists a real $\varepsilon > 0$ for which

$$(\forall n \in \mathbb{N})(|f(x_n) - f(y_n)| \geq \varepsilon).$$

Demonstrate that there is no continuous partial function $f^* : \mathbb{R} \to \mathbb{R}$ extending f such that $z \in \mathrm{dom}(f^*)$.

8. Explain why it makes no sense to introduce and consider the notion of a partial multi-function.
9. Let E be a topological space and let $f : E \to \mathbb{R}$ be a function. Define a multi-function $\Phi_f : \mathbb{R} \to \mathcal{P}(E)$ as follows:

$$\Phi_f(y) = \mathrm{cl}(f^{-1}(y)) \qquad (y \in \mathbb{R}),$$

where $\mathrm{cl}(f^{-1}(y))$ denotes the closure of the set $f^{-1}(y)$ in E.
Verify that:

    (a) if f is such that the sets $f^{-1}(y)$ are closed in E for all y in $\mathbb{R}$, then Φ_f uniquely determines f;

2 Partial Functions and Functions

(b) if $g : E \to \mathbb{R}$ and $h : E \to \mathbb{R}$ are two functions, then, in general, the equality $\Phi_g = \Phi_h$ does not imply the equality $g = h$.

10. Give an example of a connected polygonal curve in $\mathbb{R}^2$ which does not admit a continuous uniformization; moreover, prove that such a curve must have at least three sides.

11. Demonstrate effectively (i.e., without using the Axiom of Choice) that if Z is a subset of $\mathbb{R}^2$ such that, for any $x \in \mathbb{R}$, the vertical section $\{y : (x, y) \in Z\}$ is open in $\mathbb{R}$, then Z admits a uniformization.

    For this purpose, take into account the everywhere density of $\mathbb{Q}$ in $\mathbb{R}$.

12. Show effectively that if Z is a subset of $\mathbb{R}^2$, all vertical sections of which are closed, then Z admits a uniformization.

    For this purpose, first establish the following fact: within **ZF** set theory there exists a function h defined on the family of all nonempty closed subsets of $\mathbb{R}$ such that $h(X) \in X$ for any nonempty closed set X in $\mathbb{R}$.

    Remark 2.3 In **ZF** theory the following three statements hold for an arbitrary metric space E.

    (1) If there exists a function ϕ defined on the family of all nonempty closed subsets of E such that $\phi(X) \in X$ for every nonempty closed set X in E, then there exists a function ψ defined on the family of all nonempty open subsets of E such that $\psi(Y) \in Y$ for every nonempty open set Y in E.

    (2) If E is separable, then there exists a function ψ defined on the family of all nonempty open subsets of E such that $\psi(Y) \in Y$ for every nonempty open set Y in E.

    (3) If E is complete and separable, then there exists a function ϕ defined on the family of all nonempty closed subsets of E such that $\phi(X) \in X$ for every nonempty closed set X in E.

13. Let E be an arbitrary base set and let $f : E \to E$ be any bijection.

    Prove that there are two involutions $s : E \to E$ and $s' : E \to E$ such that $f = s' \circ s$.

    Argue as follows. Denote by $\mathbb{Z}$ the set of all integers and, for each point $x \in E$, introduce the set $O(x) = \{f^n(x) : n \in \mathbb{Z}\}$.

    Verify that the restriction of f to $O(x)$ is representable in the form of the composition of two involutions.

    For this purpose, consider separately two possible cases: $O(x)$ is finite and $O(x)$ is infinite. In the first case, identify $O(x)$ with the set of all vertices of a regular polygon P, on which f acts as a rotation about the center of P. In the second case, identify $O(x)$ with the group $(\mathbb{Z}, +)$ on which f acts as the shift $n \to n + 1$.

    Taking into account the above circumstance for any $O(x)$, obtain the required result.

14. Let $\{X_i : i \in I\}$ be a family of sets and denote by $\prod\{X_i : i \in I\}$ the set of all those functions $f : I \to \cup\{X_i : i \in I\}$ which satisfy the relation $f(i) \in X_i$ for each $i \in I$.

This $\prod\{X_i : i \in I\}$ is called the *product* of the family $\{X_i : i \in I\}$.
Elements of $\prod\{X_i : i \in I\}$ are often called *selectors* of the given family $\{X_i : i \in I\}$.
If $X_i = X$ for all $i \in I$, then the notation X^I is also used instead of $\prod\{X_i : i \in I\}$.
For a fixed index $i \in I$, the function

$$\mathrm{pr}_i : \prod\{X_i : i \in I\} \to X_i$$

defined by $\mathrm{pr}_i(f) = f(i)$ is called the *i-th projection* of $\prod\{X_i : i \in I\}$ to the set X_i.
Check that:

(a) if all sets X_i $(i \in I)$ are nonempty, then $\prod\{X_i : i \in I\}$ is also nonempty (one of the forms of **AC**);
(b) if $I = \emptyset$, then $\prod\{X_i : i \in I\}$ is the singleton $\{\emptyset\}$.

15. For any two sets X and Y, let $X \triangle Y$ denote the set $(X \setminus Y) \cup (Y \setminus X)$.
So $\triangle$ may be treated as a binary operation for all pairs of subsets of E, where E is a given ground set.
This operation is usually called the *symmetric difference* of two sets.
Prove that $(X \triangle Y) \triangle Z = X \triangle (Y \triangle Z)$ (the associativity of $\triangle$).
Deduce from the above fact that, for every natural number $n \geq 2$, the set $X_1 \triangle X_2 \triangle ... \triangle X_n$ is well-defined and consists of all those elements which belong to an odd number of members of the family of sets $(X_1, X_2, ..., X_n)$.
Conclude that the triple $(\mathcal{P}(E), \triangle, \cap)$ is a commutative ring in which $0 = \emptyset$ and $1 = E$ (here $\triangle$ plays the role of the addition operation $+$ and $\cap$ plays the role of the multiplication operation $\cdot$).
In this ring one has the identities $X + X = 0$ and $X \cdot X = X$.

16. Let $\{S_i : i \in I\}$ be a family of pairwise different unary relations on a ground set E.
Show that the following two assertions are equivalent:

(a) there exists a binary relation on E which is universal for the family $\{S_i : i \in I\}$;
(b) there exists an injective function acting from I into E.

17. Let n be a natural number, let a and b be distinct points of $\mathbb{R}$, and let

$$c_0, c_1, ..., c_n, \quad d_0, d_1, ..., d_n$$

be two finite sequences of real numbers.
Demonstrate that there exists a real polynomial h on $\mathbb{R}$ such that the equalities

$$h(a) = c_0, \quad h'(a) = c_1, ..., \quad h^{(n)}(a) = c_n,$$
$$h(b) = d_0, \quad h'(b) = d_1, ..., \quad h^{(n)}(b) = d_n$$

hold true for the indicated derivatives of h at the points a and b respectively. For this purpose, write h in the form

$$h(x) = f(x) + (x-a)^{n+1} g(x),$$

where both f and g are real polynomials on $\mathbb{R}$ and

$$f(a) = c_0, \qquad f'(a) = c_1, \ldots, \qquad f^{(n)}(a) = c_n.$$

Then show that, choosing an appropriate g, the required equalities hold.

18. Let n be a natural number, let $\{[a_k, b_k] : k \in K\}$ be a finite family of pairwise disjoint non-degenerate segments in $\mathbb{R}$, and let

$$f : \cup\{[a_k, b_k] : k \in K\} \to \mathbb{R}$$

be an n-times differentiable function (naturally, at a_k the right derivatives are meant and at b_k the left derivatives are meant).
Prove that there exists a function $f^* : \mathbb{R} \to \mathbb{R}$ which extends f and is also n-times differentiable on $\mathbb{R}$.
For this purpose, take into account the result of Exercise 17.

Remark 2.4 The more complicated problem of extending a partial real-valued function of several real variables (while preserving differential properties of the function) is discussed, e.g., in [44].

19. Let A and B be any two sets, let f be a mapping from A into B, and let g be a mapping from B into A. Suppose that

$$f \circ g = \mathrm{Id}_B, \qquad g \circ f = \mathrm{Id}_A.$$

Show that both f and g are bijections and $g = f^{-1}$.

20. Let E be a ground set and let f be a mapping of E into itself (in other words, let f be a self-mapping of E).
An element $x \in E$ is called a *fixed point* (or an *invariant point*) for f if the equality $f(x) = x$ holds true.
The family of all fixed points of f is sometimes denoted by the symbol $\mathrm{fix}(f)$.
Check that:

(a) if x is a fixed point of f, then for every natural number n the same x is a fixed point of the mapping f^n (here $f^0 = \mathrm{Id}_E$);
(b) if f is a bijection of E onto itself, then $\mathrm{fix}(f) = \mathrm{fix}(f^{-1})$;
(c) it may happen that both f and g are bijections of E onto itself and have fixed points, but the composition $f \circ g$ does not possess any fixed point.

For (c), consider $E = \mathbb{R}$ and take as f and g two distinct central symmetries of $\mathbb{R}$.

21. Let f and g be two partial functions and let $h = f \circ g$.

Show that:

(a) if h is a surjection, then f is also a surjection;
(b) it may happen that f is a bijection, but h is not a surjection.

22. Let again f and g be two partial mappings and let $h = f \circ g$.
Verify that:

(a) if h is an injection and g is a surjection, then f is an injection;
(b) it may happen that both g and h are injections, but f is not an injection;
(c) it may happen that h and f are injections, but g is not a surjection;
(d) it may happen that f is a bijection, g is a surjection, but h is not an injection.

23. Let h be an arbitrary mapping from a set A into a set B.
Demonstrate that $h = f \circ g$, where g is a surjective mapping and f is an injective mapping.
For this purpose, consider on A the equivalence relation R produced by h, i.e.,

$$R(x, y) \Leftrightarrow (x \in A \ \& \ y \in A \ \& \ h(x) = h(y)).$$

Then denote by A/R the corresponding quotient set and, for each element $x \in A$, put $g(x) = X$, where $X \in A/R$ and $x \in X$.
The value $g(x)$ is well-defined, because there exists a unique $X \in A/R$ such that $x \in X$.
Infer that the obtained mapping g is a surjection from A onto A/R.
Afterwards, define a mapping $f : A/R \to B$ by putting $f(X) = h(x)$, where X is any element of A/R and x is any element of X.
A straightforward verification gives that f is well-defined, is an injective function, and $h = f \circ g$.

24. Let A, B, C, D be four sets, let f be a mapping from A into B, let g be a mapping from B into C, and let h be a mapping from C into D. Suppose that both mappings $g \circ f$ and $h \circ g$ are bijections.
Prove that all three mappings f, g and h are bijections.

25. Let E be a set, let f be a mapping of E into itself, and let S be a binary relation on E defined by the formula

$$S(x, y) \Leftrightarrow (\exists m \in \mathbb{N})(\exists n \in \mathbb{N})(f^m(x) = f^n(y)).$$

Check that S is an equivalence relation on E.

26. Let n and m be any two natural numbers. Denote by $I_{n,m}$ the total number of injective mappings from a set A which contains exactly n elements into a set B which contains exactly m elements.
Show that $I_{n,m} = m(m-1) \cdots (m-(n-1))$. In particular, one has $I_{n,n} = n!$.

27. Let n and m be any two nonzero natural numbers. Denote by $S_{n,m}$ the total number of all surjective mappings from a set A which contains exactly n elements onto a set B which contains exactly m elements.
Verify the validity of the following recursive formula:

$$S_{n,m} = m(S_{n-1,m} + S_{n-1,m-1}).$$

Starting with this formula, establish the equality

$$S_{n,m} = C_m^0 m^n - C_m^1 (m-1)^n + C_m^2 (m-2)^n - \ldots + (-1)^{m-1} C_m^{m-1}.$$

In this equality all integers

$$C_m^k = m!/(k!(m-k)!) \qquad (0 \leq k \leq m-1)$$

are coefficients of the binomial $(1+x)^m$ of degree m.

Chapter 3
Elementary Facts on Cardinal Numbers

Two sets A and B are called *equinumerous* if there exists a bijection from A onto B. In this case, the notation $A \sim B$ is often used. The relation $R(A, B)$ defined by $A \sim B$ is reflexive, symmetric, and transitive, i.e., one has

$$R(A, A), \qquad R(A, B) \Rightarrow R(B, A), \qquad (R(A, B) \,\&\, R(B, C)) \Rightarrow R(A, C)$$

for any sets A, B, and C. This fact is easily verified.

So, R induces an equivalence relation on the Boolean $\mathcal{P}(E)$ of any base set E, and it can readily be seen that the validity of $R(A, B)$, for $A \subset E$ and $B \subset E$, does not depend on the choice of E.

If $R(A, B)$ holds true, then one also says that the sets A and B have the same *cardinality* (or the same *cardinal number*). This circumstance is usually expressed by the equality $\text{card}(A) = \text{card}(B)$.

Remark 3.1 The precise definition of cardinal numbers is not trivial and can be introduced in three different manners.

In **ZFC** theory (i.e., in **ZF** & **AC**), the first way to introduce the notion of a cardinal number is to define it as an ordinal α in the sense of von Neumann, such that α is not equinumerous with any proper initial subinterval of α.

Within the same theory, using the so-called Axiom of Foundation, which is not necessary for most branches of mathematics, the cardinal number of a set A may be defined as the family of all those sets which are equinumerous with A and have the least rank in the von Neumann hierarchy of sets (cf. [76, 108, 111]).

The third way of introducing cardinal numbers (cardinals) is based on *Hilbert's logical τ-operator*, which reflects a global form of the Axiom of Choice. Actually, Hilbert's τ operator associates to every nonempty class $\mathcal{K}$ of sets one of the elements of $\mathcal{K}$. In particular, for any set A, the term $\text{card}(A)$ is the value of τ on the class of all sets equinumerous with A (see [18]). This last approach cannot be realized in **ZFC** theory and needs more powerful versions of set theory.

Theorem 3.1 *The following assertions hold true:*

(1) *if $\{A_i : i \in I\}$ and $\{B_i : i \in I\}$ are any two disjoint families of sets and $A_i \sim B_i$ for all $i \in I$, then*

$$\cup\{A_i : i \in I\} \sim \cup\{B_i : i \in I\};$$

(2) *if $\{A_i : i \in I\}$ and $\{B_i : i \in I\}$ are any two families of sets and $A_i \sim B_i$ for all $i \in I$, then*

$$\prod\{A_i : i \in I\} \sim \prod\{B_i : i \in I\};$$

(3) *if $A \sim B$ and $C \sim D$, then $A^C \sim B^D$;*
(4) *if $B \cap C = \emptyset$, then $A^B \times A^C \sim A^{B \cup C}$;*
(5) *$(A^B)^C \sim A^{B \times C}$.*

The proof of Theorem 3.1 is left to the reader, because it is a direct consequence of the definition of equinumerous sets. However, it should be mentioned that the use of the Axiom of Choice (**AC**) is necessary in the proof.

It follows from (1) of Theorem 3.1 that if A is a set, $a \notin A$ and $b \notin A$, then $A \cup \{a\} \sim A \cup \{b\}$.

Also, if $a \in A$ and $b \notin A$, then $A \sim (A \setminus \{a\}) \cup \{b\}$.

Now, we introduce the fundamental concept of a finite set (in the sense of Bolzano and Dedekind).

A set A is called *finite* if $A \not\sim A \cup \{b\}$, where $b \notin A$.

Evidently, this definition does not depend on an element $b \notin A$.

Theorem 3.2 *The following two assertions hold:*

(1) *any subset of a finite set is finite;*
(2) *if A is a finite set and a is any element, then $A \cup \{a\}$ is a finite set.*

Proof To show (1), let A be finite and let $B \subset A$. Suppose to the contrary that B is not finite, i.e., $B \sim B \cup \{b\}$, where $b \notin B$. We may assume, without loss of generality, that $b \notin A$ (cf. Exercise 9). Then we get

$$A = (B \cup (A \setminus B)) \sim (B \cup \{b\}) \cup (A \setminus B) = A \cup \{b\},$$

which contradicts the finiteness of A.

To show (2), let A be again a finite set and let a be an arbitrary element. If $a \in A$, then there is nothing to prove. Let now $a \notin A$ and suppose to the contrary that $A \cup \{a\}$ is not finite. Then

$$A \cup \{a\} \sim (A \cup \{a\}) \cup \{b\}$$

for some $b \notin A \cup \{a\}$. This means that there exists a bijection

3 Elementary Facts on Cardinal Numbers

$$f : A \cup \{a\} \to A \cup \{a, b\}.$$

Only the following three cases can occur here.

1. $f(a) = a$.
   In this case, we get $A \sim f(A) = A \cup \{b\}$, which contradicts the finiteness of A.
2. $f(a) = b$.
   In this case, we get $A \sim f(A) = A \cup \{a\}$, which also contradicts the finiteness of A.
3. $f(a) \in A$.
   In this case, we have

$$f(A \cup \{a\}) = (A \setminus \{f(a)\}) \cup \{f(a)\} \cup \{a\} \cup \{b\},$$

$$A \sim f(A) = ((A \setminus \{f(a)\}) \cup \{a\}) \cup \{b\} \sim A \cup \{b\},$$

which once again yields a contradiction in view of the finiteness of A.

Theorem 3.2 has thus been proved. □

One may now define the natural numbers as the cardinalities of finite sets.

Example 3.1 Clearly, $\emptyset$ is a finite set. Theorem 3.2 implies that the set $\{a\}$ is also finite for any element a. By virtue of the same theorem, if $b \neq a$, then the set $\{a, b\}$ is finite, and so forth. To successfully continue this process, we need at every step new elements distinct from all the preceding ones.

An effective scheme for carrying out the just described procedure was suggested by von Neumann. It is as follows (for a more rigorous and complete argument, see Exercise 16).

First, let us put $a_0 = \emptyset$. If a_n has already been defined, then put

$$a_{n+1} = a_n \cup \{a_n\}.$$

It can be shown by using ordinary mathematical induction that

$$a_{n+1} = \{a_0, a_1, \ldots, a_n\}$$

for each natural index n and that all elements a_n are pairwise distinct.

These concrete objects are called the *natural numbers in the sense of von Neumann*.

Remark 3.2 Recall that a partially ordered set $(E, \leq)$ is *well-ordered* if each nonempty subset X of E possesses a least element, i.e.,

$$(\exists x \in X)(\forall y \in X)(x \leq y).$$

By using the method of transfinite induction (transfinite recursion), von Neumann generalized the above scheme of constructing natural numbers and introduced the ordinal numbers (ordinals) in his sense. Actually, he demonstrated that every well-ordered set is isomorphic to a uniquely determined ordinal in his sense (see, for instance, [76, 108, 111]).

A set A is called *infinite* (in the sense of Bolzano and Dedekind) if A is not finite, i.e., there exists a bijection of A onto the set $A \cup \{b\}$, where $b \notin A$.

Clearly, this definition does not depend on the choice of $b \notin A$ and is equivalent to the following definition:

A set A is infinite if $A \sim A \setminus \{a\}$ for some element $a \in A$.

The existence of infinite sets is guaranteed by one of the axioms of **ZF** theory (see again [76, 108, 111]).

In fact, the existence of infinite sets is equivalent to the existence of the set of all natural numbers.

In our further considerations, we will identify the natural numbers with von Neumann's natural numbers. Thus, we adopt the notation

$$0 = \emptyset, \qquad n + 1 = \{0, 1, \ldots, n\}, \qquad \mathbb{N} = \{0, 1, \ldots, n, \ldots\}.$$

Actually, $\mathbb{N}$ is the least (by inclusion) set, one of the elements of which is $\emptyset$ and which is closed under a unary operation $'$ defined by the formula $x' = x \cup \{x\}$.

The number x' is usually denoted by $x + 1$, where $1 = \{\emptyset\}$.

The infiniteness of $\mathbb{N}$ is seen from the existence of a canonical bijection

$$f : \mathbb{N} \to \mathbb{N} \setminus \{0\},$$

where $f(n) = n + 1$ for all $n \in \mathbb{N}$.

For any two sets A and B, we will write $\operatorname{card}(A) \leq \operatorname{card}(B)$ if there exists an injective function acting from A into B.

In particular, if $X \subset B$, then $\operatorname{card}(X) \leq \operatorname{card}(B)$ is trivially true.

It is also obvious that

$$\operatorname{card}(A) \leq \operatorname{card}(A),$$

$$(\operatorname{card}(A) \leq \operatorname{card}(B) \ \& \ \operatorname{card}(B) \leq \operatorname{card}(C)) \Rightarrow \operatorname{card}(A) \leq \operatorname{card}(C).$$

The above formulas show us that $\leq$ produces a certain partial quasi-ordering (or partial pre-ordering) between cardinal numbers (see Chap. 1).

Using the Axiom of Choice, it is easy to check that, for every mapping f and for every set A, the inequality $\operatorname{card}(f(A)) \leq \operatorname{card}(A)$ holds true.

The next statement of **ZF** theory, known as the Cantor–Bernstein theorem, yields that $\leq$ is, in fact, a partial ordering between cardinal numbers.

3 Elementary Facts on Cardinal Numbers

Theorem 3.3 *If A and B are two sets such that there exist an injective function $f : A \to B$ and an injective function $g : B \to A$, then there exists a bijection $h : A \to B$.*

Proof It is convenient to introduce the notation

$$X = A \setminus g(B), \qquad \phi = g \circ f.$$

Observe that, since both f and g are injections, ϕ is an injection acting from A into itself. Consider the following family of sets:

$$\mathcal{F} = \{Z \subset A : X \cup \phi(Z) \subset Z\}.$$

This family is not empty, because A belongs to it. Consequently, there exists $A' = \cap \mathcal{F}$.

Moreover, keeping in mind the injectivity of ϕ it can easily be checked that the family $\mathcal{F}$ is closed under any intersections, i.e., if $\{Z_i : i \in I\}$ is a subfamily of $\mathcal{F}$, then $\cap \{Z_i : i \in I\} \in \mathcal{F}$. Therefore, A' also belongs to $\mathcal{F}$, i.e., we have $X \cup \phi(A') \subset A'$.

Notice that the set $Y = X \cup \phi(A')$ satisfies the inclusions

$$X \subset Y, \qquad \phi(Y) \subset \phi(A') \subset Y,$$

which imply $Y \in \mathcal{F}$. Since A' is the least (by inclusion) member of $\mathcal{F}$, we infer that

$$A' \subset Y = X \cup \phi(A')$$

and, as a result, $A' = X \cup \phi(A')$. Further, let us put

$$A'' = A \setminus A', \qquad B' = f(A'), \qquad B'' = B \setminus B',$$

and let us verify that $g(B'') = A''$. Indeed, we may write

$$\begin{aligned}
g(B'') &= g(B \setminus B') = g(B) \setminus g(B') \\
&= (A \setminus X) \setminus g(f(A')) \\
&= A \setminus (X \cup \phi(A')) \\
&= A \setminus A' = A''.
\end{aligned}$$

We thus can define the required bijection $h : A \to B$ as follows:

$$h(x) = \begin{cases} f(x) & \text{if } x \in A', \\ g^{-1}(x) & \text{if } x \in A''. \end{cases}$$

This completes the proof of the Cantor–Bernstein theorem. □

Remark 3.3 The presented argument for proving Theorem 3.3 does not use **AC** and even does not rely on the concept of infinite sets. According to Theorem 3.3, the implication

$$(\operatorname{card}(A) \leq \operatorname{card}(B) \ \& \ \operatorname{card}(B) \leq \operatorname{card}(A)) \Rightarrow \operatorname{card}(A) = \operatorname{card}(B)$$

holds true. It makes sense to observe at this point that, within **ZF** theory, the formula

$$(\forall A)(\forall B)(\operatorname{card}(A) \leq \operatorname{card}(B) \ \vee \ \operatorname{card}(B) \leq \operatorname{card}(A))$$

is equivalent to the Axiom of Choice. In other words, the partial ordering $\leq$ between various cardinalities is a linear ordering if and only if **AC** is valid. Moreover, under **AC**, the same relation $\leq$ becomes a well-ordering, i.e., in any nonempty class of cardinal numbers there exists the least one.

Now, we would like to recall the following notion.

By definition, $\operatorname{card}(A) < \operatorname{card}(B)$ if there exists an injective function from A into B, but there exists no injective function from B into A.

Theorem 3.4 *If there exists an injective function from a nonempty set A into a set B, then there exists a surjective function from B onto A.*

Proof Let $A \neq \emptyset$ and let $f : A \to B$ be an injective function. Pick an element $a \in A$ and define a function $g : B \to A$ by the formula:

$$g(y) = \begin{cases} f^{-1}(y) \text{ if } y \in \operatorname{ran}(f), \\ a \text{ if } y \in B \setminus \operatorname{ran}(f). \end{cases}$$

A straightforward verification yields that g is a surjection of B onto A. $\square$

Remark 3.4 If $f : A \to B$ is a surjection, then, in general, to establish the existence of an injective function $g : B \to A$, one needs to use the Axiom of Choice (cf. (4) of Theorem 2.1 from Chap. 2).

The next theorem is Cantor's famous inequality between $\operatorname{card}(E)$ and $\operatorname{card}(\mathcal{P}(E))$ for an arbitrary set E. Like Theorem 3.3, this inequality is provable within **ZF** theory.

Theorem 3.5 $\operatorname{card}(E) < \operatorname{card}(\mathcal{P}(E))$.

Proof The inequality $\operatorname{card}(E) \leq \operatorname{card}(\mathcal{P}(E))$ follows from the existence of a canonical injective function $f : E \to \mathcal{P}(E)$ defined by

$$f(x) = \{x\} \quad (x \in E).$$

It remains to demonstrate that there is no injective function from $\mathcal{P}(E)$ into E. Taking into account that $\mathcal{P}(E) \neq \emptyset$ and keeping in mind Theorem 3.4, it suffices to show that there is no surjection from E onto $\mathcal{P}(E)$. Suppose otherwise, i.e., there exists a family $\{E_y : y \in E\}$ consisting of all subsets of E. Consider the binary relation R on E defined by the formula

$$R(x, y) \Leftrightarrow x \in E_y.$$

By definition, the y-sections of R give us all subsets of E when y ranges over E. But this contradicts the nonexistence of a binary relation universal for all unary relations on E (see Theorem 1.4 of Chap. 1).

The obtained contradiction ends the proof. □

Example 3.2 By virtue of Cantor's theorem, we have the following infinite list of inequalities:

$$\text{card}(\mathbb{N}) < \text{card}(\mathcal{P}(\mathbb{N})) < \text{card}(\mathcal{P}(\mathcal{P}(\mathbb{N}))) < \text{card}(\mathcal{P}(\mathcal{P}(\mathcal{P}(\mathbb{N})))) < \cdots.$$

Considering the union of all the sets

$$\mathbb{N}, \; \mathcal{P}(\mathbb{N}), \; \mathcal{P}(\mathcal{P}(\mathbb{N})), \; \mathcal{P}(\mathcal{P}(\mathcal{P}(\mathbb{N}))), \ldots$$

we come to a set having cardinality strictly greater than all the above cardinalities. Proceeding in such a manner (e.g., by transfinite recursion), we will obtain increasingly larger cardinal numbers.

The next theorem shows that $\text{card}(\mathbb{N})$ is the smallest infinite cardinal number.

Theorem 3.6 *If E is an infinite set, then E contains a subset E' equinumerous with $\mathbb{N}$.*

Proof Since E is infinite, there exist an element $e \in E$ and a bijection f from E onto $E \setminus \{e\}$. Consider the elements of E defined by ordinary mathematical induction as follows:

$$e_0 = e, \qquad e_{n+1} = f(e_n).$$

It is not hard to see that the sequence $\{e_n : n \in \mathbb{N}\}$ consists of pairwise different elements, so the set $E' = \{e_n : n \in \mathbb{N}\}$ is as required. This completes the proof. □

For the set $\mathbb{N}$, we have a canonical partition $\{N_0, N_1\}$, where N_0 is the set of all even natural numbers and N_1 is the set of all odd natural numbers. So we may assert that if

$$A \sim \mathbb{N}, \qquad B \sim \mathbb{N}, \qquad A \cap B = \emptyset,$$

then $A \cup B \sim \mathbb{N}$ or, equivalently,

$$(\mathbb{N} \times \{0\}) \cup (\mathbb{N} \times \{1\}) \sim \mathbb{N}.$$

In fact, for any two sets X and Y, the relation $X \sim \mathbb{N}$ & $Y \sim \mathbb{N}$ implies the relation $X \cup Y \sim \mathbb{N}$.

The next theorem is more interesting in this respect.

Theorem 3.7 *For every nonzero natural number k, the set $\mathbb{N}^k$ is equinumerous with $\mathbb{N}$.*

Proof For any pair (m, n) of natural numbers, put

$$f((m, n)) = 2^m(2n + 1) - 1.$$

A straightforward verification shows that f is a bijection from $\mathbb{N}^2$ onto $\mathbb{N}$. So we have $\mathbb{N}^2 \sim \mathbb{N}$. We then proceed by induction. Suppose that, for a natural number $k > 1$, the relation $\mathbb{N}^k \sim \mathbb{N}$ has already been established. Then we may write

$$\mathbb{N}^{k+1} \sim \mathbb{N}^k \times \mathbb{N} \sim \mathbb{N} \times \mathbb{N} \sim \mathbb{N}^2 \sim \mathbb{N},$$

which ends the proof. □

It can readily be deduced from the formula $\mathbb{N}^2 \sim \mathbb{N}$ that the set $\mathbb{Q}$ of all rational numbers is equinumerous with $\mathbb{N}$.

Remark 3.5 It was shown by A. Tarski that the following two statements are equivalent within **ZF** set theory:

(1) the Axiom of Choice (**AC**);
(2) for any infinite set E, $\text{card}(E \times E) = \text{card}(E)$.

For more details about this equivalence, see e.g. [111] and [167].

It follows from the above equivalence that, under **AC**,

$$\text{card}((E \times \{0\}) \cup (E \times \{1\})) = \text{card}(E)$$

for every infinite set E. However, the last equality does not imply **AC** in **ZF** theory.

A set E is called *countable* (sometimes, *at most countable*) if E is equinumerous with a subset of $\mathbb{N}$.

In particular, all finite sets are countable by this definition.

Working within **ZF** theory it is not hard to show that:

(i) each subset of a countable set is also countable;
(ii) the image of a countable set, under any partial function, is countable;
(iii) the union of a finite family of countable sets is countable.

However, in order to prove that the union of a countable family of countable sets is again countable, one needs to use a certain form of **AC**.

A set E is called *uncountable* if E is not countable.

In view of Cantor's inequality, the set $\mathcal{P}(\mathbb{N})$ is uncountable and there are many sets of higher cardinalities.

3 Elementary Facts on Cardinal Numbers

As is widely known, in mathematical analysis the most important example of an uncountable set is the real line $\mathbb{R}$.

card($\mathbb{R}$) is called the *cardinality of the continuum* and is denoted by **c**.

We will be dealing with **c** in the next chapter.

Exercises

1. Give a detailed proof of Theorem 3.1 and specify those steps in the argument which essentially rely on the Axiom of Choice.
2. Check that a set A is finite if and only if, for each element $a \in A$, the relation $A \not\sim A \setminus \{a\}$ holds true.
3. For the sets a_n $(n \in \mathbb{N})$ defined in Example 3.1, show that:

   (i) if $n \neq m$, then $a_n \neq a_m$;
   (ii) $a_n \notin a_n$;
   (iii) $a_n \sim n$.

4. Working in **ZF** theory, demonstrate the validity of the following assertions for finite sets in the standard sense (see Remark 3.9 below):

   (a) the image of a finite set under any partial function is a finite set;
   (b) the union of any two finite sets X and Y is a finite set;
   (c) the union of a finite family $\{X_i : i \in I\}$ of finite sets is a finite set;
   (d) the product of a finite family $\{X_i : i \in I\}$ of finite sets is a finite set.

   For (a), use the fact that a subset of a finite set is finite.
   For (b), use induction on card(Y).
   For (c) and (d), use induction on card(I).

5. Prove, within **ZF** theory, that the following two assertions are equivalent:

   (a) the Axiom of Choice (**AC**);
   (b) for any two sets E and F, if $g : E \to F$ is a surjection, then there exists a function $h : F \to E$ such that $g \circ h = \mathrm{Id}_F$.

6. Demonstrate within **ZF** theory that the following three assertions are equivalent:

   (i) A is an infinite set;
   (ii) there exist an element $a \in A$ and a bijection $f : A \to A \setminus \{a\}$;
   (iii) there exists a bijection of A onto its proper subset.

7. Verify that the following statements hold true:

   (a) every subset of a countable set is countable;
   (b) the image of a countable set under any partial mapping is countable;
   (c) the union of a finite family of countable sets is a countable set;
   (d) the product of a finite family of countable sets is a countable set;
   (e) the union of a countable family of countable sets is countable.

Check that the statements (a), (b), (c), (d) are provable in **ZF** theory.
For (e), use some weak form of **AC** and the fact that $\mathbb{N}$ is equinumerous with $\mathbb{N}^2$.

8. Let A be an uncountable set and let B be a countable subset of A.
Show that $A \sim A \setminus B$ and conclude that $A \setminus B$ is also an uncountable set.
For this purpose, argue as follows. First of all, observe that the set $A \setminus B$ is necessarily infinite, hence contains a subset C equinumerous with $\mathbb{N}$. One delicate point should be noted, namely, infinite sets in the standard sense (see Remark 3.9 below) are identified here with infinite sense in the Bolzano–Dedekind sense, so a certain weak version of **AC** is being utilized.
Then, for the set A, consider two representations

$$A = (A \setminus (B \cup C)) \cup (B \cup C),$$

$$A \setminus B = (A \setminus (B \cup C)) \cup C,$$

and keep in mind that $B \cup C \sim C$.

Remark 3.6 It can be demonstrated that, for a set E, the following two assertions are equivalent:

(1) E is uncountable;
(2) for every mapping $f : E \to E$, there exists a proper subset X of E such that $X \sim E$ and $f(X) \subset X$.

For a proof of this equivalence, see [102].

9. Using Cantor's inequality $\mathrm{card}(E) < \mathrm{card}(\mathcal{P}(E))$, infer that there exists no set X satisfying the relation $(\forall x)(x \in X)$.
In other words, no set can be universal for the class of all sets.

10. Let X be a set, let $(Y, \leq)$ be a well-ordered set, and suppose that there exists an injective function

$$g : X \times Y \to X \cup Y.$$

Work in **ZF** theory and show that the cardinal numbers $\mathrm{card}(X)$ and $\mathrm{card}(Y)$ are comparable, i.e.,

$$\mathrm{card}(X) \leq \mathrm{card}(Y) \;\vee\; \mathrm{card}(Y) \leq \mathrm{card}(X).$$

For this purpose, argue as follows. Without loss of generality assume that X and Y are disjoint and g is a bijection. In $X \times Y$ take the sets $g^{-1}(X)$ and $g^{-1}(Y)$ and consider all sets of the form $\{x\} \times Y$, where x ranges over X. If, for some $x \in X$, the set $\{x\} \times Y$ is entirely contained in $g^{-1}(X)$, then $\mathrm{card}(Y) \leq \mathrm{card}(X)$.
If, for each $x \in X$, one has

3 Elementary Facts on Cardinal Numbers   33

$$(\{x\} \times Y) \cap g^{-1}(Y) \neq \emptyset,$$

then there is an injective mapping from X into Y, which implies the inequality $\operatorname{card}(X) \leq \operatorname{card}(Y)$.

11. If $(E, \leq)$ is a well-ordered set, then the ordinal type of $(E, \leq)$ is called the ordinal number (in short, ordinal) associated with $(E, \leq)$.
   This is Cantor's classical definition of ordinal numbers (ordinals).
   Consider the following relation $\leq$ for any two ordinals α and β:

   $\alpha \leq \beta$ if and only if α is isomorphic to some initial subinterval of β.

   In this definition an initial subinterval of β can coincide with β (in other words, β itself is an improper initial subinterval of β).
   Check that:

   (a) no ordinal α is isomorphic to its proper initial subinterval;
   (b) if α and β are any two ordinals, then either $\alpha \leq \beta$ or $\beta \leq \alpha$ (so $\leq$ turns out to be a linear ordering in every set of ordinals);
   (c) the relation $\leq$ is a well-ordering (in every set of ordinals);
   (d) there does not exist a largest ordinal (equivalently, the class of all ordinals is not a set).

12. The so-called Replacement Axiom of **ZF** theory states that, for any function f and for any set X, the image $f(X)$ is a set (see [76, 108, 111]).
   Using this axiom, prove in **ZF** that if A is an arbitrary set, then there exists an ordinal (or, equivalently, a well-ordered set) which cannot be injectively mapped into A.
   For this purpose, suppose on the contrary that, for every ordinal α, there exists an injective function acting from α into A. Then every α turns out to be isomorphic to an appropriate well-ordered subset of A. Deduce from this fact that in such a case there exists a mapping from some subset of $\mathcal{P}(A \times A)$ onto the class of all ordinals. But the latter contradicts (d) of Exercise 11 and the Replacement Axiom.

13. Verify that in **ZF** theory the following two assertions are equivalent:

   (a) the Axiom of Choice;
   (b) any set can be equipped with some well-ordering.

   Argue as follows. The implication (b) $\Rightarrow$ (a) is almost trivial.
   To prove the implication (a) $\Rightarrow$ (b), take an arbitrary set A and an ordinal α which cannot be injectively mapped into A (see Exercise 12). Let

   $$\phi : \mathcal{P}(A) \to A$$

   be a choice function for $\mathcal{P}(A)$, which means that $\phi(X) \in X$ for every nonempty subset X of A. Pick any $b \notin A$.

Apply the method of transfinite recursion and define an α-sequence $\{a_\beta : \beta < \alpha\}$ as follows:

$$a_\beta = \begin{cases} \phi(A \setminus \{a_\gamma : \gamma < \beta\}) & \text{if } A \setminus \{a_\gamma : \gamma < \beta\} \neq \emptyset, \\ b & \text{if } A \setminus \{a_\gamma : \gamma < \beta\} = \emptyset. \end{cases}$$

Infer from the property of α that there exists a least $\beta_0 < \alpha$ such that $a_{\beta_0} = b$. Therefore, the β_0-sequence $\{a_\gamma : \gamma < \beta_0\}$ is injective and consists of all elements of A. This implies that A can be endowed with a well-ordering isomorphic to β_0.

Remark 3.7 Recall that the result of Exercise 13 is due to E. Zermelo (see his celebrated work [190]). Also, taking into account Exercises 10, 11, 12, and 13, it is not difficult to prove Tarski's theorem which states that in **ZF** theory the following two assertions are equivalent:

(1) the Axiom of Choice (**AC**);
(2) for any infinite set X, the product set $X \times X$ is equinumerous with X.

14. A set T is called *transitive* if, for any x and y, the relations $x \in T$ and $y \subset x$ imply $y \in T$.
    Check that:

    (a) if X is a transitive set, then $X \cup \{X\}$ is also transitive;
    (b) if $\{X_i : i \in I\}$ is an arbitrary family of transitive sets, then the set $\cup \{X_i : i \in I\}$ is transitive;
    (c) if $\{Y_j : j \in J\}$ is an arbitrary nonempty family of transitive sets, then the set $\cap \{Y_j : j \in J\}$ is transitive.

Remark 3.8 Transitive sets play an important role in those mathematical topics which are concerned with the logical foundations of **ZFC** theory, because such sets serve as appropriate models for various fragments of this theory and for its relatively consistent extensions (see, e.g., [76, 108]).

15. Let $(X, \preceq)$ be any well-ordered set. For each element $x \in X$, define the set $o(x)$ by transfinite recursion as follows:

$$o(x) = \{o(z) : z \prec x\}.$$

Verify that this definition does not depend on the original set $(X, \preceq)$. In other words, if $(Y, \preceq)$ is another well-ordered set and elements $x \in X$ and $y \in Y$ are such that the two associated initial intervals

$$X_x = \{z \in X : z \prec x\}, \qquad Y_y = \{z \in Y : z \prec y\}$$

are isomorphic, then one has the equality $o(x) = o(y)$.

3 Elementary Facts on Cardinal Numbers

The sets of the form $o(x)$ are called *von Neumann's ordinal numbers* (briefly, *von Neumann's ordinals*) and are usually denoted by Greek letters $\alpha, \beta, \gamma, \xi, \eta, \zeta, \ldots$ In the sequel, only this type of ordinal will be considered.
Demonstrate that:

(a) any ordinal α is a transitive set;
(b) if α is an ordinal, then $\alpha \cup \{\alpha\}$ is also an ordinal (called the *successor* of α and usually denoted by the symbol $\alpha + 1$);
(c) if $\{\alpha_i : i \in I\}$ is a nonempty family of ordinals, then

$$\cup \{\alpha_i : i \in I\}, \qquad \cap \{\alpha_i : i \in I\}$$

are ordinals, too;
(d) any ordinal α is a well-ordered set with respect to $\subset$; in other words, the relation

$$x \in \alpha \ \& \ y \in \alpha \ \& \ x \subset y$$

is a well-ordering of α; analogously, the relation

$$x \in \alpha \ \& \ y \in \alpha \ \& \ (x \in y \ \vee \ x = y)$$

is a well-ordering of α which coincides with the previous one;
(e) if, for some transitive set α, the above relation is a well-ordering of α, then α is an ordinal;
(f) for an arbitrary ordinal α, one has $\alpha \notin \alpha$;
(g) (α is an ordinal) $\Leftrightarrow$ (α is a transitive set and all elements of α are transitive sets).

To establish (f), apply the principle of transfinite induction. To establish (g), use the so-called Axiom of Foundation, which states that every nonempty set X has an element x such that $X \cap x = \emptyset$ (see [76, 108]).

16. As has already been mentioned, $\mathbb{N}$ is the least (by inclusion) set, one of the elements of which is $\emptyset$ and which is closed under the unary operation $'$ defined by $x' = x \cup \{x\}$. Recall that the existence of $\mathbb{N}$ is guaranteed by one of the axioms of **ZF** theory.
Check in detail that:

(a) if $n \in \mathbb{N}$, then $n' \neq \emptyset$; in addition, if $n \neq \emptyset$, then there exists an $m \in \mathbb{N}$ such that $n = m'$;
(b) if $n \in \mathbb{N}$ and $n \neq \emptyset$, then $\emptyset \in n$;
(c) all elements of $\mathbb{N}$, as well as $\mathbb{N}$ itself, are transitive sets;
(d) all elements of $\mathbb{N}$ are finite sets in the sense of Bolzano and Dedekind;
(e) if $m \in \mathbb{N}$ and $n \in \mathbb{N}$ are such that $m' = n'$, then $m = n$;
(f) $\mathbb{N}$ is infinite (in the sense of Bolzano and Dedekind);
(g) if $m \in \mathbb{N}$ and $n \in \mathbb{N}$, then $m \cup n$ and $m \cap n$ also belong to $\mathbb{N}$;

(h) for any two elements m and n from $\mathbb{N}$, the disjunction

$$m \in n \lor n \in m \lor m = n$$

holds true;

(i) $\mathbb{N}$ is well-ordered by a binary relation R defined by the formula

$$R(m, n) \Leftrightarrow m \in \mathbb{N} \,\&\, n \in \mathbb{N} \,\&\, m \subset n;$$

(j) $R(m, n)$ is equivalent to $m \in \mathbb{N} \,\&\, n \in \mathbb{N} \,\&\, (m \in n \lor m = n)$.

When checking the validity of all assertions (a)–(j), keep in mind the definition of $\mathbb{N}$.

17. Since $\mathbb{N}$ of Exercise 16 is well-ordered by R, there exists a unique isomorphism of $\mathbb{N}$ onto some ordinal ω in the sense of von Neumann (see Exercise 15). Show that this isomorphism is an identity mapping, so $\mathbb{N} = \omega$. Also, deduce from this fact that $n = \{m \in \mathbb{N} : m \in n\}$ for every $n \in \mathbb{N}$.

Remark 3.9 In **ZF** theory we have two standard definitions of finite and infinite sets.

Namely, one says that a set X is *finite in the standard sense* if there exists an element n of $\mathbb{N}$ such that $X \sim n$.

Accordingly, a set Y is *infinite in the standard sense* if Y is not finite in the standard sense.

It can easily be seen that Y is infinite in the standard sense if and only if, for any natural number n, there exists an injective function from n into Y.

Also, the sets which are finite in the standard sense are finite in the sense of Bolzano and Dedekind (equivalently, any infinite set in the sense of Bolzano and Dedekind is infinite in the standard sense).

It turns out that the converse implication is not provable within **ZF** theory but is valid under some weak versions of the Axiom of Choice. For more details about this phenomenon, we refer the reader to [66, 75, 76] (see also the next exercise).

18. Consider the following weak form of **AC**:
 For every family $\{X_n : n \in \mathbb{N}\}$ of nonempty finite sets (in the standard sense), there exists a family $\{x_n : n \in \mathbb{N}\}$ such that $(\forall n \in \mathbb{N})(x_n \in X_n)$.
 Demonstrate that, adding this statement to the axioms of **ZF** theory, any infinite set in the standard sense becomes infinite in the sense of Bolzano and Dedekind.

19. Show that the following two assertions are equivalent in **ZF** theory:

    (a) every infinite set in the standard sense is infinite in the sense of Bolzano and Dedekind;
    (b) for any set X, if there exists a surjection of X onto $\mathbb{N}$, then there exists an injective function acting from $\mathbb{N}$ into X.

Chapter 4
Some Properties of the Continuum

The real line $\mathbb{R}$ is one of the main objects in mathematical analysis.

This object can be introduced axiomatically, as a Dedekind complete linearly ordered field (see, e.g., [35]).

More concretely, $\mathbb{R}$ can be defined as the Dedekind completion (via Dedekind cuts) of the linearly ordered field $\mathbb{Q}$ of all rational numbers.

Also, $\mathbb{R}$ can be obtained by Cantor's method, as the family of all equivalence classes of Cauchy sequences of rational numbers, where two Cauchy sequences

$$\{p_n : n \in \mathbb{N}\} \subset \mathbb{Q}, \qquad \{q_n : n \in \mathbb{N}\} \subset \mathbb{Q}$$

are announced to be equivalent if $\lim_{n\to\infty}|p_n - q_n| = 0$.

The objects obtained in these three different manners turn out to be isomorphic, so one can speak of the uniqueness of $\mathbb{R}$ up to isomorphism.

Let us recall the following five important properties of $\mathbb{R}$.

1. Every nonempty subset of $\mathbb{R}$ which is bounded from above has a supremum (the *Dedekind completeness* of $\mathbb{R}$).
2. If $\{\triangle_n : n \in \mathbb{N}\}$ is a decreasing (by inclusion) sequence of segments in $\mathbb{R}$ and the length of $\triangle_n$ tends to zero as n tends to infinity, then

$$\cap \{\triangle_n : n \in \mathbb{N}\} \neq \emptyset$$

   (*Cantor's Axiom*—another form of the completeness of $\mathbb{R}$).
3. If ε and a are any two strictly positive real numbers, then there exists a natural number k such that $k\varepsilon > a$ (the *Axiom of Archimedes*).
4. Any segment $[a, b]$ in $\mathbb{R}$ is compact, i.e., every open covering of $[a, b]$ contains a finite subcovering of $[a, b]$ (the *local compactness* of $\mathbb{R}$).
5. The set $\mathbb{Q}$ of all rational numbers is everywhere dense in $\mathbb{R}$, i.e., $\mathbb{Q}$ has common points with any nonempty open interval in $\mathbb{R}$ (the *separability* of $\mathbb{R}$).

Remark 4.1 The Dedekind completeness of $\mathbb{R}$ and Cantor's Axiom are logically equivalent statements, and each of them implies the Axiom of Archimedes (see Exercises 1 and 2 in the present chapter).

The uncountability of $\mathbb{R}$ follows from its completeness and also from its local compactness.

Theorem 4.1 *Cantor's Axiom implies the uncountability of* $\mathbb{R}$.

Proof Assume Cantor's Axiom and suppose to the contrary that $\mathbb{R}$ is countable, i.e., $\mathbb{R} = \{x_n : n \in \mathbb{N}\}$. Divide the segment $[0, 1]$ into three congruent subintervals and choose the first (from the left) subsegment which does not contain the point x_0. Denote this subsegment by $\triangle_0$. Then proceed by induction. If the segment $\triangle_n$ has already been defined, then divide it into three congruent subsegments, choose the first (from the left) subsegment which does not contain the point x_{n+1} and denote it by $\triangle_{n+1}$. Proceeding in this manner, we obtain a decreasing (by inclusion) sequence of segments $\{\triangle_n : n \in \mathbb{N}\}$. By virtue of Cantor's Axiom, there exists a (unique) point

$$x \in \cap\{\triangle_n : n \in \mathbb{N}\}.$$

Since, for every $n \in \mathbb{N}$, we have $x \in \triangle_n$ and $x_n \notin \triangle_n$, we infer that $x \neq x_n$, which contradicts the equality $\mathbb{R} = \{x_n : n \in \mathbb{N}\}$. Actually, this argument shows that even the segment $[0, 1]$ is uncountable. $\square$

Theorem 4.2 *The local compactness of* $\mathbb{R}$ *(equivalently, the compactness of some non-degenerate segment in* $\mathbb{R}$*) implies the uncountability of* $\mathbb{R}$.

Proof Assume that a non-degenerate segment $[a, b] \subset \mathbb{R}$ is compact, i.e., any open covering of $[a, b]$ contains a finite subcovering. In order to show the validity of Theorem 4.2, first observe that if $\{U_i : i \in I\}$ is an arbitrary finite family of open intervals in $\mathbb{R}$ such that

$$[a, b] \subset \cup\{U_i : i \in I\},$$

then $b - a < \sum\{l(U_i) : i \in I\}$, where $l(U_i)$ denotes the length of U_i for each $i \in I$. This fact can easily be proved by induction on $\text{card}(I)$.

Now, suppose to the contrary that $\mathbb{R}$ (and hence $[a, b]$) is countable, i.e., $[a, b] = \{x_n : n \in \mathbb{N}\}$. To each x_n associate an open interval U_n whose mid-point is x_n and whose length is $(b - a)/2^{n+1}$. We thus come to an open covering $\{U_n : n \in \mathbb{N}\}$ of $[a, b]$. In view of the compactness of $[a, b]$, there exists a finite family

$$\{U_{n_1}, U_{n_2}, ..., U_{n_k}\} \subset \{U_n : n \in \mathbb{N}\}$$

such that $n_1 < n_2 < ... < n_k$ and $[a, b] \subset U_{n_1} \cup U_{n_2} \cup ... \cup U_{n_k}$. Therefore, we must have

$$b - a < l(U_{n_1}) + l(U_{n_2}) + \cdots + l(U_{n_k}).$$

On the other hand, it is easy to see that

$$l(U_{n_1}) + l(U_{n_2}) + \cdots + l(U_{n_k}) =$$
$$(b-a)(1/2^{n_1+1} + 1/2^{n_2+1} + \cdots + 1/2^{n_k+1}) < b - a.$$

The obtained contradiction ends the proof. □

As was mentioned in Chap. 3, $\mathbb{R}$ is usually treated as a canonical model of the continuum, so card($\mathbb{R}$) is called the cardinality of the continuum and is denoted by **c**.

The next theorem shows some homogeneity of the real line from the set-theoretical viewpoint.

Theorem 4.3 *All non-degenerate segments, non-degenerate open intervals, and non-degenerate half-open intervals are of cardinality continuum.*

Proof First observe that any non-degenerate segment $[a, b]$ is equinumerous with $[0, 1]$. Indeed, the affine function $f : \mathbb{R} \to \mathbb{R}$ defined by

$$f(x) = (b-a)x + a \qquad (x \in \mathbb{R})$$

induces a bijection (even homeomorphism) between $[0, 1]$ and $[a, b]$. As a trivial consequence, we obtain that $]0, 1[$ is equinumerous (homeomorphic) with $]a, b[$.

Further, as widely known, the standard trigonometric function

$$\tan \,:\,]-\pi/2, \pi/2[\, \to \mathbb{R}$$

is a homeomorphism between the open interval $]-\pi/2, \pi/2[$ and $\mathbb{R}$.

So, the required result follows now from the Cantor–Bernstein theorem (see Chap. 3). All omitted (easy) details are left to the reader. □

More precise information on card($\mathbb{R}$) gives the next theorem.

Theorem 4.4 $\mathrm{card}(\mathbb{R}) = \mathrm{card}(\mathbb{N}^{\mathbb{N}}) = \mathrm{card}(\mathcal{P}(\mathbb{N}))$.

Proof For every function $\phi : \mathbb{N} \setminus \{0\} \to \mathbb{N} \setminus \{0\}$, consider the real number

$$t(\phi) = 1/2^{\phi(1)} + 1/2^{\phi(1)+\phi(2)} + \cdots + 1/2^{\phi(1)+\phi(2)+\cdots+\phi(n)} + \cdots.$$

Then $t(\phi) \in \,]0, 1]$ and, conversely, for every $t \in \,]0, 1]$ there exists a unique function $\phi : \mathbb{N} \setminus \{0\} \to \mathbb{N} \setminus \{0\}$ such that $t = t(\phi)$. This bijective mapping (correspondence) shows us that

$$\mathbb{R} \sim \,]0, 1] \sim \mathbb{N}^{\mathbb{N}}.$$

Further, using the characteristic functions of subsets of $\mathbb{N}$, we may write

$$\mathrm{card}(P(\mathbb{N})) = \mathrm{card}(2^{\mathbb{N}}) \leq \mathrm{card}(\mathbb{N}^{\mathbb{N}}).$$

On the other hand, using Cantor's inequality, we get

$$\mathrm{card}(\mathbb{N}^\mathbb{N}) \leq \mathrm{card}((2^\mathbb{N})^\mathbb{N}) = \mathrm{card}(2^{\mathbb{N}\times\mathbb{N}})$$
$$= \mathrm{card}(2^\mathbb{N}) = \mathrm{card}(\mathcal{P}(\mathbb{N})),$$

which implies $\mathrm{card}(\mathbb{N}^\mathbb{N}) = \mathrm{card}(2^\mathbb{N})$ in view of the Cantor–Bernstein theorem. So we finally obtain

$$\mathrm{card}(\mathbb{R}) = \mathrm{card}(\mathbb{N}^\mathbb{N}) = \mathrm{card}(\mathcal{P}(\mathbb{N})),$$

and the proof is complete. □

Theorem 4.5 *The relation* $\mathbb{R} \sim \mathbb{R}^\mathbb{N}$ *holds in* **ZF** *theory. Consequently, for each nonzero* $n \in \mathbb{N}$, *one has* $\mathbb{R} \sim \mathbb{R}^n$.

Proof Obviously, we may write

$$\mathbb{R}^\mathbb{N} \sim (\mathbb{N}^\mathbb{N})^\mathbb{N} \sim \mathbb{N}^{\mathbb{N}\times\mathbb{N}} \sim \mathbb{N}^\mathbb{N} \sim \mathbb{R}.$$

Keeping in mind the Cantor–Bernstein theorem, we also get $\mathbb{R} \sim \mathbb{R}^n$, whenever n is a nonzero natural number.

Theorem 4.5 has thus been proved. □

Remark 4.2 Let us introduce the notation:

$$[\mathbb{R}]^{<\omega} = \text{ the family of all finite subsets of } \mathbb{R};$$
$$[\mathbb{R}]^{\leq\omega} = \text{ the family of all countable subsets of } \mathbb{R}.$$

Working within **ZF** theory, it is not difficult to show that

$$\mathrm{card}([\mathbb{R}]^{<\omega}) = \mathbf{c}.$$

The inequality $\mathrm{card}([\mathbb{R}]^{\leq\omega}) \geq \mathbf{c}$ is also valid in the same theory. However, the reverse inequality

$$\mathrm{card}([\mathbb{R}]^{\leq\omega}) \leq \mathbf{c}$$

is not provable within **ZF**. Moreover, in the framework of **ZF** it cannot be proved that the cardinality of Vitali's partition $\mathbb{R}/V$ of $\mathbb{R}$ (see Exercise 8 and Remark 1.1 from Chap. 1) does not exceed $\mathbf{c}$.

The next theorem is a very special case of König's inequality in general set theory (see [76, 108, 111, 167], and Exercise 7).

Theorem 4.6 *Let $\{X_n : n \in \mathbb{N}\}$ be a family of subsets of $\mathbb{R}$ satisfying the relation $\cup\{X_n : n \in \mathbb{N}\} = \mathbb{R}$.*
Then there exists an index $n \in \mathbb{N}$ such that $X_n \sim \mathbb{R}$.

Proof Since $\mathbb{R} \sim \mathbb{R}^\mathbb{N}$ (see Theorem 4.5), it suffices to demonstrate the validity of the following statement:

If $\mathbb{R}^\mathbb{N} = \cup\{Z_n : n \in \mathbb{N}\}$, then there exists an $n \in \mathbb{N}$ such that $Z_n \sim \mathbb{R}$.

Suppose otherwise, i.e., $\mathbb{R}^\mathbb{N} = \cup\{Z_n : n \in \mathbb{N}\}$ but $\text{card}(Z_n) < \text{card}(\mathbb{R})$ for all $n \in \mathbb{N}$. Represent $\mathbb{R}^\mathbb{N}$ in the form of a countable product

$$\mathbb{R}^\mathbb{N} = R_0 \times R_1 \times \cdots \times R_n \times \cdots,$$

where $(\forall n \in \mathbb{N})(R_n = \mathbb{R})$, and for each $n \in \mathbb{N}$ take the set $\text{pr}_n(Z_n)$. In view of our assumption,

$$\text{card}(\text{pr}_n(Z_n)) \leq \text{card}(Z_n) < \text{card}(\mathbb{R}).$$

Consequently, there exists a point $z_n \in \mathbb{R} \setminus \text{pr}_n(Z_n)$.

Consider now the element $z = \{z_n : n \in \mathbb{N}\}$ of $\mathbb{R}^\mathbb{N}$. Since

$$\text{pr}_n(z) = z_n \notin \text{pr}_n(Z_n)$$

for all $n \in \mathbb{N}$, we conclude that $z \notin \cup\{Z_n : n \in \mathbb{N}\}$, which contradicts the equality $\mathbb{R}^\mathbb{N} = \cup\{Z_n : n \in \mathbb{N}\}$.

The obtained contradiction ends the proof. □

Let f be an arbitrary real-valued function on $\mathbb{R}$. The following natural question arises:

Does there exist a set $X \subset \mathbb{R}$ such that $X \sim \mathbb{R}$ and the restriction $f|X$ of f to X is a bounded function?

The preceding considerations enable us to answer this question positively. Indeed, for the given function f, we may write

$$\mathbb{R} = \cup\{f^{-1}([-n, n]) : n \in \mathbb{N}\}.$$

Denoting $f^{-1}([-n, n])$ by X_n, we come to the representation

$$\mathbb{R} = \cup\{X_n : n \in \mathbb{N}\}.$$

According to Theorem 4.6, there exists an $n \in \mathbb{N}$ such that $X_n \sim \mathbb{R}$. Now, it is clear that $f|X_n$ is a bounded function. Namely, we have

$$(\forall x \in X_n)(|f(x)| \leq n).$$

In general, the descriptive structure of such X_n may be rather complicated. This is illustrated by the next theorem.

Theorem 4.7 *Within **ZF** theory, there exists a function $g : \mathbb{R} \to \mathbb{Q}$ whose graph is everywhere dense in the plane $\mathbb{R}^2$.*

In particular, the function g is unbounded from above and from below on every nonempty open subinterval of $\mathbb{R}$.

Proof Let $\{p_k : k \in \mathbb{N}\}$ be an injective enumeration of all rational numbers in $\mathbb{R}$. Since $\mathbb{Q}^3 \sim \mathbb{N}^3 \sim \mathbb{N}$, we may denote by $\{D_n : n \in \mathbb{N}\}$ the family of all those open discs in $\mathbb{R}^2$ whose centers have rational coordinates and whose radii are strictly positive rational numbers.

We first define the required g on a certain subset of $\mathbb{Q}$. At the beginning, take $(r, t) \in \mathbb{Q}^2 \cap D_0$ and put $x_0 = r$ and $g(x_0) = t$.

Suppose now that the values

$$g(x_0) \in \mathbb{Q}, \ g(x_1) \in \mathbb{Q}, \ldots, \ g(x_n) \in \mathbb{Q}$$

have already been determined for some pairwise distinct points

$$x_0 \in \mathbb{Q}, \ x_1 \in \mathbb{Q}, \ldots, \ x_n \in \mathbb{Q}$$

so that $(x_i, g(x_i)) \in D_i$, where $i = 0, 1, \ldots, n$.

Consider the disc D_{n+1} and its first projection $\mathrm{pr}_1(D_{n+1})$. Clearly, the set $\mathrm{pr}_1(D_{n+1})$ is a non-degenerate open interval in $\mathbb{R}$ and contains infinitely many rational numbers. Let p be the rational point in $\mathrm{pr}_1(D_{n+1})$ different from all points $x_0, x_1, \ldots, x_n$ and having the least index in the enumeration $\{p_k : k \in \mathbb{N}\}$. We put $x_{n+1} = p$. Further, consider the set

$$D_{n+1}(x_{n+1}) = \{y \in \mathbb{R} : (x_{n+1}, y) \in D_{n+1}\}.$$

Obviously, $D_{n+1}(x_{n+1})$ is also a non-degenerate open interval in $\mathbb{R}$. Let q be the rational point from this interval having the least index in the enumeration $\{p_k : k \in \mathbb{N}\}$. We then put $g(x_{n+1}) = q$. Evidently,

$$(x_{n+1}, g(x_{n+1})) \in D_{n+1}.$$

Proceeding in this manner, we effectively define the two sequences

$$\{x_0, x_1, \ldots, x_n, \ldots\} \subset \mathbb{Q}, \qquad \{g(x_0), g(x_1), \ldots, g(x_n), \ldots\} \subset \mathbb{Q}.$$

Finally, for each point $x \in \mathbb{R} \setminus \{x_0, x_1, \ldots, x_n, \ldots\}$, we put $g(x) = 0$.

The obtained function $g : \mathbb{R} \to \mathbb{Q}$ is as required.

Indeed, let D be an arbitrary open disc in $\mathbb{R}^2$. There exists an index $n \in \mathbb{N}$ such that $D_n \subset D$. By virtue of our construction, $(x_n, g(x_n)) \in D_n$ and, consequently, $(x_n, g(x_n)) \in D$. This shows us that the graph of g is everywhere dense in $\mathbb{R}^2$.

Theorem 4.7 has thus been proved. $\square$

Exercises

1. Argue within **ZF** theory and establish that the Dedekind completeness of $\mathbb{R}$ is equivalent to Cantor's Axiom.
2. Prove within **ZF** theory that the Dedekind completeness of $\mathbb{R}$ implies the Axiom of Archimedes.
3. Demonstrate that the Axiom of Archimedes does not imply the completeness of $\mathbb{R}$.
4. Work again in **ZF** theory and show that $\mathbb{R}$ cannot be represented as the union of a countable family of finite sets.
   Moreover, verify that this statement follows from Cantor's Axiom.
5. Prove, within **ZF** theory, that the following two assertions are equivalent:

   (a) if $\mathbb{R} = X \cup Y$, then either $\mathbb{R} \sim X$ or $\mathbb{R} \sim Y$;
   (b) for every natural number n, if $\mathbb{R} = X_1 \cup X_2 \cup ... \cup X_n$, then there exists a natural index $i \in [1, n]$ such that $\mathbb{R} \sim X_i$.

   To prove the implication (a) $\Rightarrow$ (b), use induction on n.
6. Let $\{X_i : i \in I\}$ and $\{Y_i : i \in I\}$ be two families of sets such that, for each index $i \in I$, the set X_i is a proper subset of Y_i (so $X_i \neq \emptyset$ and $\text{card}(Y_i) \geq 2$). Let $P = \prod\{Y_i : i \in I\}$.
   Demonstrate that there exists a disjoint family $\{Z_i : i \in I\}$ of subsets of P such that $Z_i \sim X_i$ for any $i \in I$. Conclude from this fact that the inequality

   $$\text{card}(\cup\{X_i : i \in I\}) \leq \text{card}(P)$$

   holds true.
7. Show the validity of König's inequality:
   If $\{X_i : i \in I\}$ and $\{Y_i : i \in I\}$ are any two families of sets such that

   $$(\forall i \in I)(\text{card}(X_i) < \text{card}(Y_i)),$$

   then

   $$\text{card}(\cup\{X_i : i \in I\}) < \text{card}(\prod\{Y_i : i \in I\}).$$

   For this purpose, assume without loss of generality that all sets X_i, where $i \in I$, are nonempty and argue analogously to the proof of Theorem 4.6, keeping in mind Exercise 6.
8. Deduce from König's inequality Cantor's inequality.
   In order to establish the required implication, consider a disjoint family of sets $\{X_i : i \in I\}$, where $\text{card}(X_i) = 1$ for all $i \in I$, and take another family of sets $\{Y_i : i \in I\}$, where $\text{card}(Y_i) = 2$ for all $i \in I$. Conclude that

   $$\text{card}(I) < \text{card}(2^I),$$

which is equivalent to Cantor's inequality.

Remark 4.3 A weak side of König's inequality is that its proof needs the Axiom of Choice, while Cantor's inequality is provable within the framework of **ZF** set theory.

9. A model of **ZF** theory was constructed by A. Levy in which the real line $\mathbb{R}$ is representable as the union of a countable family of countable sets (see, e.g., [66, 75, 76]).
   Check that:

   (a) in this model there exists a function $f : \mathbb{R} \to \mathbb{R}$ such that, for every uncountable subset X of $\mathbb{R}$, the restriction $f|X$ is unbounded;
   (b) any nonempty open subset of $\mathbb{R}$ contains infinitely many discontinuity points of f;
   (c) in the same model no uncountable subset of $\mathbb{R}$ is equinumerous with a well-ordered set.

   Remark 4.4 Actually, within **ZF** theory, $\mathbb{R}$ is not a union of countably many countable sets if and only if any real-valued function on $\mathbb{R}$ is bounded on an appropriate uncountable subset of $\mathbb{R}$. Using the Axiom of Choice, Sierpiński and Zygmund [168] were able to construct a function $g : \mathbb{R} \to \mathbb{R}$ having the property that, for each set $X \subset \mathbb{R}$ with $\text{card}(X) = \mathbf{c}$, the restricted function $g|X$ is not continuous on X. Notice that such a g can be a bounded function. Some additional interesting facts about Sierpiński–Zygmund type functions may be found in [32, 51, 93, 94, 136, 147].

10. Denote by $\mathcal{H}$ the family of all those functions $h : \mathbb{R} \to \mathbb{R}$ whose graphs are everywhere dense in the plane $\mathbb{R}^2$.
    Work in **ZF** theory and prove that $\text{card}(\mathcal{H}) = \text{card}(\mathbb{R}^{\mathbb{R}}) = 2^{\mathbf{c}}$.

11. Let E be a ground set and let X be a set for which the relation $E^X \sim E$ holds true. Suppose also that E contains some well-orderable subset Y.
    Show in **ZF** theory that there exists no family of sets $\{Z_i : i \in I\}$ satisfying the following three conditions:

    (a) $\text{card}(I) = \text{card}(X)$;
    (b) $\cup \{Z_i : i \in I\} = E$;
    (c) $\text{card}(Z_i) < \text{card}(Y)$ for each $i \in I$.

12. Demonstrate that there is no countably infinite partition of the segment $[0, 1]$ into closed subsets of $[0, 1]$.
    For this purpose, use Cantor's Axiom and the fact that $[0, 1]$ is connected.

13. Let $\mathcal{F}$ be a family of functions acting from $\mathbb{R}$ into itself and let g be a function also acting from $\mathbb{R}$ into itself.
    This g is called *universal* for $\mathcal{F}$ (with respect to the family of all everywhere dense subsets of $\mathbb{R}$) if, for any function $f \in \mathcal{F}$, there exists an everywhere dense set $X = X_f$ in $\mathbb{R}$ such that $f|X = g|X$.

Work in **ZF** theory and prove that if card($\mathcal{F}$) $\leq$ **c**, then there exists a universal function for $\mathcal{F}$.

For this purpose, use an effective partition of $\mathbb{R}$ into continuum many everywhere dense countable subsets of $\mathbb{R}$.

On the other hand, check that if $\mathcal{F}$ is the family of all functions acting from $\mathbb{R}$ into $\mathbb{R}$, then there exists no universal function for $\mathcal{F}$.

Remark 4.5 According to the result of Exercise 13, there exists a universal function g for the family $C(\mathbb{R}, \mathbb{R})$ of all continuous mappings acting from $\mathbb{R}$ into itself.

Analogously, there exists a universal function h for the family $B(\mathbb{R}, \mathbb{R})$ of all Borel mappings acting from $\mathbb{R}$ into itself (recall that $f \in B(\mathbb{R}, \mathbb{R})$ if and only if $f^{-1}(U)$ is a Borel subset of $\mathbb{R}$ whenever U is an open subset of $\mathbb{R}$).

It is not hard to show that g cannot be a continuous function and h cannot be a Borel function.

14. Work in **ZF** theory and demonstrate that if X is a set of cardinality **c** and Y is a countable subset of X, then card($X \setminus Y$) = **c**.

    For this purpose, argue as follows. First, without loss of generality, identify X with the real line $\mathbb{R}$ and represent its countable subset Y in the form $Y = \{y_n : n \in \mathbb{N}\}$. Then recursively construct a family

    $$\{\Delta_{i_1, i_2, \ldots, i_n} : (i_1, i_2, \ldots, i_n) \in 2^n, \ n \in \mathbb{N}\}$$

    of non-degenerate segments in $\mathbb{R}$ with rational endpoints such that:

    (a) for each $n \in \mathbb{N}$, the family $\{\Delta_{i_1, i_2, \ldots, i_n} : (i_1, i_2, \ldots, i_n) \in 2^n\}$ is disjoint;
    (b) for each $n \in \mathbb{N}$, the set $\cup \{\Delta_{i_1, i_2, \ldots, i_n} : (i_1, i_2, \ldots, i_n) \in 2^n\}$ does not contain y_n;
    (c) the length of any segment $\Delta_{i_1, i_2, \ldots, i_n}$ does not exceed $1/2^n$;
    (d) for every $n \in \mathbb{N}$, the set $\Delta_{i_1, i_2, \ldots, i_n, 0} \cup \Delta_{i_1, i_2, \ldots, i_n, 1}$ is entirely contained in the interior of $\Delta_{i_1, i_2, \ldots, i_n}$.

    Finally, verify that the set

    $$Z = \cap \{\cup \{\Delta_{i_1, i_2, \ldots, i_n} : (i_1, i_2, \ldots, i_n) \in 2^n\} : n \in \mathbb{N}\}$$

    has cardinality continuum and is contained in $\mathbb{R} \setminus Y$ (actually, Z is homeomorphic to the classical Cantor set on $\mathbb{R}$).

Remark 4.6 It makes sense to compare the above exercise with Exercise 8 from Chap. 3. In this connection, let us recall that in Exercise 8 from Chap. 3 it is necessary to use a certain weak form of **AC**.

A much stronger statement than that given in Exercise 14 was obtained by Sierpiński. His result is formulated as follows.

It is provable within **ZF** theory that if **a** is an infinite cardinal in the Bolzano–Dedekind sense, X is a set with $\text{card}(X) = 2^{\mathbf{a}}$, and Y is a subset of X with $\text{card}(Y) = \mathbf{a}$, then $\text{card}(X \setminus Y) = 2^{\mathbf{a}}$.

For a detailed proof of this statement, see e.g. Sierpiński's well-known monograph [167].

15. Show that the following assertion does not contradict **ZF** theory:

There exists a function $\phi : \mathbb{R} \to \mathbb{R}$ which is unbounded on every uncountable subset of $\mathbb{R}$ and whose restriction to some uncountable set X in $\mathbb{R}$ is the identical mapping (hence $\phi|X$ is trivially continuous).

For this purpose, suppose that $\mathbb{R}$ is representable as a union of a countable family of countable sets and check that in $\mathbb{R}$ there exists an uncountable locally countable set X. Then consider the set $\mathbb{R} \setminus X$ and keep in mind Exercise 9.

Chapter 5
The Oscillation of a Real-Valued Function at a Point

Let E be a topological space. In this chapter, we will be dealing with various real-valued functions defined on the whole space E or on a subset of E.

In the sequel we adopt the convention that the empty subset $\emptyset$ of $\mathbb{R}$ is bounded in $\mathbb{R}$ (because $\emptyset$ is contained in any bounded subinterval of $\mathbb{R}$).

A partial function $f : E \to \mathbb{R}$ is called *locally bounded from above (from below)* at a point $x \in E$ if there exists a neighborhood $U(x)$ of this point such that the set $f(U(x))$ is bounded from above (from below) in $\mathbb{R}$.

In this definition we do not assume that $x \in \text{dom}(f)$.

Accordingly, we shall say that a partial function $f : E \to \mathbb{R}$ is *locally bounded* at $x \in E$ if there exists a neighborhood $U(x)$ of x such that the set $f(U(x))$ is bounded in $\mathbb{R}$.

Recall that a topological space E is *quasi-compact* if any open covering of E contains a finite subcovering. If, in addition, E is a Hausdorff space, then E is called a *compact space* (see [19]).

Theorem 5.1 *Let E be a quasi-compact topological space and let*

$$f : E \to \mathbb{R}$$

be a partial function locally bounded from above (from below) at all points of E.

Then f is bounded from above (from below) on the entire E.

Consequently, if $f : E \to \mathbb{R}$ is locally bounded at all points of E, then f is bounded on the entire E.

Proof It suffices to consider the case when f is locally bounded from above at each point of E.

For any $x \in E$, there exist an open neighborhood $U(x)$ and a real constant $t(x)$ such that

$$(\forall y)(y \in U(x) \cap \text{dom}(f) \Rightarrow f(y) \leq t(x)).$$

We thus come to the open covering $\{U(x) : x \in E\}$ of E and, by virtue of the quasi-compactness of E, there exist finitely many points $x_1, x_2, \ldots, x_n$ in E such that

$$E \subset U(x_1) \cup U(x_2) \cup \cdots \cup U(x_n).$$

Now, defining

$$t = \max(t(x_1), t(x_2), \ldots, t(x_n)),$$

we easily conclude that $f(x) \leq t$ for every point $x \in \text{dom}(f)$. This completes the proof. □

It immediately follows from Theorem 5.1 that if $f : \mathbb{R} \to \mathbb{R}$ is a partial function locally bounded from above (from below) at all points of $\mathbb{R}$, then the restriction of f to any subsegment of $\mathbb{R}$ is bounded from above (from below). Consequently, if f is locally bounded at all points of $\mathbb{R}$, then its restriction to any subsegment of $\mathbb{R}$ is bounded.

Let us consider more thoroughly the behavior of a given partial function $f : E \to \mathbb{R}$ at a fixed point $x_0 \in E$ which belongs to the closure of $\text{dom}(f)$.

In any case, there exist two values associated with f and x_0:

$$\limsup\nolimits_{x \to x_0} f(x), \qquad \liminf\nolimits_{x \to x_0} f(x).$$

These values are defined in a similar manner. For example, let us recall the definition of $\limsup_{x \to x_0} f(x)$. For this purpose, introduce the set

$$T = \{t \in \mathbb{R} : (\forall U(x_0))(\exists x \in U(x_0) \cap \text{dom}(f))(f(x) \geq t)\}$$

and stipulate

$$\limsup\nolimits_{x \to x_0} f(x) = \sup(T).$$

Note that if $\limsup_{x \to x_0} f(x)$ is finite and ε is an arbitrary strictly positive real number, then the following two conditions are satisfied:

$$(\forall U(x_0))(\exists z \in U(x_0) \cap \text{dom}(f))(f(z) > \limsup\nolimits_{x \to x_0} f(x) - \varepsilon),$$

$$(\exists V(x_0))(\forall z \in V(x_0) \cap \text{dom}(f))(f(z) < \limsup\nolimits_{x \to x_0} f(x) + \varepsilon).$$

Of course, the cases

$$\limsup\nolimits_{x \to x_0} f(x) = +\infty \ (-\infty), \qquad \liminf\nolimits_{x \to x_0} f(x) = -\infty \ (+\infty)$$

can take place. Clearly, we have the inequality

5 The Oscillation of a Real-Valued Function at a Point

$$\liminf_{x \to x_0} f(x) \le \limsup_{x \to x_0} f(x).$$

Also, it should be noticed that if $x_0 \in \mathrm{dom}(f)$, then

$$\liminf_{x \to x_0} f(x) \le f(x_0) \le \limsup_{x \to x_0} f(x),$$

so in this situation we cannot have

$$\limsup_{x \to x_0} f(x) = -\infty \ \vee\ \liminf_{x \to x_0} f(x) = +\infty.$$

Obviously, f is not locally bounded from above at x_0 if and only if

$$\limsup_{x \to x_0} f(x) = +\infty.$$

Analogously, f is not locally bounded from below at x_0 if and only if

$$\liminf_{x \to x_0} f(x) = -\infty.$$

So the values $\limsup_{x \to x_0} f(x)$ and $\liminf_{x \to x_0} f(x)$ yield a certain characterization of the behavior of f at x_0.

It makes sense to introduce the following value:

$$O_f(x_0) = \limsup_{x \to x_0} f(x) - \liminf_{x \to x_0} f(x).$$

This value is called the *oscillation* of f at x_0 (cf. [19, 35, 40, 84, 102, 141]). Notice that if

$$\limsup_{x \to x_0} f(x) = \liminf_{x \to x_0} f(x) = +\infty$$

or

$$\limsup_{x \to x_0} f(x) = \liminf_{x \to x_0} f(x) = -\infty,$$

then we stipulate $O_f(x_0) = +\infty$.

Some simple properties of the value $O_f(x_0)$ are presented in the next theorem.

Theorem 5.2 *The following five assertions hold:*

(1) $O_f(x_0) \ge 0$;
(2) $O_f(x_0)$ *is finite if and only if both values*

$$\limsup_{x \to x_0} f(x), \qquad \liminf_{x \to x_0} f(x)$$

are finite;

(3) *for any $t \in \mathbb{R}$, one has $O_{tf}(x_0) = |t| O_f(x_0)$; in particular,*

$$O_{-f}(x_0) = O_f(x_0);$$

(4) *if $x_0 \in \mathrm{dom}(f)$, then the equality $O_f(x_0) = 0$ is equivalent to the continuity of f at x_0;*

(5) *if, for a function $f : E \to \mathbb{R}$ and for some real $t \geq 0$, one has*

$$||f|| = \sup\{|f(x)| : x \in E\} \leq t,$$

then $O_f(x) \leq 2t$ for all $x \in E$.

The proof of Theorem 5.2 directly follows from the definition of $O_f(x_0)$ and is left to the reader.

Let us consider a more general situation when a partial function f from E to M is given, where (M, d) is a metric space (d denotes a metric on M).

In this case, for a point x_0 belonging to the closure of $\mathrm{dom}(f)$, the values $\limsup_{x \to x_0} f(x)$ and $\liminf_{x \to x_0} f(x)$ are, in general, senseless.

However, the oscillation of f at x_0 can be defined by the following formula:

$$O_f(x_0) = \lim_{U(x_0) \to x_0} \mathrm{diam}(f(U(x_0))).$$

Here $U(x_0)$ is any neighborhood of x_0, the symbol $\mathrm{diam}(f(U(x_0)))$ denotes the diameter of the set $f(U(x_0))$, and the limit is understood with respect to the filter $\mathcal{U}(x_0)$ of all neighborhoods of x_0. Equivalently, we may write

$$O_f(x_0) = \inf\{\mathrm{diam}(f(U)) : U \in \mathcal{U}(x_0)\}.$$

As earlier, we do not exclude the infinite value from the above formula, so the equality $O_f(x_0) = +\infty$ can take place for various metric spaces (M, d).

Fortunately, for the real line $\mathbb{R}$, the two definitions of the oscillation of a partial function $f : E \to \mathbb{R}$ at a point x_0 belonging to the closure of $\mathrm{dom}(f)$ are completely compatible.

Theorem 5.3 *If $M = \mathbb{R}$ is equipped with the standard metric of $\mathbb{R}$, then the two definitions of the oscillation of a function $f : E \to M$ at a point $x_0 \in E$ are equivalent, i.e., both definitions give the same value.*

Proof To demonstrate this equivalence, let us introduce the notation:

$$O_f(x_0) = \limsup_{x \to x_0} f(x) - \liminf_{x \to x_0} f(x);$$

$$O'_f(x_0) = \lim_{U(x_0) \to x_0} \mathrm{diam}(f(U(x_0))).$$

Our goal is to show that $O_f(x_0) = O'_f(x_0)$.

5 The Oscillation of a Real-Valued Function at a Point

First of all, it is not difficult to verify that

$$O_f(x_0) = +\infty \Leftrightarrow O'_f(x_0) = +\infty.$$

So we may restrict our further argument to the case when the relations $O_f(x_0) < +\infty$ and $O'_f(x_0) < +\infty$ are fulfilled simultaneously.

Take any real $\varepsilon > 0$. By the definitions of the values

$$\limsup_{x \to x_0} f(x), \qquad \liminf_{x \to x_0} f(x),$$

there exists a neighborhood $U(x_0)$ such that

$$(\forall z \in U(x_0))(f(z) < \limsup_{x \to x_0} f(x) + \varepsilon/2),$$

and there exists a neighborhood $V(x_0)$ such that

$$(\forall y \in V(x_0))(f(y) > \liminf_{x \to x_0} f(x) - \varepsilon/2).$$

Consider the neighborhood $W(x_0) = U(x_0) \cap V(x_0)$. Then, for any two points z and y from $W(x_0)$, we have

$$|f(z) - f(y)| < (\limsup_{x \to x_0} f(x) + \varepsilon/2) - (\liminf_{x \to x_0} f(x) - \varepsilon/2)$$
$$= O_f(x_0) + \varepsilon.$$

Consequently, we may write

$$\operatorname{diam}(f(W(x_0))) = \sup\{|f(z) - f(y)| : \{z, y\} \subset W(x_0)\} \leq O_f(x_0) + \varepsilon,$$
$$O'_f(x_0) \leq \operatorname{diam}(f(W(x_0))) \leq O_f(x_0) + \varepsilon.$$

Since $\varepsilon > 0$ was arbitrarily small, we deduce that

$$O'_f(x_0) \leq O_f(x_0).$$

To show the validity of the reverse inequality, take again any real $\varepsilon > 0$ and consider any neighborhood $G(x_0)$ of x_0. Again, by the definitions of the values $\limsup_{x \to x_0} f(x)$ and $\liminf_{x \to x_0} f(x)$, there exists a point $z \in G(x_0)$ such that

$$f(z) > \limsup_{x \to x_0} f(x) - \varepsilon/2,$$

and there exists a point $y \in G(x_0)$ such that

$$f(y) < \liminf_{x \to x_0} f(x) + \varepsilon/2.$$

Consequently, we have

$$\mathrm{diam}(f(G(x_0))) \geq |f(z) - f(y)| \geq f(z) - f(y) \geq O_f(x_0) - \varepsilon,$$

whence it follows that

$$O'_f(x_0) = \lim_{G(x_0) \to x_0} \mathrm{diam}(f(G(x_0))) \geq O_f(x_0) - \varepsilon.$$

Finally, keeping in mind the arbitrary smallness of ε, we conclude that

$$O'_f(x_0) \geq O_f(x_0)$$

and so we obtain the desired equality $O_f(x_0) = O'_f(x_0)$.

Theorem 5.3 has thus been proved. □

Theorem 5.4 *Let E be a topological space, let x_0 be a point of E, and let f and g be real-valued functions on E.*

The following two assertions hold:

(1) $O_{f+g}(x_0) \leq O_f(x_0) + O_g(x_0)$;
(2) if $O_f(x_0) < +\infty$ and $O_g(x_0) < +\infty$, then the inequality

$$O_{f-g}(x_0) \geq |O_f(x_0) - O_g(x_0)|$$

is fulfilled.

Proof It suffices to demonstrate the validity of (1), because (2) is a logical corollary of (1). Suppose, without loss of generality, that

$$O_f(x_0) + O_g(x_0) < +\infty.$$

Evidently, in this case we have $O_{f+g}(x_0) < +\infty$.

Let $U(x_0)$ be an arbitrary neighborhood of x_0. For any real $\varepsilon > 0$, there exist two points x and y in $U(x_0)$ such that

$$\mathrm{diam}((f+g)(U(x_0))) - \varepsilon \leq |(f+g)(x) - (f+g)(y)|.$$

Obviously, we may write

$$|(f+g)(x) - (f+g)(y)| \leq |f(x) - f(y)| + |g(x) - g(y)|$$
$$\leq \mathrm{diam}(f(U(x_0))) + \mathrm{diam}(g(U(x_0))).$$

Consequently,

$$\mathrm{diam}((f+g)(U(x_0))) - \varepsilon \leq \mathrm{diam}(f(U(x_0))) + \mathrm{diam}(g(U(x_0))).$$

5 The Oscillation of a Real-Valued Function at a Point

The limits of both sides of the latter inequality, when $U(x_0) \to x_0$, exist and we get

$$O_{f+g}(x_0) - \varepsilon \leq O_f(x_0) + O_g(x_0).$$

Since $\varepsilon > 0$ can be taken arbitrarily small, we conclude that

$$O_{f+g}(x_0) \leq O_f(x_0) + O_g(x_0),$$

which completes the proof. □

Theorem 5.4 easily implies that if the three non-negative values

$$O_f(x_0), \qquad O_g(x_0), \qquad O_{f+g}(x_0)$$

are finite, then they may be considered as the side lengths of a triangle (notice that in some cases this triangle can be degenerate).

Also, by using the method of induction on n, it can easily be deduced from Theorem 5.4 that

$$O_{f_1+f_2+\cdots+f_n}(x_0) \leq O_{f_1}(x_0) + O_{f_2}(x_0) + \cdots + O_{f_n}(x_0).$$

Actually, this formula expresses the finite subadditivity of the mapping

$$f \to O_f(x_0)$$

defined on the family of all those real-valued functions f on E which have finite oscillation at $x_0 \in E$.

The next theorem shows, under an appropriate assumption, the countable subadditivity of the same mapping $f \to O_f(x_0)$. Naturally, the countable subadditivity is a much deeper property than the finite subadditivity.

Theorem 5.5 *Let E be a topological space, $\{f_n : n \in \mathbb{N}\}$ be a countable family of real-valued functions on E, and let x_0 be a point of E.*

Suppose also that:

(1) all functions f_n are locally bounded at x_0;
(2) the series $\sum\{f_n : n \in \mathbb{N}\}$ converges uniformly on E to some function $f : E \to \mathbb{R}$.

Then the inequality $O_f(x_0) \leq \sum\{O_{f_n}(x_0) : n \in \mathbb{N}\}$ holds true.

Proof It suffices to restrict our argument to the case when

$$\sum\{O_{f_n}(x_0) : n \in \mathbb{N}\} < +\infty.$$

In this case, according to Theorem 5.4, we may write

$$O_f(x_0) \leq O_{f_0}(x_0) + \cdots + O_{f_n}(x_0) + O_{\phi_n}(x_0),$$

where

$$\phi_n = f_{n+1} + f_{n+2} + \cdots \quad (n \in \mathbb{N}).$$

Take any real $\varepsilon > 0$. Because of (2), there exists an $n \in \mathbb{N}$ such that

$$||\phi_n|| = \sup\{|\phi_n(x)| : x \in E\} \leq \varepsilon/2.$$

In view of (5) of Theorem 5.2, we have the inequality $O_{\phi_n}(x_0) \leq \varepsilon$. Consequently,

$$O_f(x_0) \leq O_{f_0}(x_0) + \cdots + O_{f_n}(x_0) + \varepsilon$$

or, equivalently,

$$O_f(x_0) - \varepsilon \leq O_{f_0}(x_0) + \cdots + O_{f_n}(x_0) \leq \sum \{O_{f_m}(x_0) : m \in \mathbb{N}\}.$$

Since $\varepsilon > 0$ can be arbitrarily small, we obtain the required result.

Theorem 5.5 has thus been proved. □

If $g : E \to \mathbb{R}$ is a function locally bounded at all points of a topological space E, then we can consider the associated oscillation function

$$O_g : E \to [0, +\infty[$$

whose value at any point $x \in E$ is equal to the oscillation of g at x.

Theorem 5.6 *Let E be a topological space and let $\{f_n : n \in \mathbb{N}\}$ be a sequence of real-valued locally bounded functions on E which converges uniformly on E to some function $f : E \to \mathbb{R}$.*

Then the sequence of oscillation functions $\{O_{f_n} : n \in \mathbb{N}\}$ converges uniformly to the oscillation function O_f.

The proof of this theorem is left to the reader.

Actually, Theorem 5.6 is a generalization of the following well-known theorem in classical mathematical analysis:

If a sequence $\{f_n : n \in \mathbb{N}\}$ of real-valued continuous functions on a topological space E converges uniformly on E to some function f, then f is also continuous on E (see, e.g., [35, 44, 152]).

We would like to end this chapter with one remark. Suppose that E is a topological space, (M, d) is a metric space, and $f : E \to M$ is a partial function locally bounded at all points of the closure of $\mathrm{dom}(f)$. Then the real value $O_f(x)$ is defined for any point x belonging to $\mathrm{cl}(\mathrm{dom}(f))$. Thus, we may introduce the non-negative function

$$O_f : \mathrm{cl}(\mathrm{dom}(f)) \to \mathbb{R}$$

which carries significant information about the behavior of f. It is natural to call O_f the oscillation function associated with f.

Exercises

1. Let E be a topological space and let $f : E \to \mathbb{R}$, $g : E \to \mathbb{R}$ be two functions locally bounded at a point $x_0 \in E$.
   Prove that the inequalities

   $$\mathrm{limsup}_{x \to x_0}(f(x) + g(x)) \leq \mathrm{limsup}_{x \to x_0} f(x) + \mathrm{limsup}_{x \to x_0} g(x);$$
   $$\mathrm{liminf}_{x \to x_0}(f(x) + g(x)) \geq \mathrm{liminf}_{x \to x_0} f(x) + \mathrm{liminf}_{x \to x_0} g(x)$$

   hold true for these f, g and x_0.
   Using the above inequalities, give another proof of the relation

   $$O_{f+g}(x_0) \leq O_f(x_0) + O_g(x_0)$$

   and infer from it the finite subadditivity of the mapping $f \to O_f(x_0)$.

2. Give a detailed proof of Theorem 5.2.
3. Verify that if E is a topological space, $[a, b]$ is a segment in $\mathbb{R}$, and $f : E \to [a, b]$ is a function, then $O_f(x) \leq b - a$ for all points x of E.
4. Let $f = \chi_\mathbb{Q}$ denote the Dirichlet function.
   Check that $O_f(x) = 1$ for all points $x \in \mathbb{R}$.
5. Work in **ZF** theory and prove that there are $2^\mathfrak{c}$ many functions $\phi : \mathbb{R} \to [0, 1]$ such that the graph of ϕ is everywhere dense in the strip $\mathbb{R} \times [0, 1]$.
   In particular, for any ϕ with the above property, infer that:

   (a) $\mathrm{limsup}_{x \to x_0} \phi(x) = 1$ for each $x_0 \in \mathbb{R}$;
   (b) $\mathrm{liminf}_{x \to x_0} \phi(x) = 0$ for each $x_0 \in \mathbb{R}$;
   (c) $O_\phi(x_0) = 1$ for each $x_0 \in \mathbb{R}$.

6. Show that in the formulation of Theorem 5.5 the uniform convergence of a series of functions is essential and cannot be replaced by the weaker pointwise convergence.
7. Give a detailed proof of Theorem 5.6 and demonstrate that in its formulation the uniform convergence of a sequence of functions is essential and cannot be replaced by the weaker pointwise convergence.
   Also, infer Theorem 5.5 from Theorem 5.6.
8. Give an example of a function $\psi : \mathbb{R} \to [0, 1]$ such that there exists no function $\phi : \mathbb{R} \to \mathbb{R}$ for which $O_\phi = \psi$.
9. It was proved in Chap. 4 that there exists a function

$$g : \mathbb{R} \to \mathbb{Q}$$

whose graph is everywhere dense in the plane $\mathbb{R}^2$ (see Theorem 4.7 therein). Obviously, such a g satisfies the equality $O_g(x) = +\infty$ for any point x of $\mathbb{R}$. Work in **ZF** theory and present an example of a function $f : \mathbb{R} \to \mathbb{Q}$ such that:

(a) f is non-negative, so the graph of f is not everywhere dense in $\mathbb{R}^2$;
(b) the equality $O_f(x) = +\infty$ holds true for all points $x \in \mathbb{R}$.

10. Let D be a nonempty open subinterval of $\mathbb{R}$ and let $f : D \to \mathbb{R}$ be a bounded function. According to the standard definition of mathematical analysis, the oscillation $O(f, D)$ of f on D is expressed by the formula

$$O(f, D) = \sup\{f(x) : x \in D\} - \inf\{f(x) : x \in D\}.$$

Starting with this definition, verify that, for each point $x_0 \in D$, the equality

$$O_f(x_0) = \inf\{O(f, \Delta) : \Delta \text{ is an open subinterval of } D \text{ and } x_0 \in \Delta\}$$

holds.

Chapter 6
Points of Continuity and Discontinuity of Real-Valued Functions

Let E be a topological space and let $g : E \to \mathbb{R}$ be a partial function.

In Chap. 5, for each point x belonging to $\mathrm{cl}(\mathrm{dom}(g))$, the following value has been defined: $O_g(x)$ = the oscillation of g at x.

It will be convenient to extend this concept to all points from the set $E \setminus \mathrm{cl}(\mathrm{dom}(g))$ by putting

$$O_g(x) = 0 \qquad (x \in E \setminus \mathrm{cl}(\mathrm{dom}(g))).$$

As has already been mentioned in the same chapter, if $x \in \mathrm{dom}(g)$, then the continuity of g at x is equivalent to the equality $O_g(x) = 0$.

Now, let us introduce the following two sets canonically associated with the given partial function $g : E \to \mathbb{R}$:

$C(g) =$ the set of all those points $x \in \mathrm{dom}(g)$ at which g is continuous;

$D(g) =$ the set of all those points $x \in \mathrm{dom}(g)$ at which g is discontinuous

(i.e., is not continuous).

Clearly, we have the equalities:

$$C(g) \cap D(g) = \emptyset, \qquad C(g) \cup D(g) = \mathrm{dom}(g).$$

In this chapter we will be dealing with the descriptive structure of the sets $C(g)$ and $D(g)$ (cf. [40, 109, 141]).

Theorem 6.1 *For any real number $r > 0$, the set of all those points $x \in E$ which satisfy the inequality $O_g(x) \geq r$ is closed in E.*

Proof Denote the above-mentioned set by $D_r(g)$. Obviously, we have

$$(E \setminus \mathrm{cl}(\mathrm{dom}(g))) \cap D_r(g) = \emptyset.$$

Consequently, the inclusion $D_r(g) \subset \mathrm{cl}(\mathrm{dom}(g))$ holds true.

Take any point x belonging to $\mathrm{cl}(D_r(g))$ and consider any open neighborhood $U(x)$ of x. Clearly, there exists a point

$$y \in U(x) \cap D_r(g).$$

From the inequality $O_g(y) \geq r$, we easily infer that

$$\mathrm{diam}(g(U(x))) \geq r.$$

Since $U(x)$ was taken arbitrarily, we conclude that

$$O_g(x) \geq r, \qquad x \in D_r(g), \qquad \mathrm{cl}(D_r(g)) = D_r(g),$$

which completes the proof. □

As a trivial corollary of Theorem 6.1, we obtain

Theorem 6.2 *For any real number $r > 0$, the set of all those points $x \in E$ which satisfy the inequality $O_g(x) < r$ is open in E.*

Proof Indeed, since the above-mentioned set coincides with $E \setminus D_r(g)$, we immediately get the required result. □

The next theorem describes the topological structure of $D(g)$.

Theorem 6.3 *The set $D(g)$ is representable in the form*

$$D(g) = \cup\{\mathrm{dom}(g) \cap F_n : n \in \mathbb{N}\},$$

where $\{F_n : n \in \mathbb{N}\}$ is some sequence of closed sets in E.

Proof Evidently, we may write

$$D(g) = \{x \in \mathrm{dom}(g) : O_g(x) > 0\} = \cup\{\mathrm{dom}(g) \cap F_n : n \in \mathbb{N}\},$$

where

$$F_n = \{x \in E : O_g(x) \geq 1/2^n\}$$

for each natural number n.

In view of Theorem 6.1, all sets F_n are closed in E, which ends the proof of Theorem 6.3. □

As a straightforward consequence of Theorem 6.3, we have

Theorem 6.4 *If $g : E \to \mathbb{R}$ is a function, then the set $D(g)$ is representable as the union of a countable family of closed subsets of E.*

The following theorem describes the topological structure of $C(g)$.

Theorem 6.5 *The set $C(g)$ is representable in the form*

$$C(g) = \cap\{\mathrm{dom}(g) \cap U_n : n \in \mathbb{N}\},$$

where $\{U_n : n \in \mathbb{N}\}$ is some sequence of open sets in E.

Proof Evidently, we may write

$$C(g) = \mathrm{dom}(g) \setminus D(g) = \mathrm{dom}(g) \setminus \cup\{\mathrm{dom}(g) \cap F_n : n \in \mathbb{N}\},$$

where all the sets

$$F_n = \{x \in E : O_g(x) \geq 1/2^n\}$$

are closed in E. Consequently,

$$C(g) = \cap\{\mathrm{dom}(g) \cap U_n : n \in \mathbb{N}\},$$

where all $U_n = E \setminus F_n$ are open subsets of E. In fact, we have

$$U_n = \{x \in E : O_g(x) < 1/2^n\}$$

for every natural number n. Theorem 6.5 has thus been proved. $\square$

As a trivial corollary of Theorem 6.5, we get

Theorem 6.6 *If $g : E \to \mathbb{R}$ is a function, then the set $C(g)$ is representable as the intersection of a countable family of open subsets of E.*

According to the two dual Theorems 6.4 and 6.6, if we have an arbitrary real-valued function g on a topological space E, then $D(g)$ is the union of countably many closed sets in E and $C(g)$ is the intersection of countably many open sets in E.

In this connection, the following two questions naturally arise:

(*) If a set $X \subset E$ is representable as the union of countably many closed subsets of E, does there exist a function $g : E \to \mathbb{R}$ such that $D(g) = X$?
(**) If a set $Y \subset E$ is representable as the intersection of countably many open subsets of E, does there exist a function $h : E \to \mathbb{R}$ such that $C(h) = Y$?

It can easily be seen that these two questions are equivalent in the following sense:

If a topological space E is given, then a positive answer to (*) implies a positive answer to (**) and, conversely, a positive answer to (**) implies a positive answer to (*).

We are now going to consider question (*) in more detail.

We will try to give a positive solution to (*) for a sufficiently wide class of topological spaces E. Let us introduce this class.

A topological space E is called *resolvable* (in the sense of E. Hewitt) if there exists a partition $\{A, B\}$ of E such that both sets A and B are everywhere dense in E (see [67]).

Equivalently, one may say that a space E is resolvable if there exist two nonempty sets $A \subset E$ and $B \subset E$ such that

$$A \cap B = \emptyset, \qquad \mathrm{cl}(A) = \mathrm{cl}(B) = E.$$

It immediately follows from the above definition that a resolvable space does not have isolated points.

Naturally, a topological space E is called *irresolvable* if E is not resolvable.

All classical spaces in mathematical analysis (nonempty metric spaces without isolated points, various spaces of functions, non-discrete locally compact topological groups, etc.) are resolvable.

More information on resolvable and irresolvable topological spaces is presented in Appendix E.

Lemma 6.1 *Every nonempty open subspace of a resolvable topological space is also resolvable.*

Proof Let E be a resolvable space and let U be an arbitrary nonempty open set in E. Consider a partition $\{A, B\}$ of E such that

$$\mathrm{cl}(A) = \mathrm{cl}(B) = E.$$

This partition of E produces the partition $\{U \cap A, U \cap B\}$ of U. Obviously, both sets $U \cap A$ and $U \cap B$ are everywhere dense in U, which immediately implies the desired result. □

Lemma 6.2 *If E is a resolvable topological space and F is a closed subset of E, then there exists a function*

$$\psi : E \to \{0, 1\}$$

such that $\psi|(E \setminus F) = 0$ and $O_\psi(x) = 1$ for each point $x \in F$.

Proof Here only two cases are possible.

1. The given closed set F is nowhere dense in E.
    In this case, we put $\psi = \chi_F$, where χ_F denotes the characteristic function of F. Clearly, $O_\psi(x) = 1$ for all points $x \in F$ and $\psi|(E \setminus F) = 0$ by definition of ψ.

2. The interior int(F) of F is not empty.
   In this case, let us put $U = \text{int}(F)$ and consider a partition $\{A, B\}$ of U into two subsets each of which is everywhere dense in U (see Lemma 6.1).
   Then define the required function $\psi : E \to \{0, 1\}$ as follows:

   (a) $\psi(x) = 0$ if $x \in A$,
   (b) $\psi(x) = 1$ if $x \in B$,
   (c) $\psi(x) = 1$ if $x \in F \setminus U$,
   (d) $\psi(x) = 0$ if $x \in E \setminus F$.

   A straightforward verification shows that $O_\psi(x) = 1$ for any point $x \in F$.

Lemma 6.2 has thus been proved. □

Now, we are ready to establish the main result of this chapter.

Theorem 6.7 *Let E be a resolvable topological space.*

Then any subset X of E which admits a representation in the form of the union of countably many closed sets in E coincides with the set $D(g)$ of some bounded non-negative function $g : E \to \mathbb{R}$ (depending on X).

Proof Without loss of generality, we may represent the set X in the form

$$X = F_0 \cup F_1 \cup \cdots \cup F_n \cup \cdots,$$

where all sets F_n are closed, $F_0 = \emptyset$, and $F_n \subset F_{n+1}$ for each index $n \in \mathbb{N}$.

Applying Lemma 6.2 to every set F_n, we may define a function

$$\psi_n : E \to \{0, 1\}$$

such that $\psi_n|(E \setminus F_n) = 0$ and $O_{\psi_n}(x) = 1$ for all points $x \in F_n$.

Further, let us take a series $\sum \{a_n : n \in \mathbb{N}\}$ of strictly positive real numbers such that

$$a_{n+1} + a_{n+2} + \cdots + a_k + \cdots < a_n \qquad (n \in \mathbb{N}).$$

For instance, in order to be more concrete, we can put

$$a_n = 1/3^n \qquad (n \in \mathbb{N}).$$

Define a function $g : E \to \mathbb{R}$ by the formula

$$g = a_0 \psi_0 + a_1 \psi_1 + \cdots + a_n \psi_n + \cdots.$$

This function is well-defined in view of the uniform convergence on E of the written function series.

Let us demonstrate that g is continuous at all points of $E \setminus X$.

Indeed, take any point x from $E \setminus X$. By definition, $g(x) = 0$. Fix a real $\varepsilon > 0$ and choose $n \in \mathbb{N}$ so large that $a_n < \varepsilon$. Since $x \notin F_n$, there exists a neighborhood $U(x)$ such that

$$U(x) \cap F_n = \emptyset.$$

For all points $y \in U(x)$, we may write

$$g(y) = a_{n+1}\psi_{n+1}(y) + a_{n+2}\psi_{n+2}(y) + \cdots + a_k\psi_k(y) + \cdots < a_n < \varepsilon,$$

which shows us that g is continuous at x.

Further, let us demonstrate that g is discontinuous at each point $x \in X$. Obviously, there exists an $n \in \mathbb{N}$ satisfying $x \in F_n \setminus F_{n-1}$. We have

$$g(x) = a_n\psi_n(x) + a_{n+1}\psi_{n+1}(x) + \cdots + a_k\psi_k(x) + \cdots$$

and, analogously,

$$g(y) = a_n\psi_n(y) + a_{n+1}\psi_{n+1}(y) + \cdots + a_k\psi_k(y) + \cdots$$

for all points $y \in E \setminus F_{n-1}$, where $E \setminus F_{n-1}$ is an open subset of E containing the point x.

For the remainder of the proof we will be dealing with the restriction of g to $E \setminus F_{n-1}$, but we will preserve the same notation g for this restricted function. Defining

$$\phi_n = a_{n+1}\psi_{n+1} + a_{n+2}\psi_{n+2} + \cdots + a_k\psi_k + \cdots,$$

let us express g (on the set $E \setminus F_{n-1}$) in the form

$$g = a_n\psi_n + \phi_n$$

or, equivalently,

$$a_n\psi_n = g - \phi_n = g + (-\phi_n).$$

By virtue of Theorem 5.4 from Chap. 5, we can write

$$a_n = O_{a_n\psi_n}(x) \leq O_g(x) + O_{\phi_n}(x).$$

Also, according to Theorem 5.5 from the same Chap. 5,

$$O_{\phi_n}(x) \leq a_{n+1} + a_{n+2} + \cdots + \cdots + a_k + \cdots.$$

Consequently,

$$a_n \leq O_g(x) + a_{n+1} + a_{n+2} + \cdots + \cdots + a_k + \cdots$$

and, therefore,

$$0 < a_n - (a_{n+1} + a_{n+2} + \cdots + \cdots + a_k + \cdots) \leq O_g(x),$$

which establishes the relation $x \in D(g)$ and completes the proof of Theorem 6.7. □

It should be mentioned that, in general, the assertion of Theorem 6.7 is not true for irresolvable topological spaces (in this connection, see e.g. Exercise 9 of Appendix E).

Exercises

1. Give an example of a function $f : \mathbb{R} \to \mathbb{R}$ such that the set $D(f)$ is not closed in $\mathbb{R}$.
2. Show that if a function $g : \mathbb{R}^2 \to \mathbb{R}$ is an injection, then the set $D(g)$ is everywhere dense in $\mathbb{R}^2$.
3. Verify that no countable everywhere dense subset of $\mathbb{R}$ can be represented as the intersection of countably many open sets in $\mathbb{R}$.
4. Check that there exists no function $f : \mathbb{R} \to \mathbb{R}$ such that the set $C(f)$ is countable and everywhere dense in $\mathbb{R}$.
5. Show that a nonempty subspace X of $\mathbb{R}$ is resolvable if and only if X does not have isolated points.
6. Let E be a nonempty topological space satisfying the following conditions:

    (a) E does not contain isolated points;
    (b) every singleton in E is closed;
    (c) there exists a countable base of E.

    Generalizing the result of Exercise 5, demonstrate that there are two disjoint countable everywhere dense subsets of E (so E is a resolvable space).
    For this purpose, construct recursively the required subsets of E.
7. Using the Kuratowski–Zorn lemma, prove that if each member of some base of a topological space E is resolvable, then E is resolvable.
8. For every infinite cardinal number **a**, give an example of an irresolvable topological space E with card$(E) = $ **a**, which does not have isolated points.
    For this purpose, take into account the fact that if X is an arbitrary infinite set, then there exists a nontrivial ultrafilter $\mathcal{F}$ of subsets of X (see, e.g., [19, 40, 75, 76, 84, 111]). The nontriviality of $\mathcal{F}$ means that $\{x\} \notin \mathcal{F}$ for any element x of X.
9. Let E be a topological space, (M, d) be a metric space, and let $g : E \to M$ be a partial function. Define:

$C(g)$ = the set of all those points $x \in \text{dom}(g)$ at which g is continuous;

$D(g)$ = the set of all those points $x \in \text{dom}(g)$ at which g is discontinuous.

Show that Theorems 6.1–6.6 of this chapter remain valid for the introduced sets $C(g)$ and $D(g)$.

Chapter 7
Real-Valued Monotone Functions

Let $(E, \leq)$ and $(F, \leq)$ be two partially ordered sets and let $g : E \to F$ be a mapping (function).

Recall that g is an *increasing* mapping if

$$(\forall x \in E)(\forall y \in E)(x \leq y \Rightarrow g(x) \leq g(y)).$$

By the dual definition, g is a *decreasing* mapping if

$$(\forall x \in E)(\forall y \in E)(x \leq y \Rightarrow g(y) \leq g(x)).$$

Accordingly, g is called a *monotone* mapping if g is either increasing or decreasing.

In a similar manner *strictly increasing*, *strictly decreasing*, and *strictly monotone* mappings from $(E, \leq)$ into $(F, \leq)$ are introduced. For instance, $g : E \to F$ is strictly increasing if

$$(\forall x \in E)(\forall y \in E)(x < y \Rightarrow g(x) < g(y)).$$

Let X be a subset of $(E, \leq)$. Then the partial ordering $\leq$ of E induces the partial ordering of X (denoted usually by the same symbol $\leq$).

Observe that if $g : E \to F$ is an increasing (decreasing, strictly increasing, strictly decreasing) mapping, then the restriction $g|X$ is also an increasing (decreasing, strictly increasing, strictly decreasing) mapping from $(X, \leq)$ into $(F, \leq)$. In this context, the following question is of interest:

Let X be a subset of $(E, \leq)$ and let $h : X \to F$ be an increasing mapping from X into $(F, \leq)$. Does there exist an increasing mapping $h^* : E \to F$ which is an extension of h?

Simple examples show that, in general, the answer to this question is negative. The theorem presented below gives a necessary and sufficient condition for the existence of a desired extension h^* in the special case when $E = F = \mathbb{R}$ and $\mathbb{R}$ is equipped with its standard linear ordering.

Theorem 7.1 *For a partial increasing function $h : \mathbb{R} \to \mathbb{R}$, the following two assertions are equivalent:*

(1) h is locally bounded at all points of $\mathbb{R}$;
(2) there exists an increasing function $h^* : \mathbb{R} \to \mathbb{R}$ extending h.

Proof To demonstrate the validity of (1) $\Rightarrow$ (2), suppose (1). According to Theorem 5.1 from Chap. 5, the given h is bounded on every subsegment of $\mathbb{R}$. If $\text{dom}(h) = \emptyset$, then any real-valued constant function on $\mathbb{R}$ is an increasing extension of h. If $\text{dom}(h) \neq \emptyset$, then we pick a point $t \in \text{dom}(h)$ and define a function $h^* : \mathbb{R} \to \mathbb{R}$ as follows:

$$h^*(x) = \begin{cases} \sup\{h(y) : t \leq y \leq x, \ y \in \text{dom}(h)\} & \text{if } x \in [t, +\infty[, \\ \inf\{h(y) : x \leq y \leq t, \ y \in \text{dom}(h)\} & \text{if } x \in]-\infty, t]. \end{cases}$$

A straightforward verification shows that h^* extends h and is an increasing function.

To demonstrate the validity of (2) $\Rightarrow$ (1), suppose (2). Let x be an arbitrary point of $\mathbb{R}$. Take any two real numbers a and b such that $a < x < b$. For each point $y \in \text{dom}(h) \cap]a, b[$, we may write

$$h^*(a) \leq h^*(y) = h(y) = h^*(y) \leq h^*(b),$$

which immediately implies that h is locally bounded at x.

This completes the proof of Theorem 7.1. □

For a slightly more general statement, see Exercise 5 of the present chapter.
The following assertions are almost trivial:

(i) if a partial function $g : \mathbb{R} \to \mathbb{R}$ is increasing (decreasing) and a real t is non-negative (non-positive), then the partial function tg is increasing;
(ii) if a partial function $g : \mathbb{R} \to \mathbb{R}$ is decreasing (increasing) and a real t is non-negative (non-positive), then the partial function tg is decreasing;
(iii) if two given partial functions

$$g : (E, \leq) \to \mathbb{R}, \qquad h : (E, \leq) \to \mathbb{R}$$

have a common domain and both are increasing (decreasing), then the partial function

$$g + h : (E, \leq) \to \mathbb{R}$$

is also increasing (decreasing).

7 Real-Valued Monotone Functions

In connection with (iii), it will be proved in the sequel that there exist two increasing functions from $\mathbb{R}$ into $\mathbb{R}$ such that their difference is not monotone on any non-degenerate subinterval of $\mathbb{R}$.

Now, let us turn our attention to the discontinuity points of monotone partial real-valued functions on $\mathbb{R}$. In view of Theorem 7.1, we may restrict our further considerations to the case of monotone (e.g., increasing) functions acting from $\mathbb{R}$ into itself.

Observe that if an increasing function $g : \mathbb{R} \to \mathbb{R}$ is given and x_0 is any point of $\mathbb{R}$, then there exist two one-sided limits

$$\lim_{x \to x_0-} g(x), \qquad \lim_{x \to x_0+} g(x),$$

i.e., the left-sided limit of f at x_0 and the right-sided limit of f at x_0. For them the inequalities

$$\lim_{x \to x_0-} g(x) \leq g(x_0) \leq \lim_{x \to x_0+} g(x)$$

trivially hold true, and we obviously have

$$O_g(x_0) = \lim_{x \to x_0+} g(x) - \lim_{x \to x_0-} g(x),$$

where $O_g(x_0)$ denotes, as usual, the oscillation of g at x_0 (see Chap. 5). Consequently, g is continuous at x_0 if and only if

$$\lim_{x \to x_0+} g(x) = \lim_{x \to x_0-} g(x) = g(x_0),$$

and g is discontinuous at x_0 if and only if

$$\lim_{x \to x_0+} g(x) > \lim_{x \to x_0-} g(x).$$

In Chap. 6 it was proved that if a topological space E is resolvable and $f : E \to \mathbb{R}$ is a function, then the set $D(f)$ of all discontinuity points of f is representable as the union of some countable family of closed subsets of E. Evidently, $\mathbb{R}$ is a resolvable space. So the above-mentioned result is applicable to $\mathbb{R}$ and to any real-valued monotone function on $\mathbb{R}$. But in this special case one has a much stronger result. In order to establish it, let us first formulate and prove the following simple lemma.

Lemma 7.1 *Let $\{\Delta_i : i \in I\}$ be any family of non-degenerate subintervals of $\mathbb{R}$, which pairwise do not have common interior points.*

*Then effectively (i.e., without using **AC**) one has* $\mathrm{card}(I) \leq \mathrm{card}(\mathbb{N})$.

Proof In fact, the assertion of Lemma 7.1 is a direct consequence of the separability of $\mathbb{R}$. Let $\{q_n : n \in \mathbb{N}\}$ be an injective enumeration of all rational numbers. Since each interval Δ_i is non-degenerate, it contains in its interior infinitely many rational numbers. Denote by $n(i)$ the least natural number such that $q_{n(i)} \in \text{int}(\Delta_i)$. In view of the assumption on the given family of intervals, we have

$$(\forall i \in I)(\forall j \in I)(i \neq j \Rightarrow n(i) \neq n(j)),$$

i.e., there is an injection from I into $\mathbb{N}$. Therefore, the set I is at most countable, which completes the proof. □

Theorem 7.2 *The set $D(f)$ of all discontinuity points of an arbitrary monotone function $f : \mathbb{R} \to \mathbb{R}$ is at most countable.*

Proof It suffices to consider the case when f is an increasing function. To each point $x \in D(f)$ associate the non-degenerate interval

$$\Delta(x) = [\lim_{t \to x-} f(t), \lim_{t \to x+} f(t)].$$

Let us check that the family $\{\Delta(x) : x \in D(f)\}$ of intervals satisfies the assumption of Lemma 7.1. For this purpose, take any two distinct points x and y from $D(f)$. Without loss of generality, we may suppose that $x < y$. Let z be such that $x < z < y$. Then we can write

$$\lim_{t \to x+} f(t) \leq f(z) \leq \lim_{t \to y-} f(t),$$

which shows us that the intervals $\Delta(x)$ and $\Delta(y)$ do not possess common interior points. Applying Lemma 7.1, we conclude that the set $D(f)$ is at most countable, as required. □

We also have the next, more profound theorem, which is converse to Theorem 7.2 (see, e.g., [152]).

Theorem 7.3 *Let $X = \{x_i : i \in I\}$ be an arbitrary countable subset of the real line $\mathbb{R}$.*

Then there exists a non-negative increasing bounded function f from $\mathbb{R}$ into itself such that $D(f) = X$.

Proof The case when I is a finite set is not difficult and is left to the reader. Suppose now that $\text{card}(I) = \text{card}(\mathbb{N})$ and represent the given set X in the form of an injective sequence $\{x_n : n \in \mathbb{N}\}$. Further, for each point $x \in \mathbb{R}$, let us introduce the notation

$$N_x = \{n \in \mathbb{N} : x_n < x\}.$$

It is clear that the implication $x < y \Rightarrow N_x \subset N_y$ holds true.

Let $\{a_n : n \in \mathbb{N}\}$ be an injective sequence of strictly positive real numbers such that

$$\sum \{a_n : n \in \mathbb{N}\} < +\infty.$$

7 Real-Valued Monotone Functions

Define a function $f : \mathbb{R} \to \mathbb{R}$ by the formula

$$f(x) = \sum \{a_n : n \in N_x\} \qquad (x \in \mathbb{R}).$$

This function is well-defined in view of the convergence of the series

$$\sum \{a_n : n \in N_x\}.$$

Notice, by the way, that f is increasing and if $N_x = \emptyset$ for some $x \in \mathbb{R}$, then $f(x) = 0$.

It remains to demonstrate that $D(f) = X$. For this purpose, take an arbitrary point $t \in \mathbb{R}$. Only two cases are possible.

1. $t \notin X$.

In this case, we will show that f is continuous at t. Indeed, it suffices to establish that

$$\lim_{x \to t-} f(x) = \lim_{x \to t+} f(x).$$

Pick any real $\varepsilon > 0$ and find a natural number m such that

$$\sum \{a_n : n \in \mathbb{N},\ n > m\} < \varepsilon.$$

Then consider the points $x_0, x_1, \ldots, x_m$ of X. Evidently, there exist two points $x < t$ and $y > t$ satisfying the equality

$$[x, y] \cap \{x_0, x_1, \ldots, x_m\} = \emptyset.$$

By virtue of the definition of f, we may write

$$f(x) = \sum \{a_n : n \in N_x\}, \qquad f(y) = \sum \{a_n : n \in N_y\}.$$

At the same time, it is easy to see that

$$N_y \setminus N_x \subset \{n : n \in \mathbb{N},\ n > m\}$$

and, consequently,

$$f(y) - f(x) \leq \sum \{a_n : n \in \mathbb{N},\ n > m\} < \varepsilon,$$

which implies the continuity of f at t.

2. $t \in X$.

In this case $t = x_m$ for some $m \in \mathbb{N}$. Again, we may write

$$f(t) = \sum \{a_n : n \in N_t\}$$
$$= \sum \{a_n : x_n < t\}.$$

If $y \in \mathbb{R}$ is any point such that $t < y$, then

$$f(y) = \sum \{a_n : n \in N_y\}$$
$$= \sum \{a_n : x_n < y\}$$
$$\geq f(t) + a_m,$$

which immediately implies that f is discontinuous at t (because of $a_m > 0$).

Theorem 7.3 has thus been proved. $\square$

Lemma 7.2 *Let X be a finite set of points in $\mathbb{R}$ and let, for each point $x \in X$, a non-degenerate segment $\triangle_x$ be given so that the following two conditions are satisfied:*

(1) $(\forall x \in X)(\triangle_x \subset]0, 1[)$;
(2) $(\forall x \in X)(\forall y \in X)(x < y \Rightarrow \sup \triangle_x < \inf \triangle_y)$.

Let p, q, and r be any three distinct points in $\mathbb{R}$ not belonging to X.

Then there exist three non-degenerate segments $\triangle_p$, $\triangle_q$, $\triangle_r$ such that the set $X' = X \cup \{p, q, r\}$ satisfies the analogs of (1) and (2), i.e.,

(1') $(\forall x \in X')(\triangle_x \subset]0, 1[)$;
(2') $(\forall x \in X')(\forall y \in X')(x < y \Rightarrow \sup \triangle_x < \inf \triangle_y)$.

The easy proof of this lemma is left to the reader.

Theorem 7.4 *There exist two increasing functions f and g acting from $\mathbb{R}$ into $[0, 1]$ such that, for any nonempty open subinterval U of $\mathbb{R}$, the difference $f - g$ is not monotone on U.*

Proof Pick an injective enumeration $\{U_n : n \in \mathbb{N}\}$ of all those nonempty open subintervals of $\mathbb{R}$ whose endpoints are rational numbers.

Using the method of mathematical induction, we are going to construct an increasing (by inclusion) sequence $\{X_n : n \in \mathbb{N}\}$ of finite subsets of $\mathbb{R}$, a sequence of partial functions $\{f_n : n \in \mathbb{N}\}$ and a sequence of partial functions $\{g_n : n \in \mathbb{N}\}$ such that:

(a) for every set X_n, there is a sequence $\{\triangle_x : x \in X_n\}$ of non-degenerate segments satisfying conditions (1) and (2) of Lemma 7.2 (where $X_n = X$);
(b) for any $n \in \mathbb{N}$, one has $X_{n+1} \setminus X_n = \{p_n, q_n, r_n\}$, where p_n, q_n, r_n are some points in $\mathbb{R}$ and $p_n < q_n < r_n$;
(c) $\{p_n, q_n, r_n\} \subset U_n$;
(d) the functions $f_n : X_n \to]0, 1[$ and $g_n : X_n \to]0, 1[$ are defined so that $f_n \subset f_{n+1}$ and $g_n \subset g_{n+1}$ for each $n \in \mathbb{N}$;

7 Real-Valued Monotone Functions

(e) the functions $f_{n+1} : X_{n+1} \to \,]0, 1[$ and $g_{n+1} : X_{n+1} \to \,]0, 1[$ are defined so that

$$f_{n+1}(p_n) = \sup\triangle_{p_n}, \qquad g_{n+1}(p_n) = \inf\triangle_{p_n},$$
$$f_{n+1}(q_n) = \inf\triangle_{q_n}, \qquad g_{n+1}(q_n) = \sup\triangle_{q_n},$$
$$f_{n+1}(r_n) = \sup\triangle_{r_n}, \qquad g_{n+1}(r_n) = \inf\triangle_{r_n}.$$

At the beginning of our construction we take

$$X_0 = f_0 = g_0 = \emptyset$$

and choose distinct points p_0, q_0, r_0 from U_0 so that $p_0 < q_0 < r_0$. Put $X_1 = \{p_0, q_0, r_0\}$ and define non-degenerate segments

$$\triangle_{p_0} \subset \,]0, 1[, \qquad \triangle_{q_0} \subset \,]0, 1[, \qquad \triangle_{r_0} \subset \,]0, 1[$$

satisfying condition (2) of Lemma 7.2 for $X = X_1$. Then define two functions

$$f_1 : X_1 \to \,]0, 1[, \qquad g_1 : X_1 \to \,]0, 1[$$

fulfilling property (e) for $n = 0$.

Further proceed by induction in the analogous manner. Namely, suppose that the finite set $X_n \subset \mathbb{R}$ has already been constructed with the corresponding increasing functions

$$f_n : X_n \to \,]0, 1[, \qquad g_n : X_n \to \,]0, 1[.$$

Consider the interval U_n and choose in it three distinct points p_n, q_n, r_n such that

$$p_n < q_n < r_n \qquad [p_n, r_n] \cap X_n = \emptyset.$$

Keeping in mind Lemma 7.2, it is not difficult to show that, for appropriate non-degenerate segments

$$\triangle_{p_n} \subset \,]0, 1[, \qquad \triangle_{q_n} \subset \,]0, 1[, \qquad \triangle_{r_n} \subset \,]0, 1[$$

and for the expanded finite set $X_{n+1} = X_n \cup \{p_n, q_n, r_n\}$, there exist two increasing functions

$$f_{n+1} : X_{n+1} \to \,]0, 1[, \qquad g_{n+1} : X_{n+1} \to \,]0, 1[$$

which extend, respectively, f_n and g_n and fulfil the property (e).

We then put

$$X = \cup\{X_n : n \in \mathbb{N}\}, \qquad f = \cup\{f_n : n \in \mathbb{N}\}, \qquad g = \cup\{g_n : n \in \mathbb{N}\}.$$

Obviously, both f and g are increasing partial functions from $\mathbb{R}$ into $]0, 1[$. According to Theorem 7.1, these two partial functions can be extended to two increasing functions acting from $\mathbb{R}$ into $\mathbb{R}$. For the sake of brevity, we denote those extensions by the same symbols f and g.

It is not hard to see that $0 \leq f(x) \leq 1$ and $0 \leq g(x) \leq 1$ for all $x \in \mathbb{R}$.

Finally, by virtue of our construction, for any interval U_n and its points p_n, q_n, r_n, we have

$$f(p_n) - g(p_n) > 0, \qquad f(q_n) - g(q_n) < 0, \qquad f(r_n) - g(r_n) > 0.$$

The above inequalities directly imply that the difference $f - g$ cannot be monotone on U_n, which completes the proof of Theorem 7.4. □

Remark 7.1 A much deeper result than Theorem 7.4 is known. Namely, it can be demonstrated that there exists an everywhere differentiable function $h : \mathbb{R} \to \mathbb{R}$ which is not monotone on any non-degenerate segment of $\mathbb{R}$ and whose derivative is bounded on $\mathbb{R}$ (see [32, 34, 57, 81, 184]). For this h, it can easily be deduced that there are two increasing continuous functions $f : \mathbb{R} \to \mathbb{R}$ and $g : \mathbb{R} \to \mathbb{R}$ such that $h = f - g$.

Remark 7.2 In classical mathematical analysis, numerous interesting and important facts have been established concerning the structure of monotone functions. Let us mention only few of them:

(1) Every real-valued monotone function on $[0, 1]$ has a finite derivative at almost all points of $[0, 1]$ (with respect to the standard Lebesgue measure λ on $\mathbb{R}$).
(2) Any real-valued increasing function on $[0, 1]$ can be considered as the derivative of some convex function on $[0, 1]$ (excepting countably many points of $[0, 1]$).
(3) There exists a real-valued strictly increasing continuous function on $\mathbb{R}$ whose derivative is zero at almost all points of $\mathbb{R}$ (with respect to the same measure λ on $\mathbb{R}$).

The above-mentioned facts are discussed in many textbooks devoted to real analysis and Lebesgue measure theory (see, for instance, [23, 44, 70, 71, 100, 104, 134, 153]).

Exercises

1. Give an example of two partially ordered sets $(E, \leq)$ and $(F, \leq)$ for which there exists a bijective strictly increasing mapping $g : E \to F$ such that the inverse mapping $g^{-1} : F \to E$ is not monotone.
2. Show that if $(E, \leq)$ and $(F, \leq)$ are any two linearly ordered sets and $g : E \to F$ is a bijective strictly increasing mapping, then the inverse mapping $g^{-1} :$

7 Real-Valued Monotone Functions

$F \to E$ is strictly increasing, too. Consequently, in this case g turns out to be an isomorphism between $(E, \leq)$ and $(F, \leq)$.

3. Let $\{g_i : i \in I\}$ be a family of partial functions acting from a partially ordered set $(E, \leq)$ into a partially ordered set $(F, \leq)$, and suppose that all g_i ($i \in I$) are increasing (decreasing) on their domains. Suppose also that, for any two indices $i \in I$ and $j \in I$, there exists a $k \in I$ such that g_k is a common extension of g_i and g_j.
   Demonstrate that the partial function $g = \cup\{g_i : i \in I\}$ is increasing (decreasing) on its domain.

4. Let E be a topological space and let $h : E \to \mathbb{R}$ be a function.
   Verify that the set of all those points of E at which h is locally bounded from above (from below) is open in E.

5. Generalize Theorem 7.1 in the following manner:
   Any monotone partial function $h : \mathbb{R} \to \mathbb{R}$ can be extended to a monotone partial function $h^* : \mathbb{R} \to \mathbb{R}$ defined on a maximal (by inclusion) open subinterval of $\mathbb{R}$ (which may coincide with $\mathbb{R}$).
   Also, show that h^* cannot be locally bounded at the endpoints of the above-mentioned maximal interval (if it differs from $\mathbb{R}$).

6. Let $f : \mathbb{R} \to [a, b]$ be a partial monotone function.
   Verify that there exists a monotone function $f^* : \mathbb{R} \to [a, b]$ which extends f.

7. Check that the composition of any two monotone partial functions is a monotone partial function.
   Also, check that the pointwise limit of a sequence of monotone functions acting from a partially ordered set $(E, \preceq)$ into $\mathbb{R}$ is a monotone function.

8. Give a detailed proof of Lemma 7.2.

9. Let $g : \mathbb{R} \to \mathbb{R}$ be a function and let x be a point in $\mathbb{R}$.
   This x is called a *simple discontinuity point* of g if there exist both one-sided limits

$$\lim_{y \to x, y < x} g(y) = g_-(x), \qquad \lim_{y \to x, y > x} g(y) = g_+(x)$$

   and either $g_-(x) \neq g(x)$ or $g_+(x) \neq g(x)$.
   In view of the above definition, if g is a monotone function, then all discontinuity points of g are simple discontinuity points.
   Generalizing Theorem 7.2, show that if f is an arbitrary function acting from $\mathbb{R}$ into $\mathbb{R}$, then the set of all simple discontinuity points of f is at most countable. For this purpose, denote by $D_s(f)$ the set of all simple discontinuity points of f and also stipulate

$$D_1 = \{x \in D_s(f) : f_-(x) < f_+(x)\},$$
$$D_2 = \{x \in D_s(f) : f_-(x) > f_+(x)\},$$
$$D_3 = \{x \in D_s(f) : f(x) < f_-(x) = f_+(x)\},$$
$$D_4 = \{x \in D_s(f) : f(x) > f_-(x) = f_+(x)\}.$$

Verify that
$$D_s(f) = D_1 \cup D_2 \cup D_3 \cup D_4,$$

so it suffices to demonstrate that each set D_i ($i = 1, 2, 3, 4$) is at most countable. Actually, the argument is similar for all D_i.

For instance, show that $\text{card}(D_1) \leq \text{card}(\mathbb{N})$. In order to do this, define a mapping
$$x \to (p(x), q(x), r(x)) \qquad (x \in D_1)$$

such that
$$\{p(x), q(x), r(x)\} \subset \mathbb{Q}, \qquad x \in \,]q(x), r(x)[,$$
$$(\forall t \in \,]q(x), x[)(f(t) < p(x)), \qquad (\forall t \in \,]x, r(x)[)(f(t) > p(x)).$$

The existence of such a mapping is obvious, because of the inequality
$$f_-(x) < f_+(x).$$

Now, it is not hard to verify that this mapping is injective. Indeed, suppose to the contrary that, for some distinct points x_1 and x_2 from D_1, one has the equality
$$(p(x_1), q(x_1), r(x_1)) = (p(x_2), q(x_2), r(x_2)).$$

Without loss of generality one may assume that $x_1 < x_2$. Choose any point t from $]x_1, x_2[$. Then one must have simultaneously
$$f(t) < p(x_2) = p(x_1), \qquad f(t) > p(x_1) = p(x_2),$$

which is impossible. The contradiction obtained establishes that the described mapping is injective, so
$$\text{card}(D_1) \leq \text{card}(\mathbb{Q} \times \mathbb{Q} \times \mathbb{Q}) = \text{card}(\mathbb{N}).$$

Analogously argue for the sets D_2, D_3 and D_4.

Remark 7.3 Let $g : \mathbb{R} \to \mathbb{R}$ be an everywhere differentiable function. Then the derivative g' of g may be discontinuous at some points of $\mathbb{R}$. However, as is known, there are no simple discontinuity points of g'. The latter fact easily follows from the so-called Darboux property of g' (see, e.g., [24, 44]).

10. Demonstrate that there exist two real-valued strictly positive functions ϕ and ψ on $\mathbb{R}$ which satisfy the following three conditions:

7 Real-Valued Monotone Functions 75

(a) ϕ is increasing;
(b) ψ is decreasing;
(c) the function $\phi \cdot \psi$ is not monotone on any non-degenerate subinterval of $\mathbb{R}$.

For this purpose, use the result formulated in Theorem 7.4.

11. For a partial function $f : \mathbb{R} \to \mathbb{R}$, consider the following two properties:

    (a) there exists an uncountable set $X \subset \mathrm{dom}(f)$ such that $f|X$ is monotone;
    (b) there exists an uncountable set $Y \subset \mathrm{dom}(f)$ such that $f|Y$ is continuous.

    Verify that the implication (a) $\Rightarrow$ (b) holds.

    Remark 7.4 It is consistent with **ZFC** set theory that there exists a partial function g from $\mathbb{R}$ into $\mathbb{R}$ such that:

    (1) $\mathrm{dom}(g)$ is uncountable and g is continuous on $\mathrm{dom}(g)$;
    (2) for each uncountable subset Z of $\mathrm{dom}(g)$, the restriction $g|Z$ is not monotone.

    More details on partial functions similar to this g can be found in [100].
    At the same time, if a partial function f from $\mathbb{R}$ into itself is defined on an uncountable closed subset of $\mathbb{R}$ and is continuous, then there exists a nonempty perfect set P in $\mathbb{R}$ entirely contained in $\mathrm{dom}(f)$ and such that the restriction $f|P$ is monotone (see again [100]).

12. Let $(X, \leq)$ and $(Y, \leq)$ be two linearly ordered sets and let h be a partial monotone function from X to Y.

    Supposing that $(Y, \leq)$ is Dedekind complete, formulate and prove for h an appropriate analog of Theorem 7.1.

Chapter 8
Real-Valued Convex Functions

It is well known that monotone functions (briefly considered in the previous chapter) are closely connected with real-valued convex functions (see, e.g., [44, 134], and Exercise 5 of this chapter). Here we would like to touch upon convex functions and some of their elementary properties.

First of all, it makes sense to recall the precise definition of a convex function on a non-degenerate subsegment $[a, b]$ of $\mathbb{R}$.

Let f be a function acting from $[a, b]$ into $\mathbb{R}$.

We shall say that f is *mid-point convex* if the inequality

$$f((x + y)/2) \leq (f(x) + f(y))/2$$

holds for all points x and y of $[a, b]$.

Further, we shall say that f is *lower convex* (in short, *convex*) if the inequality

$$f(tx + (1 - t)y) \leq tf(x) + (1 - t)f(y)$$

holds for all points x and y of $[a, b]$ and for every $t \in [0, 1]$.

Accordingly, we shall say that f is *upper convex* (in short, *concave*) if the inequality

$$f(tx + (1 - t)y) \geq tf(x) + (1 - t)f(y)$$

holds for all points x and y of $[a, b]$ and for every $t \in [0, 1]$.

Remark 8.1 If one has the strict inequality

$$f(tx + (1 - t)y) < tf(x) + (1 - t)f(y)$$

for any two distinct points x and y of $[a, b]$ and for each t from the open interval $]0, 1[$, then f is called a *strictly convex* function. A *strictly concave* function is defined in a similar manner.

Obviously, if a function f is convex, then it is also mid-point convex and $-f$ is concave. In fact, f is convex if and only if $-f$ is concave.

If a function f is simultaneously convex and concave on $[a, b]$, then f is an *affine* function (i.e., the graph of f is a line segment).

In general, mid-point convexity of f does not imply convexity of f. This implication is true under certain additional assumptions on f (see Exercise 3 and Remark 8.6 of the present chapter).

Theorem 8.1 *The following three assertions hold:*

(1) if $f : [a, b] \to \mathbb{R}$ is convex and a real r is non-negative, then rf is also convex;
(2) if f and g are both convex on $[a, b]$, then $f + g$ is also convex on $[a, b]$;
(3) if $f : [a, b] \to \mathbb{R}$ is convex and $g : \mathbb{R} \to \mathbb{R}$ is convex and increasing, then the composition $g \circ f$ is convex on $[a, b]$.

The easy proof of this theorem is left to the reader.

Remark 8.2 It may happen that a function f is strictly convex, but its square $f^2 = f \cdot f$ is strictly concave. So the family of convex functions is not closed under the usual multiplication operation (cf. Exercise 11 of this chapter).

Theorem 8.2 *If a function $f : [a, b] \to \mathbb{R}$ is convex, then the following inequality (due to J. Jensen) holds true:*

$$f(t_1 x_1 + t_2 x_2 + \cdots + t_n x_n) \leq t_1 f(x_1) + t_2 f(x_2) + \cdots + t_n f(x_n),$$

where n is a nonzero natural number, $t_1, t_2, \ldots, t_n$ are non-negative real numbers such that $\sum \{t_i : 1 \leq i \leq n\} = 1$, and $x_1, x_2, \ldots, x_n$ are any points of $[a, b]$.

Proof We argue by induction on n. For $n = 1$ and $n = 2$ there is nothing to prove. Suppose that $n \geq 2$ and Jensen's inequality holds true for n. Take any non-negative real numbers $t_1, t_2, \ldots, t_n, t_{n+1}$ such that

$$\sum \{t_i : 1 \leq i \leq n + 1\} = 1$$

and choose arbitrarily points $x_1, x_2, \ldots, x_n, x_{n+1}$ from $[a, b]$. Since $n \geq 2$, there exists $t_i < 1$. We may assume, without loss of generality, that $t_{n+1} < 1$. Consider the point x in $[a, b]$ defined by

$$x = (t_1 x_1 + t_2 x_2 + \cdots + t_n x_n)/(1 - t_{n+1}).$$

Clearly, we can write

$$f(t_{n+1} x_{n+1} + (1 - t_{n+1}) x) \leq t_{n+1} f(x_{n+1}) + (1 - t_{n+1}) f(x).$$

8 Real-Valued Convex Functions

By virtue of the inductive assumption for n, we get

$$f(x) \leq (t_1 f(x_1) + t_2 f(x_2) + \cdots + t_n f(x_n))/(1 - t_{n+1}),$$

whence it readily follows that

$$f(t_1 x_1 + t_2 x_2 + \cdots + t_{n+1} x_{n+1}) \leq t_1 f(x_1) + t_2 f(x_2) + \cdots + t_{n+1} f(x_{n+1}).$$

Jensen's inequality has thus been proved by the induction method. □

Remark 8.3 Obviously, the analogous inequality for real-valued concave functions on $[a, b]$ also holds. Namely, if $f : [a, b] \to \mathbb{R}$ is concave, then

$$f(t_1 x_1 + t_2 x_2 + \cdots + t_n x_n) \geq t_1 f(x_1) + t_2 f(x_2) + \cdots + t_n f(x_n),$$

where n is a nonzero natural number, $t_1, t_2, \ldots, t_n$ are non-negative real numbers such that $\sum \{t_i : 1 \leq i \leq n\} = 1$, and $x_1, x_2, \ldots, x_n$ are any points of $[a, b]$.

Remark 8.4 Jensen's inequality for convex functions and its dual for concave functions are extremely useful for obtaining many other inequalities in mathematical analysis. For instance, the classical inequality between the geometrical and arithmetical means of any given positive real numbers $x_1, x_2, \ldots, x_n$, i.e., the relation

$$(x_1 x_2 \cdots x_n)^{1/n} \leq (x_1 + x_2 + \cdots + x_n)/n \qquad (n > 0)$$

is a straightforward consequence of the fact that the function $x \to \ln(x)$ is concave on the unbounded interval $]0, +\infty[$.

Theorem 8.3 *If $g : [a, b] \to \mathbb{R}$ is a convex function and differs from a constant, then g cannot attain its supremum at an interior point of the segment $[a, b]$.*

Proof Suppose on the contrary that $\sup(g) = g(z)$, where z is an interior point of $[a, b]$. Let x and y be any two points in $[a, b]$ such that $x < z < y$. Then, for some $t \in]0, 1[$, we have $z = tx + (1 - t)y$. In view of the convexity of g, we may write

$$g(z) \leq tg(x) + (1 - t)g(y) \leq g(z),$$

whence it immediately follows that $g(x) = g(y) = g(z)$, i.e., g is a constant function.

The obtained contradiction completes the proof. □

Consider now an arbitrary convex function $f : [a, b] \to \mathbb{R}$.

Let x, y, and z be three interior points of $[a, b]$ such that $x < y < z$. Evidently, y admits a representation in the form

$$y = ((y - x)/(z - x))z + ((z - y)/(z - x))x.$$

Consequently, we get

$$f(y) \leq ((y-x)/(z-x))f(z) + ((z-y)/(z-x))f(x).$$

It is not difficult to check that the above relation implies

$$(f(y) - f(x))/(y-x) \leq (f(z) - f(x))/(z-x) \leq (f(z) - f(y))/(z-y).$$

The latter inequalities enable us to infer that:

(i) if y tends to x, then the value $(f(y) - f(x))/(y-x)$ decreases and is bounded from below; therefore, there exists a right derivative $f'_r(x)$;
(ii) the relation

$$f'_r(x) \leq (f(z) - f(x))/(z-x) \leq f'_r(z)$$

holds, so the function $x \to f'_r(x)$ is increasing on $]a, b[$.

The analogous two assertions hold true for the left derivative f'_l of f.

Since $f'_r(x)$ and $f'_l(x)$ exist at each interior point x of $[a, b]$, one can conclude that f is continuous at all points of the open interval $]a, b[$ (cf. Exercise 4).

It should also be mentioned that the equality

$$f'_r(x) = f'_l(x)$$

is valid for all, except countably many, points of $[a, b]$. Therefore, f is differentiable on a co-countable subset of $[a, b]$ (in this connection, see Exercise 7).

Remark 8.5 In our lecture course we do not intend to consider many other important properties of convex and mid-point convex functions. They are discussed in various monographs and textbooks (see especially [107]). Obviously, of more interest are the situations where real-valued convex functions are defined on convex subsets of a topological vector space over $\mathbb{R}$ (in particular, on convex subsets of the Euclidean space $\mathbb{R}^m$, where $m \in \mathbb{N}$). Recall that a subset Z of a real vector space E is convex if

$$(\forall x \in Z)(\forall y \in Z)(\{tx + (1-t)y : 0 < t < 1\} \subset Z).$$

The Lebesgue theorem stating the differentiability almost everywhere (with respect to the standard Lebesgue measure λ on $\mathbb{R}$) of an arbitrary real-valued monotone function on $\mathbb{R}$ readily implies that any convex function on $\mathbb{R}$ is twice differentiable almost everywhere. The analogous much deeper result holds true for any convex function defined on $\mathbb{R}^m$, in terms of the m-dimensional Lebesgue measure λ_m on $\mathbb{R}^m$.

8 Real-Valued Convex Functions

Exercises

1. Let $h : [a, b] \to \mathbb{R}$ be a mid-point convex function and let $n \geq 1$ be a natural number.
   Demonstrate that the inequality
   $$h((x_1 + x_2 + \cdots + x_n)/n) \leq (h(x_1) + h(x_2) + \cdots + h(x_n))/n$$
   holds true for any points $x_1, x_2, \ldots, x_n$ from $[a, b]$.
   For this purpose, use Cauchy type induction on n.
   In other words, prove that:

   (i) if the above inequality is valid for n, then it is also valid for $2n$;
   (ii) if the above inequality is valid for $n + 1$, then it is also valid for n.

   Conclude that if, in addition, h is continuous, then h is a convex function on the segment $[a, b]$.

2. Let f be a function acting from a segment $[a, b]$ into $\mathbb{R}$.
   Show that the following two assertions are equivalent:

   (i) f is a convex function;
   (ii) the set $\{(x, t) \in [a, b] \times \mathbb{R} : f(x) \leq t\}$ is convex in the plane $\mathbb{R}^2$.

   Infer from the above equivalence that if $\{f_i : i \in I\}$ is a family of convex (respectively, concave) functions on $[a, b]$ and $f = \sup\{f_i : i \in I\}$ (respectively, $f = \inf\{f_i : i \in I\}$), then f is a convex (respectively, concave) function on $[a, b]$.

3. Prove that:

   (i) there exists a mid-point convex function $f : \mathbb{R} \to \mathbb{R}$ which is not convex;
   (ii) if a mid-point convex function $g : [a, b] \to \mathbb{R}$ is bounded from above on $[a, b]$, then g is convex.

   For (i), consider $\mathbb{R}$ as a vector space over $\mathbb{Q}$ and, using a basis of this space (a Hamel basis), define an additive function $f : \mathbb{R} \to \mathbb{Q}$ which trivially is mid-point convex, but is not convex.
   For (ii), argue as follows. First, consider the special case where $[a, b]$ is a neighborhood of 0 and the equality $g(0) = 0$ holds for a mid-point convex and bounded from above function g on $[a, b]$. In this case, prove that g is continuous at 0.
   Then consider the general case of a mid-point convex and bounded from above function $g : [a, b] \to \mathbb{R}$, where $[a, b]$ is an arbitrary non-degenerate segment in $\mathbb{R}$. Take any interior point x_0 of $[a, b]$ and introduce the function
   $$h(x) = g(x + x_0) - g(x_0) \qquad (x \in [a, b] - x_0).$$

Verify that h is a mid-point convex and bounded from above function on some neighborhood of 0, such that $h(0) = 0$. Consequently, h is continuous at 0, which means that g is continuous at x_0. Infer that g is a convex function on $[a, b]$ (cf. Exercise 1).

Remark 8.6 It was demonstrated by W. Sierpiński that if a mid-point convex function $h : [a, b] \to \mathbb{R}$ is Lebesgue measurable, then h is convex. For a proof of this highly nontrivial fact, see e.g. [161].

4. Give an example of a convex function $f : [a, b] \to \mathbb{R}$ such that f is discontinuous at the endpoints of $[a, b]$.
5. Let $g : [r, +\infty[\to \mathbb{R}$ be a function, where r is a real number.
   Show that the following two assertions are equivalent:

   (a) g is a convex function and has a right derivative at r;
   (b) there exists an increasing function $\phi : [r, +\infty[\to \mathbb{R}$ such that

   $$g(x) = g(r) + \int_r^x \phi(t)\,dt \qquad (x \in [r, +\infty[).$$

   Here the integration is meant in the Riemann sense (see Appendix C).

   Conclude that if (a) holds true, then g is differentiable on a co-countable subset of $[r, +\infty[$.
   For this purpose, keep in mind the following facts:

   (i) the function ϕ of (b) is continuous at all points of a co-countable subset of $[r, +\infty[$;
   (ii) the derivative $g'(x)$ exists and is equal to $\phi(x)$ whenever x belongs to $[r, +\infty[$ and is a continuity point of ϕ.

6. Give an example of a continuous function $h : [0, 1] \to \mathbb{R}$ such that:

   (i) h is a convex function;
   (ii) h is not representable in the form

   $$h(x) = h(0) + \int_0^x \psi(t)\,dt \qquad (x \in [0, 1]),$$

   where $\psi : [0, 1] \to \mathbb{R}$ is an increasing function.

7. A widely known theorem of elementary geometry states that the sum of all external angles of a convex polygon in $\mathbb{R}^2$ is equal to 2π.
   Starting with this theorem, give another proof of the fact that any convex function $f : [a, b] \to \mathbb{R}$ is differentiable at each point of a co-countable subset of $[a, b]$.
8. Let $\{f_n : n \in \mathbb{N}\}$ be a sequence of real-valued convex functions on a segment $[a, b] \subset \mathbb{R}$ and suppose that

$$f(x) = \lim_{n\to\infty} f_n(x) \qquad (x \in [a.b]).$$

Demonstrate that $f : [a, b] \to \mathbb{R}$ is also a convex function.
In addition, assuming that all functions f_n ($n \in \mathbb{N}$) and the function f are differentiable at a point x_0 of $]a, b[$, show that the equality

$$f'(x_0) = \lim_{n\to\infty} f'_n(x_0)$$

is fulfilled.

9. Consider the vector space $C([0, 1])$ ($= C([0, 1], \mathbb{R})$) of all real-valued continuous functions f defined on the unit segment $[0, 1]$. Equip this space with the topology of uniform convergence by using the standard sup-norm

$$||f|| = \sup\{|f(x)| : x \in [0, 1]\}.$$

As is well known, $C([0, 1])$ becomes a separable Banach space with respect to $||\cdot||$.

Check that one of the closed subsets of $C([0, 1])$ is the complete separable metric space $\mathrm{Conv}([0, 1])$ formed by all continuous convex functions on $[0, 1]$. Further, introduce the following notation:

$F_0 = $ the family of all those functions from $C([0, 1])$ which are differentiable at least at one point of $[0, 1]$;

$F_1 = $ the family of all those functions from $\mathrm{Conv}([0, 1])$ which are differentiable at each interior point of $[0, 1]$.

Show that:

(a) F_0 is a first category subset of the space $C([0, 1])$ (the theorem of Banach and Mazurkiewicz);
(b) F_1 is a co-meager subset of $\mathrm{Conv}([0, 1])$, i.e., the set $\mathrm{Conv}([0, 1]) \setminus F_1$ is of first category in $\mathrm{Conv}([0, 1])$.

Argue step by step. Begin by establishing (a). For every natural number $n > 0$ and for every rational number $q > 0$, define

$$K_{n,q} = \{(f, x) \in C([0, 1]) \times [0, 1] :$$
$$(\forall t)(0 < |t| < q \Rightarrow (|f(x+t) - f(x)|)/|t| \leq n)\}.$$

Verify that the set $K_{n,q}$ is closed in the product space $C([0, 1]) \times [0, 1]$. Then, using the compactness of $[0, 1]$, prove that the set $\mathrm{pr}_1(K_{n,q})$ is closed and nowhere dense in $C([0, 1])$. Finally, check the inclusion

$$F_0 \subset \cup\{\mathrm{pr}_1(K_{n,q}) : n \in \mathbb{N}, q \in \mathbb{Q}, q > 0\},$$

from which (a) immediately follows.

Concerning (b), argue in a similar manner. First, for any natural number $n > 0$, let $a_n = 1/3^n$ and $b_n = 1 - 1/3^n$.

If $t \in [a_n, b_n]$ and $f \in \operatorname{Conv}([0, 1])$, consider two straight lines $L_1(t, f)$ and $L_2(t, f)$ in $\mathbb{R}^2$ defined by their canonical equations in terms of the standard coordinates $(x, y) \in \mathbb{R}^2$. Namely, the equation corresponding to $L_1(t, f)$ is

$$y - f(t) = f'_r(t)(x - t)$$

and the equation corresponding to $L_2(t, f)$ is

$$y - f(t) = f'_l(t)(x - t).$$

Also, denote by $\alpha(t, f)$ the value of the convex angle which is formed by $L_1(t, f)$ and $L_2(t, f)$ and which contains the graph of f.

Now, take a nonzero natural number m and define

$$P_{m,n} = \{(f, t) \in \operatorname{Conv}([0, 1]) \times [a_n, b_n] : \alpha(t, f) \leq \pi - 1/m\}.$$

Show that the set $P_{m,n}$ is closed in the product space $\operatorname{Conv}([0, 1]) \times [0, 1]$. Then, using again the compactness of $[0, 1]$, establish the closedness and nowhere density of $\operatorname{pr}_1(P_{m,n})$ in the space $\operatorname{Conv}([0, 1])$. Finally, take into account the relation

$$F_1 = \operatorname{Conv}([0, 1]) \setminus \cup \{\operatorname{pr}_1(P_{m,n}) : m \in \mathbb{N} \setminus \{0\}, \ n \in \mathbb{N} \setminus \{0\}\},$$

which trivially implies the validity of (b).

Remark 8.7 Let us define:

$$F_2 = \text{the family of all those functions from } Conv([0, 1]) \text{ which}$$
$$\text{are twice differentiable everywhere on } [0, 1].$$

In connection with (b) of Exercise 9, it should be noted that according to Gruber's theorem, the family F_2 is of first category in the space $\operatorname{Conv}([0, 1])$.

10. Let $[a, b]$ be a non-degenerate segment in $\mathbb{R}$ and let f be a real-valued convex function on $[a, b]$.

    Prove that the following two assertions are equivalent:

    (i) there exists a convex function $f^* : \mathbb{R} \to \mathbb{R}$ extending f;
    (ii) there exists a right derivative of f at the point a and there exists a left derivative of f at the point b.

11. Verify that the function

8 Real-Valued Convex Functions

$$x \to -x^{1/3} \quad (x > 0)$$

is strictly convex and the function

$$x \to x^{2/3} \quad (x > 0)$$

is strictly concave.

Conclude that the family of all convex functions on $]0, +\infty[$ is not closed under the standard multiplication operation.

Also, check that there exists a real-valued strictly convex function g on the same interval $]0, +\infty[$ such that the composition $g \circ g$ is strictly concave (e.g., put $g(x) = x^{-1/3}$ for $x > 0$).

12. Let $f : [a, b] \to \mathbb{R}$ be a convex function.

    Show that if this f is differentiable at all interior points of $[a, b]$, then the derivative f' is continuous on $]a, b[$.

Remark 8.8 It is widely known that a function $g : \mathbb{R} \to \mathbb{R}$ can be differentiable at all points of $\mathbb{R}$, but the derivative g' of g can be discontinuous at many points of $\mathbb{R}$. In this context, Exercise 12 indicates that the derivatives of convex functions have a certain nice property.

Chapter 9
Semicontinuity of a Real-Valued Function at a Point

Let E be a topological space, x_0 be a point of E, and let $f : E \to \mathbb{R}$ be a function. We would like to begin this chapter by recalling two standard definitions (see, e.g., [4, 19, 40, 104, 134]).

One says that f is *lower semicontinuous* at x_0 if, for any real $\varepsilon > 0$, there exists a neighborhood $U(x_0)$ of x_0 such that

$$(\forall x \in U(x_0))(f(x) > f(x_0) - \varepsilon).$$

In a similar manner, one says that f is *upper semicontinuous* at x_0 if, for any real $\varepsilon > 0$, there exists a neighborhood $U(x_0)$ of x_0 such that

$$(\forall x \in U(x_0))(f(x) < f(x_0) + \varepsilon).$$

Remark 9.1 In the above definitions one can replace, respectively, the symbol $>$ by $\geq$, and the symbol $<$ by $\leq$. Clearly, after this replacement we get the same notions of lower semicontinuity and upper semicontinuity of a function $f : E \to \mathbb{R}$ at a point x_0. Also, it is easy to see that the lower semicontinuity of f at x_0 can equivalently be expressed by the formula

$$\liminf\nolimits_{x \to x_0} f(x) = f(x_0),$$

and the upper semicontinuity of f at x_0 can equivalently be expressed by the formula

$$\limsup\nolimits_{x \to x_0} f(x) = f(x_0).$$

It will be seen in the sequel that there is a close connection between lower semicontinuous functions at x_0 and upper semicontinuous functions at x_0.

A function $f : E \to \mathbb{R}$ is called *semicontinuous* at $x_0 \in E$ if f is either lower semicontinuous at x_0 or upper semicontinuous at x_0.

© The Author(s), under exclusive license to Springer Nature Switzerland AG 2025
A. Kharazishvili, *Lectures on Real-valued Functions*,
https://doi.org/10.1007/978-3-031-95369-9_9

Remark 9.2 It directly follows from the above definitions that, for a function $g : E \to \mathbb{R}$, the following two statements are equivalent:

(a) g is continuous at a point $x_0 \in E$;
(b) g is simultaneously lower and upper semicontinuous at $x_0 \in E$.

Theorem 9.1 *The following four assertions hold:*

(1) *a function $f : E \to \mathbb{R}$ is lower (upper) semicontinuous at $x_0 \in E$ if and only if the function $-f$ is upper (lower) semicontinuous at x_0;*
(2) *if two functions $f : E \to \mathbb{R}$ and $g : E \to \mathbb{R}$ are lower (respectively, upper) semicontinuous at $x_0 \in E$, then the functions*

$$f + g, \quad \sup(f, g), \quad \inf(f, g)$$

are lower (respectively, upper) semicontinuous at x_0;
(3) *if $f : E \to \mathbb{R}$ is lower (upper) semicontinuous at $x_0 \in E$ and a real number t is non-negative, then the function tf is lower (upper) semicontinuous at x_0;*
(4) *if $f : E \to \mathbb{R}$ and $g : E \to \mathbb{R}$ are both non-negative and lower (upper) semicontinuous functions at $x_0 \in E$, then $f \cdot g$ is also a lower (upper) semicontinuous function at x_0.*

Proof Assertion (1) immediately follows from the definitions of lower semicontinuous and upper semicontinuous functions at x_0.

To show the validity of (2), it suffices (in view of (1)) to consider the case when both $f : E \to \mathbb{R}$ and $g : E \to \mathbb{R}$ are lower semicontinuous at a point $x_0 \in E$. Pick a real $\varepsilon > 0$. For $\varepsilon/2$, there exists a neighborhood $U(x_0)$ such that

$$(\forall x \in U(x_0))(f(x) > f(x_0) - \varepsilon/2).$$

Similarly, for the same $\varepsilon/2$, there exists a neighborhood $V(x_0)$ such that

$$(\forall x \in V(x_0))(g(x) > g(x_0) - \varepsilon/2).$$

Let $W(x_0) = U(x_0) \cap V(x_0)$. Then, for the neighborhood $W(x_0)$, we have

$$(\forall x \in W(x_0))(f(x) + g(x) > (f(x_0) - \varepsilon/2) + (g(x_0) - \varepsilon/2)$$
$$= (f(x_0) + g(x_0)) - \varepsilon),$$

which establishes the lower semicontinuity of $f + g$ at x_0.

An analogous argument works for the functions $\sup(f, g)$ and $\inf(f, g)$.

The easy proof of assertion (3) is left to the reader.

In order to demonstrate (4), suppose first that both functions $f \geq 0$ and $g \geq 0$ are upper semicontinuous at x_0 and take a real $\varepsilon > 0$. There exist a neighborhood $U(x_0)$ and a neighborhood $V(x_0)$ such that

9 Semicontinuity of a Real-Valued Function at a Point

$$(\forall x \in U(x_0))(f(x) < f(x_0) + \varepsilon),$$
$$(\forall x \in V(x_0))(g(x) < g(x_0) + \varepsilon).$$

Again letting $W(x_0) = U(x_0) \cap V(x_0)$ and keeping in mind the non-negativity of f and g, we may write

$$f(x)g(x) < (f(x_0) + \varepsilon)(g(x_0) + \varepsilon) = f(x_0)g(x_0) + \varepsilon(f(x_0) + g(x_0)) + \varepsilon^2$$

for all points $x \in W(x_0)$, whence it readily follows that $f \cdot g$ is upper semicontinuous at x_0.

Finally, let both non-negative functions f and g be lower semicontinuous at x_0. If $f(x_0) = 0$ or $g(x_0) = 0$, then $(f \cdot g)(x_0) = 0$ and $f \cdot g$ is automatically lower semicontinuous at x_0 (cf. Exercise 1 of this chapter). It remains to consider the case when $f(x_0) > 0$ and $g(x_0) > 0$. Let a real $\varepsilon > 0$ be arbitrarily small. In particular, we may assume that

$$f(x_0) - \varepsilon > 0, \qquad g(x_0) - \varepsilon > 0.$$

There exist a neighborhood $U(x_0)$ and a neighborhood $V(x_0)$ such that

$$(\forall x \in U(x_0))(f(x) > f(x_0) - \varepsilon > 0),$$
$$(\forall x \in V(x_0))(g(x) > g(x_0) - \varepsilon > 0).$$

Once again defining $W(x_0) = U(x_0) \cap V(x_0)$, we may write

$$f(x)g(x) > (f(x_0) - \varepsilon)(g(x_0) - \varepsilon)$$
$$= f(x_0)g(x_0) - \varepsilon(f(x_0) + g(x_0)) + \varepsilon^2$$
$$> f(x_0)g(x_0) - \varepsilon(f(x_0) + g(x_0))$$

for all points $x \in W(x_0)$, which implies the lower semicontinuity of $f \cdot g$ at the given point x_0.

Theorem 9.1 has thus been proved. $\square$

Exercise 3 of the present chapter shows that the assumption of the non-negativity of both functions f and g in assertion (4) of Theorem 9.1 is essential for the validity of (4).

Remark 9.3 Let L be a vector space over $\mathbb{R}$ and let 0 stand for the neutral element of L. Recall that a nonempty set $C \subset L$ is a *cone* in L (whose vertex is 0) if the following condition is fulfilled:

For any $x \in C$ and for each real number $t \geq 0$, one has $tx \in C$.

In the geometric language this means that if $x \in C \setminus \{0\}$, then the ray which passes through x and whose endpoint coincides with 0 is entirely contained in C.

A cone C is called a *convex cone* if, for any two points $x \in C$ and $y \in C$, one has $x + y \in C$.

This definition of a convex cone is compatible with the usual definition of convex sets in real vector spaces:

A subset Z of L is called *convex* if, for any two points $x \in Z$ and $y \in Z$, the line segment $[x, y] = \{\tau x + (1 - \tau)y : 0 \leq \tau \leq 1\}$ is entirely contained in Z.

Indeed, it can readily be verified that a cone C is convex in the usual sense if and only if it is convex in the sense of the above-mentioned definition.

It should be noticed that convex cones play an important role in various topics of modern functional analysis. In particular, such cones are very useful in those questions which are concerned with some extension theorems in analysis (see, e.g., Chap. 18).

Now, consider the real vector space $\mathbb{R}^E$ of all functions acting from a topological space E into $\mathbb{R}$. Actually, assertions (2) and (3) of Theorem 9.1 state that the functions on E which are lower (respectively, upper) semicontinuous at a fixed point $x_0 \in E$ form some convex cone C (respectively, $-C$) in $\mathbb{R}^E$, and $C \cap (-C)$ coincides with the vector subspace of $\mathbb{R}^E$ consisting of all those real-valued functions on E which are continuous at x_0.

Moreover, assertion (2) of the same theorem shows us that C (respectively, $-C$) is a lattice with respect to the standard operations

$$(f, g) \to \inf(f, g), \quad (f, g) \to \sup(f, g)$$

in the vector space $\mathbb{R}^E$.

The following theorem presents one important property of this lattice (its so-called *conditional completeness*).

Theorem 9.2 *Let E be a topological space, x_0 be a point of E, and let $\{f_i : i \in I\}$ be a family of real-valued functions on E such that $\sup\{f_i(x) : i \in I\}$ exists for each $x \in E$ (respectively, $\inf\{f_i(x) : i \in I\}$ exists for each $x \in E$).*

If all functions f_i ($i \in I$) are lower (respectively, upper) semicontinuous at x_0, then the function $\sup\{f_i : i \in I\}$ is lower semicontinuous at x_0 (respectively, the function $\inf\{f_i : i \in I\}$ is upper semicontinuous at x_0).

Proof It suffices to consider the case when all f_i ($i \in I$) are lower semicontinuous at x_0. Let $f = \sup\{f_i : i \in I\}$ and take an arbitrary real $\varepsilon > 0$. Since we have

$$f(x_0) = \sup\{f_i(x_0) : i \in I\},$$

there is an index $j \in I$ such that $f(x_0) < f_j(x_0) + \varepsilon/2$. Since the function f_j is lower semicontinuous at x_0, there exists a neighborhood $U(x_0)$ such that

$$(\forall x \in U(x_0))(f_j(x) > f_j(x_0) - \varepsilon/2).$$

9 Semicontinuity of a Real-Valued Function at a Point

Consequently, for any point $x \in U(x_0)$, we have

$$f(x) \geq f_j(x) > f_j(x_0) - \varepsilon/2 > f(x_0) - \varepsilon/2 - \varepsilon/2 = f(x_0) - \varepsilon,$$

which shows us that the function f is lower semicontinuous at x_0. □

Remark 9.4 Let $(I, \preceq)$ be a partially ordered set, E be a topological space, x_0 be a point in E, and let $\{f_i : i \in I\}$ be an increasing (respectively, decreasing) family of real-valued functions on E such that $\sup\{f_i(x) : i \in I\}$ exists for each $x \in E$ (respectively, $\inf\{f_i(x) : i \in I\}$ exists for each $x \in E$). Then $\{f_i : i \in I\}$ converges pointwise on E to the function

$$f : E \to \mathbb{R}$$

defined by the formula $f(x) = \sup\{f_i(x) : i \in I\}$ (respectively, defined by the formula $f(x) = \inf\{f_i(x) : i \in I\}$).

Theorem 9.2 immediately implies that if all functions f_i ($i \in I$) are lower (respectively, upper) semicontinuous at x_0, then f is lower semicontinuous at x_0 (respectively, f is upper semicontinuous at x_0).

In the special case $(I, \preceq) = (\mathbb{N}, \leq)$, where $\leq$ is the standard well-ordering of $\mathbb{N}$, we obtain the following assertion:

If $\{f_n : n \in \mathbb{N}\}$ is an increasing (respectively, decreasing) sequence of real-valued functions on E, all of which are lower (respectively, upper) semicontinuous at $x_0 \in E$, then the function $f = \lim_{n \to \infty} f_n$ is lower (respectively, upper) semicontinuous at the same x_0.

For any ground set E (not necessarily endowed with a topology) and for any function $f : E \to \mathbb{R}$, let us introduce the notation

$$\Gamma^*(f) = \{(x, t) \in E \times \mathbb{R} : f(x) \leq t\},$$
$$\Gamma_*(f) = \{(x, t) \in E \times \mathbb{R} : f(x) \geq t\}.$$

Evidently, we have the equality

$$\Gamma(f) = \Gamma^*(f) \cap \Gamma_*(f),$$

where $\Gamma(f)$ denotes the graph of f. It is also easy to see that the equivalence

$$(x, t) \in \Gamma^*(f) \Leftrightarrow (x, -t) \in \Gamma_*(-f)$$

is true.

Theorem 9.3 *Let E be a topological space, x_0 be a point of E, and let $f : E \to \mathbb{R}$ be a function.*
The following two assertions hold:

(1) if f is lower semicontinuous at x_0 and (x_0, t_0) is an adherent point of $\Gamma^*(f)$ in the topological product space $E \times \mathbb{R}$, then (x_0, t_0) belongs to $\Gamma^*(f)$;
(2) if f is upper semicontinuous at x_0 and (x_0, t_0) is an adherent point of $\Gamma_*(f)$ in the topological product space $E \times \mathbb{R}$, then (x_0, t_0) belongs to $\Gamma_*(f)$.

Proof It suffices to demonstrate the validity of (1). Suppose that the given f is lower semicontinuous at x_0 and (x_0, t_0) is an adherent point of $\Gamma^*(f)$. We must show that $f(x_0) \leq t_0$. Suppose otherwise, i.e., $f(x_0) > t_0$. Take a real $\varepsilon > 0$ such that $f(x_0) > t_0 + 2\varepsilon$. For this ε, there exists a neighborhood $U(x_0)$ such that

$$(\forall x \in U(x_0))(f(x) > f(x_0) - \varepsilon > t_0 + \varepsilon).$$

In the topological space $E \times \mathbb{R}$ consider the set $U(x_0) \times \,]t_0 - \varepsilon, t_0 + \varepsilon[$. Clearly, this set is a neighborhood of the point (x_0, t_0) and it can easily be verified that

$$(U(x_0) \times \,]t_0 - \varepsilon, t_0 + \varepsilon[) \cap \Gamma^*(f) = \emptyset.$$

But the last formula contradicts the condition that (x_0, t_0) is an adherent point of $\Gamma^*(f)$. The obtained contradiction ends the proof. □

Notice that the statement converse to Theorem 9.3 is also true (in this connection, see Exercise 9).

Exercises

1. Let E be a topological space and let $f : E \to \mathbb{R}$ be a function. Suppose that f has a local minimum (respectively, a local maximum) at the point $x_0 \in E$.
   Check that f is lower semicontinuous (respectively, upper semicontinuous) at x_0.
   Infer from this fact that if $E = \mathbb{R}$, then a function $f : \mathbb{R} \to \mathbb{R}$ can be semicontinuous at $x_0 \in \mathbb{R}$ and not have one-sided limits at x_0.
2. Give a detailed proof of the fact that if E is a topological space, x_0 is a point of E, and $f : E \to \mathbb{R}$ and $g : E \to \mathbb{R}$ are two functions which are upper (lower) semicontinuous at x_0, then the functions $\sup(f, g)$ and $\inf(f, g)$ are also upper (lower) semicontinuous at the same point x_0.
3. Let two functions $f : \mathbb{R} \to \mathbb{R}$ and $g : \mathbb{R} \to \mathbb{R}$ be defined as follows:

$$f(x) = \begin{cases} 1 & \text{if } x = 0, \\ 0 & \text{if } x \neq 0, \end{cases} \qquad g(x) = \begin{cases} -1 & \text{if } x = 0, \\ -2 & \text{if } x \neq 0. \end{cases}$$

Verify that:

(a) both functions f and g are upper semicontinuous at 0;

9 Semicontinuity of a Real-Valued Function at a Point

(b) the function $f \cdot g$ is lower semicontinuous at 0, but is not upper semicontinuous at 0.

Conclude that the assumption of the non-negativity of both functions f and g in assertion (4) of Theorem 9.1 is essential for the validity of (4).

4. Check that the definition of a convex cone in a real vector space L, which was formulated in Remark 9.3, is compatible with the standard definition of a convex set in L.
5. Give an example of a monotone function f acting from $\mathbb{R}$ into $\mathbb{R}$ such that f is not lower semicontinuous at 0 and is not upper semicontinuous at 0.
6. Let two functions $f : \mathbb{R} \to \mathbb{R}$ and $g : \mathbb{R} \to \mathbb{R}$ be defined as follows:

$$f = \text{the characteristic function of } \{0\};$$

$$g = \text{the characteristic function of } \{1\}.$$

Check that:

(a) both f and g are upper semicontinuous at all points of $\mathbb{R}$;
(b) the composition $f \circ g$ is not upper semicontinuous at 1;
(c) the composition $f \circ g$ is lower semicontinuous at 1.

7. Let E be a topological space, x_0 be a point of E, and let

$$f : \mathbb{R} \to \mathbb{R}, \qquad g : E \to \mathbb{R}$$

be two functions satisfying the following conditions:

(a) f is an increasing function and is upper semicontinuous at the point $g(x_0)$;
(b) g is upper semicontinuous at the point x_0.

Prove that the composition $f \circ g$ is upper semicontinuous at the point x_0.
For this purpose, argue as follows. Since f is upper semicontinuous at $g(x_0)$, for any real $\varepsilon > 0$ there exists a real $\delta > 0$ such that

(i) $(\forall y)(g(x_0) - \delta < y < g(x_0) + \delta \Rightarrow f(y) < f(g(x_0)) + \varepsilon)$.

Further, since g is upper semicontinuous at x_0, for the same $\delta > 0$, there exists a neighborhood $U(x_0)$ of x_0 such that

(ii) $(\forall x)(x \in U(x_0) \Rightarrow g(x) < g(x_0) + \delta)$.

Let now x be an arbitrary point from $U(x_0)$. If one has

$$g(x_0) - \delta < g(x) < g(x_0) + \delta,$$

then $f(g(x)) < f(g(x_0)) + \varepsilon$ in view of (i). If

$$g(x) \notin \,]g(x_0) - \delta, g(x_0) + \delta[,$$

then, in view of (ii), one has $g(x) \leq g(x_0) - \delta < g(x_0)$. Remembering that f is an increasing function, one obtains

$$f(g(x)) \leq f(g(x_0)) < f(g(x_0)) + \varepsilon,$$

which establishes the upper semicontinuity of $f \circ g$ at x_0.

8. Let E be a ground set and let $f : E \to \mathbb{R}$ be a function. Demonstrate that:

   (a) each of the sets $\Gamma^*(f)$ and $\Gamma_*(f)$ completely determines f;
   (b) each of the sets

   $$\{(x,t) \in E \times \mathbb{R} : f(x) < t\}, \qquad \{(x,t) \in E \times \mathbb{R} : f(x) > t\}$$

   also completely determines f.

9. Let E be a topological space, x_0 be a point of E, and let $f : E \to \mathbb{R}$ be a function.
   Show that the following two assertions hold true:

   (a) if any adherent point of $\Gamma^*(f) \subset E \times \mathbb{R}$, having the form (x_0, t), belongs to $\Gamma^*(f)$, then f is lower semicontinuous at x_0;
   (b) if any adherent point of $\Gamma_*(f) \subset E \times \mathbb{R}$, having the form (x_0, t), belongs to $\Gamma_*(f)$, then f is upper semicontinuous at x_0.

   For this purpose, argue similarly to the proof of Theorem 9.3.

10. Let E be a topological space, x_0 be a point of E, let $\{f_n : n \in \mathbb{N}\}$ be a sequence of functions from E to $\mathbb{R}$, and let f be a function also acting from E to $\mathbb{R}$. Suppose that the following two conditions are fulfilled:

    (a) for every real $\varepsilon > 0$, there exists a natural number $n = n(\varepsilon)$ such that $(\forall x \in E)(|f_n(x) - f(x)| < \varepsilon)$;
    (b) all functions f_n ($n \in \mathbb{N}$) are lower (upper) semicontinuous at x_0.

    Prove that the function f is lower (upper) semicontinuous at x_0.

Chapter 10
Semicontinuous Real-Valued Functions on Quasi-Compact Spaces

Suppose that a topological space E is given with a function f acting from E to $\mathbb{R}$.

In the previous chapter the notion of lower (respectively, upper) semicontinuity of f at a point x_0 of E was introduced.

In accordance with this definition, it is natural to call f *lower* (respectively, *upper*) *semicontinuous* on E if f is lower (respectively, upper) semicontinuous at all points of E.

Remark 10.1 Observe that a function $g : E \to \mathbb{R}$ is continuous on E if and only if g is simultaneously lower and upper semicontinuous on E.

Example 10.1 Denoting by $\lfloor x \rfloor$ the biggest integer not exceeding a real number x, it can easily be seen that the function

$$x \to \lfloor x \rfloor \qquad (x \in \mathbb{R})$$

is upper semicontinuous on $\mathbb{R}$. Consequently, the function

$$x \to x - \lfloor x \rfloor \qquad (x \in \mathbb{R})$$

is lower semicontinuous on $\mathbb{R}$. Evidently, both these functions are discontinuous at infinitely many points of $\mathbb{R}$.

Example 10.2 If E is a topological space and X is an open set in E, then the characteristic function of X is lower semicontinuous on E (but, in general, is not upper semicontinuous). Consequently, if Y is a closed set in E, then the characteristic function of Y is upper semicontinuous on E (but, in general, is not lower semicontinuous).

Remark 10.2 It immediately follows from Theorem 9.3 of Chap. 9 that if a function $f : E \to \mathbb{R}$ is lower (respectively, upper) semicontinuous on E, then the set $\Gamma^*(f)$ (respectively, the set $\Gamma_*(f)$) is closed in the topological product space $E \times \mathbb{R}$. The converse assertion is also true (see Exercise 3 of this chapter).

Theorem 10.1 *For a function* $f : E \to \mathbb{R}$, *the following three assertions are equivalent:*

(1) f *is lower semicontinuous on* E;
(2) for any real t, *the set* $\{x \in E : f(x) > t\}$ *is open in* E;
(3) for any real t, *the set* $\{x \in E : f(x) \le t\}$ *is closed in* E.

Proof The equivalence of (2) and (3) is trivial. Let us demonstrate the equivalence of (1) and (2).

Supposing (1), take $t \in \mathbb{R}$ and consider the set $\{x \in E : f(x) > t\}$. If x_0 is a point from this set, then $f(x_0) > t$. Choose a real $\varepsilon > 0$ so small that $f(x_0) > t + \varepsilon$. Since f is lower semicontinuous at x_0, there exists a neighborhood $U(x_0)$ such that $f(x_0) \le f(x) + \varepsilon$ for all points $x \in U(x_0)$. This implies that $f(x) > t$ for all $x \in U(x_0)$. Therefore,

$$U(x_0) \subset \{x \in E : f(x) > t\},$$

which shows us that $\{x \in E : f(x) > t\}$ is open in E, i.e., (2) holds true.

Conversely, suppose (2) and let x_0 be a point of E. Take a real $\varepsilon > 0$ and put $t = f(x_0) - \varepsilon$. According to (2), the set $\{x \in E : f(x) > t\}$ is open in E. Since x_0 belongs to $\{x \in E : f(x) > t\}$, there exists a neighborhood $U(x_0)$ entirely contained in $\{x \in E : f(x) > t\}$. Clearly, we have $f(x) > f(x_0) - \varepsilon$ for all $x \in U(x_0)$, so f is lower semicontinuous at x_0, i.e., (1) holds true.

This completes the proof of Theorem 10.1. □

Remark 10.3 It directly follows from the above theorem that, for a function $f : E \to \mathbb{R}$, the following three assertions are equivalent:

(1) f is upper semicontinuous on E;
(2) for any real t, the set $\{x \in E : f(x) < t\}$ is open in E;
(3) for any real t, the set $\{x \in E : f(x) \ge t\}$ is closed in E.

Consequently, every real-valued upper (lower) semicontinuous function on E is a Borel mapping from E into $\mathbb{R}$ (general Borel mappings will be introduced and discussed later in this lecture course).

Theorem 10.2 *Let* E *be an arbitrary nonempty quasi-compact topological space and let* $f : E \to \mathbb{R}$ *be a lower semicontinuous function on* E.

Then there exists a point $z \in E$ *such that*

$$f(z) = \inf\{f(x) : x \in E\}.$$

In other words, f *attains its infimum on the space* E.

Proof Let $r = \inf(\operatorname{ran}(f)) = \inf\{f(x) : x \in E\}$.

Observe first that $r > -\infty$. Indeed, suppose for a moment otherwise. Then, for every natural number n, the closed set

$$X_n = \{x \in E : f(x) \leq -n\}$$

is nonempty, $X_{n+1} \subset X_n$, and

$$\cap \{X_n : n \in \mathbb{N}\} = \emptyset,$$

which is impossible for the quasi-compact space E. So, $r > -\infty$.

Further, in view of the definition of the infimum of a set of real numbers, for each nonzero natural number n, the closed set

$$Z_n = \{x \in E : f(x) \leq r + 1/n\}$$

is also nonempty. Therefore, we have a decreasing (by inclusion) sequence $\{Z_n : n \in \mathbb{N} \setminus \{0\}\}$ of nonempty closed sets in the quasi-compact space E. This implies that

$$\cap \{Z_n : n \in \mathbb{N} \setminus \{0\}\} \neq \emptyset.$$

Taking any point $z \in \cap\{Z_n : n \in \mathbb{N} \setminus \{0\}\}$, we easily conclude that $f(z) = r$, which ends the proof. □

Remark 10.4 Obviously, we have a statement dual to Theorem 10.2, which is formulated in terms of upper semicontinuous functions. Namely, if E is a nonempty quasi-compact topological space and $f : E \to \mathbb{R}$ is an upper semicontinuous function on E, then there exists a point z in E such that

$$f(z) = \sup\{f(x) : x \in E\}.$$

In other words, f attains its supremum on the space E. As a consequence of the above-mentioned fact and of Theorem 10.2, we obtain that every real-valued continuous function defined on a nonempty quasi-compact topological space attains its infimum and attains its supremum. Of course, the latter fact is also directly implied by the following well-known theorem of general topology: any continuous image of a quasi-compact space is again quasi-compact.

Theorem 10.3 *Let E be an arbitrary quasi-compact topological space and let $\{f_n : n \in \mathbb{N}\}$ be an increasing sequence of real-valued functions on E such that the following two conditions are satisfied:*

(1) all f_n ($n \in \mathbb{N}$) are lower semicontinuous;
(2) for each $x \in E$, the limit $\lim_{n \to \infty} f_n(x)$ exists.

Let $f : E \to \mathbb{R}$ be defined by the formula

$$f(x) = \lim_{n \to \infty} f_n(x) \qquad (x \in E).$$

If the function f is upper semicontinuous on E, then the given sequence $\{f_n : n \in \mathbb{N}\}$ converges to f uniformly.

Proof Suppose that f is upper semicontinuous at all points of E. Pick a real $\varepsilon > 0$. For any $x \in E$, there exists an open neighborhood $U(x, \varepsilon)$ of x such that

$$(\forall y \in U(x, \varepsilon))(f(y) \leq f(x) + \varepsilon/3).$$

Also, there exists a natural number $n(x, \varepsilon)$ such that

$$f(x) \leq f_{n(x,\varepsilon)}(x) + \varepsilon/3.$$

Further, since the function $f_{n(x,\varepsilon)}$ is lower semicontinuous, there exists an open neighborhood $V(x, \varepsilon)$ of x such that

$$(\forall y \in V(x, \varepsilon))(f_{n(x,\varepsilon)}(x) \leq f_{n(x,\varepsilon)}(y) + \varepsilon/3).$$

Let $W(x, \varepsilon) = U(x, \varepsilon) \cap V(x, \varepsilon)$. Then, for the neighborhood $W(x, \varepsilon)$, we have

$$(\forall y \in W(x, \varepsilon))(f(y) \leq f_{n(x,\varepsilon)}(y) + \varepsilon).$$

Using the quasi-compactness of E, we can find a finite family of open neighborhoods

$$W(x_1, \varepsilon), W(x_2, \varepsilon), \ldots, W(x_k, \varepsilon)$$

of this type, which collectively cover E. Simultaneously, we get the corresponding finite family of natural numbers

$$n(x_1, \varepsilon), n(x_2, \varepsilon), \ldots, n(x_k, \varepsilon).$$

Defining

$$n(\varepsilon) = \max\{n(x_1, \varepsilon), n(x_2, \varepsilon), \ldots, n(x_k, \varepsilon)\},$$

we see that, for every point $y \in E$, the relation

$$f(y) \leq f_{n(\varepsilon)}(y) + \varepsilon$$

holds true. Since the sequence of functions $\{f_n : n \in \mathbb{N}\}$ is increasing, we deduce that

$$(\forall y \in E)(\forall n \geq n(\varepsilon))(f_n(y) \leq f(y) \leq f_n(y) + \varepsilon),$$

which gives us the uniform convergence of $\{f_n : n \in \mathbb{N}\}$ to f, and the proof is complete. □

Naturally, we have a direct analog of the previous theorem in terms of a decreasing sequence of upper semicontinuous real-valued functions on E.

Theorem 10.4 *Let E be an arbitrary quasi-compact topological space and let $\{f_n : n \in \mathbb{N}\}$ be a decreasing sequence of real-valued functions on E such that the following two conditions are satisfied:*

(1) all f_n ($n \in \mathbb{N}$) are upper semicontinuous;
(2) for each $x \in E$, the limit $\lim_{n\to\infty} f_n(x)$ exists.

Let $f : E \to \mathbb{R}$ be defined by the formula

$$f(x) = \lim_{n\to\infty} f_n(x) \qquad (x \in E).$$

If the function f is lower semicontinuous on E, then the given sequence $\{f_n : n \in \mathbb{N}\}$ converges to f uniformly.

Proof It suffices to consider the increasing sequence $\{-f_n : n \in \mathbb{N}\}$ of lower semicontinuous real-valued functions on E and to apply Theorem 10.3 to this sequence. □

The two preceding theorems enable us to demonstrate a particularly useful theorem of mathematical analysis, which is due to Dini (see, e.g., [35, 40, 44]).

Theorem 10.5 *Let E be an arbitrary quasi-compact topological space and let $\{f_n : n \in \mathbb{N}\}$ be a monotone sequence of real-valued functions on E such that:*

(1) all f_n ($n \in \mathbb{N}$) are continuous;
(2) for each $x \in E$, the limit $\lim_{n\to\infty} f_n(x)$ exists.

Let $f : E \to \mathbb{R}$ be defined by the formula

$$f(x) = \lim_{n\to\infty} f_n(x) \qquad (x \in E).$$

Then the following two assertions are equivalent:

(a) the sequence $\{f_n : n \in \mathbb{N}\}$ converges uniformly to f;
(b) the function f is continuous on E.

Proof The implication (a) $\Rightarrow$ (b) is trivial in view of condition (1).

Suppose now that (b) is valid.

First, consider the case of an increasing sequence $\{f_n : n \in \mathbb{N}\}$. In this case, all f_n ($n \in \mathbb{N}$) are continuous (hence lower semicontinuous) functions and f is simultaneously lower and upper semicontinuous. By virtue of Theorem 10.3, the sequence $\{f_n : n \in \mathbb{N}\}$ converges uniformly to f, i.e., (a) holds true.

Lastly, consider the case of a decreasing sequence $\{f_n : n \in \mathbb{N}\}$. In this case, all f_n ($n \in \mathbb{N}$) are continuous (hence upper semicontinuous) functions and f is simultaneously lower and upper semicontinuous. By virtue of Theorem 10.4, the sequence $\{f_n : n \in \mathbb{N}\}$ converges uniformly to f, i.e., (a) again holds true.

This ends the proof of Dini's theorem. □

Exercises

1. Recalling that $\lfloor x \rfloor$ denotes the biggest integer not exceeding a real number x, check that the so-called *floor function*

$$x \to \lfloor x \rfloor \qquad (x \in \mathbb{R})$$

is upper semicontinuous on $\mathbb{R}$. Deduce from this fact that the function

$$x \to x - \lfloor x \rfloor \qquad (x \in \mathbb{R})$$

(the *fractional part* of x) is lower semicontinuous on $\mathbb{R}$.

In addition, verify that the set of all discontinuity points of each of these two functions is identical with the set $\mathbb{Z}$ of all integers.

2. Let E be a topological space and let X be a subset of E.
Demonstrate that:

   (a) X is an open set in E if and only if the characteristic function of X is lower semicontinuous on E;

   (b) X is a closed set in E if and only if the characteristic function of X is upper semicontinuous on E.

3. Let E be a topological space and let $f : E \to \mathbb{R}$ be a function.
Prove that the following two assertions hold:

   (a) f is lower semicontinuous on E if and only if the set $\Gamma^*(f)$ is closed in the topological product space $E \times \mathbb{R}$;

   (b) f is upper semicontinuous on E if and only if the set $\Gamma_*(f)$ is closed in the topological product space $E \times \mathbb{R}$.

   For this purpose, use Theorem 9.3 and Exercise 9 from Chap. 9.

4. Let E be a ground set and let $\{f_i : i \in I\}$ be a family of real-valued functions on E.
Demonstrate that:

   (a) if $\sup\{f_i : i \in I\}$ exists, then

$$\Gamma^*(\sup\{f_i : i \in I\}) = \cap\{\Gamma^*(f_i) : i \in I\};$$

(b) if, for some function $g : E \to \mathbb{R}$, one has the equality

$$\Gamma^*(g) = \cap \{\Gamma^*(f_i) : i \in I\},$$

then $g = \sup\{f_i : i \in I\}$;
(c) if $\inf\{f_i : i \in I\}$ exists, then

$$\Gamma_*(\inf\{f_i : i \in I\}) = \cap\{\Gamma_*(f_i) : i \in I\};$$

(d) if, for some function $h : E \to \mathbb{R}$, one has the equality

$$\Gamma_*(h) = \cap\{\Gamma_*(f_i) : i \in I\},$$

then $h = \inf\{f_i : i \in I\}$.

5. Let E be a topological space and let $\{f_i : i \in I\}$ be a family of real-valued functions on E.
   Verify that:

   (a) if all functions f_i ($i \in I$) are lower semicontinuous and $f = \sup\{f_i : i \in I\}$ exists, then f is also lower semicontinuous;
   (b) if all functions f_i ($i \in I$) are upper semicontinuous and $g = \inf\{f_i : i \in I\}$ exists, then g is also upper semicontinuous.

   For this purpose, either use Theorem 9.2 of Chap. 9 or apply the results of Exercises 3 and 4 formulated above.

6. A topological space E is called *countably quasi-compact* if every countable open covering of E contains a finite subcovering (equivalently, if every countable centered family of closed sets in E has a nonempty intersection).
   Check that any quasi-compact space is countably quasi-compact and give an example of a countably quasi-compact space which is not quasi-compact.

7. For a topological space E, prove that the following two assertions are equivalent:

   (a) E is countably quasi-compact;
   (b) every countably infinite subset of E has at least one accumulation point in E.

   In order to show the implication (a) $\Rightarrow$ (b), suppose (a) and consider any countably infinite set $X = \{x_n : n \in \mathbb{N}\}$ in E. For each $k \in \mathbb{N}$, define

   $$F_k = \mathrm{cl}(\{x_n : n \in \mathbb{N} \ \& \ n \geq k\}).$$

   Then all sets F_k are closed, nonempty, and the relations

   $$F_0 \supset F_1 \supset \cdots \supset F_k \supset \cdots$$

are satisfied. In view of (a), there exists a point $x \in \cap\{F_k : k \in \mathbb{N}\}$. Verify that this x is an accumulation point for X.

In order to show the implication (b) $\Rightarrow$ (a), suppose (b) and take any countable centered family $\{F_n : n \in \mathbb{N}\}$ of closed sets in E. Assume, without loss of generality, that

$$F_0 \supset F_1 \supset \cdots \supset F_n \supset \cdots, \qquad (\forall n \in \mathbb{N})(F_n \setminus F_{n+1} \neq \emptyset).$$

Further, for each $n \in \mathbb{N}$, choose a point $x_n \in F_n \setminus F_{n+1}$ and consider the countably infinite set

$$X = \{x_n : n \in \mathbb{N}\}.$$

In view of (b), there exists an accumulation point x of this X. Verify that $x \in \cap\{F_n : n \in \mathbb{N}\}$.

8. Check that:
   (a) a closed subspace of a countably quasi-compact space is countably quasi-compact;
   (b) a continuous image of a countably quasi-compact space is countably quasi-compact.

Remark 10.5 As is widely known, the topological product of any family of quasi-compact spaces is quasi-compact (this is one of the logical equivalents of the Axiom of Choice; see [83]). At the same time, there are even two countably quasi-compact spaces such that their topological product is not countably quasi-compact (in this connection, see e.g. [40]).

9. Let E be an arbitrary nonempty countably quasi-compact topological space and let $f : E \to \mathbb{R}$ be a lower semicontinuous function on E.
   Demonstrate that there exists a point $z \in E$ such that

   $$f(z) = \inf\{f(x) : x \in E\}.$$

   In other words, f attains its infimum on E.
   Formulate and prove the corresponding statement for an arbitrary nonempty countably quasi-compact topological space E and for any upper semicontinuous function $g : E \to \mathbb{R}$.

10. Let E be a quasi-compact topological space and let $(I, \preceq)$ be a directed partially ordered set (i.e., for any $i \in I$ and $j \in I$, there exists a $k \in I$ such that $i \preceq k$ and $j \preceq k$).
    Formulate and prove generalized versions of Theorems 10.3, 10.4, and 10.5 for the case of monotone families $\{f_i : i \in I\}$ of real-valued semicontinuous functions on E.

11. Let E be a topological space, $g : E \to \mathbb{R}$ be a lower semicontinuous function, and let $f : g(E) \to \mathbb{R}$ be an increasing lower semicontinuous function.

Show that $f \circ g : E \to \mathbb{R}$ is a lower semicontinuous function (cf. Exercise 7 from Chap. 9).

12. Let E be a topological space and let $g : E \to \mathbb{R}$ be a lower semicontinuous function. Define

$$g_1 = g/(1 + |g|)$$

and verify that the following two assertions hold:

(a) g_1 is a bounded lower semicontinuous function;
(b) the set $C(g)$ of all continuity points of g coincides with the set $C(g_1)$ of all continuity points of g_1.

For (a), use the result of Exercise 11.

13. Let E be a ground set and let $\mathcal{K}$ be a class of subsets of E. Denote by $\mathcal{F}(\mathcal{K})$ the family of all finite unions of members of $\mathcal{K}$.

Suppose that $\mathcal{K}$ is a countably quasi-compact class in Marczewski's sense, i.e., every countable centered family of sets from $\mathcal{K}$ has nonempty intersection.

Prove that the class $\mathcal{F}(\mathcal{K})$ is also countably quasi-compact in Marczewski's sense.

For this purpose, reduce the exercise to a certain statement of infinite combinatorics, which is equivalent to some weak form of the Axiom of Choice.

Chapter 11
The Banach–Steinhaus Theorem

As is mentioned in [38], there are three fundamental principles of modern linear functional analysis:

(i) the Hahn–Banach theorem on a continuous linear extension of a partial continuous linear functional;
(ii) the Banach–Steinhaus theorem (also known as the *principle of condensation of singularities*);
(iii) Banach's theorem on open linear mappings (or on the closed graph of a linear mapping).

In our lecture course we will touch upon each of the statements (i), (ii), and (iii). Extensive information about these principles can be found e.g. in [8, 10, 38, 78, 104], and in many other textbooks on modern mathematical analysis.

In the present chapter, we will be concerned with the Banach–Steinhaus theorem and with some of its immediate consequences.

Below, we would like to give a result which is more general than the classical Banach–Steinhaus theorem.

First of all, recall that a subset X of a topological space E is of *first Baire category* (simply, of *first category* or *meager*) in E if X can be represented as the union of countably many nowhere dense subsets of E.

Accordingly, a subset Y of E is of *second Baire category* (simply, of *second category* or *non-meager*) in E if Y is not of first Baire category in E.

A subset Z of E is called *co-meager* (or *residual*) in E if the set $E \setminus Z$ is of first category in E.

Theorem 11.1 *Let $(G, +)$ be a commutative (abelian) topological group and let $\{f_i : i \in I\}$ be a family of lower semicontinuous mappings acting from G into $\mathbb{R}$. Suppose that this family satisfies the following two conditions:*

(1) *each mapping f_i ($i \in I$) is subadditive, i.e.,*

$$f_i(x+y) \leq f_i(x) + f_i(y)$$

for all $x \in G$ and for all $y \in G$;
(2) *there is an everywhere dense set $Z \subset G$ of second category in G such that*

$$\sup\{f_i(z) : i \in I\} < +\infty$$

for any point $z \in Z$.

Then the family $\{f_i : i \in I\}$ is locally bounded from above, i.e., for each point $x \in G$, there exists a neighborhood $V(x)$ of x such that the relation

$$\sup\{f_i(t) : i \in I, \ t \in V(x)\} < +\infty$$

holds true.

Proof For every natural number n, let us define

$$Y_n = \{y \in G : (\forall i \in I)(f_i(y) \leq n)\}.$$

Since all f_i ($i \in I$) are lower semicontinuous functions, the set Y_n is closed in G. By virtue of condition (2), we may write the inclusion

$$Z \subset \cup\{Y_n : n \in \mathbb{N}\}.$$

Since the set Z is of second category, at least one Y_n is not nowhere dense in G, so must have nonempty interior. Let $m \in \mathbb{N}$ be such that $\text{int}(Y_m) \neq \emptyset$.

Now, take any point $x \in G$. Using the everywhere density of Z in G, we can find an element $z \in Z$ such that $Y_m + z$ is a neighborhood of x. Obviously, each point t from $Y_m + z$ admits a representation in the form $t = y + z$, where $y \in Y_m$. So, for any index $j \in I$, we may write

$$f_j(t) = f_j(y+z) \leq f_j(y) + f_j(z) \leq m + a,$$

where $a = \sup\{f_i(z) : i \in I\} < +\infty$. Defining $V(x) = Y_m + z$, we finally have

$$\sup\{f_i(t) : i \in I, \ t \in V(x)\} \leq m + a < +\infty.$$

This completes the proof of the theorem. □

Remark 11.1 In Theorem 11.1 the assumption of commutativity of the group $(G, +)$ is not essential. The reader can easily generalize the above result to the case of non-commutative topological groups (after an appropriate reformulation of it). Also, the proof of Theorem 11.1 does not rely on the continuity of the mapping

11 The Banach–Steinhaus Theorem

$$(x, y) \to x + y \quad (x \in G, \ y \in G).$$

Actually, in our argument we only use the fact that all translations of G and the symmetry of G are homeomorphisms of G onto itself.

As a direct consequence of Theorem 11.1, we get the following classical theorem of Banach and Steinhaus [10]. It should be mentioned here that this theorem was called by its authors "the principle of condensation of singularities". Below, we will show by relevant examples why such an abbreviation is justified and meaningful.

Theorem 11.2 *Let $(E, || \cdot ||)$ be a real Banach space, and let $\{f_i : i \in I\}$ be a family of semicontinuous linear functionals on E such that, for some everywhere dense set $Z \subset E$ of second category in E, one has*

$$(\forall z \in Z)(\sup\{f_i(z) : i \in I\} < +\infty).$$

Then all functionals from the given family are continuous and their norms are uniformly bounded.

Proof First, the condition that each linear functional f_i is semicontinuous implies that f_i is continuous. Actually, a much stronger result is true, namely, if an additive functional $f : E \to \mathbb{R}$ is Borel measurable (more generally, if this f has the Baire property), then f is continuous. For details, see Chap. 30.

Further, applying Theorem 11.1, we infer that

$$\sup\{f_i(t) : i \in I, \ t \in V(0)\} < +\infty$$

for some symmetric neighborhood $V(0)$ of the neutral element 0 of E. But this circumstance readily gives us that $\sup\{||f_i|| : i \in I\} < +\infty$, which ends the proof. □

Another consequence of Theorem 11.1 is formulated as follows.

Theorem 11.3 *Let $(E, || \cdot ||)$ be a real Banach space, let $(F, || \cdot ||)$ be a real normed vector space, and let $\{A_i : i \in I\}$ be a family of continuous linear operators acting from E into F. Suppose that, for some everywhere dense set $Z \subset E$ of second category in E, one has*

$$(\forall z \in Z)(\sup\{||A_i(z)|| : i \in I\} < +\infty).$$

Then the following two assertions hold:

(1) *the norms of all operators A_i ($i \in I$) are uniformly bounded, i.e., $\sup\{||A_i|| : i \in I\} \leq b$ for some real constant $b > 0$;*
(2) *the family $\{A_i : i \in I\}$ is uniformly equicontinuous, i.e., for any real $\varepsilon > 0$, there exists a real $\delta > 0$ such that the formula*

$$(\forall i \in I)(\forall x \in E)(\forall y \in E)(||x - y|| < \delta \Rightarrow ||A_i(x) - A_i(y)|| < \varepsilon)$$

holds true.

Proof For each index $i \in I$, let us define

$$p_i(x) = ||A_i(x)|| \qquad (x \in E).$$

So, we come to the family $\{p_i : i \in I\}$ of sublinear continuous functionals on E. Keeping in mind the condition

$$(\forall z \in Z)(\sup\{||A_i(z)|| : i \in I\} < +\infty)$$

and Theorem 11.1, we deduce that

$$\sup\{||A_i(t)|| : i \in I, \ t \in V(0)\} < +\infty$$

for a certain symmetric neighborhood $V(0)$ of 0 in E. This immediately yields the validity of (1) for some strictly positive real number b.

Further, if a real $\varepsilon > 0$ is chosen arbitrarily and δ is taken to be equal to ε/b, then

$$||x - y|| < \delta \Rightarrow ||A_i(x) - A_i(y)|| \leq ||A_i|| ||x - y|| < b\delta = \varepsilon$$

for all $i \in I$ and for all $\{x, y\} \subset E$, which establishes the uniform equicontinuity of the given family $\{A_i : i \in I\}$.

This completes the proof of Theorem 11.3. $\qquad \square$

Example 11.1 Let $(E, ||\cdot||)$ and $(F, ||\cdot||)$ be two real Banach spaces and let $\{A_n : n \in \mathbb{N}\}$ be a sequence of continuous linear operators acting from E into F. Suppose, in addition, that there is a set $Z \subset E$ of second category in E such that the sequence of restrictions $\{A_n|Z : n \in \mathbb{N}\}$ converges pointwise on Z (this assumption implies that the sequence $\{A_n : n \in \mathbb{N}\}$ converges pointwise on E; cf. Chap. 30).

Then $\sup\{||A_n|| : n \in \mathbb{N}\} < +\infty$ and there exists a unique continuous linear operator $A : E \to F$ which satisfies the following two conditions:

(1) $(\forall x \in E)(A(x) = \lim_{n \to \infty} A_n(x))$;
(2) $||A|| \leq \limsup\{||A_n|| : n \in \mathbb{N}\}$.

Notice also that, in general, the equality

$$\lim_{n \to \infty} ||A_n|| = ||A||$$

fails to be true. Moreover, it may happen that $||A_n|| = 1$ for all $n \in \mathbb{N}$, but $||A|| = 0$.

Example 11.2 Let E be a topological space and let f be a function acting from E into $\mathbb{R}$. Recall that f belongs to the *first Baire class* $B_1(E, \mathbb{R})$ if and only if there exists a sequence $\{f_n : n \in \mathbb{N}\}$ of real-valued continuous functions on E pointwise converging to f, i.e.,

$$f(x) = \lim_{n \to \infty} f_n(x) \qquad (x \in E).$$

The class $B_1(E, \mathbb{R})$ plays an important role in many topics of mathematical analysis, because the functions belonging to $B_1(E, \mathbb{R})$ possess more or less nice structural properties (see, e.g., [4, 100, 109, 141]). Notice that if $E = \mathbb{R}$, then all derivatives on E, all monotone functions on E, and all lower (upper) semicontinuous functions on E belong to $B_1(E, \mathbb{R})$.

Remark 11.2 The Banach–Steinhaus theorem is often used for constructing various examples of real-valued functions with extraordinary (singular) properties. In this context, see [10]. It can be shown that in some cases the Banach–Steinhaus theorem is implied by a purely topological theorem due to Kuratowski and Ulam [113]. The latter theorem is a topological analog of the classical Fubini theorem from measure theory (cf. [141]; see also Appendix B).

Exercises

1. Prove the classical Baire theorem stating that no nonempty complete metric space (M, d) is of first category in itself.
   To obtain the required result, take any countable family $\{F_n : n \in \mathbb{N}\}$ of nowhere dense closed subsets of M and define by recursion a sequence $\{B_n : n \in \mathbb{N}\}$ of closed balls in M such that:
   (a) $(\forall n \in \mathbb{N})(B_{n+1} \subset B_n)$;
   (b) $\lim_{n \to \infty} \operatorname{diam}(B_n) = 0$;
   (c) $(\forall n \in \mathbb{N})(B_n \cap F_n = \emptyset)$.

   In view of the completeness of M, there is a point $x \in \cap\{B_n : n \in \mathbb{N}\}$. Verify that $x \notin \cup\{F_n : n \in \mathbb{N}\}$.

2. Prove another classical Baire theorem stating that no nonempty locally compact topological space is of first category on itself.
   Use a method similar to the argument presented in Exercise 1.

3. The *Axiom of Dependent Choice* (denoted by **DC**) is the following set-theoretical statement:
   If X is a nonempty set and S is a binary relation on X satisfying

   $$(\forall x \in X)(\exists y \in X) S(x, y),$$

   then there exists a sequence $\{x_n : n \in \mathbb{N}\} \subset X$ such that

   $$(\forall n \in \mathbb{N}) S(x_n, x_{n+1}).$$

   Demonstrate that the following three assertions are valid in **ZF** & **DC** set theory:
   (a) if $\{X_i : i \in I\}$ is an arbitrary countable family of nonempty sets, then $\prod\{X_i : i \in I\} \neq \emptyset$;

(b) the union of any countable family of countable sets is a countable set;

(c) if a given set X has the property that, for each $n \in \mathbb{N}$, a subset of X can be found which is equinumerous with n, then X is infinite in the sense of Bolzano and Dedekind.

4. Following C. Blair [13], show that the following two assertions are equivalent within **ZF** set theory:

(a) the Axiom of Dependent Choice (**DC**);

(b) every nonempty complete metric space is of second category.

The implication (a) $\Rightarrow$ (b) can be established by using a fairly standard argument (cf. Exercise 1).

To prove the converse implication (b) $\Rightarrow$ (a), suppose (b) and take an arbitrary nonempty set X with a binary relation S on X satisfying

$$(\forall x \in X)(\exists y \in X)S(x, y).$$

Equip X with the discrete metric (topology) and consider the topological product $X^{\mathbb{N}}$, which is a complete metrizable space. For each natural number k, define

$$U_k = \{\{x_n : n \in \mathbb{N}\} \in X^{\mathbb{N}} : (\exists n > k) S(x_k, x_n)\}.$$

Check that any set U_k ($k \in \mathbb{N}$) is open and everywhere dense in $X^{\mathbb{N}}$. Consequently, there exists

$$\{x_n : n \in \mathbb{N}\} \in \cap\{U_k : k \in \mathbb{N}\}.$$

Using the sequence $\{x_n : n \in \mathbb{N}\}$, define by recursion its subsequence $\{x_{n(m)} : m \in \mathbb{N}\}$ such that $S(x_{n(m)}, x_{n(m+1)})$ for all $m \in \mathbb{N}$.

5. Work in **ZF** theory and demonstrate that any nonempty complete separable metric space is of second category.

6. Formulate and prove an appropriate analog of Theorem 11.1 for non-commutative topological groups $(G, \cdot)$.

7. Verify that the proof of Theorem 11.1 does not use the continuity of the mapping

$$(x, y) \to x + y \qquad (x \in G, y \in G)$$

and, actually, it suffices to require in the formulation of this theorem that all translations of G and the symmetry of G are homeomorphisms of G onto itself.

8. Give a proof of the statement formulated in Example 11.1 of this chapter.

For this purpose, take into account the following nontrivial fact:

if $(G, +)$ is a commutative topological group and X is a subset of G having the Baire property and of second category, then the difference set

$$X - X = \{x - y : x \in X, \, y \in X\}$$

is a neighborhood of 0 in G (for more details, see Chap. 30).

9. Let $(E, ||\cdot||)$ and $(F, ||\cdot||)$ be two Banach spaces (over the field $\mathbb{R}$), and let $\{A_n : n \in \mathbb{N}\}$ be a sequence of continuous linear operators from E into F such that

$$\sup\{||A_n|| : n \in \mathbb{N}\} = +\infty.$$

Show that the set X of all those points $x \in E$ at which the sequence $\{A_n(x) : n \in \mathbb{N}\}$ does not converge in F is everywhere dense and of second category in E.

For this purpose, keep in mind that $E \setminus X$ is a Borel vector subspace of E and use the hint of Exercise 8.

10. Let E be a topological space and let $\{f_n : n \in \mathbb{N}\}$ be a sequence of real-valued lower semicontinuous functions on E. Denote by X the set of all those points $x \in E$ for which the corresponding sequence $\{f_n(x) : n \in \mathbb{N}\}$ is not bounded from above in $\mathbb{R}$.

Check that X is of type G_δ in E, i.e., X is representable in the form of the intersection of a countable family of open sets in E.

11. Demonstrate that there exists a continuous 2π-periodic function

$$g : \mathbb{R} \to \mathbb{R}$$

whose trigonometric Fourier series is divergent at a point $x_0 \in \,]-\pi, \pi[$.

For this purpose, denote by $s_n(f, x_0)$ the n-th Fourier sum at x_0 of any 2π-periodic continuous function $f : \mathbb{R} \to \mathbb{R}$, and consider the sequence $\{s_n(\cdot, x_0) : n \in \mathbb{N}\}$ of continuous linear functionals defined on the space $\mathcal{F}$ of all such functions f, equipped with the standard sup-norm.

Verify that the norms of these functionals are not bounded from above and infer from Theorem 11.2 of this chapter that there exists a function $g \in \mathcal{F}$ for which the sequence $\{s_n(g, x_0) : n \in \mathbb{N}\}$ is unbounded, so is divergent.

12. Generalize the result of the previous exercise and prove that there is a residual set $\mathcal{F}_0$ in the space $\mathcal{F}$ such that, for every function $f \in \mathcal{F}_0$, the trigonometric Fourier series of f is divergent on a residual subset of $\mathbb{R}$.

For this purpose, take into account Exercise 11 and the fact formulated in Exercise 10.

Remark 11.3 Answering a question posed many years ago by N. Luzin, it was established by L. Carleson [28] that the trigonometric Fourier series of any real-valued square-integrable (in particular, continuous) function on $[-\pi, \pi]$ is convergent almost everywhere with respect to the standard Lebesgue measure λ on $[-\pi, \pi]$. This deep result and Exercise 12 show a radical difference between the Lebesgue measure and Baire category in problems concerning the pointwise convergence of the trigonometric Fourier series of real-valued 2π-periodic continuous functions on $\mathbb{R}$.

13. Prove that the vector space of all bounded functions g from the Baire class $B_1(\mathbb{R}, \mathbb{R})$, equipped with the standard sup-norm

$$||g|| = \sup\{|g(x)| : x \in \mathbb{R}\},$$

is a real Banach space.

Chapter 12
A Characterization of Oscillation Functions

Let E be a topological space and let $f : E \to \mathbb{R}$ be a function. Suppose that this f is locally bounded from above at a point $x_0 \in E$. Then one can consider the real value

$$f^*(x_0) = \limsup_{x \to x_0} f(x).$$

Moreover, if f is locally bounded from above at all points $x \in E$, then we come to the corresponding function $f^* : E \to \mathbb{R}$ defined by the formula

$$f^*(x) = \limsup_{y \to x} f(y) \qquad (x \in E).$$

The following theorem is readily obtained from the definition of f^*.

Theorem 12.1 *The real-valued function f^* is upper semicontinuous on the entire space E.*

Proof Take any point $x_0 \in E$ and consider the value $f^*(x_0)$. By definition of $f^*(x_0)$, for every real $\varepsilon > 0$, there exists an open neighborhood $U(x_0)$ such that

$$(\forall x \in U(x_0))(f(x) < f^*(x_0) + \varepsilon).$$

This formula trivially implies

$$(\forall x \in U(x_0))(f^*(x) \leq f^*(x_0) + \varepsilon),$$

which shows us that the function f^* is upper semicontinuous at x_0. Since x_0 was taken arbitrarily in E, we come to the desired result.

Theorem 12.1 has thus been proved. □

Similarly, if a function $g : E \to \mathbb{R}$ is locally bounded from below at a point $x_0 \in E$, then one can consider the real value

$$g_*(x_0) = \liminf_{x \to x_0} g(x).$$

Moreover, if g is locally bounded from below at all points $x \in E$, then we come to the corresponding function $g_* : E \to \mathbb{R}$ defined by the formula

$$g_*(x) = \liminf_{y \to x} g(y).$$

The next theorem is completely analogous to Theorem 12.1.

Theorem 12.2 *The real-valued function g_* is lower semicontinuous on the entire space E.*

Proof Consider the function $-g$, which is locally bounded from above on E. According to Theorem 12.1, the function $(-g)^*$ is upper semicontinuous on E. Now, since we have the equality

$$-((-g)^*) = g_*,$$

the required result immediately follows, which ends the proof. $\square$

Let now $h : E \to \mathbb{R}$ be a function which is locally bounded at all points of the topological space E. Then we may simultaneously introduce two real-valued functions h^* and h_* canonically associated with h by the formulas

$$h^*(x) = \limsup_{y \to x} h(y), \qquad h_*(x) = \liminf_{y \to x} h(y),$$

where $x \in E$. As has already been mentioned, h^* is upper semicontinuous on E and h_* is lower semicontinuous on E.

We also have the function O_h defined by the formula

$$O_h = h^* - h_*.$$

Recall that O_h is the oscillation function of h (see Chap. 5).

Theorem 12.3 *The following three assertions hold:*

(1) the function h is continuous at $x_0 \in E$ if and only if $O_h(x_0) = 0$;
(2) the function O_h is non-negative and upper semicontinuous on E;
(3) if $t \in \mathbb{R}$ and $x \in E$, then $O_{th}(x) = |t| O_h(x)$ (consequently, one has $O_{-h}(x) = O_h(x)$).

Proof The straightforward verification of (1) and (3) is left to the reader. Let us check the validity of assertion (2). The non-negativity of O_h is trivial. Further, we may write

$$O_h = h^* + (-(h_*)).$$

By virtue of Theorem 12.2, the function h_* is lower semicontinuous, so the function $-(h_*)$ is upper semicontinuous. We know that the sum of any two upper semicontinuous functions on E is upper semicontinuous on E (see Theorem 9.1 from Chap. 9). Consequently, O_h is an upper semicontinuous function on E.

Theorem 12.3 has thus been proved. □

Now, our goal is to demonstrate that, for any non-negative upper semicontinuous function $O : \mathbb{R} \to \mathbb{R}$, there exists a function $h : \mathbb{R} \to \mathbb{R}$ such that $O = O_h$.

For this purpose, we need a useful notion from classical point set theory.

Let Z be a subset of the plane $\mathbb{R}^2$ and let z be a point in $\mathbb{R}^2$.

This z is called a *condensation point* of Z if every neighborhood of z contains uncountably many points of Z.

More generally, suppose that E is an arbitrary topological space and x is a point in E.

This x is called a condensation point of E if any neighborhood of x is uncountable (cf. [19, 35, 40, 84, 109]).

Lemma 12.1 *If Z is a subset of $\mathbb{R}^2$, then Z can be represented in the form*

$$Z = Z_0 \cup Z_1 \quad (Z_0 \cap Z_1 = \emptyset),$$

where Z_0 is at most countable and Z_1 coincides with the set of all those condensation points of Z which belong to Z.

Consequently, if Z is uncountable, then Z_1 is also uncountable.

The proof of Lemma 12.1 is left to the reader. Actually, this lemma follows from the fact that the subspace Z of $\mathbb{R}^2$ has a countable base (cf. Exercise 6 of the present chapter, where a more general result is formulated).

As a corollary of Lemma 12.1, we get the next lemma.

Lemma 12.2 *Let $f : \mathbb{R} \to \mathbb{R}$ be a partial function whose domain is an uncountable subset of $\mathbb{R}$.*

Then $\mathrm{dom}(f)$ admits a representation $\mathrm{dom}(f) = X \cup Y$, where:

(1) *for any $x \in X$, the point $(x, f(x))$ is not a condensation point of the graph of f;*
(2) *for any $y \in Y$, the point $(y, f(y))$ is a condensation point of the graph of f.*

In particular, $X \cap Y = \emptyset$ and the inequalities

$$\mathrm{card}(X) \leq \mathrm{card}(\mathbb{N}), \quad \mathrm{card}(Y) > \mathrm{card}(\mathbb{N})$$

hold.

We now are ready to formulate and prove the main theorem of this chapter.

Theorem 12.4 *For every non-negative upper semicontinuous function* $O : \mathbb{R} \to \mathbb{R}$, *there exists a Borel function* $f : \mathbb{R} \to \mathbb{R}$ *such that* $O = O_f$.

Proof According to Lemma 12.2, we have a representation $\mathbb{R} = X \cup Y$ such that:

(a) $X \cap Y = \emptyset$;
(b) for each $x \in X$, the point $(x, O(x))$ is not a condensation point of the graph of O;
(c) for each $y \in Y$, the point $(y, O(y))$ is a condensation point of the graph of O;
(d) $\text{card}(X) \le \text{card}(\mathbb{N})$ and $\text{card}(Y) = \mathbf{c} > \text{card}(\mathbb{N})$.

Let A be a countable everywhere dense subset of $\mathbb{R}$ not intersecting the set X. The existence of such A is evident, because of the countability of X.

We define a function $f : \mathbb{R} \to \mathbb{R}$ as follows:

(i) $f(x) = 0$ for all $x \in A$;
(ii) $f(x) = O(x)$ for all $x \in \mathbb{R} \setminus A$.

Observe that f is a Borel function, because the set

$$\{x \in \mathbb{R} : f(x) \neq O(x)\} \subset A$$

is at most countable. Now, we assert that the non-negative Borel function f satisfies the equality $O_f = O$.

Indeed, take an arbitrary point x in $\mathbb{R}$ and consider the two possible cases for this x.

1. $x \in X$.

Since A is everywhere dense in $\mathbb{R}$ and $f(a) = 0$ for each $a \in A$, we have

$$\liminf_{y \to x} f(y) = 0.$$

At the same time, using the upper semicontinuity of O at x and the relations $A \cap X = \emptyset$ and $O(x) = f(x)$, we infer

$$\limsup_{y \to x} f(y) = f(x) = O(x).$$

Therefore,

$$O_f(x) = f(x) - 0 = O(x) - 0 = O(x).$$

2. $x \in \mathbb{R} \setminus X = Y$.

Again, in view of the everywhere density of A in $\mathbb{R}$, we have

$$\liminf_{y \to x} f(y) = 0.$$

At the same time, keeping in mind that $(x, O(x))$ is a condensation point of the graph of O, the set $\{(a, f(a)) : a \in A\}$ is countable, and O is upper semicontinuous at x, it is not hard to see that

$$\limsup_{y \to x} f(y) = O(x).$$

Therefore, in this case we also get

$$O_f(x) = O(x) - 0 = O(x),$$

which completes the proof of Theorem 12.4. □

Exercises

1. Let E be a topological space and let $f : E \to \mathbb{R}$ be a function locally bounded from above at all points of E.
   Check that:

   (a) $f \leq f^*$;
   (b) f is upper semicontinuous at $x \in E$ if and only if $f(x) = f^*(x)$.

2. Let E be a topological space and let $g : E \to \mathbb{R}$ be a function locally bounded from below at all points of E.
   Verify that:

   (a) $g_* \leq g$;
   (b) g is lower semicontinuous at $x \in E$ if and only if $g(x) = g_*(x)$.

3. Let E be a topological space and let $h : E \to \mathbb{R}$ be a function locally bounded at all points of E.
   Check that:

   (a) $h_* \leq h \leq h^*$;
   (b) h is continuous at $x \in E$ if and only if $h^*(x) = h_*(x)$.

4. Let E be a topological space and let $f : E \to \mathbb{R}$ be a function locally bounded from above (respectively, from below) at all points of E.
   Show that $(f^*)^* = f^*$ (respectively, $(f_*)_* = f_*$).

5. Generalize Theorem 5.6 from Chap. 5 to the case when, instead of real-valued locally bounded functions on a topological space E, there are locally bounded functions on E taking their values in some nonempty metric space (M, d).

6. A topological space E is called *Lindelöf* if every open covering of E contains a countable subcovering (see, e.g., [19, 40, 84]).
   Accordingly, a topological space E is called *hereditarily Lindelöf* if any subspace of E is Lindelöf.

Check that these five assertions hold true:

(a) each closed subspace of a Lindelöf space is also Lindelöf;
(b) any quasi-compact space is Lindelöf;
(c) there exists a compact space with an open everywhere dense non-Lindelöf subspace;
(d) a topological space is hereditarily Lindelöf if and only if any open subspace of it is Lindelöf;
(e) every topological space having a countable base is hereditarily Lindelöf.

Further, let E be an arbitrary uncountable hereditarily Lindelöf topological space.
Prove that there exists a decomposition of E in the form

$$E = X \cup Y \qquad (X \cap Y = \emptyset),$$

where X and Y satisfy the following two conditions:

(i) $\operatorname{card}(X) \leq \operatorname{card}(\mathbb{N})$ and no point of the set X is a condensation point of E;
(ii) $\operatorname{card}(Y) > \operatorname{card}(\mathbb{N})$ and all points of the set Y are condensation points of E.

Conclude that X is an open subset of E and Y is a closed subset of E.

7. Give a detailed proof of Lemma 12.2.
8. Let E be a topological space satisfying the following two conditions:

   (a) there exists an infinite base $\mathcal{B}$ of E whose cardinality is strictly less than $\operatorname{card}(E)$;
   (b) for each nonempty open set $U \subset E$, the equality

   $$\operatorname{card}(U) = \operatorname{card}(E)$$

   holds true (in other words, E is an *isodyne topological space*; see Appendix E).

   Let $O : E \to \mathbb{R}$ be any non-negative upper semicontinuous function.
   Demonstrate that there exists a function $f : E \to \mathbb{R}$ such that $O = O_f$.
   For this purpose, argue similarly to the proof of Theorem 12.4.

9. Check in detail that the function f defined in the proof of Theorem 12.4 is a Borel mapping from $\mathbb{R}$ into itself.
10. Consider the topological space $(E, \mathcal{T}_\Phi)$ described in Exercise 9 of Appendix E.
    Let $O : E \to \mathbb{R}$ be any strictly positive constant function.
    Verify that there exists no function $f : E \to \mathbb{R}$ satisfying the equality $O_f = O$.
    Conclude that Theorem 12.4 of this chapter cannot be extended to the class of all topological spaces.

12 A Characterization of Oscillation Functions

Remark 12.1 Let E be an isodyne resolvable topological space and let

$$O : E \to \mathbb{R}$$

be a non-negative upper semicontinuous function. The following situation can occur:

(a) there exists a function $f : E \to \mathbb{R}$ such that $O = O_f$;
(b) there exists no Borel function $g : E \to \mathbb{R}$ such that $O = O_g$.

For more details about this fact, see [102], Chapter 2, Exercise 27.

11. Work in **ZF** theory and show that each uncountable bounded subset of $\mathbb{R}$ has at least one condensation point in $\mathbb{R}$.
12. Prove that in **ZF** theory the following two assertions are equivalent:

    (a) the union of any countable family of countable subsets of $\mathbb{R}$ is countable;
    (b) every uncountable bounded subset of $\mathbb{R}$ has at least two condensation points in $\mathbb{R}$.

13. Let E be a topological space. For each subset X of E, denote by $d(X)$ the set of all accumulation points of X (in E).
    Check that:

    (a) the set $d(X)$ is closed in E;
    (b) the equality

    $$d(X_1 \cup X_2) = d(X_1) \cup d(X_2)$$

    holds and, consequently, the inclusion $X_1 \subset X_2$ implies the inclusion $d(X_1) \subset d(X_2)$;
    (c) in general, the formula

    $$d(\cup\{X_n : n \in \mathbb{N}\}) = \cup\{d(X_n) : n \in \mathbb{N}\}$$

    fails to be true.

14. Let E be a metrizable separable topological space without isolated points and let F be an arbitrary closed set in E.
    Demonstrate that there exists a subset Y of E satisfying the following two conditions:

    (a) Y is discrete in E (hence is at most countable) and $Y \cap F = \emptyset$;
    (b) $d(F \cup Y) = F$ (here the notation of Exercise 13 is used).

    Argue as follows. Denote by Z the set of all isolated points of F. Since F is separable, Z is at most countable. Assume, without loss of generality, that Z is represented in the form

    $$Z = \{z_n : n \in \mathbb{N}\},$$

where all z_n ($n \in \mathbb{N}$) are pairwise distinct. Then construct by recursion a disjoint family $\{B_n : n \in \mathbb{N}\}$ of closed balls in E such that:

(i) for any $n \in \mathbb{N}$, the center of B_n is z_n and $F \cap B_n = \{z_n\}$;
(ii) for any $n \in \mathbb{N}$, the radius of B_n is less than $1/(n+1)$.

Let Y_n be a discrete subset of $\text{int}(B_n) \setminus \{z_n\}$ having the property that z_n is a unique accumulation point for Y_n. Put

$$Y = \cup \{Y_n : n \in \mathbb{N}\}$$

and verify that the set Y is as required.

Remark 12.2 In fact, the result of Exercise 14 is essentially due to Cantor.

15. Recall that, by definition, a *Peano curve* is any (effectively constructed) continuous surjective mapping acting from the unit segment $[0, 1]$ onto the unit square $[0, 1]^2$.
Show that the existence of a Peano curve implies the existence of a subset X of $[0, 1]$ such that:

(a) X is homeomorphic to the standard Cantor set on $\mathbb{R}$;
(b) the Lebesgue measure of X is zero (so X is nowhere dense in $[0, 1]$).

For this purpose, take some Peano curve $f : [0, 1] \to [0, 1]^2$ and consider the disjoint family of closed sets

$$\{f^{-1}(\{x\} \times [0, 1]) : x \in [0, 1]\}.$$

Check that there exists a point $x_0 \in [0, 1]$ for which the set

$$T = f^{-1}(\{x_0\} \times [0, 1])$$

is nowhere dense in $[0, 1]$ and, simultaneously, is of Lebesgue measure zero. Then consider the set X of all condensation points of T and demonstrate that X is as required.

Chapter 13
Semicontinuity Versus Continuity

In this chapter we consider some nontrivial connections between real-valued continuous functions and real-valued semicontinuous functions.

We begin with several simple auxiliary lemmas from general topology (see, e.g., [40, 84, 109]).

Let $(E_1, \mathcal{T}_1)$ and $(E_2, \mathcal{T}_2)$ be two topological spaces and let $E = E_1 \times E_2$.

Recall that, by definition, the *product topology* of $\mathcal{T}_1$ and $\mathcal{T}_2$ is the least (with respect to inclusion) topology $\mathcal{T}$ on E such that both canonical projections

$$\mathrm{pr}_1 : (E, \mathcal{T}) \to (E_1, \mathcal{T}_1) \qquad \mathrm{pr}_2 : (E, \mathcal{T}) \to (E_2, \mathcal{T}_2)$$

are continuous mappings.

Lemma 13.1 *Let $\mathcal{B}_1$ be a base of $(E_1, \mathcal{T}_1)$ and let $\mathcal{B}_2$ be a base of $(E_2, \mathcal{T}_2)$. Then the family*

$$\mathcal{B} = \{U \times V : U \in \mathcal{B}_1, \ V \in \mathcal{B}_2\}$$

is a base of the topological product space $(E, \mathcal{T}) = (E_1, \mathcal{T}_1) \times (E_2, \mathcal{T}_2)$.

In particular, one has $\mathrm{card}(\mathcal{B}) = \mathrm{card}(\mathcal{B}_1) \cdot \mathrm{card}(\mathcal{B}_2)$.

This lemma directly follows from the definition of the product of two topological spaces.

Example 13.1 Let E be a topological space possessing a base of infinite cardinality **a** and let F be a topological space with a countable base. Then, according to Lemma 13.1, the product topological space $E \times F$ has a base of cardinality not exceeding **a**.

Lemma 13.2 *Let $(E, \mathcal{T})$ be a topological space and let $\mathcal{B}$ be a base of $(E, \mathcal{T})$. Suppose that an open set $U \subset E$ is representable in the form*

$$U = \cup\{U_i : i \in I\},$$

where $\{U_i : i \in I\}$ *is some family of open subsets of* E.
Then there exists a set $J \subset I$ *such that:*

(1) $\mathrm{card}(J) \leq \mathrm{card}(\mathcal{B})$;
(2) $U = \cup\{U_j : j \in J\}$.

Proof Define a family $\mathcal{B}' \subset \mathcal{B}$ as follows:

$$\mathcal{B}' = \{V \in \mathcal{B} : (\exists i \in I)(V \subset U_i)\}.$$

For each $V \in \mathcal{B}'$, choose an index $j(V) \in I$ satisfying $V \subset U_{j(V)}$ and put

$$J = \{j(V) : V \in \mathcal{B}'\}.$$

A straightforward verification shows that the set J is as required. This ends the proof. $\square$

Example 13.2 Let $(E, \mathcal{T})$ be a topological space and let $\mathcal{B}$ be a base of $(E, \mathcal{T})$. Suppose that a closed set $Z \subset E$ is representable in the form

$$Z = \cap\{Z_i : i \in I\},$$

where $\{Z_i : i \in I\}$ is some family of closed subsets of E. Then, by virtue of Lemma 13.2, there exists a set $J \subset I$ such that $\mathrm{card}(J) \leq \mathrm{card}(\mathcal{B})$ and $Z = \cap\{Z_j : j \in J\}$.

Recall that a Hausdorff topological space E is *completely regular* if, for each point $x \in E$ and for every closed set $X \subset E$ not containing x, there exists a continuous function $f : E \to [0, 1]$ satisfying the conditions $f|X = 0$ and $f(x) = 1$.

It can readily be seen that the above definition is equivalent to the following:

A Hausdorff topological space E is *completely regular* if, for each point $x \in E$ and for every closed set $X \subset E$ not containing x, there exists a continuous function $f : E \to [a, b]$ such that $f|X = a$ and $f(x) = b$ (here a and b are any two given real numbers satisfying $a < b$).

It is well known that all metric spaces are completely regular (and even perfectly normal). Moreover, all Hausdorff completely regular spaces can be characterized as the subspaces of normal spaces or the subspaces of compact spaces (see, e.g., [40, 84, 109]).

Recall that, for a topological space E, its *(topological) weight* is defined as the infimum of the cardinalities of all bases of E. This infimum is usually denoted by the symbol $w(E)$. Since the cardinal numbers are well-ordered, there exists a base $\mathcal{B}$ of E such that

$$\mathrm{card}(\mathcal{B}) = w(E).$$

13 Semicontinuity Versus Continuity

Obviously, $w(E)$ is one of the main cardinal-valued invariants for the class of all topological spaces, i.e., if E and F are any two homeomorphic topological spaces, then $w(E) = w(F)$.

Lemma 13.3 *Let E be a completely regular topological space, let U be an open subset of E, and let χ_U denote the characteristic function of U.*

Then $\chi_U = \sup\{f_i : i \in I\}$ for some family $\{f_i : i \in I\}$ of continuous functions on E such that $\mathrm{ran}(f_i) \subset [0,1]$ whenever $i \in I$.

Proof The case $U = \emptyset$ is trivial. So suppose that $U \neq \emptyset$.

Take an arbitrary point $x \in U$ and consider the closed set $E \setminus U$. Since E is completely regular, there exists a function

$$f_x : E \to [0,1]$$

such that $f_x(x) = 1$ and $f_x|(E \setminus U) = 0$. Clearly, we have

$$(\forall y \in E)(f_x(y) \leq \chi_U(y)), \qquad f_x(x) = \chi_U(x),$$

whence it immediately follows that $\chi_U = \sup\{f_x : x \in U\}$.

Lemma 13.3 has thus been proved. □

Lemma 13.4 *Let E be a completely regular space and let $f : E \to \mathbb{R}$ be a non-negative lower semicontinuous function.*

Then there exists a family $\{f_i : i \in I\}$ of real-valued non-negative functions on E satisfying the following two conditions:

(1) $f = \sup\{f_i : i \in I\}$;
(2) *for each index $i \in I$, the function f_i is representable in the form $f_i = t_i \chi_{U_i}$, where $t_i \geq 0$ and U_i is an open subset of E.*

Proof It suffices to demonstrate that if $x \in E$ and $f(x) > 0$, then, for any real ε from the interval $]0, f(x)[$, there exists an open set $U_{\varepsilon,x} \subset E$ such that

$$x \in U_{\varepsilon,x}, \qquad (f(x) - \varepsilon)\chi_{U_{\varepsilon,x}} \leq f.$$

Since the given function f is lower semicontinuous at x, there is an open neighborhood $V(x)$ such that

$$(\forall y \in V(x))(f(y) \geq f(x) - \varepsilon).$$

So we may put $U_{\varepsilon,x} = V(x)$, which completes the proof. □

Theorem 13.1 *Let E be a completely regular space with $w(E) \geq \mathrm{card}(\mathbb{N})$ and let $f : E \to \mathbb{R}$ be a lower semicontinuous function such that, for some real a, one has*

$$(\forall x \in E)(f(x) \geq a).$$

Then there exists a family $\{f_j : j \in J\}$ of real-valued continuous functions on E satisfying the following three conditions:

(1) $\operatorname{card}(J) \leq w(E)$;
(2) $f = \sup\{f_j : j \in J\}$;
(3) $(\forall j \in J)(\forall x \in E)(f_j(x) \geq a)$.

Proof Without loss of generality, we may assume that $a = 0$. Lemmas 13.3 and 13.4 readily imply that

$$f = \sup\{f_i : i \in I\}$$

for some family $\{f_i : i \in I\}$ of real-valued non-negative continuous functions on E.

Now, consider the topological product space $E \times \mathbb{R}$. According to Lemma 13.1, the topological weight of $E \times \mathbb{R}$ does not exceed $w(E)$, because $\mathbb{R}$ possesses a countable base. We also have the equality

$$\Gamma^*(f) = \cap\{\Gamma^*(f_i) : i \in I\}.$$

Since all the sets

$$\Gamma^*(f), \qquad \Gamma^*(f_i) \qquad (i \in I)$$

are closed in $E \times \mathbb{R}$, we can apply to them the result formulated in Example 13.2. Consequently, there exists a set $J \subset I$ such that:

(a) $\operatorname{card}(J) \leq w(E)$;
(b) $\Gamma^*(f) = \cap\{\Gamma^*(f_j) : j \in J\}$.

Finally, taking into account Exercise 4 from Chap. 10, we conclude that the equality $f = \sup\{f_j : j \in J\}$ holds true.

This ends the proof of Theorem 13.1. □

Remark 13.1 The dual version of the previous theorem is formulated as follows. Let E be again a completely regular space with $w(E) \geq \operatorname{card}(\mathbb{N})$ and let $g : E \to \mathbb{R}$ be an upper semicontinuous function such that, for some real b, one has

$$(\forall x \in E)(g(x) \leq b).$$

Then there exists a family $\{g_j : j \in J\}$ of real-valued continuous functions on E satisfying the following conditions:

(1) $\operatorname{card}(J) \leq w(E)$;
(2) $g = \inf\{g_j : j \in J\}$;
(3) $(\forall j \in J)(\forall x \in E)(g_j(x) \leq b)$.

Remark 13.2 Suppose in Theorem 13.1 that $w(E) = \operatorname{card}(\mathbb{N})$. Then, for any lower semicontinuous function $f : E \to [a, +\infty[$, we may assert that there exists a countable family of functions $\{f_n : n \in \mathbb{N}\}$ such that every f_n acts from E into $[a, +\infty[$, is continuous on E, and

$$f = \sup\{f_n : n \in \mathbb{N}\}.$$

Moreover, introducing for each index $n \in \mathbb{N}$ the function

$$\phi_n = \sup\{f_k : 0 \leq k \leq n\},$$

we come to the increasing sequence of continuous functions $\{\phi_n : n \in \mathbb{N}\}$ such that

$$f = \lim_{n \to \infty} \phi_n.$$

Analogously, under the same assumption $w(E) = \mathrm{card}(\mathbb{N})$, for any upper semicontinuous function $g : E \to \,]-\infty, b]$, we may assert that there exists a countable family of functions $\{g_n : n \in \mathbb{N}\}$ such that every g_n acts from E into $]-\infty, b]$, is continuous on E, and

$$g = \inf\{g_n : n \in \mathbb{N}\}.$$

Moreover, introducing for each index $n \in \mathbb{N}$ the function

$$\psi_n = \inf\{g_k : 0 \leq k \leq n\},$$

we come to the decreasing sequence of continuous functions $\{\psi_n : n \in \mathbb{N}\}$ such that

$$g = \lim_{n \to \infty} \psi_n.$$

These two facts are often used in various applications of real-valued lower (upper) semicontinuous functions.

For a partial function $\phi : E \to \mathbb{R}$, let the symbol ϕ_+ denote the function on $\mathrm{dom}(\phi)$ defined as follows:

$$\phi_+(x) = \begin{cases} \phi(x) & \text{if } \phi(x) \geq 0, \\ 0 & \text{if } \phi(x) < 0 \end{cases}$$

Using the above notation, we have the next lemma.

Lemma 13.5 *The following two assertions hold:*

(1) *if $\phi \leq \psi$, then $\phi_+ \leq \psi_+$;*
(2) *if $\phi = \lim_{n \to \infty} \phi_n$, then $\phi_+ = \lim_{n \to \infty} (\phi_n)_+$.*

The proof of Lemma 13.5 is quite easy and is left to the reader.

We now present a simple version of the so-called separation theorem for two functions, one of which is upper semicontinuous and the other is lower semicontinuous.

Theorem 13.2 *Let E be any topological space, let $\{f_n : n \in \mathbb{N}\}$ be an increasing sequence of real-valued continuous functions on E, and let $\{g_n : n \in \mathbb{N}\}$ be a decreasing sequence of real-valued continuous functions on E, such that there exist two pointwise limits*

$$f = \lim_{n \to \infty} f_n, \qquad g = \lim_{n \to \infty} g_n,$$

which satisfy the inequality $g \leq f$.

Then there exists a continuous function $h : E \to \mathbb{R}$ separating f and g, i.e., the relation $g \leq h \leq f$ holds true.

Proof For any $n \in \mathbb{N}$, it will be convenient to introduce the notation

$$\phi_n = (g_n - f_n)_+, \qquad \psi_n = (g_n - f_{n+1})_+.$$

Since all the functions f_n and g_n are continuous, the produced functions ϕ_n and ψ_n are also continuous.

Using this notation and keeping in mind Lemma 13.5, it is easy to check that:

(a) $\phi_n \geq \psi_n \geq \phi_{n+1} \geq 0$ for each $n \in \mathbb{N}$;
(b) $\lim_{n \to \infty} \phi_n = \lim_{n \to \infty} \psi_n = 0$.

Now, for every point $x \in E$, let us define

$$h(x) = f_0(x) + \phi_0(x) - \psi_0(x) + \phi_1(x) - \psi_1(x) + \phi_2(x) - \psi_2(x) + \cdots.$$

In view of (a) and (b), the above alternating series converges and the value $h(x)$ is well-defined. So we come to the function $h : E \to \mathbb{R}$.

From the representation of h one easily infers that h is a pointwise limit of an increasing sequence of real-valued continuous functions on E and, simultaneously, h is a pointwise limit of a decreasing sequence of real-valued continuous functions on E. Therefore, h is lower semicontinuous and upper semicontinuous, which implies that h is continuous on E.

It remains to verify that $g(x) \leq h(x) \leq f(x)$ whenever $x \in E$.

For this purpose, consider the two possible cases.

1. $x \in E$ is such that $g(x) = f(x)$.

In this case, we have $g_n(x) \geq f_m(x)$ for all natural numbers n and m, so $h(x)$ is equal to

$$f_0(x) + (g_0(x) - f_0(x)) - (g_0(x) - f_1(x)) + (g_1(x) - f_1(x)) - (g_1(x) - f_2(x)) + \cdots,$$

13 Semicontinuity Versus Continuity

which immediately gives us

$$h(x) = \lim_{n \to \infty} f_n(x) = \lim_{n \to \infty} g_n(x) = f(x) = g(x).$$

2. $x \in E$ is such that $g(x) < f(x)$.

In this case, for sufficiently large $m \in \mathbb{N}$, we have $g_m(x) < f_m(x)$, so in the expansion of $h(x)$ only finitely many terms differ from zero. In other words, there is a least $n \in \mathbb{N}$ such that either $\phi_n(x) = 0$ or $\psi_n(x) = 0$.

Let this n satisfy the relation

$$\phi_0(x) > 0, \ \psi_0(x) > 0, \ldots, \ \psi_{n-1}(x) > 0, \ \phi_n(x) = 0.$$

Then we may write

$$h(x) = f_0(x) + (g_0(x) - f_0(x)) - \cdots - (g_{n-1}(x) - f_n(x)) = f_n(x),$$

whence it follows that $g(x) \leq g_n(x) \leq f_n(x) = h(x) \leq f(x)$.

Let now n satisfy the relation

$$\phi_0(x) > 0, \ \psi_0(x) > 0, \ldots, \ \phi_n(x) > 0, \ \psi_n(x) = 0.$$

Then we may write

$$h(x) = f_0(x) + (g_0(x) - f_0(x)) - \cdots + (g_n(x) - f_n(x)) = g_n(x),$$

whence it follows that $g(x) \leq g_n(x) = h(x) \leq f_{n+1}(x) \leq f(x)$.

This completes the proof of Theorem 13.2. □

Remark 13.3 We would like to especially mention that, in the formulation of Theorem 13.2, E is an arbitrary topological space.

Remark 13.4 It can easily be seen that Theorem 13.2 implies the following result.

Let E be a separable metric space and let f and g be two real-valued functions on E such that:

(a) $g \leq f$;
(b) f is a lower semicontinuous function;
(c) g is an upper semicontinuous function.

Then there exists a continuous function $h : E \to \mathbb{R}$ which separates f and g.

Remark 13.5 In the literature there are known several results on separation of real-valued semicontinuous functions, which are similar to Theorem 13.2 (in this connection, see e.g. [40, 148]). Actually, Theorem 13.2 and its analogs belong to the group of statements concerning the existence of various kinds of selectors with prescribed nice properties. We mean here theorems on the existence of continuous selectors, measurable selectors, and selectors having certain algebraic properties.

This topic is very important for many branches of contemporary mathematics, because the results of the above-mentioned type have found numerous applications in theoretical and practical questions (see [112, 130, 131, 148]).

Exercises

1. Give a detailed proof of Lemma 13.1.
2. Prove the assertion formulated in Example 13.2.
3. Demonstrate the equivalence of the two definitions of completely regular spaces.
4. Give a detailed proof of the statement formulated in Remark 13.4.
5. Recall that a (Hausdorff) topological space is *normal* if, for any two disjoint closed sets X and Y in E, there exist open sets U and V in E such that

$$X \subset U, \quad Y \subset V, \quad U \cap V = \emptyset.$$

A (Hausdorff) topological space E is called *perfectly normal* if, for every closed subset X of E, there exists a continuous function

$$f : E \to [0, 1]$$

such that $f|X = 0$ and $f(x) > 0$ for all $x \in E \setminus X$.
Verify that:

(a) every perfectly normal space is normal (hence completely regular);
(b) every metric space is perfectly normal.

6. Show, for a topological space E, that the following two assertions are equivalent:

(a) E is perfectly normal;
(b) E is normal and every closed set in E is of type G_δ (i.e., can be represented as the intersection of a countable family of open subsets of E).

7. Recall that a (Hausdorff) topological space E is a *regular space* if, for each point x of E and for every neighborhood $U(x)$ of x there exists a closed neighborhood $V(x)$ of x such that $V(x) \subset U(x)$.
Let E be a regular hereditarily Lindelöf space.
Demonstrate that:

(a) E is a normal space;
(b) E is a perfectly normal space.

For (a), argue as follows. Take any two disjoint closed sets X and Y in E. For any two points $x \in X$ and $y \in Y$, choose two open neighborhoods $U(x)$ and $V(y)$ satisfying the relations

13 Semicontinuity Versus Continuity

$$U(x) \subset \mathrm{cl}(U(x)) \subset E \setminus Y, \qquad V(y) \subset \mathrm{cl}(V(y)) \subset E \setminus X.$$

Let $\mathcal{U}$ be the family of all such $U(x)$ and let $\mathcal{V}$ be the family of all such $V(y)$. Since E is hereditarily Lindelöf, there exist a countable subset X_0 of X and a countable subset Y_0 of Y for which

$$X \subset \cup\{U(x) : x \in X_0\}, \qquad Y \subset \cup\{V(y) : y \in Y_0\}.$$

Further, let

$$X_0 = \{x_0, x_1, \ldots, x_n, \ldots\}, \qquad Y_0 = \{y_0, y_1, \ldots, y_n, \ldots\},$$

$$U_n = U(x_n) \setminus \cup\{\mathrm{cl}(V(y_i)) : 0 \leq i \leq n\} \qquad (n \in \mathbb{N}),$$

$$V_n = V(y_n) \setminus \cup\{\mathrm{cl}(U(x_i)) : 0 \leq i \leq n\} \qquad (n \in \mathbb{N}),$$

$$U = \cup\{U_n : n \in \mathbb{N}\}, \qquad V = \cup\{V_n : n \in \mathbb{N}\}.$$

Finally, verify that the sets U and V are open in E and

$$X \subset U, \qquad Y \subset V, \qquad U \cap V = \emptyset,$$

which establishes that E is a normal space.

For (b), argue as follows. Take an arbitrary open subset W of E and, for each point z of W, choose an open neighborhood $O(z)$ of z satisfying the relation

$$O(z) \subset \mathrm{cl}(O(z)) \subset W.$$

Again, since E is hereditarily Lindelöf, there exists a countable subset Z of W such that

$$W = \cup\{O(z) : z \in W\} = \cup\{O(z) : z \in Z\}.$$

Infer from the last equality that

$$W = \cup\{\mathrm{cl}(O(z)) : z \in Z\}$$

and conclude that E is perfectly normal.

As an immediate consequence of (a) and (b), obtain the following result: every regular topological space with a countable base is perfectly normal.

Remark 13.6 In fact, every Hausdorff regular topological space with a countable base is metrizable (see, e.g., [19, 40, 84, 109]).

8. Let E be a topological space, U be an open subset of E, and suppose that

$$\chi_U = \sup\{f_i : i \in I\}$$

for some countable family $\{f_i : i \in I\}$ of real-valued upper semicontinuous functions on E.

Check that U is of type F_σ in E (i.e., U admits a representation in the form of a union of countably many closed subsets of E).

Chapter 14
Outer Measures

Let E be a nonempty ground (base) set and let $\{X_n : n \in \mathbb{N}\}$ be a sequence of subsets of E.

We introduce the following fairly standard notation:

$$\limsup\{X_n : n \in \mathbb{N}\} = \bigcap_{n \in \mathbb{N}} \bigcup_{m \geq n} X_m,$$

$$\liminf\{X_n : n \in \mathbb{N}\} = \bigcup_{n \in \mathbb{N}} \bigcap_{m \geq n} X_m.$$

The set $\limsup\{X_n : n \in \mathbb{N}\}$ is called the *upper limit* of this sequence and the set $\liminf\{X_n : n \in \mathbb{N}\}$ is called the *lower limit* of this sequence.

Clearly, the sets $\limsup\{X_n : n \in \mathbb{N}\}$ and $\liminf\{X_n : n \in \mathbb{N}\}$ do not depend on the choice of a set E containing all X_n ($n \in \mathbb{N}$).

Also, it is easy to check that:

(i) $\limsup\{X_n : n \in \mathbb{N}\}$ consists of all those elements of E which belong to infinitely many members of $\{X_n : n \in \mathbb{N}\}$;
(ii) $\liminf\{X_n : n \in \mathbb{N}\}$ consists of all those elements of E which belong to all, except finitely many, members of $\{X_n : n \in \mathbb{N}\}$.

Therefore, we always have the inclusion

$$\liminf\{X_n : n \in \mathbb{N}\} \subset \limsup\{X_n : n \in \mathbb{N}\}.$$

Moreover, changing finitely many members of $\{X_n : n \in \mathbb{N}\}$, we do not change the sets $\liminf\{X_n : n \in \mathbb{N}\}$ and $\limsup\{X_n : n \in \mathbb{N}\}$.

The same conclusion remains true if we remove finitely many members from $\{X_n : n \in \mathbb{N}\}$ or if we add to $\{X_n : n \in \mathbb{N}\}$ finitely many members.

In the very special case when the equality

$$\liminf\{X_n : n \in \mathbb{N}\} = \limsup\{X_n : n \in \mathbb{N}\}$$

holds true, we say that the sequence $\{X_n : n \in \mathbb{N}\}$ *converges* or is *convergent* (in the purely set-theoretical sense). In such a case we put

$$\lim\{X_n : n \in \mathbb{N}\} = \liminf\{X_n : n \in \mathbb{N}\} = \limsup\{X_n : n \in \mathbb{N}\}$$

and we say that the given sequence $\{X_n : n \in \mathbb{N}\}$ converges to its limit $\lim\{X_n : n \in \mathbb{N}\}$.

Example 14.1 Let $\{X_n : n \in \mathbb{N}\}$ be an increasing (by inclusion) sequence of subsets of E. It can readily be checked that $\{X_n : n \in \mathbb{N}\}$ converges to the set $\cup \{X_n : n \in \mathbb{N}\}$, so

$$\lim\{X_n : n \in \mathbb{N}\} = \cup\{X_n : n \in \mathbb{N}\}.$$

Analogously, if $\{Y_n : n \in \mathbb{N}\}$ is a decreasing (by inclusion) sequence of subsets of E, then $\{Y_n : n \in \mathbb{N}\}$ converges to the set $\cap \{Y_n : n \in \mathbb{N}\}$, so

$$\lim\{Y_n : n \in \mathbb{N}\} = \cap\{Y_n : n \in \mathbb{N}\}.$$

Theorem 14.1 *Let $\{X_n : n \in \mathbb{N}\}$ and $\{Y_n : n \in \mathbb{N}\}$ be any two sequences of subsets of E.*

The following formulas hold:

$$\liminf\{X_n : n \in \mathbb{N}\} = E \setminus \limsup\{E \setminus X_n : n \in \mathbb{N}\};$$
$$\limsup\{X_n : n \in \mathbb{N}\} \cup \limsup\{Y_n : n \in \mathbb{N}\} = \limsup\{X_n \cup Y_n : n \in \mathbb{N}\};$$
$$\limsup\{X_n : n \in \mathbb{N}\} \cap \limsup\{Y_n : n \in \mathbb{N}\} \supset \limsup\{X_n \cap Y_n : n \in \mathbb{N}\};$$
$$\liminf\{X_n : n \in \mathbb{N}\} \cap \liminf\{Y_n : n \in \mathbb{N}\} = \liminf\{X_n \cap Y_n : n \in \mathbb{N}\};$$
$$\liminf\{X_n : n \in \mathbb{N}\} \cup \liminf\{Y_n : n \in \mathbb{N}\} \subset \liminf\{X_n \cup Y_n : n \in \mathbb{N}\}.$$

Moreover, if $\{X_n : n \in \mathbb{N}\}$ converges to X and $\{Y_n : n \in \mathbb{N}\}$ converges to Y, then $\{X_n \cup Y_n : n \in \mathbb{N}\}$ converges to $X \cup Y$ and $\{X_n \cap Y_n : n \in \mathbb{N}\}$ converges to $X \cap Y$.
In addition, if $Z \subset E$, then

$$\limsup\{X_n : n \in \mathbb{N}\} \setminus Z = \limsup\{X_n \setminus Z : n \in \mathbb{N}\};$$
$$\liminf\{X_n : n \in \mathbb{N}\} \setminus Z = \liminf\{X_n \setminus Z : n \in \mathbb{N}\};$$
$$Z \setminus \liminf\{X_n : n \in \mathbb{N}\} = \limsup\{Z \setminus X_n : n \in \mathbb{N}\};$$
$$Z \setminus \limsup\{X_n : n \in \mathbb{N}\} = \liminf\{Z \setminus X_n : n \in \mathbb{N}\}.$$

14 Outer Measures

Proof A straightforward verification. □

A family $\mathcal{A}$ of subsets of E is an *algebra* (in E) if $\mathcal{A}$ is closed under finite unions of its members and under taking the complements of its members.

An equivalent definition is as follows: $\mathcal{A} \neq \emptyset$ and

$$(X \in \mathcal{A}\ \&\ Y \in \mathcal{A}) \Rightarrow (E \setminus X \in \mathcal{A}\ \&\ X \cup Y \in \mathcal{A}).$$

Each of these two definitions implies that $\emptyset \in \mathcal{A}$ and $E \in \mathcal{A}$.

Actually, the smallest (by inclusion) algebra in E is exactly $\{\emptyset, E\}$ and the largest (by inclusion) algebra in E is $\mathcal{P}(E)$.

Also, any algebra in E is closed under finite intersections of its members.

An algebra $\mathcal{A}$ of subsets of E is called a σ-*algebra* (in E) if $\mathcal{A}$ is closed under countable unions of its members or, formally,

$$\{X_n : n \in \mathbb{N}\} \subset \mathcal{A} \Rightarrow \cup\{X_n : n \in \mathbb{N}\} \in \mathcal{A}.$$

It follows from this definition that any σ-algebra in E is closed under countable intersections of its members.

Example 14.2 Let $E = [t_1, t_2[$ be a half-open (from the right) interval on the real line $\mathbb{R}$ and let $\mathcal{A}$ be the family of all those sets in $[t_1, t_2[$ which are representable in the form of a finite union of half-open (from the right) subintervals of $[t_1, t_2[$. Then $\mathcal{A}$ is an algebra in $[t_1, t_2[$.

The family $\mathcal{B}([t_1, t_2[)$ of all Borel subsets of $[t_1, t_2[$ is a σ-algebra in $[t_1, t_2[$. In fact, $\mathcal{B}([t_1, t_2[)$ is generated by $\mathcal{A}$, i.e., $\mathcal{B}([t_1, t_2[)$ is the smallest (by inclusion) σ-algebra containing $\mathcal{A}$.

Remark 14.1 Borel σ-algebras of topological spaces play an important role in many topics of descriptive set theory, general topology, functional analysis, and probability theory (see, for instance, [17, 19, 20, 33, 38, 40, 48, 63, 69, 82]).

Recall that a family $\mathcal{I}$ of subsets of E is an *ideal* (in E) if $E \notin \mathcal{I}$ and $\mathcal{I}$ is closed under finite unions of its members and under taking subsets of its members. Formally, this means that

$$\emptyset \in \mathcal{I}, \quad E \notin \mathcal{I},$$
$$(X \in \mathcal{I}\ \&\ Y \in \mathcal{I}) \Rightarrow (X \cup Y \in \mathcal{I}),$$
$$(X \in \mathcal{I}\ \&\ Y \subset X) \Rightarrow Y \in \mathcal{I}.$$

If $E \neq \emptyset$, then $\{\emptyset\}$ is the smallest (by inclusion) ideal in E. If E contains at least two distinct elements, then there does not exist a largest (by inclusion) ideal in E.

Example 14.3 Let $\mathcal{I}$ be an ideal in E. Consider the family of sets

$$\mathcal{F} = \{E \setminus X : X \in \mathcal{I}\}.$$

This family is called the *filter* in E dual to the given ideal I.
Clearly, $\mathcal{F}$ has the following properties:

(a) $\emptyset \notin \mathcal{F}$ and $E \in \mathcal{F}$;
(b) $X \in \mathcal{F}$ and $Y \in \mathcal{F}$ imply $X \cap Y \in \mathcal{F}$;
(c) $X \in \mathcal{F}$ and $X \subset Y \subset E$ imply $Y \in \mathcal{F}$.

It can readily be checked that the family of sets $\mathcal{A} = I \cup \mathcal{F}$ is an algebra in E.

All elements of I may be treated as small (little) subsets of E and all elements of $\mathcal{F}$ may be treated as large (big) subsets of E.

An ideal I (respectively, a filter $\mathcal{F}$) is called *countably additive* or a σ-*ideal* (respectively, *countably complete* or a δ-*filter*) if I is closed under countable unions of its members (respectively, if $\mathcal{F}$ is closed under countable intersections of its members).

Example 14.4 If I is an arbitrary σ-ideal in E and $\mathcal{F}$ is its dual filter, then $\mathcal{F}$ is countably complete and $\mathcal{A} = I \cup \mathcal{F}$ is a σ-algebra in E.

Let E be a ground set. By definition (see, e.g., [17, 38, 63, 104, 134, 141]), an *outer measure* on E is any function

$$\mu^* : \mathcal{P}(E) \to [0, +\infty]$$

which satisfies the following three conditions:

(1) $\mu^*(\emptyset) = 0$;
(2) μ^* is increasing, i.e., $\mu^*(X) \leq \mu^*(Y)$ whenever $X \subset Y \subset E$;
(3) μ^* is countably subadditive, i.e.,

$$\mu^*(\cup\{X_n : n \in \mathbb{N}\}) \leq \Sigma\{\mu^*(X_n) : n \in \mathbb{N}\}$$

for every countable family $\{X_n : n \in \mathbb{N}\}$ of subsets of E.

Notice that the conjunction of the two conditions (2) and (3) can be replaced by one equivalent condition:

If $X \subset E$ and $\{X_n : n \in \mathbb{N}\} \subset \mathcal{P}(E)$ are such that $X \subset \cup\{X_n : n \in \mathbb{N}\}$, then

$$\mu^*(X) \leq \Sigma\{\mu^*(X_n) : n \in \mathbb{N}\}.$$

Notice also that (1) and (3) trivially imply the finite subadditivity of μ^*, i.e., one has

$$\mu^*(\cup\{X_n : 0 \leq n \leq m\}) \leq \Sigma\{\mu^*(X_n) : 0 \leq n \leq m\}$$

for an arbitrary finite family $\{X_n : 0 \leq n \leq m\}$ of subsets of E.

If $X \subset E$, then the value $\mu^*(X)$ is called the μ^*-*measure* of X.

If $\mu^*(X) < +\infty$, then it is natural to say that X is of finite μ^*-measure.

14 Outer Measures

For any two sets $X \subset E$ and $Y \subset E$, which are both of finite μ^*-measure, let us introduce the notation

$$d(X, Y) = \mu^*(X \triangle Y),$$

where, as usual,

$$X \triangle Y = (X \setminus Y) \cup (Y \setminus X)$$

is the symmetric difference of X and Y.

By virtue of the monotonicity and finite subadditivity of μ^*, we may write

$$d(X, Y) = \mu^*(X \triangle Y) \leq \mu^*(X) + \mu^*(Y) < +\infty,$$

so we come to the real-valued non-negative function

$$d : \mathcal{F}(E, \mu^*) \times \mathcal{F}(E, \mu^*) \to [0, +\infty[,$$

where

$$\mathcal{F}(E, \mu^*) = \{X \subset E : \mu^*(X) < +\infty\}.$$

The function d is a *quasi-metric* (or *pseudo-metric*) on $\mathcal{F}(E, \mu^*)$, i.e.,

$$d(X, X) = 0, \qquad d(X, Y) = d(Y, X) \qquad (X, Y \in \mathcal{F}(E, \mu^*));$$
$$d(X, Y) \leq d(X, Z) + d(Z, Y) \qquad (X, Y, Z \in \mathcal{F}(E, \mu^*)).$$

We only will check the validity of $d(X, Y) \leq d(X, Z) + d(Z, Y)$, because the other two formulas presented above are trivial.

Since we have the obvious inclusion

$$X \triangle Y \subset (X \triangle Z) \cup (Z \triangle Y),$$

the monotonicity and finite subadditivity of μ^* yields at once

$$d(X, Y) \leq d(X, Z) + d(Z, Y).$$

More generally, for a finite sequence $\{X_k : 1 \leq k \leq n\}$ of members of $\mathcal{F}(E, \mu^*)$, where $n \geq 2$, we may write

$$d(X_1, X_n) \leq d(X_1, X_2) + d(X_2, X_3) + \ldots + d(X_{n-1}, X_n).$$

Further, for any two sets X and Y from $\mathcal{F}(E, \mu^*)$, the inequality

$$|\mu^*(X) - \mu^*(Y)| \leq \mu^*(X \triangle Y)$$

holds true (therefore, if $\mu^*(X \triangle Y) = 0$, then $\mu^*(X) = \mu^*(Y)$).

To prove this, we first write the following two simple inclusions:

$$X \subset Y \cup (X \triangle Y), \qquad Y \subset X \cup (X \triangle Y).$$

Applying the monotonicity and finite subadditivity of μ^*, we get

$$\mu^*(X) \leq \mu^*(Y) + \mu^*(X \triangle Y), \qquad \mu^*(Y) \leq \mu^*(X) + \mu^*(X \triangle Y)$$

or, equivalently,

$$\mu^*(X) - \mu^*(Y) \leq \mu^*(X \triangle Y), \qquad \mu^*(Y) - \mu^*(X) \leq \mu^*(X \triangle Y),$$

whence it follows that $|\mu^*(X) - \mu^*(Y)| \leq \mu^*(X \triangle Y)$.

Example 14.5 Let μ^* be an outer measure on E not identically equal to zero. We say that a set $X \subset E$ is of μ^*-measure zero if $\mu^*(X) = 0$. It is easy to see that the family of all sets of μ^*-measure zero is a σ-ideal in E.

An outer measure μ^* on E is called *regular* if, for any increasing (by inclusion) sequence $\{X_n : n \in \mathbb{N}\}$ of subsets of E, the equality

$$\mu^*(\cup\{X_n : n \in \mathbb{N}\}) = \lim_{n \to \infty} \mu^*(X_n)$$

holds (in some sense, this equality can be treated as the lower semicontinuity of μ^*).

Theorem 14.2 *If μ^* is a regular outer measure on E, then the quasi-metric d is complete. In other words, every Cauchy sequence in the quasi-metric space $(\mathcal{F}(E, \mu^*), d)$ is convergent.*

Proof Let $\{X_n : n \in \mathbb{N}\}$ be a Cauchy sequence in $(\mathcal{F}(E, \mu^*), d)$. From a well-known fact of the general theory of quasi-metric spaces, in order to show the convergence of a Cauchy sequence, it suffices to establish the convergence of at least one of its subsequences. So, replacing the given sequence by an appropriate subsequence, we may assume, without loss of generality, that

$$d(X_n, X_{n+1}) \leq 1/2^{n+1} \qquad (n \in \mathbb{N}).$$

Observe now that, for each $m \in \mathbb{N}$, we have the inclusion

$$\limsup\{X_n : n \in \mathbb{N}\} \setminus \liminf\{X_n : n \in \mathbb{N}\} \subset \cup\{X_n \triangle X_{n+1} : m \leq n\}.$$

Indeed, take an arbitrary element

$$x \in \limsup\{X_n : n \in \mathbb{N}\} \setminus \liminf\{X_n : n \in \mathbb{N}\}.$$

Obviously, there exists a natural number $i \geq m$ such that $x \in X_i$ and there exists a natural number $j > i$ such that $x \notin X_j$. Therefore, there is a natural number k in the interval $[i, j[$ such that

$$x \in X_k \setminus X_{k+1} \subset X_k \triangle X_{k+1},$$

which shows the validity of the required inclusion. So, for each $m \in \mathbb{N}$, we may write

$$\mu^*(\limsup\{X_n : n \in \mathbb{N}\} \setminus \liminf\{X_n : n \in \mathbb{N}\})$$
$$\leq \Sigma\{\mu^*(X_n \triangle X_{n+1}) : m \leq n\}$$
$$\leq 1/2^{m+1} + 1/2^{m+2} + \cdots$$
$$= 1/2^m.$$

The latter implies at once that

$$\mu^*(\limsup\{X_n : n \in \mathbb{N}\} \setminus \liminf\{X_n : n \in \mathbb{N}\}) = 0$$

or, equivalently,

$$\mu^*(\limsup\{X_n : n \in \mathbb{N}\} \triangle \liminf\{X_n : n \in \mathbb{N}\}) = 0.$$

Further, applying Theorem 14.1, we get

$$X_m \setminus \limsup\{X_n : n \in \mathbb{N}\} = \liminf\{X_m \setminus X_n : n \in \mathbb{N}\}.$$

Consider the sequence of sets $\{X_m \setminus X_k : k \in \mathbb{N}\}$. For each natural number $n > m$, we obviously have

$$\cap\{X_m \setminus X_k : n \leq k\} \subset X_m \setminus X_n \subset X_m \triangle X_n,$$

which implies that

$$\mu^*(\cap\{X_m \setminus X_k : n \leq k\}) \leq \mu^*(X_m \triangle X_n)$$
$$\leq \mu^*(X_m \triangle X_{m+1}) + \cdots + \mu^*(X_{n-1} \triangle X_n)$$
$$< 1/2^m.$$

Remembering the formula

$$\liminf\{X_m \setminus X_n : n \in \mathbb{N}\} = \bigcup_{n \in \mathbb{N}} \bigcap_{k \geq n} (X_m \setminus X_k)$$

and keeping in mind the regularity of μ^*, we get

$$\mu^*(X_m \setminus \limsup\{X_n : n \in \mathbb{N}\}) = \mu^*(\liminf\{X_m \setminus X_n : n \in \mathbb{N}\}) \leq 1/2^m.$$

Applying Theorem 14.1 once again, we may write

$$\liminf\{X_n : n \in \mathbb{N}\} \setminus X_m = \liminf\{X_n \setminus X_m : n \in \mathbb{N}\}$$

and the argument completely analogous to the previous one gives us

$$\mu^*(\liminf\{X_n : n \in \mathbb{N}\} \setminus X_m) = \mu^*(\liminf\{X_n \setminus X_m : n \in \mathbb{N}\}) \leq 1/2^m.$$

Finally, taking into account the equality

$$\mu^*(\limsup\{X_n : n \in \mathbb{N}\} \triangle \liminf\{X_n : n \in \mathbb{N}\}) = 0,$$

we conclude that $\{X_n : n \in \mathbb{N}\}$ converges in the quasi-metric d to each of the two sets $\liminf\{X_n : n \in \mathbb{N}\}$ and $\limsup\{X_n : n \in \mathbb{N}\}$.

This completes the proof of Theorem 14.2. □

Using the standard factorization operation, i.e., identifying all those sets $X \in \mathcal{F}(E, \mu^*)$ and $Y \in \mathcal{F}(E, \mu^*)$ which satisfy $d(X, Y) = 0$, we obtain the metric space

$$(\mathcal{F}(E, \mu^*)/_\equiv, d/_\equiv),$$

where the equivalence relation $\equiv$ is naturally defined as follows:

$$X \equiv Y \Leftrightarrow d(X, Y) = 0.$$

For any two $\equiv$-equivalence classes $[Z_1]$ and $[Z_2]$, where $Z_1 \in \mathcal{F}(E, \mu^*)$ and $Z_2 \in \mathcal{F}(E, \mu^*)$, we have

$$d/_\equiv([Z_1], [Z_2]) = d(Z_1, Z_2).$$

The function $d/_\equiv$ is well-defined and, for a regular outer measure μ^*, the metric $d/_\equiv$ is complete in view of Theorem 14.2.

Exercises

1. Let $(I, \leq)$ be a directed partially ordered set, i.e.,

$$(\forall i \in I)(\forall j \in I)(\exists k \in I)(i \leq k \ \& \ j \leq k),$$

14 Outer Measures

and let $\{X_i : i \in I\}$ be a family of subsets of a base set E. Define

$$\mathrm{limsup}\{X_i : i \in I\} = \bigcap_{i \in I} \bigcup_{j \geq i} X_j,$$

$$\mathrm{liminf}\{X_i : i \in I\} = \bigcup_{i \in I} \bigcap_{j \geq i} X_j.$$

The set $\mathrm{limsup}\{X_i : i \in I\}$ is called the *upper limit* of this family and the set $\mathrm{liminf}\{X_i : i \in I\}$ is called the *lower limit* of this family.

Verify that:

(a) $x \in \mathrm{limsup}\{X_i : i \in I\}$ if and only if x belongs to all those X_i whose indices constitute a cofinal subset of $(I, \leq)$;
(b) $x \in \mathrm{liminf}\{X_i : i \in I\}$ if and only if x belongs to all those X_i whose indices are greater than or equal to some $j \in I$;
(c) $\mathrm{liminf}\{X_i : i \in I\} \subset \mathrm{limsup}\{X_i : i \in I\}$.

In the special case when the equality

$$\mathrm{liminf}\{X_i : i \in I\} = \mathrm{limsup}\{X_i : i \in I\}$$

holds true, one says that $\{X_i : i \in I\}$ *converges* or is *convergent* (in the purely set-theoretical sense). More precisely, in such a case the notation

$$\lim\{X_i : i \in I\} = \mathrm{liminf}\{X_i : i \in I\} = \mathrm{limsup}\{X_i : i \in I\}$$

is used and one says that the family $\{X_i : i \in I\}$ converges to its limit $\lim\{X_i : i \in I\}$.

Try to establish some analogues of Example 14.1 and Theorem 14.1 for the just introduced sets $\mathrm{limsup}\{X_i : i \in I\}$ and $\mathrm{liminf}\{X_i : i \in I\}$.

2. Check that if $E = \emptyset$, then there is no ideal (filter) of subsets of E.
3. Let E be a ground set and let $\mathcal{L}$ be a family of subsets of E such that $\emptyset \in \mathcal{L}$ and there exists a countable subfamily of $\mathcal{L}$ which covers E.

Let $f : \mathcal{L} \to [0, +\infty]$ be a function such that $f(\emptyset) = 0$.

For any set $X \subset E$, consider the value

$$\inf\{\Sigma\{f(X_n) : n \in \mathbb{N}\} : \{X_n : n \in \mathbb{N}\} \subset \mathcal{L}, \ X \subset \cup\{X_n : n \in \mathbb{N}\}\}$$

and denote it by $\mu^*(X)$.

Verify that the function $\mu^* : \mathcal{P}(E) \to [0, +\infty]$ is an outer measure on E.

4. Let μ^* be an outer measure on E and let $\{X_n : n \in \mathbb{N}\}$ be a sequence of subsets of E such that

$$\mu^*(X_0) + \mu^*(X_1) + \cdots + \mu^*(X_n) + \cdots < +\infty.$$

Show that the formula $\mu^*(\mathrm{limsup}\{X_n : n \in \mathbb{N}\}) = 0$ holds.

Remark 14.2 This formula is an easy part of the classical Borel–Cantelli lemma (see, e.g., [17, 63]).

5. Give an example of an outer measure which is not regular.
   For this purpose, use Exercise 3 with appropriate ground set E, a countable family $\mathcal{L}$ of subsets of E, and a function $f : \mathcal{L} \to [0, +\infty]$.
6. Let (M, ρ) be a quasi-metric (metric) space and let $\{x_n : n \in \mathbb{N}\}$ be a Cauchy sequence in M such that some subsequence of $\{x_n : n \in \mathbb{N}\}$ converges, with respect to ρ, to a point x of M.
   Prove that the given sequence $\{x_n : n \in \mathbb{N}\}$ also converges to the same point x.

Chapter 15
Finitely Additive and Countably Additive Measures

By definition, a *finite finitely additive measure* on a ground set E is an arbitrary function $\mu : \mathcal{A} \to [0, +\infty[$ which satisfies the following two conditions:

(i) $\mathcal{A}$ is an algebra of subsets of E;
(ii) for any two disjoint sets $X \in \mathcal{A}$ and $Y \in \mathcal{A}$, one has

$$\mu(X \cup Y) = \mu(X) + \mu(Y).$$

Putting $X = Y = \emptyset$, we get from (ii) that $\mu(\emptyset) = 0$.

Also, it follows from (ii) by induction on n that

$$\mu(X_1 \cup X_2 \cup \cdots \cup X_n) = \mu(X_1) + \mu(X_2) + \cdots + \mu(X_n)$$

for every finite disjoint family $\{X_1, X_2, \ldots, X_n\} \subset \mathcal{A}$.

The same condition (ii) gives us the formula

$$\mu(X \cup Y) + \mu(X \cap Y) = \mu(X) + \mu(Y)$$

for any two sets X and Y from $\mathcal{A}$ (which are not, in general, disjoint).

An essential generalization of the latter formula, the so-called principle of inclusion and exclusion, is presented in Exercise 1 of this chapter.

Further, if X and Y are in $\mathcal{A}$ and $X \subset Y$, then $Y = (Y \setminus X) \cup X$, so

$$\mu(Y) = \mu(Y \setminus X) + \mu(X) \geq \mu(X),$$

which means the monotonicity of μ (or, more precisely, μ is an increasing function with respect to the inclusion relation).

If, in addition, one has the equality

$$\mu(\cup\{X_n : n \in \mathbb{N}\}) = \Sigma\{\mu(X_n) : n \in \mathbb{N}\}$$

whenever $\{X_n : n \in \mathbb{N}\}$ is a countable disjoint family of members of $\mathcal{A}$ such that $\cup\{X_n : n \in \mathbb{N}\} \in \mathcal{A}$, then μ is called a finite *countably additive measure* on the algebra $\mathcal{A}$.

It is easy to see that the countable additivity of μ on $\mathcal{A}$ is equivalent to the following condition:

For any decreasing (by inclusion) sequence $\{Y_n : n \in \mathbb{N}\} \subset \mathcal{A}$ such that $\cap \{Y_n : n \in \mathbb{N}\} = \emptyset$, the equality $\lim_{n \to \infty} \mu(Y_n) = 0$ holds true.

The latter condition is sometimes called the *upper semicontinuity* of μ at the empty set $\emptyset$ (cf. Chap. 9).

A natural question arises: how do we extend a given finitely additive (respectively, countably additive) measure to a finitely additive (respectively, to a countably additive) measure defined on a maximally large algebra (respectively, σ-algebra) of subsets of E?

Our goal is to show the validity of the following three basic facts concerning countably additive measures and finitely additive measures.

1. Every finite countably additive measure defined on an algebra $\mathcal{A}$ of subsets of E is uniquely extendible to a countably additive measure defined on the σ-algebra $\sigma(\mathcal{A})$ generated by $\mathcal{A}$ (Carathéodory's theorem).
2. For any set $X \subset E$ and for any finite countably additive measure μ on E, there exists a countably additive measure μ' on E which extends μ and for which $X \in \operatorname{dom}(\mu')$; at the same time, it may happen that the above-mentioned measure μ' is not uniquely determined (Marczewski's theorem).
3. Every finite finitely additive measure on E is extendible to a finitely additive measure defined on the power set $\mathcal{P}(E)$ of E (Tarski's theorem).

For proving Carathéodory's theorem and also for many other purposes, we need the notion of the *outer measure* μ^* associated with (or produced by) a given finitely additive measure μ on a ground set E.

Let μ be a finite finitely additive measure on E whose domain is an algebra $\mathcal{A} \subset \mathcal{P}(E)$ and let X be an arbitrary subset of E. We put

$$\mu^*(X) =$$
$$\inf\{\Sigma\{\mu(A_n) : n \in \mathbb{N}\} : \{A_n : n \in \mathbb{N}\} \subset \mathcal{A},\ X \subset \cup\{A_n : n \in \mathbb{N}\}\}.$$

It can readily be seen that in the above formula it suffices to take only those disjoint countable families $\{A_n : n \in \mathbb{N}\} \subset \mathcal{A}$ which cover the set X.

Thus, we have defined by this formula the function

$$\mu^* : \mathcal{P}(E) \to [0, +\infty[\,.$$

15 Finitely Additive and Countably Additive Measures

Obviously, if $Y \in \mathcal{A}$ and $\mu(Y) = 0$, then $\mu^*(Y) = 0$. Also, for some $Z \in \mathcal{A}$, it may happen that $\mu(Z) > 0$ and $\mu^*(Z) = 0$ (see Exercise 4 of this chapter).

It is absolutely clear that $\mu^*(\emptyset) = 0$.

If $X \subset Y \subset E$, then $\mu^*(X) \leq \mu^*(Y)$ (the monotonicity of μ^*).

A less trivial fact is presented in the following theorem.

Theorem 15.1 *The function μ^* is countably subadditive, so μ^* is a finite outer measure on E.*

Proof Take an arbitrary sequence $\{X_n : n \in \mathbb{N}\}$ of subsets of E and pick a real $\varepsilon > 0$.

For each index $n \in \mathbb{N}$, there is a countable family $\{A_{n,i} : i \in I(n)\} \subset \mathcal{A}$ such that

$$X_n \subset \cup\{A_{n,i} : i \in I(n)\}, \qquad \Sigma\{\mu(A_{n,i}) : i \in I(n)\} \leq \mu^*(X_n) + \varepsilon/2^{n+1}.$$

Consider the countable double family of sets

$$\{A_{n,i} : n \in \mathbb{N}, \ i \in I(n)\} \subset \mathcal{A}.$$

Obviously,

$$\cup\{X_n : n \in \mathbb{N}\} \subset \cup\{A_{n,i} : n \in \mathbb{N}, \ i \in I(n)\}.$$

Consequently, we get

$$\mu^*(\cup\{X_n : n \in \mathbb{N}\}) \leq \Sigma\{\mu(A_{n,i}) : n \in \mathbb{N}, \ i \in I(n)\}$$
$$\leq \Sigma\{\mu^*(X_n) : n \in \mathbb{N}\} + \varepsilon.$$

Since $\varepsilon > 0$ can be taken arbitrarily small, we finally obtain

$$\mu^*(\cup\{X_n : n \in \mathbb{N}\}) \leq \Sigma\{\mu^*(X_n) : n \in \mathbb{N}\},$$

which completes the proof (cf. Exercise 3 from Chap. 14). □

Summarizing the above, we can consider μ^* as the outer measure canonically associated with μ, or as the outer measure produced by μ.

In connection with this produced outer measure, it is natural to ask whether μ^* always extends μ.

The answer to this question is given by the next theorem.

Theorem 15.2 *The following two assertions are equivalent:*

(1) μ^* *is an extension of* μ;
(2) μ *is countably additive on* $\mathcal{A} = \mathrm{dom}(\mu)$.

Proof Assume (1) and let $\{A_n : n \in \mathbb{N}\}$ be an arbitrary disjoint countable family of members of $\mathcal{A}$ such that $\cup \{A_n : n \in \mathbb{N}\} \in \mathcal{A}$.

Since $\mu = \mu^*|\mathcal{A}$ and μ^* is countably subadditive in view of Theorem 15.1, we have

$$\mu(\cup\{A_n : n \in \mathbb{N}\}) \leq \Sigma\{\mu(A_n) : n \in \mathbb{N}\}.$$

On the other hand, by virtue of the finite additivity and monotonicity of μ, we may write

$$\Sigma\{\mu(A_n) : 0 \leq n \leq k\}) \leq \mu(\cup\{A_n : n \in \mathbb{N}\})$$

for each $k \in \mathbb{N}$. From this formula it immediately follows that

$$\mu(\cup\{A_n : n \in \mathbb{N}\}) \geq \Sigma\{\mu(A_n) : n \in \mathbb{N}\}.$$

Therefore, we obtain

$$\mu(\cup\{A_n : n \in \mathbb{N}\}) = \Sigma\{\mu(A_n) : n \in \mathbb{N}\},$$

i.e., assertion (2) is valid.

Now, suppose that (2) holds true and take any set $X \in \mathcal{A}$. Consider an arbitrary countable covering $\{A_n : n \in \mathbb{N}\} \subset \mathcal{A}$ of X. It is not hard to show that there exists a disjoint family $\{A'_n : n \in \mathbb{N}\} \subset \mathcal{A}$ such that

$$(\forall n \in \mathbb{N})(A'_n \subset A_n), \qquad X = \cup\{A'_n : n \in \mathbb{N}\}.$$

Consequently, using the countable additivity of μ, we get

$$\mu(X) = \Sigma\{\mu(A'_n) : n \in \mathbb{N}\} \leq \Sigma\{\mu(A_n) : n \in \mathbb{N}\},$$

which gives us the inequality $\mu(X) \leq \mu^*(X)$. On the other hand, taking the trivial covering $\{X\}$ of X, we have $\mu^*(X) \leq \mu(X)$, whence it follows that $\mu^*(X) = \mu(X)$. We thus see that (1) is valid, which ends the proof of Theorem 15.2. $\square$

Remark 15.1 As mentioned above, for a finite finitely additive measure μ on $\mathcal{A}$, the inequality $\mu^*(X) \leq \mu(X)$ is true whenever $X \in \mathcal{A}$, and in some situations the strict inequality $\mu^*(X) < \mu(X)$ can take place. According to Theorem 15.2, the equality $\mu^*(X) = \mu(X)$ for all $X \in \mathcal{A}$ means the countable additivity of μ on $\mathcal{A}$. In this case the term "outer measure" is justified and relevant.

Theorem 15.3 *Suppose that a finite measure μ on an algebra of subsets of E is countably additive. Let $\{A_n : n \in \mathbb{N}\}$ be an increasing (by inclusion) countable family of sets from $\mathrm{dom}(\mu)$ and let μ^* denote the outer measure produced by μ.*

Then, with the notation $A = \cup\{A_n : n \in \mathbb{N}\}$, one has:

15 Finitely Additive and Countably Additive Measures

(1) $\lim_{n\to\infty} \mu(A_n) = \mu^*(A)$;
(2) *for each $k \in \mathbb{N}$, the equality*

$$\mu^*(A \setminus A_k) = \mu^*(A) - \mu(A_k)$$

holds true.

Proof Let us establish (1). Since $\{A_n : n \in \mathbb{N}\}$ is increasing and μ is monotone, the limit $\lim_{n\to\infty} \mu(A_n)$ exists. Further, we may write

$$A = A_0 \cup (A_1 \setminus A_0) \cup (A_2 \setminus A_1) \cup \cdots \cup (A_{n+1} \setminus A_n) \cup \cdots.$$

In view of the countable subadditivity of μ^*, we get

$$\mu^*(A) \leq \mu(A_0) + (\mu(A_1) - \mu(A_0)) + (\mu(A_2) - \mu(A_1)) + \cdots$$
$$+ (\mu(A_{n+1}) - \mu(A_n)) + \cdots$$
$$= \lim_{n\to\infty} \mu(A_{n+1})$$
$$= \lim_{n\to\infty} \mu(A_n).$$

On the other hand, the monotonicity of μ^* gives us $\mu(A_n) \leq \mu^*(A)$ for each $n \in \mathbb{N}$. So we infer that $\lim_{n\to\infty} \mu(A_n) \leq \mu^*(A)$. Summarizing the above, we obtain the equality

$$\mu^*(A) = \lim_{n\to\infty} \mu(A_n),$$

which ends the proof of (1).

To demonstrate (2), fix $k \in \mathbb{N}$ and express the set $A \setminus A_k$ in the form

$$A \setminus A_k = (A_{k+1} \setminus A_k) \cup (A_{k+2} \setminus A_{k+1}) \cup \cdots \cup (A_{n+1} \setminus A_n) \cup \cdots.$$

Utilizing once again the countable subadditivity of μ^*, we get

$$\mu^*(A \setminus A_k) \leq (\mu(A_{k+1}) - \mu(A_k)) + (\mu(A_{k+2}) - \mu(A_{k+1})) + \cdots$$
$$+ (\mu(A_{n+1}) - \mu(A_n)) + \cdots$$
$$= \lim_{n\to\infty} (\mu(A_{n+1}) - \mu(A_k))$$
$$= \mu^*(A) - \mu(A_k).$$

On the other hand,

$$\mu^*(A \setminus A_k) \geq \mu^*(A) - \mu^*(A_k) = \mu^*(A) - \mu(A_k).$$

Therefore,
$$\mu^*(A \setminus A_k) = \mu^*(A) - \mu(A_k),$$

i.e., (2) is satisfied, too, and Theorem 15.3 has thus been proved. □

Remark 15.2 In general, the set $A = \cup\{A_n : n \in \mathbb{N}\}$ of Theorem 15.3 does not belong to the algebra $\mathrm{dom}(\mu)$. Actually, if for any increasing (by inclusion) family $\{B_n : n \in \mathbb{N}\} \subset \mathrm{dom}(\mu)$, one has

$$\cup \{B_n : n \in \mathbb{N}\} \in \mathrm{dom}(\mu),$$

then $\mathrm{dom}(\mu)$ is a σ-algebra of subsets of E.

Exercises

1. Let $\mathcal{L}$ be a lattice of subsets of a base set E, i.e.,

$$X \cup Y \in \mathcal{L}, \qquad X \cap Y \in \mathcal{L}$$

whenever $X \in \mathcal{L}$ and $Y \in \mathcal{L}$. Suppose also that $\emptyset \in \mathcal{L}$.
A function $\nu : \mathcal{L} \to \mathbb{R}$ is called *modular* if $\nu(\emptyset) = 0$ and

$$\nu(X \cup Y) + \nu(X \cap Y) = \nu(X) + \nu(Y)$$

for any two sets $X \in \mathcal{L}$ and $Y \in \mathcal{L}$.
   Verify that every modular function ν on $\mathcal{L}$ is finitely additive.
   Let k be a nonzero natural number, let $j \in \{1, 2, \ldots, k\}$, and let

$$s = \{i_1, i_2, \ldots, i_j\}$$

be a sequence of pairwise distinct indices from $\{1, 2, \ldots, k\}$. In other words, s is a j-element subset of $\{1, 2, \ldots, k\}$. The set-theorists usually express this circumstance in the form $s \in [\{1, 2, \ldots, k\}]^j$. For the sake of brevity, we also denote $s \in \mathcal{S}_j$, where $\mathcal{S}_j = [\{1, 2, \ldots, k\}]^j$ is the family of all j-element subsets of $\{1, 2, \ldots, k\}$.
   Further, suppose that a k-sequence of sets $(X_1, X_2, \ldots, X_k)$ is given. For any set

$$s = \{i_1, i_2, \ldots, i_j\} \in \mathcal{S}_j,$$

where $j \in \{1, 2, \ldots, k\}$, consider the corresponding sets $X_{i_1}, X_{i_2}, \ldots, X_{i_j}$ from the family $\{X_1, X_2, \ldots, X_k\}$ and define $X_s = X_{i_1} \cap X_{i_2} \cap \cdots \cap X_{i_j}$.

15 Finitely Additive and Countably Additive Measures

With the introduced notation, prove the following *principle of inclusion and exclusion* for any modular function v on $\mathcal{L}$:

$$v(X_1 \cup X_2 \cup \cdots \cup X_k) = \sum\{v(X_s) : s \in \mathcal{S}_1\} - \sum\{v(X_s) : s \in \mathcal{S}_2\}$$
$$+ \cdots + (-1)^{k+1} \sum\{v(X_s) : s \in \mathcal{S}_k\}.$$

For this purpose, use induction on k.

2. For an arbitrary finite finitely additive measure μ on an algebra of subsets of E, show the equivalence of the following two conditions:

   (a) μ is countably additive;
   (b) μ is upper semicontinuous at $\emptyset$.

3. Let μ be a finite finitely additive measure on E whose domain is an algebra $\mathcal{A} \subset \mathcal{P}(E)$ and let X be an arbitrary subset of E. The value $\mu^*(X)$ was defined as follows:

$$\mu^*(X) =$$
$$\inf\{\Sigma\{\mu(A_n) : n \in \mathbb{N}\} : \{A_n : n \in \mathbb{N}\} \subset \mathcal{A},\ X \subset \cup\{A_n : n \in \mathbb{N}\}\}.$$

Verify that the same value is obtained if in the above formula one takes only those disjoint countable families $\{A_n : n \in \mathbb{N}\} \subset \mathcal{A}$ which cover the set X.

4. Give a concrete example of a finite finitely additive measure μ on $\mathbb{N}$ satisfying the following two conditions:

   (a) $\text{dom}(\mu)$ is the algebra of all finite and co-finite subsets of $\mathbb{N}$;
   (b) $\mu(\mathbb{N}) = 1$ and $\mu^*(\mathbb{N}) = 0$.

Remark 15.3 In connection with Exercise 4, the following interesting and somewhat surprising fact should be mentioned:

In **ZF & DC** theory it is impossible to define on a σ-algebra of sets a finite finitely additive measure which is not countably additive.

For more information about this fact, see e.g. Appendix 2 of [100].

5. Specify a form of the Axiom of Choice that was used in the proof of Theorem 15.1 of this chapter.

6. Let an algebra $\mathcal{A}$ of subsets of a ground set E be such that, for every increasing (by inclusion) family $\{A_n : n \in \mathbb{N}\} \subset \mathcal{A}$, one has

$$\cup\{A_n : n \in \mathbb{N}\} \in \mathcal{A}.$$

Check that $\mathcal{A}$ is a σ-algebra of subsets of E.

7. Let $\mathcal{A}$ be an algebra of subsets of E and let $\{\mu_n : n \in \mathbb{N}\}$ be a sequence of finite finitely additive measures on $\mathcal{A}$. Suppose that, for each member A of $\mathcal{A}$, the limit $\lim_{n \to \infty} \mu_n(A)$ exists, and denote it by $\mu(A)$.

Verify that the obtained real-valued function μ is a finite finitely additive measure on $\mathcal{A}$.

8. Let $\mathcal{A}$ be a σ-algebra of subsets of E and let $\{\mu_n : n \in \mathbb{N}\}$ be a sequence of finite countably additive measures on $\mathcal{A}$. Suppose that, for each member A of $\mathcal{A}$, the limit $\lim_{n\to\infty} \mu_n(A)$ exists, and denote it by $\mu(A)$.

Verify that the obtained real-valued function μ is a finite countably additive measure on $\mathcal{A}$.

For this purpose, argue as follows. First of all, since the limit

$$\lim_{n\to\infty} \mu_n(E) = \mu(E)$$

exists, the finitely additive measure μ (see Exercise 7) is finite. In order to show the countable additivity of μ it suffices to prove the upper semicontinuity of μ at $\emptyset$.

Suppose otherwise, i.e., there exists a decreasing (by inclusion) sequence $\{A_m : m \in \mathbb{N}\}$ of members of $\mathcal{A}$, converging to $\emptyset$, such that $\mu(A_m) > \varepsilon$ for some real constant $\varepsilon > 0$ and for all $m \in \mathbb{N}$.

Then, for any natural number m, the sequence

$$\{\mu_0(A_m), \mu_1(A_m), \ldots, \mu_n(A_m), \ldots\}$$

contains infinitely many terms strictly exceeding ε.

Now, take a natural number $m_1 > 0$.

There is a natural number $m_2 > m_1$ satisfying $\mu_{m_2}(A_{m_1}) > \varepsilon$. Since μ_{m_2} is upper semicontinuous at $\emptyset$, there exists a natural number $m_3 > m_2$ such that $\mu_{m_2}(A_{m_3}) < \varepsilon/2$. Consequently,

$$\mu_{m_2}(A_{m_1} \setminus A_{m_3}) > \varepsilon/2.$$

Proceeding in this manner, one can define by induction two subsequences

$$\{\mu_{m_{2k}} : k = 1, 2, \ldots\}, \qquad \{A_{m_{2k-1}} : k = 1, 2, \ldots\}$$

which have the following properties:

$$m_1 < m_2 < m_3 < m_4 < \cdots,$$

$$(\forall k \geq 1)(\mu_{m_{2k}}(A_{m_{2k-1}} \setminus A_{m_{2k+1}}) > \varepsilon/2).$$

For each natural number $k \geq 1$, define

$$B_k = A_{m_{2k-1}} \setminus A_{m_{2k+1}}.$$

Clearly, the sets B_k ($k = 1, 2, \ldots$) are pairwise disjoint.

Let $\{K_j : j \in \mathbb{N}\}$ be a partition of $\mathbb{N} \setminus \{0\}$ into infinite subsets, and let

$$C_j = \cup\{B_k : k \in K_j\} \qquad (j \in \mathbb{N}).$$

The family of sets $\{C_j : j \in \mathbb{N}\} \subset \mathcal{A}$ is disjoint and the relations

$$\mu(C_j) = \lim_{k \in K_j, k \to \infty} \mu_{m_{2k}}(C_j) \geq \varepsilon/2 \qquad (j \in \mathbb{N})$$

are easily verified. Therefore, μ cannot be a finite measure.

The obtained contradiction yields the required result.

Remark 15.4 Actually, the above argument also shows the validity of the following statement:

For an arbitrary real $\varepsilon > 0$, there exists a natural number $p = p(\varepsilon)$ such that $\mu_n(A_m) \leq \varepsilon$ whenever $n > p$ and $m > p$.

9. Let E be a ground set and let $\mathcal{R}$ be a nonempty family of some subsets of E. This $\mathcal{R}$ is called a *ring of sets* in E if the following condition is fulfilled:

$$(\forall X \in \mathcal{R})(\forall Y \in \mathcal{R})(X \cup Y \in \mathcal{R} \,\&\, X \setminus Y \in \mathcal{R}).$$

For example, any ideal of sets in E is a ring of sets in E (the converse assertion is not true in general).

Supposing that $\mathcal{R}$ is a ring of subsets of E, check that:

(a) $\emptyset \in \mathcal{R}$;
(b) if X and Y belong to $\mathcal{R}$, then $X \cap Y$ also belongs to $\mathcal{R}$;
(c) the family of sets

$$\mathcal{A} = \{Z \subset E : Z \in \mathcal{R} \vee E \setminus Z \in \mathcal{R}\}$$

is the algebra of subsets of E generated by $\mathcal{R}$, i.e., $\mathcal{A}$ is the smallest (by inclusion) algebra of sets in E containing $\mathcal{R}$.

10. Let $\mathcal{R}$ be a ring of subsets of E. Define two binary operations $\cdot$ and $+$ in $\mathcal{R}$ by the formulas:

$$X \cdot Y = X \cap Y \qquad (X \in \mathcal{R}, Y \in \mathcal{R}),$$
$$X + Y = X \triangle Y \qquad (X \in \mathcal{R}, Y \in \mathcal{R}),$$

where $\triangle$ stands, as usual, for the operation of symmetric difference of two sets.

Show that the algebraic structure $(\mathcal{R}, +, \cdot)$ is a commutative ring, in which $+$ is an addition operation and $\cdot$ is a multiplication operation.

Moreover, in this ring one has $0_\mathcal{R} = \emptyset$, and the equalities

$$X \cdot X = X, \qquad X + X = 0_\mathcal{R}$$

hold for all $X \in \mathcal{R}$.

11. Let $\mathcal{R}$ be a ring of subsets of E and let $\mu : \mathcal{R} \to [0, +\infty]$ be a function. This μ is called a *finitely additive measure* on $\mathcal{R}$ if $\mu(\emptyset) = 0$ and

$$\mu(\cup\{X_i : i \in I\}) = \sum\{\mu(X_i) : i \in I\}$$

for any disjoint finite family $\{X_i : i \in I\}$ of sets from $\mathcal{R}$.

Suppose that $E \notin \mathcal{R}$ (i.e., the ring $\mathcal{R}$ is not an algebra of subsets of E) and suppose that μ is a finitely additive measure on $\mathcal{R}$.

For each set $Z \in \mathcal{R}$, put

$$\mu'(Z) = \mu(Z), \qquad \mu'(E \setminus Z) = +\infty.$$

Verify that μ' is a finitely additive functional (in fact, measure) on the algebra $\mathcal{A}(\mathcal{R})$ of subsets of E generated by $\mathcal{R}$, and μ' trivially extends μ.

12. Let $\mathcal{R}$ be again a ring of subsets of E and let $\mu : \mathcal{R} \to [0, +\infty]$ be a function. This μ is called a *countably additive measure* on $\mathcal{R}$ if $\mu(\emptyset) = 0$ and

$$\mu(\cup\{X_i : i \in I\}) = \sum\{\mu(X_i) : i \in I\}$$

for any disjoint countable family $\{X_i : i \in I\}$ of sets from $\mathcal{R}$ such that $\cup\{X_i : i \in I\} \in \mathcal{R}$.

Clearly, every countably additive measure on $\mathcal{R}$ is simultaneously finitely additive.

Suppose that there exists a countable family $\{Z_n : n \in \mathbb{N}\}$ of sets from $\mathcal{R}$ such that $E = \cup\{Z_n : n \in \mathbb{N}\}$. Since $\mathcal{R}$ is a ring, one may assume below that this family is disjoint.

Let μ be a countably additive measure on $\mathcal{R}$ and let $\mathcal{A}(\mathcal{R})$ denote the algebra of sets in E generated by $\mathcal{R}$. For each $Z \in \mathcal{A}(\mathcal{R})$, define

$$\mu'(Z) = \sum\{\mu(Z \cap Z_n) : n \in \mathbb{N}\}.$$

Demonstrate that:

(a) μ' is well-defined for all members of $\mathcal{A}(\mathcal{R})$;
(b) μ' is a countably additive measure on $\mathcal{A}(\mathcal{R})$;
(c) μ' extends μ.

13. Let E be a ground set and let $\mathcal{S}$ be a nonempty family of some subsets of E.

This family is called a *semiring of sets* in E if the following two conditions are satisfied:

(i) $X \cap Y \in \mathcal{S}$ whenever $X \in \mathcal{S}$ and $Y \in \mathcal{S}$;
(ii) if X and Y belong to $\mathcal{S}$ and $X \subset Y$, then there exists a nonempty finite disjoint family $\{Z_i : i \in I\}$ of members of $\mathcal{S}$ such that

$$Y \setminus X = \cup \{Z_i : i \in I\}.$$

For a given semiring $\mathcal{S}$, show that:

(a) $\emptyset \in \mathcal{S}$;
(b) the ring $\mathcal{R}(\mathcal{S})$ generated by $\mathcal{S}$ coincides with the family of all those subsets of E which admit a representation in the form

$$Z_1 \cup Z_2 \cup \cdots \cup Z_n,$$

where n is a natural number and the sets Z_i ($i = 1, 2, \ldots, n$) are some pairwise disjoint members of $\mathcal{S}$.

14. Let $\mathcal{S}$ be a semiring of subsets of a base set E and let

$$\mu : \mathcal{S} \to [0, +\infty]$$

be a function.

This μ is called a *finitely additive measure* on $\mathcal{S}$ if $\mu(\emptyset) = 0$ and

$$\mu(\cup\{X_i : i \in I\}) = \sum \{\mu(X_i) : i \in I\}$$

for every disjoint finite family $\{X_i : i \in I\}$ of sets belonging to $\mathcal{S}$ such that $\cup \{X_i : i \in I\} \in \mathcal{S}$.

Prove that any finitely additive measure on $\mathcal{S}$ admits a unique extension to a finitely additive measure which is defined on the ring $\mathcal{R}(\mathcal{S})$ generated by $\mathcal{S}$.

For this purpose, use (b) of Exercise 13.

15. Let $\mathcal{S}$ be again a semiring of subsets of a base set E and let

$$\mu : \mathcal{S} \to [0, +\infty]$$

be a function.

This μ is called a *countably additive measure* on $\mathcal{S}$ if $\mu(\emptyset) = 0$ and

$$\mu(\cup\{X_j : j \in J\}) = \sum \{\mu(X_j) : j \in J\}$$

for every disjoint countable family $\{X_j : j \in J\}$ of sets from $\mathcal{S}$ such that $\cup \{X_j : j \in J\} \in \mathcal{S}$.

Prove that any countably additive measure on $\mathcal{S}$ admits a unique extension to a countably additive measure which is defined on the ring $\mathcal{R}(\mathcal{S})$ generated by $\mathcal{S}$.

16. Let E be a ground set of cardinality ω_1, where ω_1 is the least uncountable ordinal (cardinal) number.

Check that there exists a ring $\mathcal{R}$ of subsets of E such that:

(a) card($\mathcal{R}$) = ω_1;
(b) $E = \cup \{X : X \in \mathcal{R}\}$;
(c) there is no countable covering of E with elements of $\mathcal{R}$.

For this purpose, take as $\mathcal{R}$ the family of all finite subsets of E.

Let now $\mathcal{R}'$ be the ring consisting of all at most countable subsets of E. Examine whether the analogs of (a), (b), and (c) are valid for this $\mathcal{R}'$.

17. Let E be a ground set, let $\mathcal{R}$ be a ring of subsets of E, and let $\{X_i : i \in I\}$ be a disjoint infinite family of members of $\mathcal{R}$ satisfying the equality

$$E = \cup\{X_i : i \in I\}.$$

Consider the family $\mathcal{A}$ of all those sets X in E, for which the relation

$$(\forall i \in I)(X \cap X_i \in \mathcal{R})$$

holds true.

Verify that $\mathcal{A}$ is an algebra of subsets of E containing $\mathcal{R}$ and, therefore, containing the algebra $\mathcal{A}(\mathcal{R})$ generated by $\mathcal{R}$.

Suppose, in addition, that μ is a nonzero countably additive measure on $\mathcal{R}$ and, for any set $X \in \mathcal{A}$, define

$$\mu'(X) = \sum\{\mu(X \cap X_i) : i \in I\}.$$

Demonstrate that:

(a) the function μ' is a countably additive measure on $\mathcal{A}$;
(b) if the set I is countably infinite, then μ' extends μ;
(c) if the set I is uncountable, then it may happen that μ' is identically equal to zero, so does not extend μ.

Chapter 16
Extensions of Measures

In the present chapter we would like to discuss a fairly standard method for obtaining certain extensions of a given finite measure (cf. [17, 63, 104, 141, 152]). For this purpose, we first need to introduce an important auxiliary notion.

Let E be a ground set, μ be a finite finitely additive measure defined on some algebra $\mathcal{A}$ of subsets of E, and let μ^* denote the outer measure produced by μ (see Chap. 15).

We shall say that a set $X \subset E$ is μ^*-*measurable* if, for every real $\varepsilon > 0$, there exists a set A belonging to $\mathcal{A}$ and satisfying the inequality

$$\mu^*(X \triangle A) \leq \varepsilon.$$

Obviously, the definition just formulated is equivalent to the following:

A set $X \subset E$ is μ^*-*measurable* if there exists a sequence $\{A_n : n \in \mathbb{N}\}$ of members of $\mathcal{A}$ such that

$$\lim_{n \to \infty} \mu^*(X \triangle A_n) = 0.$$

We will denote by $\mathcal{A}^*$ the family of all μ^*-measurable subsets of E.

The following two assertions are trivially true:

(i) every set of μ^*-measure zero is μ^*-measurable, i.e., belongs to $\mathcal{A}^*$;
(ii) if $X \in \mathcal{A}^*$ and $\mu^*(X \triangle Y) = 0$, then $Y \in \mathcal{A}^*$.

The following lemma is less trivial.

Lemma 16.1 *The family $\mathcal{A}^*$ is an algebra of subsets of E containing $\mathcal{A}$ and the function $\mu^*|\mathcal{A}^*$ is finitely additive.*

Proof The inclusion $\mathcal{A} \subset \mathcal{A}^*$ is obvious.

Let X be any member of $\mathcal{A}^*$. For an arbitrary real $\varepsilon > 0$, there is a set $A \in \mathcal{A}$ such that $\mu^*(X \triangle A) \leq \varepsilon$. Keeping in mind the equality

$$(E \setminus X) \triangle (E \setminus A) = X \triangle A,$$

we have

$$\mu^*((E \setminus X) \triangle (E \setminus A)) = \mu^*(X \triangle A) \leq \varepsilon,$$

which gives us $E \setminus X \in \mathcal{A}^*$.

Further, let A and B be any two sets from $\mathcal{A}^*$. There exist two sequences

$$\{A_n : n \in \mathbb{N}\} \subset \mathcal{A}, \qquad \{B_n : n \in \mathbb{N}\} \subset \mathcal{A}$$

such that

$$\lim_{n \to \infty} \mu^*(A \triangle A_n) = 0, \qquad \lim_{n \to \infty} \mu^*(B \triangle B_n) = 0.$$

Using the obvious inclusion

$$(A \cup B) \triangle (A_n \cup B_n) \subset (A \triangle A_n) \cup (B \triangle B_n),$$

we obtain

$$\mu^*((A \cup B) \triangle (A_n \cup B_n)) \leq \mu^*(A \triangle A_n) + \mu^*(B \triangle B_n)$$

and, therefore,

$$\lim_{n \to \infty} \mu^*((A \cup B) \triangle (A_n \cup B_n)) = 0,$$

which shows us that $A \cup B \in \mathcal{A}^*$. Hence $\mathcal{A}^*$ is an algebra of sets.

In particular, $A \cap B \in \mathcal{A}^*$ and, keeping in mind the formulas

$$(A \cap B) \triangle (A_n \cap B_n) \subset (A \triangle A_n) \cup (B \triangle B_n),$$

$$\mu^*((A \cap B) \triangle (A_n \cap B_n)) \leq \mu^*(A \triangle A_n) + \mu^*(B \triangle B_n),$$

it is easy to see that

$$\lim_{n \to \infty} \mu^*((A \cap B) \triangle (A_n \cap B_n)) = 0.$$

Further, for any natural number n, we have

$$\mu(A_n \cup B_n) + \mu(A_n \cap B_n) = \mu(A_n) + \mu(B_n).$$

Letting n tend to infinity in the above formula, we come to the equality

$$\mu^*(A \cup B) + \mu^*(A \cap B) = \mu^*(A) + \mu^*(B).$$

16 Extensions of Measures

If $A \cap B = \emptyset$, then $\mu^*(A \cap B) = 0$, and in this case we get

$$\mu^*(A \cup B) = \mu^*(A) + \mu^*(B),$$

which establishes the finite additivity of $\mu^*|\mathcal{A}^*$.

Lemma 16.1 has thus been proved. □

Lemma 16.2 *Suppose that a finite measure μ on an algebra $\mathcal{A} \subset \mathcal{P}(E)$ is countably additive.*

Then the family $\mathcal{A}^*$ is a σ-algebra of subsets of E containing $\mathcal{A}$, and $\mu^*|\mathcal{A}^*$ is a finite countably additive measure on $\mathcal{A}^*$.

Proof Let $\{X_n : n \in \mathbb{N}\}$ be a countable family of members of $\mathcal{A}^*$. Pick a real number $\varepsilon > 0$. For each $n \in \mathbb{N}$, there exists a set $A_n \in \mathcal{A}$ such that

$$\mu^*(X_n \triangle A_n) \leq \varepsilon/2^{n+2}.$$

Using the trivial inclusion

$$(\cup\{X_n : n \in \mathbb{N}\}) \triangle (\cup\{A_n : n \in \mathbb{N}\}) \subset \cup\{X_n \triangle A_n : n \in \mathbb{N}\}$$

and the countable subadditivity of μ^*, we obtain that

$$\mu^*((\cup\{X_n : n \in \mathbb{N}\}) \triangle (\cup\{A_n : n \in \mathbb{N}\})) \leq \varepsilon/2.$$

Let us define

$$A = \cup\{A_n : n \in \mathbb{N}\}.$$

Continuing our argument, we can take an increasing (by inclusion) sequence $\{B_n : n \in \mathbb{N}\}$ of sets from $\mathcal{A}$ satisfying the relation

$$A = \cup\{A_n : n \in \mathbb{N}\} = \cup\{B_n : n \in \mathbb{N}\}.$$

Then, for each $k \in \mathbb{N}$, we may write

$$(\cup\{X_n : n \in \mathbb{N}\}) \triangle B_k \subset ((\cup\{X_n : n \in \mathbb{N}\}) \triangle A) \cup (A \triangle B_k).$$

Therefore,

$$\mu^*((\cup\{X_n : n \in \mathbb{N}\}) \triangle B_k) \leq \mu^*((\cup\{X_n : n \in \mathbb{N}\}) \triangle A) + \mu^*(A \triangle B_k).$$

Let a natural number k be such that

$$\mu^*(A \triangle B_k) = \mu^*(A \setminus B_k) \leq \varepsilon/2.$$

The existence of k with this property immediately follows from Theorem 15.3 of Chap. 15. Then, keeping in mind the above, we get

$$\mu^*((\cup\{X_n : n \in \mathbb{N}\}) \triangle B_k) \leq \varepsilon,$$

which establishes the countable additivity of $\mathcal{A}^*$.

Finally, suppose that $\{X_n : n \in \mathbb{N}\}$ is a countable disjoint family of sets from $\mathcal{A}^*$ and let

$$X = \cup\{X_n : n \in \mathbb{N}\}.$$

The countable subadditivity of μ^* gives us

$$\mu^*(X) \leq \Sigma\{\mu^*(X_n) : n \in \mathbb{N}\}.$$

On the other hand, since $\mu^*|\mathcal{A}^*$ is finitely additive and μ^* is monotone, we have

$$\mu^*(X) \geq \Sigma\{\mu^*(X_n) : 0 \leq n \leq m\}$$

for any natural number m. Therefore,

$$\mu^*(X) \geq \Sigma\{\mu^*(X_n) : n \in \mathbb{N}\}$$

and we obtain that $\mu^*(X) = \Sigma\{\mu^*(X_n) : n \in \mathbb{N}\}$.

This completes the proof of Lemma 16.2. □

Summarizing the preceding lemmas, we can formulate the following result, essentially due to Carathéodory, who established it by using a different argument (see [17, 63], and the next chapter).

Theorem 16.1 *Let E be a base set, $\mathcal{A}$ be an algebra of subsets of E, and let $\sigma(\mathcal{A})$ denote the σ-algebra generated by $\mathcal{A}$.*

Then every finite countably additive measure μ with $\mathrm{dom}(\mu) = \mathcal{A}$ admits a unique extension to a countably additive measure μ' with $\mathrm{dom}(\mu') = \sigma(\mathcal{A})$.

Proof Denote by μ^* the outer measure canonically associated with a given measure μ. Since $\mathcal{A}^*$ is a σ-algebra containing $\mathcal{A}$ (see Lemma 16.2), we have the trivial inclusion $\sigma(\mathcal{A}) \subset \mathcal{A}^*$. So, according to the same Lemma 16.2, the restriction of μ^* to $\sigma(\mathcal{A})$ is a countably additive measure μ' on $\sigma(\mathcal{A})$.

Suppose now that ν is some countably additive measure on $\sigma(\mathcal{A})$ which extends μ. Consider the family $\mathcal{S}$ of all those sets X from $\sigma(\mathcal{A})$ for which the equality $\nu(X) = \mu'(X)$ holds. A straightforward verification shows that $\mathcal{S}$ is a σ-algebra of sets in E and $\mathcal{S}$ contains the initial algebra $\mathcal{A}$ (keep in mind Theorem 15.2 of Chap. 15).

By definition of $\sigma(\mathcal{A})$, we conclude that $\sigma(\mathcal{A}) \subset \mathcal{S}$. Therefore,

16 Extensions of Measures

$$\sigma(\mathcal{A}) = \mathcal{S},$$

which immediately yields the uniqueness of μ'.

Theorem 16.1 has thus been proved. □

Remark 16.1 In general, the σ-algebra $\mathcal{A}^*$ is much bigger than the σ-algebra $\sigma(\mathcal{A})$. For instance, if $E = \mathbf{R}$ and an algebra $\mathcal{A}$ is at most countable, then the cardinality of $\sigma(\mathcal{A})$ does not exceed card($\mathbf{R}$) (= the cardinality of the continuum denoted usually by $\mathbf{c}$). On the other hand, if μ is a finite countably additive measure on the same $\mathcal{A}$ and μ^* is the associated outer measure, then the cardinality of $\mathcal{A}^*$ may be equal to $2^{\mathbf{c}}$, which is strictly greater than $\mathbf{c}$ by virtue of Cantor's classical theorem. For more details, see Exercise 1 of this chapter.

Actually, Theorem 16.1 states that when dealing with finite countably additive measures on a ground set E, it suffices to consider only those measures which are defined on various σ-algebras of subsets of E.

In connection with Theorem 16.1, the following question arises:

Is it possible to further extend the obtained measure $\mu^*|\mathcal{A}^*$ to a wider class of subsets of E?

It turns out that the answer to this question is positive, so the process of extending an initial measure μ can be continued. More detailed information concerning the posed question will be given later.

Now, we introduce two auxiliary notions.

Let μ be a finite countably additive measure on an algebra $\mathcal{A} \subset \mathcal{P}(E)$ and let μ^* be the outer measure produced by μ. We define

$$\mathcal{A}' = \sigma(\mathcal{A}), \qquad \mu' = \mu^*|\mathcal{A}'.$$

Let X be a subset of E. It is easy to check that

$$\mu^*(X) = \inf\{\mu'(A) : X \subset A, \ A \in \mathcal{A}'\}$$

and there exists a set $A \in \mathcal{A}'$ such that $X \subset A$ and $\mu^*(X) = \mu'(A)$.

Any set A having this property is called a μ'-*measurable hull* (or a μ'-*measurable envelope*) of X.

Observe that, in general, a μ'-measurable hull of X is not unique. Indeed, if A is some μ'-measurable hull of X and Z is a set of μ'-measure zero, then the set $A \cup Z$ is also a μ'-measurable hull of X.

At the same time, if A_1 and A_2 are any two μ'-measurable hulls of X, then we obviously have

$$\mu'(A_1 \triangle A_2) = 0.$$

The set $A_1 \cap A_2$ is a μ'-measurable hull of X, too, and

$$\mu^*(X) = \mu'(A_1 \cap A_2) = \mu'(A_1) = \mu'(A_2).$$

Let $\{X_n : n \in \mathbb{N}\}$ be a countable family of sets in E and let, for each $n \in \mathbb{N}$, the set A_n be a μ'-measurable hull of X_n.

Then it is not hard to show that $\cup \{A_n : n \in \mathbb{N}\}$ is a μ'-measurable hull of the set $\cup \{X_n : n \in \mathbb{N}\}$.

In a similar way, the dual notion of a μ'-measurable kernel of an arbitrary set $X \subset E$ can be introduced. Namely, we first put

$$\mu_*(X) = \sup\{\mu'(B) : B \subset X, \ B \in \mathcal{A}'\}.$$

The obtained real-valued function

$$\mu_* : \mathcal{P}(E) \to [0, +\infty[$$

is called the *inner measure canonically associated with* μ, or the *inner measure produced by* μ.

Clearly, the inequality $\mu_*(X) \leq \mu^*(X)$ is satisfied for all subsets X of E.

Observe also that, for every set $X \subset E$, there exists a set $B \in \mathcal{A}'$ such that $B \subset X$ and $\mu_*(X) = \mu'(B)$.

Any set B having this property is called a μ'-*measurable kernel* of X.

In general, a μ'-measurable kernel of X is not unique. Indeed, if B is some μ'-measurable kernel of X and Z is a set of μ'-measure zero, then the set $B \setminus Z$ is also a μ'-measurable kernel of X.

At the same time, if B_1 and B_2 are any two μ'-measurable kernels of X, then we have

$$\mu'(B_1 \triangle B_2) = 0.$$

The set $B_1 \cup B_2$ is a μ'-measurable kernel of X, too, and

$$\mu_*(X) = \mu'(B_1 \cup B_2) = \mu'(B_1) = \mu'(B_2).$$

Starting with the definitions of a μ'-measurable hull and of a μ'-measurable kernel of a set $X \subset E$, we easily get

$$\mu_*(A \setminus X) = \mu_*(X \setminus B) = 0,$$

where A is an arbitrary μ'-measurable hull of X and B is an arbitrary μ'-measurable kernel of X. In addition, the equality

$$\mu_*(X) + \mu^*(E \setminus X) = \mu(E)$$

holds true, which is not difficult to verify.

It becomes possible to give a characterization of the measurability of sets in terms of the outer and inner measures. Namely, one can formulate and prove the next theorem.

16 Extensions of Measures

Theorem 16.2 *Let μ be a finite countably additive measure on a base set E, let μ^* denote the outer measure produced by μ, and let μ_* denote the inner measure produced by μ.*

For every set $X \subset E$, the following two assertions are equivalent:

(1) *X is μ^*-measurable;*
(2) *the equality $\mu^*(X) = \mu_*(X)$ holds.*

Proof Suppose (1). Since $\mathcal{A}^*$ is an algebra of sets, we have $E \setminus X \in \mathcal{A}^*$. Let A be some μ'-measurable hull of $E \setminus X$, i.e.,

$$E \setminus X \subset A, \qquad \mu^*(E \setminus X) = \mu'(A).$$

Then we have $E \setminus A \subset X$ and

$$\mu_*(X) \geq \mu'(E \setminus A) = \mu'(E) - \mu'(A) = \mu^*(E) - \mu^*(E \setminus X) = \mu^*(X)$$

in view of the finite additivity of $\mu^*|\mathcal{A}^*$. Hence we get $\mu_*(X) \geq \mu^*(X)$. But we also have $\mu_*(X) \leq \mu^*(X)$. Consequently, $\mu_*(X) = \mu^*(X)$, which proves (2).

Now, suppose (2). Then there exist two sets A and B from $\mathcal{A}'$ such that

$$X \subset A, \qquad B \subset X, \qquad \mu'(A) = \mu^*(X) = \mu_*(X) = \mu'(B).$$

Indeed, we may take as the set A any μ'-measurable hull of X and as the set B any μ'-measurable kernel of X. The above relations imply that

$$B \subset A, \qquad \mu'(A \setminus B) = \mu^*(A \setminus B) = 0.$$

We thus see that X is representable in the form $X = B \cup Z$, where Z is a certain set of μ^*-measure zero. From this representation it directly follows that X is a μ^*-measurable set, i.e., (1) holds.

This ends the proof of Theorem 16.2. □

The next theorem shows that the outer measures produced by finite countably additive measures possess the regularity property (see Chap. 14 for the definition of a regular outer measure).

Theorem 16.3 *Let μ be a finite countably additive measure on an algebra of subsets of a ground set E.*

Then the outer measure μ^* canonically associated with μ is regular.

Proof Take any increasing (by inclusion) sequence $\{X_n : n \in \mathbb{N}\}$ of subsets of E. We have to demonstrate that

$$\mu^*(\cup\{X_n : n \in \mathbb{N}\}) = \lim_{n \to \infty} \mu^*(X_n).$$

For each $n \in \mathbb{N}$, denote by A_n a μ'-measurable hull of X_n. It is easy to check that

$$\mu'(A_n \setminus A_{n+1}) = 0 \qquad (n \in \mathbb{N}).$$

So $\mu'(A_n) \leq \mu'(A_{n+1})$ and, in view of the trivial relation

$$A_0 \cup A_1 \cup \cdots \cup A_n = A_n \cup (A_0 \setminus A_1) \cup (A_1 \setminus A_2) \cup \cdots \cup (A_{n-1} \setminus A_n),$$

we obtain

$$\mu'(A_0 \cup A_1 \cup \cdots \cup A_n) = \mu'(A_n) = \mu^*(X_n).$$

Consider the set

$$A = \cup\{A_n : n \in \mathbb{N}\}.$$

A straightforward verification shows that A is a μ'-measurable hull of the set $\cup\{X_n : n \in \mathbb{N}\}$. Therefore,

$$\mu^*(\cup\{X_n : n \in \mathbb{N}\}) = \mu'(A) = \lim_{n \to \infty} \mu'(A_n) = \lim_{n \to \infty} \mu^*(X_n).$$

This completes the proof of Theorem 16.3. $\square$

Remark 16.2 In connection with Theorem 16.3, the natural question arises whether the outer measure μ^* canonically associated with a finite countably additive measure μ possesses the analogous regularity property for decreasing sequences of subsets of E. More precisely, suppose that $\{X_n : n \in \mathbb{N}\}$ is a decreasing (by inclusion) family of sets in E. Is it true that

$$\mu^*(\cap\{X_n : n \in \mathbb{N}\}) = \lim_{n \to \infty} \mu^*(X_n)?$$

Unfortunately, the answer to this question is negative (see, for instance, Exercise 6). However, a certain analog of Theorem 16.3 is valid for the inner measure μ_* associated with the same μ and for decreasing (by inclusion) sequences of subsets of E (see Exercise 7).

Remark 16.3 Let E be a ground set and let μ be a finite countably additive measure given on some σ-algebra $\mathcal{A}$ of subsets of E. Obviously, the pair $(\mathcal{A}, d)$ may be treated as a quasi-metric space with respect to the function d defined in Chap. 14 by

$$d(X, Y) = \mu^*(X \triangle Y) \qquad (X \in \mathcal{A}, \ Y \in \mathcal{A}),$$

where μ^* stands for the outer measure produced by μ. By repeating the argument of the proof of Theorem 14.2 from Chap. 14, one can infer that this quasi-metric space is complete. At the same time, $(\mathcal{A}, d)$ is a subspace of $(\mathcal{P}(E), d)$. Thus, by virtue of a general fact from the theory of quasi-metric and metric spaces, $(\mathcal{A}/_{\equiv}, d/_{\equiv})$ turns out to be a closed subspace of the space $(\mathcal{P}(E)/_{\equiv}, d/_{\equiv})$.

Exercises

1. Let $E = [0, 1[\subset \mathbf{R}$ and consider the algebra $\mathcal{A}$ formed by all finite unions of half-open (from the right) intervals in $[0, 1[$ with rational endpoints. Let λ^* denote the outer Lebesgue measure produced by the standard Lebesgue measure λ on $[0, 1[$.
   Show that:

   (a) the cardinality of $\sigma(\mathcal{A})$ is equal to the cardinality of the continuum $\mathbf{c}$;
   (b) the cardinality of the σ-algebra $\mathcal{A}^*$ of all λ^*-measurable sets is equal to $2^{\mathbf{c}}$.

   Using Cantor's theorem, conclude that $\text{card}(\sigma(\mathcal{A})) < \text{card}(\mathcal{A}^*)$.

2. Let μ be a finite countably additive measure on an algebra $\mathcal{A}$ of subsets of a ground set E, let X be a set in E, and let μ^* denote the outer measure produced by μ.
   Prove the existence of a μ'-measurable hull of X, i.e., prove that there exists a set $A \in \sigma(\mathcal{A})$ such that $X \subset A$ and $\mu^*(X) = \mu'(A)$.

3. Let μ be a finite countably additive measure on an algebra $\mathcal{A}$ of subsets of E, let $\{X_n : n \in \mathbb{N}\}$ be a countable family of sets in E and let, for each $n \in \mathbb{N}$, the set A_n be a μ'-measurable hull of X_n.
   Verify that $\cup \{A_n : n \in \mathbb{N}\}$ is a μ'-measurable hull of $\cup \{X_n : n \in \mathbb{N}\}$.

4. Let μ be a finite countably additive measure on an algebra $\mathcal{A}$ of subsets of E, let Y be any set in E, and let μ_* denote the inner measure produced by μ.
   Prove the existence of a μ'-measurable kernel of Y, i.e., show that there exists a set $B \in \sigma(\mathcal{A})$ such that $B \subset Y$ and $\mu_*(Y) = \mu'(B)$.
   Further, let Y_1 and Y_2 be two subsets of E, let B_1 be a μ'-measurable kernel of Y_1, and let B_2 be a μ'-measurable kernel of Y_2.
   Can one assert that $B_1 \cup B_2$ is a μ'-measurable kernel of $Y_1 \cup Y_2$?

5. Let μ be a finite countably additive measure on an algebra $\mathcal{A}$ of subsets of E, let X be any set in E, and let μ^* (respectively, μ_*) denote the outer measure (respectively, the inner measure) produced by μ.
   Demonstrate that:

   (i) $\mu_*(A \setminus X) = \mu_*(X \setminus B) = 0$, where A is an arbitrary μ'-measurable hull of X and B is an arbitrary μ'-measurable kernel of X;
   (ii) a set $A \subset E$ is a μ'-measurable hull of X if and only if the set $E \setminus A$ is a μ'-measurable kernel of $E \setminus X$;
   (iii) $\mu_*(X) + \mu^*(E \setminus X) = \mu(E)$.

6. As in Exercise 1, let λ denote the standard Lebesgue measure on the interval $[0, 1[$.

Construct a decreasing (by inclusion) sequence $\{Y_n : n \in \mathbb{N}\}$ of subsets of $[0, 1[$ such that:

(a) $\lambda^*(Y_n) = 1$ for any $n \in \mathbb{N}$;
(b) $\cap\{Y_n : n \in \mathbb{N}\} = \emptyset$.

Conclude that the outer measure λ^* is not upper semicontinuous at $\emptyset$.

7. Let E be a ground set, μ be a finite countably additive measure on an algebra of subsets of E, and let $\{X_n : n \in \mathbb{N}\}$ be a decreasing (by inclusion) sequence of sets in E.

 Prove that
$$\mu_*(\cap\{X_n : n \in \mathbb{N}\}) = \lim_{n \to \infty} \mu_*(X_n),$$
where μ_* denotes, as usual, the inner measure produced by μ.

8. Let E be a ground set, let μ be a finite countably additive measure on an algebra of subsets of E, and let μ^* (respectively, μ_*) denote the outer measure (respectively, the inner measure) produced by μ.

 Show that, for any two subsets X and Y of E, the following inequalities hold true:

(a) $\mu^*(X \cup Y) + \mu^*(X \cap Y) \leq \mu^*(X) + \mu^*(Y)$;
(b) $\mu_*(X \cup Y) + \mu_*(X \cap Y) \geq \mu_*(X) + \mu_*(Y)$.

Give an example of some sets $X \subset E$ and $Y \subset E$ such that both inequalities in (a) and (b) are strong.

Chapter 17
Caratheodory's and Marczewski's Extension Theorems

There is another classical approach to the notion of a μ^*-measurable set for any outer measure μ^*. This approach was suggested by C. Carathéodory and differs from the method described in Chap. 16.

Let μ^* be an outer measure on a ground set E (see Chap. 14).

We say that a set $A \subset E$ is μ^*-*measurable in the sense of Carathéodory* if, for every $X \subset E$, the equality

$$\mu^*(X \cap A) + \mu^*(X \cap (E \setminus A)) = \mu^*(X)$$

holds true or, equivalently, if the inequality

$$\mu^*(X \cap A) + \mu^*(X \cap (E \setminus A)) \leq \mu^*(X)$$

is satisfied.

We denote by $C(\mu^*)$ the family of all μ^*-measurable subsets of E in the sense of Carathéodory. Obviously, we have

$$\emptyset \in C(\mu^*), \qquad E \in C(\mu^*).$$

First, let us demonstrate that $C(\mu^*)$ is an algebra of sets in E.

From the definition it is clear that if $A \in C(\mu^*)$, then $E \setminus A \in C(\mu^*)$.

Take now any two sets $A \in C(\mu^*)$ and $B \in C(\mu^*)$. For an arbitrary set $X \subset E$, we may write

$$\mu^*(X \cap (A \cup B)) + \mu^*(X \cap (E \setminus (A \cup B)))$$
$$= \mu^*((X \cap A \cap B) \cup (X \cap A \cap (E \setminus B)) \cup (X \cap B \cap (E \setminus A)))$$
$$+ \mu^*(X \cap (E \setminus (A \cup B)))$$

$$\leq \mu^*(X \cap A \cap B) + \mu^*(X \cap A \cap (E \setminus B)) + \mu^*(X \cap (E \setminus A) \cap B)$$
$$+ \mu^*(X \cap (E \setminus A) \cap (E \setminus B))$$
$$\leq \mu^*(X \cap A) + \mu^*(X \cap (E \setminus A))$$
$$= \mu^*(X),$$

which implies that the family $C(\mu^*)$ is closed under finite unions. Thus, $C(\mu^*)$ is an algebra of subsets of E.

Let $A \in C(\mu^*)$ and $B \in C(\mu^*)$ be any two disjoint sets. In the formula

$$\mu^*(X \cap A) + \mu^*(X \cap (E \setminus A)) = \mu^*(X)$$

we put $X = A \cup B$, and so we get that

$$\mu^*(A) + \mu^*(B) = \mu^*(A \cup B),$$

which gives us the finite additivity of μ^* on $C(\mu^*)$.

Similarly, if in the same formula we replace X by $X \cap (A \cup B)$, then we obtain

$$\mu^*(X \cap A) + \mu^*(X \cap B) = \mu^*(X \cap (A \cup B)).$$

By induction on n, it follows from the last equality that

$$\Sigma\{\mu^*(X \cap A_k) : 1 \leq k \leq n\} = \mu^*(X \cap (\cup\{A_k : 1 \leq k \leq n\}))$$

for each set $X \subset E$ and for every disjoint finite family $\{A_k : 1 \leq k \leq n\}$ of members of $C(\mu^*)$.

Keeping in mind the monotonicity and countable subadditivity of μ^*, we infer from the above that

$$\Sigma\{\mu^*(X \cap A_n) : n \in \mathbb{N}\} = \mu^*(X \cap (\cup\{A_n : n \in \mathbb{N}\}))$$

for any $X \subset E$ and for every disjoint countable family $\{A_n : n \in \mathbb{N}\}$ of members of $C(\mu^*)$.

In particular, putting $X = E$, we come to the equality

$$\Sigma\{\mu^*(A_n) : n \in \mathbb{N}\} = \mu^*(\cup\{A_n : n \in \mathbb{N}\}).$$

The results just presented enable us to prove the following lemma.

Lemma 17.1 *Let $\{A_n : n \in \mathbb{N}\}$ be an increasing (by inclusion) sequence of sets from $C(\mu^*)$ and let $\{B_n : n \in \mathbb{N}\}$ be a decreasing (by inclusion) sequence of sets from $C(\mu^*)$. Let*

$$A = \lim\{A_n : n \in \mathbb{N}\}, \qquad B = \lim\{B_n : n \in \mathbb{N}\}.$$

17 Caratheodory's and Marczewski's Extension Theorems

Then, for any set $X \subset E$, the following two assertions hold:
(1) $\mu^*(X \cap A) = \lim_{n \to \infty} \mu^*(X \cap A_n)$;
(2) *if* $\mu^*(X) < +\infty$, *then* $\mu^*(X \cap B) = \lim_{n \to \infty} \mu^*(X \cap B_n)$.

Proof Both assertions (1) and (2) are proved by the same scheme. We will only establish the validity of (2). Clearly, we may write

$$X \cap B_0 = (X \cap B) \cup (\cup \{X \cap (B_n \setminus B_{n+1}) : n \in \mathbb{N}\}),$$

which gives us

$$+\infty > \mu^*(X \cap B_0) = \mu^*(X \cap B) + \Sigma\{\mu^*(X \cap (B_n \setminus B_{n+1})) : n \in \mathbb{N}\}$$

or, equivalently,

$$+\infty > \mu^*(X \cap B_0) = \mu^*(X \cap B) + \lim_{n \to \infty} (\mu^*(X \cap B_0) - \mu^*(X \cap B_{n+1}))$$
$$= \mu^*(X \cap B) + \mu^*(X \cap B_0) - \lim_{n \to \infty} \mu^*(X \cap B_n).$$

The last formula obviously implies

$$\mu^*(X \cap B) = \lim_{n \to \infty} \mu^*(X \cap B_n),$$

and the proof of Lemma 17.1 is thus complete. □

Lemma 17.2 *The union of any countable family of members of $C(\mu^*)$ is also a member of $C(\mu^*)$.*

Proof Let $\{A_n : n \in \mathbb{N}\}$ be an arbitrary countable family of sets from $C(\mu^*)$. Let

$$A = \cup\{A_n : n \in \mathbb{N}\}$$

and let us check that, for each set $X \subset E$, the equality

$$\mu^*(X) = \mu^*(X \cap A) + \mu^*(X \cap (E \setminus A))$$

is fulfilled.

If $\mu^*(X) = +\infty$, then the above equality trivially holds. So, suppose that $\mu^*(X) < +\infty$. Keeping in mind that $C(\mu^*)$ is an algebra of subsets of E, we may assume without loss of generality that $\{A_n : n \in \mathbb{N}\}$ is an increasing (by inclusion) family of sets. For any natural number n, we have

$$\mu^*(X) = \mu^*(X \cap A_n) + \mu^*(X \cap (E \setminus A_n)).$$

Applying Lemma 17.1, we get

$$\lim_{n \to \infty} \mu^*(X \cap A_n) = \mu^*(X \cap A),$$

$$\lim_{n \to \infty} \mu^*(X \cap (E \setminus A_n)) = \mu^*(X \cap (E \setminus A)),$$

which yields the required equality

$$\mu^*(X) = \mu^*(X \cap A) + \mu^*(X \cap (E \setminus A)).$$

Lemma 17.2 has thus been proved. □

Summarizing all the established results, we come to the following fundamental theorem first obtained by Carathéodory.

Theorem 17.1 *The family $C(\mu^*)$ is a σ-algebra of subsets of E and the restriction of μ^* to this family is a countably additive measure on E (not necessarily finite).*

The restriction $\mu^*|C(\mu^*)$ is called the *Carathéodory measure induced by the outer measure μ^** (or the *measure produced by μ^**, or the *measure derived from μ^**).

In connection with extensions of measures, it makes sense to compare Carathéodory's method described above and the method presented in Chap. 16. It turns out that these two methods are compatible, as the next theorem shows.

Theorem 17.2 *Let μ be a finite countably additive measure on an algebra $\mathcal{A}$ of subsets of E, let μ^* denote the outer measure associated with μ, and let $\mathcal{A}^*$ denote the σ-algebra of all μ^*-measurable subsets of E in the sense of Chap. 16.*

If μ_c stands for the Carathéodory measure derived from μ^*, then the equality $\mu^*|\mathcal{A}^* = \mu_c$ holds true.

Proof First, let us check that every μ^*-measurable set A is measurable in the sense of Carathéodory, i.e., for each $X \subset E$, the formula

$$\mu^*(X \cap A) + \mu^*(X \cap (E \setminus A)) = \mu^*(X)$$

holds. For this purpose, denote by D_1 some μ'-measurable hull of $X \cap A$ and denote by D_2 some μ'-measurable hull of $X \cap (E \setminus A)$ (here μ' is defined as in Chap. 16). It is easy to see that

$$\mu^*(D_1 \setminus A) = \mu^*(D_2 \setminus (E \setminus A)) = 0,$$

whence it follows that

$$\mu'(D_1 \cap D_2) = 0.$$

The set $D_1 \cup D_2$ is a μ'-measurable hull of X and

$$\mu'(D_1) + \mu'(D_2) = \mu'(D_1 \cup D_2).$$

At the same time, we have

$$\mu^*(X \cap A) = \mu'(D_1), \qquad \mu^*(X \cap (E \setminus A)) = \mu'(D_2),$$
$$\mu^*(X) = \mu'(D_1 \cup D_2) = \mu'(D_1) + \mu'(D_2),$$

which implies the measurability of A in Carathéodory's sense.

Conversely, suppose that a set $A \subset E$ is measurable in Carathéodory's sense. Then, in particular, the equality

$$\mu^*(A) + \mu^*(E \setminus A) = \mu^*(E)$$

holds true. Now, denote by D_1 a μ'-measurable hull of A and denote by D_2 a μ'-measurable hull of $E \setminus A$. Clearly, we may write

$$\mu'(D_1) + \mu'(D_2) = \mu^*(E) = \mu'(E),$$
$$\mu'(E \setminus D_2) = \mu'(D_1).$$

Since $E \setminus D_2 \subset A$, we obtain

$$\mu'(D_1) = \mu'(E \setminus D_2) \leq \mu_*(A) \leq \mu^*(A) = \mu'(D_1),$$

whence it follows that $\mu_*(A) = \mu^*(A)$. The latter immediately implies that the set A is μ^*-measurable (see Theorem 16.2 from Chap. 16).

Consequently, the equality $\mathcal{A}^* = C(\mu^*)$ holds. Finally, we have by definition

$$\mu^*(A) = \mu_c(A) \qquad (A \in \mathcal{A}^*),$$

which ends the proof of Theorem 17.2. □

The next important theorem is due to E. Marczewski (E. Szpilrajn) and provides a certain method of extending countably additive measures with the aid of appropriate σ-ideals of subsets of a ground set E (cf. [173, 175]).

Theorem 17.3 *Let μ be a finite countably additive measure defined on a σ-algebra of subsets of E, let μ_* be the inner measure produced by μ, and let $\mathcal{I}$ be a σ-ideal of subsets of E such that $\mu_*(Z) = 0$ for each set $Z \in \mathcal{I}$.*

Then there exists a countably additive measure ν on E satisfying the following two conditions:

(1) *ν extends μ;*
(2) *$\mathcal{I} \subset \mathrm{dom}(\nu)$ and $\nu(Z) = 0$ for any set $Z \in \mathcal{I}$.*

Proof Consider the σ-algebra $\mathcal{S}$ generated by $\mathrm{dom}(\mu) \cup \mathcal{I}$. It is not difficult to verify that $\mathcal{S}$ consists of all those sets X which can be represented in the form

$$X = (Y \cup Z') \setminus Z'',$$

where $Y \in \text{dom}(\mu)$ and $\{Z', Z''\} \subset \mathcal{I}$ (in general, such a representation of X is not unique).

Indeed, if X admits the indicated representation, then $X \in \mathcal{S}$ and, as can easily be shown, $E \setminus X$ also admits an analogous representation, so $E \setminus X \in \mathcal{S}$.

Now, let $\{X_n : n \in \mathbb{N}\}$ be a countable family of sets, each of which can be represented in the above-mentioned form, i.e.,

$$X_n = (Y_n \cup Z'_n) \setminus Z''_n \quad (n \in \mathbb{N}),$$

where $Y_n \in \text{dom}(\mu)$ and $\{Z'_n, Z''_n\} \subset \mathcal{I}$. Then we have the inclusions

$$(\cup\{Y_n : n \in \mathbb{N}\}) \setminus (\cup\{Z''_n : n \in \mathbb{N}\}) \subset \cup\{X_n : n \in \mathbb{N}\}$$

$$\subset (\cup\{Y_n : n \in \mathbb{N}\}) \cup (\cup\{Z'_n : n \in \mathbb{N}\}),$$

whence it follows that $\cup\{X_n : n \in \mathbb{N}\}$ is representable in the form

$$\cup\{X_n : n \in \mathbb{N}\} = ((\cup\{Y_n : n \in \mathbb{N}\}) \cup T') \setminus T'',$$

where $\{T', T''\} \subset \mathcal{I}$. Consequently,

$$\cup\{X_n : n \in \mathbb{N}\} \in \mathcal{S}.$$

Starting with a representation $X = (Y \cup Z') \setminus Z''$ of any $X \in \mathcal{S}$, we put $\nu(X) = \mu(Y)$.

First of all, we need to check that the introduced function

$$\nu : \mathcal{S} \to [0, +\infty[$$

is well-defined. For this purpose, take any two representations of X in the above-mentioned form:

$$X = (Y_1 \cup Z'_1) \setminus Z''_1, \quad X = (Y_2 \cup Z'_2) \setminus Z''_2.$$

From these representations it is not difficult to infer the formula

$$Y_1 \triangle Y_2 \subset (Z'_1 \cup Z''_1 \cup Z'_2 \cup Z''_2) \in \mathcal{I}.$$

Keeping in mind the condition on $\mathcal{I}$, we get

$$\mu(Y_1 \triangle Y_2) = 0,$$

whence it follows that $\mu(Y_1) = \mu(Y_2)$. Therefore, ν is well-defined.

17 Caratheodory's and Marczewski's Extension Theorems

Further, if $Y \in \text{dom}(\mu)$, then Y can be expressed in the form

$$Y = (Y \cup \emptyset) \setminus \emptyset,$$

which implies at once that $\nu(Y) = \mu(Y)$, so ν is an extension of μ.

If $Z \in \mathcal{I}$, then Z may be written as

$$Z = (\emptyset \cup Z) \setminus \emptyset,$$

which gives us that $\nu(Z) = \mu(\emptyset) = 0$.

It remains to check the countable additivity of ν. Let $\{X_n : n \in \mathbb{N}\}$ be an arbitrary disjoint countable family of sets from $\mathcal{S}$. For each $n \in \mathbb{N}$, we have a representation

$$X_n = (Y_n \cup Z'_n) \setminus Z''_n,$$

where $Y_n \in \text{dom}(\mu)$ and $\{Z'_n, Z''_n\} \subset \mathcal{I}$. As above, it can readily be verified that

$$Y_n \cap Y_m \subset (X_n \cap X_m) \cup (Z'_n \cup Z''_n \cup Z'_m \cup Z''_m)$$

for any two distinct $n \in \mathbb{N}$ and $m \in \mathbb{N}$. Consequently, keeping in mind that

$$X_n \cap X_m = \emptyset, \qquad \mu_*(Z'_n \cup Z''_n \cup Z'_m \cup Z''_m) = 0,$$

we obtain $\mu(Y_n \cap Y_m) = 0$ and, therefore,

$$\nu(\cup\{X_n : n \in \mathbb{N}\}) = \mu(\cup\{Y_n : n \in \mathbb{N}\})$$
$$= \Sigma\{\mu(Y_n) : n \in \mathbb{N}\}$$
$$= \Sigma\{\nu(X_n) : n \in \mathbb{N}\}.$$

This gives us the desired result and ends the proof. □

Remark 17.1 A measure ν on an algebra of subsets of a ground set E is called *complete* if, for every set X of ν-measure zero, one has

$$(\forall Y)(Y \subset X \Rightarrow Y \in \text{dom}(\nu)).$$

Applying Marczewski's method to any nonzero finite countably additive measure μ on E and to the σ-ideal generated by all μ-measure zero sets, we easily conclude that there always exists a complete countably additive measure ν on E which extends μ. The least (by inclusion) such extension is usually called the *completion* of μ.

Theorem 17.4 *Let μ be a finite countably additive measure on a σ-algebra $\mathcal{A}$ of subsets of E and let X be an arbitrary subset of E.*

There exists a countably additive measure ν on some σ-algebra of subsets of E which extends μ and satisfies the relation $X \in \mathrm{dom}(\nu)$.

Proof As we know, two non-negative numbers are canonically associated with X, namely, the outer measure $\mu^*(X)$ of X and the inner measure $\mu_*(X)$ of X. Only two cases are possible.

1. $\mu_*(X) = \mu^*(X)$.
    In this case, the set X is μ^*-measurable (see Theorem 16.2 from Chap. 16). Hence we may take $\mu^*|\mathcal{A}^*$ as the required ν.
2. $\mu_*(X) < \mu^*(X)$.

In this case, the set X is not μ^*-measurable. Let B denote some μ-measurable kernel of X. Consider the set

$$Z = X \setminus B.$$

Obviously, we have the relation $0 = \mu_*(Z) < \mu^*(Z)$.

Further, denote by $\mathcal{I}$ the family of all subsets of Z. This $\mathcal{I}$ is a σ-ideal of sets in E and satisfies all the conditions of Theorem 17.3. So, according to Marczewski's result, there exists a measure ν on E extending μ and such that $\mathcal{I} \subset \mathrm{dom}(\nu)$. In particular, $Z \in \mathrm{dom}(\nu)$. But we also have

$$X = B \cup Z, \qquad B \in \mathrm{dom}(\mu) \subset \mathrm{dom}(\nu),$$

whence it follows that $X \in \mathrm{dom}(\nu)$, as required.

Theorem 17.4 is thus proved. □

Remark 17.2 The above argument shows us that if a set $X \subset E$ fulfils the equality $\mu_*(X) = \mu^*(X)$, then the value $\nu(X)$ is uniquely determined for any measure ν on E extending μ and satisfying $X \in \mathrm{dom}(\nu)$. In this context, we may say that the set X possesses the *uniqueness property* with respect to the class of all those measures on E which extend μ.

On the other hand, it can be demonstrated that if, for a set $Y \subset E$, the inequality $\mu_*(Y) < \mu^*(Y)$ holds true, then there are at least two measures ν_1 and ν_2 on E such that:

(a) both ν_1 and ν_2 extend μ;
(b) $Y \in \mathrm{dom}(\nu_1) \cap \mathrm{dom}(\nu_2)$;
(c) $\nu_1(Y) \neq \nu_2(Y)$.

So, according to (c), the set Y does not possess the uniqueness property with respect to the class of all those measures on E which extend μ. A much stronger result is formulated in Exercise 5 of this chapter.

A finitely additive measure μ on a ground set E is called *σ-finite* if there exists a countable family $\{X_i : i \in I\}$ of μ-measurable subsets of E such that

$$E = \cup\{X_i : i \in I\}, \qquad (\forall i \in I)(\mu(X_i) < +\infty).$$

Clearly, any finite measure is σ-finite.

Most countably additive measures considered and studied in mathematical analysis are σ-finite. Many statements presented in this and the preceding chapters, concerning properties of finite measures, remain true for σ-finite measures. We leave to the reader the reformulations of those statements for the case of σ-finite measures.

At the end of this chapter, it makes sense to recall the notion of a real-valued measurable function (see, e.g., [17, 45, 53, 63, 104, 134, 141]).

Let μ be a σ-finite countably additive measure on a ground set E and let $f : E \to \mathbb{R}$ be a function (or a partial function).

One says that f is *measurable with respect to* μ (or f is μ-*measurable*) if, for each open subset U of $\mathbb{R}$, the set $f^{-1}(U)$ is μ-measurable.

Obviously, in this definition it suffices to require that all sets of the form $f^{-1}(V)$ are μ-measurable, where V ranges over some countable base of the topology of $\mathbb{R}$.

As is well known, the family of all real-valued μ-measurable functions on E is closed under the standard algebraic operations and also under taking the pointwise limits of sequences of real-valued functions on E.

For our further purposes, it is convenient to expand the notion of measurability of real-valued functions.

Let $\mathcal{M}$ be a family (class) of σ-finite countably additive measures on E (in general, the domains of members of $\mathcal{M}$ are various σ-algebras of subsets of E).

We shall say that a function $f : E \to \mathbb{R}$ is *absolutely measurable* (or *universally measurable*) with respect to $\mathcal{M}$ if f is measurable with respect to every measure from $\mathcal{M}$.

We shall say that a function $g : E \to \mathbb{R}$ is *relatively measurable* with respect to $\mathcal{M}$ if there exists at least one measure μ from $\mathcal{M}$ such that g is μ-measurable.

We shall say that a function $h : E \to \mathbb{R}$ is *absolutely nonmeasurable* with respect to $\mathcal{M}$ if h is nonmeasurable with respect to every measure from $\mathcal{M}$.

Accordingly, a subset X of E will be called *absolutely measurable* (or *universally measurable*) with respect to $\mathcal{M}$ if the characteristic function χ_X is absolutely measurable with respect to $\mathcal{M}$.

A subset Y of E will be called *relatively measurable* with respect to $\mathcal{M}$ if the characteristic function χ_Y is relatively measurable with respect to $\mathcal{M}$.

A subset Z of E will be called *absolutely nonmeasurable* with respect to $\mathcal{M}$ if the characteristic function χ_Z is absolutely nonmeasurable with respect to $\mathcal{M}$.

For example, using this terminology, Theorem 17.4 of the present chapter can be reformulated in the following manner.

Theorem 17.5 *Let E be a ground set, μ be a finite countably additive measure on some σ-algebra of subsets of E, and let $\mathcal{M}_\mu$ denote the family (class) of all those countably additive measures on E which extend μ.*

Then any set $X \subset E$ is relatively measurable with respect to $\mathcal{M}_\mu$.

Exercises

1. Let E be a ground set, G be a group of transformations of E, and let $\mathcal{A}$ be an algebra of subsets of E, which is G-invariant, i.e.,

$$(\forall X \in \mathcal{A})(\forall g \in G)(g(X) \in \mathcal{A}).$$

   Verify that the σ-algebra $\sigma(\mathcal{A})$ generated by $\mathcal{A}$ is also G-invariant, i.e., the formula

$$(\forall X \in \sigma(\mathcal{A}))(\forall g \in G)(g(X) \in \sigma(\mathcal{A}))$$

   holds.

2. Using the notation and assumption of Exercise 1, suppose in addition that μ is a σ-finite countably additive measure with $\mathrm{dom}(\mu) = \mathcal{A}$ and that μ is G-invariant, i.e.,

$$(\forall X \in \mathcal{A})(\forall g \in G)(\mu(g(X)) = \mu(X)).$$

   Show that the measure μ' produced by μ and defined on $\sigma(\mathcal{A})$ is also G-invariant, i.e., the relation

$$(\forall X \in \sigma(\mathcal{A}))(\forall g \in G)(\mu(g(X)) = \mu(X))$$

   is true.

3. Let E be a ground set and let G be a group of transformations of E.

   Formulate and prove the extension theorem for a σ-finite G-invariant countably additive measure given on a G-invariant algebra of subsets of E.

   For this purpose, use the results of Exercises 1 and 2.

4. Let E be a ground set and let G be a group of transformations of E.

   Formulate and prove the analog of Marczewski's extension theorem for a σ-finite countably additive G-invariant measure μ defined on a σ-algebra of subsets of E and for a G-invariant σ-ideal $\mathcal{I}$ of subsets of E such that $\mu_*(Z) = 0$ whenever $Z \in \mathcal{I}$.

5. Let E be a ground set, μ be a finite countably additive measure on E, and let Y be a subset of E such that $\mu_*(Y) < \mu^*(Y)$. Pick a real number

$$t \in [\mu_*(Y), \mu^*(Y)].$$

   Demonstrate that there exists a countably additive measure μ_t on E extending μ and satisfying the relations $Y \in \mathrm{dom}(\mu_t)$ and $\mu_t(Y) = t$.

   For this purpose, consider a μ-measurable kernel B of Y and a μ-measurable hull of $Y \setminus B$.

17 Caratheodory's and Marczewski's Extension Theorems

6. Let E be a ground set, let $\mathcal{R}$ be a ring of subsets of E, and let μ be a countably additive measure (not necessarily finite) with $\text{dom}(\mu) = \mathcal{R}$.

    Prove that there exists a countably additive measure μ' which extends μ and is defined on the σ-algebra $\sigma(\mathcal{R})$ generated by $\mathcal{R}$.

7. Using the notation and assumptions of Exercise 6, show that if μ is σ-finite on $\mathcal{R}$, then an extension μ' is uniquely determined on $\sigma(\mathcal{R})$.

    Also, give an example of a countably additive non-σ-finite measure μ on $\mathcal{R}$ such that there are two different countably additive measures, both defined on $\sigma(\mathcal{R})$ and extending μ.

8. Prove that a non-finite countably additive measure μ defined on a σ-algebra of subsets of a ground set E is σ-finite if and only if the following two conditions are satisfied:

    (a) for any μ-measurable set X with $\mu(X) = +\infty$, there exists a μ-measurable set Y contained in X such that
    $$0 < \mu(Y) < +\infty;$$

    (b) there exists no uncountable disjoint family $\{X_i : i \in I\}$ of μ-measurable sets such that $\mu(X_i) > 0$ for all indices $i \in I$.

    To establish this result, use the method of transfinite induction over countable ordinals.

9. Let μ be a σ-finite countably additive measure on a σ-algebra of subsets of a ground set E.

    Verify that:

    (a) if f and g are any two real-valued μ-measurable functions on E, then the functions
    $$|f|, \quad f+g, \quad f \cdot g, \quad tf \ (t \in \mathbb{R}), \quad \sup\{f, g\}, \quad \inf\{f, g\}$$
    are also μ-measurable;

    (b) if $\{f_n : n \in \mathbb{N}\}$ is a pointwise convergent sequence of real-valued μ-measurable functions on E, then the function
    $$f = \lim\{f_n : n \in \mathbb{N}\}$$
    is also μ-measurable;

    (c) if $\{g_n : n \in \mathbb{N}\}$ is a sequence of real-valued μ-measurable functions on E, then the set of all those elements x of E for which the limit $\lim_{n \to \infty} g_n(x)$ exists is necessarily μ-measurable.

    More generally, let $\mathcal{M}$ be a family (class) of σ-finite countably additive measures on E.

Formulate and prove the analogues of (a), (b) and (c) for those real-valued functions on E which are absolutely measurable (universally measurable) with respect to $\mathcal{M}$.

10. Give an example of two real-valued functions f and g on $\mathbb{R}$ which are Lebesgue measurable, but their composition $f \circ g$ is not Lebesgue measurable.

11. Let $h : \mathbb{R} \to \mathbb{R}$ be a bijective Lebesgue measurable function.

    Can one assert that the inverse function h^{-1} is also Lebesgue measurable?

Chapter 18
Positive Linear Functionals

We turn our attention to finite finitely additive measures on a ground set X. As we have already said, any such measure admits an extension to a finitely additive measure defined on $\mathcal{P}(X)$. We would like to establish this classical fact (essentially due to Tarski) in a more general setting. Actually, we will connect the problem of extending a finite finitely additive measure with the problem of extending partial positive linear functionals.

Let E be a real vector space (i.e., E is a vector space over the standard field $(\mathbb{R}, +, \cdot)$).

As usual, we say that E is a *partially pre-ordered space*, with respect to a binary relation $\leq$ on E, if the following four conditions hold:

(1) $x \leq x$ for all $x \in E$ (the reflexivity of $\leq$);
(2) $x \leq y$ and $y \leq z$ imply $x \leq z$ (the transitivity of $\leq$);
(3) $x \leq y$ and $z \in E$ imply $x+z \leq y+z$ (the invariance of $\leq$ under the translations of E);
(4) $x \leq y$ and $t \in [0, +\infty[$ imply $tx \leq ty$ (the invariance of $\leq$ under the positive homotheties of E with center $0 \in E$).

From these conditions it immediately follows that:

(a) $x \leq x_1$ and $y \leq y_1$ imply $x + y \leq x_1 + y_1$;
(b) $0 \leq x$ and $0 \leq y$ imply $0 \leq x + y$.

Thus, one can conclude that the set

$$C = \{x \in E : 0 \leq x\}$$

is a convex cone in E whose vertex is 0.

Conversely, if C is a convex cone in E with vertex 0, then the binary relation $\leq$ defined by the formula

$$x \leq y \Leftrightarrow y - x \in C$$

is a partial pre-ordering of E determined by the cone C.

In other words, to speak of a real partially pre-ordered vector space is to speak of a real vector space in which some convex cone with vertex 0 is given.

If a pre-ordering $\leq$ of E is such that $x \leq y$ and $y \leq x$ imply $x = y$, then $\leq$ is a partial ordering of E, and we come to a real *partially ordered vector space* $(E, \leq)$.

Obviously, a partial pre-ordering determined by a convex cone $C \subset E$ is a partial ordering of E if and only if the equality $C \cap (-C) = \{0\}$ holds.

We have the following deep result due to M. Riesz [149].

Theorem 18.1 *Let E be a real vector space, let U be a vector subspace of E, and let C be a convex cone in E with vertex 0. Suppose, in addition, that:*

(1) $C + U = E$;
(2) $f : U \to \mathbb{R}$ *is a linear functional positive on* $C \cap U$, *i.e., for all* $x \in C \cap U$ *one has* $f(x) \geq 0$.

Then there exists a linear functional $f^* : E \to \mathbb{R}$ *extending* f *and such that* $f^*|C$ *is positive on* C, *i.e., for all* $x \in C$ *one has* $f^*(x) \geq 0$.

Proof Denote by $\preceq$ the partial pre-ordering of E determined by C. First, observe that condition (1) is equivalent to the following:

For any vector $x \in E$, there exists a vector $u_1 \in U$ such that $u_1 \preceq x$ and there exists a vector $u_2 \in U$ such that $x \preceq u_2$.

In other words, the latter condition tells us that the set U is simultaneously coinitial and cofinal in the pre-ordered set $(E, \preceq)$.

Actually, by using the standard method of transfinite induction (or the well-known Kuratowski–Zorn lemma), it suffices to demonstrate that if x belongs to $E \setminus U$, then there exists a linear functional $f^* : V \to \mathbb{R}$ such that:

(a) V is a vector subspace of E generated by $\{x\} \cup U$;
(b) the restriction $f^*|(C \cap V)$ is positive.

To prove the existence of f^* with properties (a) and (b), consider the two sets:

$$A = \{y \in U : y \preceq x\}, \quad B = \{z \in U : x \preceq z\}.$$

As mentioned above, both A and B are nonempty sets. If $y \in A$ and $z \in B$, then $z - y \in C \cap U$ and $f(y) \leq f(z)$ in view of (2). Therefore,

$$\sup\{f(y) : y \in A\} \leq \inf\{f(z) : z \in B\}.$$

Choose any τ from the line segment

18 Positive Linear Functionals

$$[\sup\{f(y) : y \in A\}, \inf\{f(z) : z \in B\}]$$

and define a linear functional $f^* : V \to \mathbb{R}$ by putting

$$f^*(tx + u) = t \cdot \tau + f(u) \qquad (t \in \mathbb{R},\ u \in U).$$

We now assert that the functional f^* is as required.

Indeed, take any vector $tx + u \in C \cap V$. Suppose first that $t > 0$. Then we have $-u/t \preceq x$, whence it follows that

$$f^*(-u/t) = f(-u/t) \le \tau = f^*(x), \qquad f^*(tx+u) \ge 0.$$

Analogously, supposing that $t < 0$, we may write $x \preceq -u/t$, whence it follows that

$$f^*(x) = \tau \le f(-u/t) = f^*(-u/t), \qquad f^*(tx+u) \ge 0.$$

We thus see that $f^*|(C \cap V)$ is positive, which completes the proof. □

Example 18.1 The condition (1) in Theorem 18.1 is very essential. To see why, put $E = \mathbb{R}^2$ and let U and C be defined as follows:

$$U = \{(x, y) \in \mathbb{R}^2 : y = 0\} = \mathbb{R} \times \{0\},$$
$$C = \{(x, y) \in \mathbb{R}^2 : y \ge 0,\ (\forall t < 0)((x, y) \ne (t, 0))\}.$$

Consider also the linear functional $f : U \to \mathbb{R}$ defined by

$$f((x, 0)) = x \qquad ((x, 0) \in U).$$

It is not difficult to check that f is positive on the convex cone $C \cap U$, but the same f does not admit a linear extension $f^* : \mathbb{R}^2 \to \mathbb{R}$ positive on C.

As a direct consequence of Theorem 18.1, we obtain the following result of Krein.

Theorem 18.2 *Let E be a real vector space, C be a convex cone in E with vertex 0, and let a vector $x \in E \setminus (-C)$ be such that $\mathbb{R}x + C = E$.*

Then there exists a linear functional $f^* : E \to \mathbb{R}$ positive on C and satisfying the equality $f^*(x) = 1$.

Proof Consider the vector subspace $U = \mathbb{R}x$ of E. Let $f : U \to \mathbb{R}$ be a linear functional such that $f(tx) = t$ for $t \in \mathbb{R}$. Let us verify that the restriction $f|(C \cap U)$ is positive. Suppose otherwise, i.e., $f(tx) < 0$ for some $tx \in C \cap U$. By virtue of the definition of f, we must have $t < 0$. Consequently, $-1/t > 0$ and

$$-x = (-1/t)(tx) \in C, \qquad x \in -C,$$

which contradicts the assumption of the theorem. Now, applying Theorem 18.1 to these U, f, and C, we obtain the required result. □

One more consequence of Theorem 18.1 is the celebrated Hahn–Banach extension theorem. To formulate it, recall the notion of a sublinear functional.

Let E be again a real vector space and let $p : E \to \mathbb{R}$ be a functional.
By definition, this p is *sublinear* if the following two conditions hold:

(i) $p(tx) = tp(x)$ for any real number $t \geq 0$ and for any $x \in E$;
(ii) $p(x + y) \leq p(x) + p(y)$ for all $x \in E$ and for all $y \in E$.

These conditions immediately imply that $p(0) = 0$ and

$$0 = p(0) = p(x + (-x)) \leq p(x) + p(-x),$$
$$-p(-x) \leq p(x), \qquad -p(x) \leq p(-x).$$

It also follows from the above conditions that every sublinear functional p is a convex functional, i.e., we have

$$p(tx + (1-t)y) \leq tp(x) + (1-t)p(y) \qquad (0 \leq t \leq 1,\ x \in E,\ y \in E).$$

Below, we give the standard formulation of the Hahn–Banach extension theorem. For several geometric versions of this theorem, see the exercises of the present chapter (especially, Exercise 2).

Theorem 18.3 *Let E be a real vector space with a sublinear functional $p : E \to \mathbb{R}$, let U be a vector subspace of E, and let $f : U \to \mathbb{R}$ be a linear functional such that $f(u) \leq p(u)$ for all $u \in U$.*

Then there exists a linear functional $f^* : E \to \mathbb{R}$ which extends f and satisfies $f^*(x) \leq p(x)$ for any $x \in E$.

Proof In the product vector space $\mathbb{R} \times E$ consider the two sets

$$C = \{(t, x) : t \in \mathbb{R},\ x \in E,\ p(x) \leq t\},$$
$$V = \mathbb{R} \times U.$$

A direct verification gives us that C is a convex cone in $\mathbb{R} \times E$ with vertex $(0, 0)$ and V is a vector subspace of $\mathbb{R} \times E$. Moreover, in view of the relation

$$(p(e), e) + (t - p(e), 0) = (t, e) \qquad (t \in \mathbb{R},\ e \in E),$$

we trivially have the equality

$$C + V = \mathbb{R} \times E.$$

Further, define a linear functional $g : V \to \mathbb{R}$ by putting

$$g((t, u)) = t - f(u) \qquad (t \in \mathbb{R}, \ u \in U).$$

Obviously, the relation $f(u) \leq p(u)$ for any $u \in U$ implies that the restriction $g|(C \cap V)$ is positive. So, in our case we can apply the Riesz extension theorem. Consequently, we obtain that there exists a linear functional

$$g^* : \mathbb{R} \times E \to \mathbb{R}$$

which extends g and is positive on the cone C.

Now, consider the linear functional $f^* : E \to \mathbb{R}$ defined by

$$f^*(x) = -g^*((0, x)) \qquad (x \in E).$$

Let us check that f^* is as required. Clearly, f^* is an extension of f. Finally, take any vector $x \in E$. By the definition of C, we have $(p(x), x) \in C$, so $g^*((p(x), x)) \geq 0$ or, keeping in mind the linearity of g^*, the relation

$$g^*((p(x), x)) = g^*((0, x)) + g^*((p(x), 0))$$
$$= g^*((0, x)) + g((p(x), 0))$$
$$= -f^*(x) + p(x) \geq 0$$

must be fulfilled, which is equivalent to $f^*(x) \leq p(x)$.

This ends the proof of the Hahn–Banach theorem. $\square$

In practice the most typical situation is when a given sublinear functional $p : E \to \mathbb{R}$ is a *seminorm* on E. This means that, for all $t \in \mathbb{R}$ and for all $x \in E$, the relation

$$p(tx) = |t| p(x)$$

is fulfilled. In its turn, the latter equality implies that $p(-x) = p(x)$ and, therefore,

$$0 = p(x + (-x)) \leq p(-x) + p(x) = 2p(x), \qquad p(x) \geq 0.$$

It is easy to see that in the case of a seminorm p on E, the relation

$$(\forall x \in E)(f(x) \leq p(x))$$

for a linear functional f is equivalent to the relation

$$(\forall x \in E)(|f(x)| \leq p(x)).$$

We are going to apply the above profound results to the problem of extending finite finitely additive measures.

Consider a ground set X equipped with some algebra $\mathcal{A}$ of its subsets and with a finitely additive measure $\mu : \mathcal{A} \to [0, +\infty[$.

We may associate with this measure the vector space U of all those real-valued functions on X which are finite linear combinations of the characteristic functions of sets belonging to $\mathcal{A}$.

Let E denote the vector space of all real-valued bounded functions on X. Clearly, we have the inclusion $U \subset E$.

Let C stand for the convex cone in E consisting of all non-negative functions w from E, i.e.,

$$w \in C \Leftrightarrow (\forall x \in X)(w(x) \geq 0).$$

Further, take any vector $w \in E$. Since w is a real-valued bounded function on X and all real-valued constant functions on X belong to U, we can find $u \in U$ such that

$$(\forall x \in X)((u + w)(x) \geq 0).$$

In other words, $u + w \in C$, whence it follows that $w \in C + U$ or, equivalently,

$$E = C + U.$$

Now, define a linear functional $f : U \to \mathbb{R}$ by putting

$$f(u) = \Sigma \{t_k \mu(X_k) : 1 \leq k \leq n\},$$

where $\{X_k : 1 \leq k \leq n\}$ is a finite partition of X into sets belonging to $\mathcal{A}$,

$$u = \Sigma \{t_k u_k : 1 \leq k \leq n\},$$

and each u_k ($1 \leq k \leq n$) is the characteristic function of the set X_k.

A straightforward verification shows that the value $f(u)$ does not depend on the above-mentioned representation of u, so f is well-defined. Also, f is positive on $C \cap U$.

According to Theorem 18.1, there exists a linear functional $f^* : E \to \mathbb{R}$ extending f and such that $f^*|C$ is positive.

Let Z be any subset of X. Remembering the notation χ_Z for the characteristic function of Z, we define $\nu(Z) = f^*(\chi_Z)$.

Theorem 18.4 *With the above notation, the produced mapping*

$$\nu : \mathcal{P}(X) \to \mathbb{R}$$

is a finitely additive measure on X which extends the initial measure μ.

Proof Obviously, for every set $Z \subset X$, the inequality $v(Z) \geq 0$ holds true in view of the positivity of f^*.

Let now Z_1 and Z_2 be any two disjoint subsets of X. Then we have

$$\chi_{Z_1} + \chi_{Z_2} = \chi_{Z_1 \cup Z_2}.$$

The additivity of f^* immediately implies the equality

$$f^*(\chi_{Z_1}) + f^*(\chi_{Z_2}) = f^*(\chi_{Z_1 \cup Z_2}).$$

Consequently,

$$v(Z_1 \cup Z_2) = v(Z_1) + v(Z_2),$$

which demonstrates the finite additivity of v.

Finally, the fact that v is an extension of μ follows at once from the analogous fact that f^* is an extension of f.

Theorem 18.4 has thus been proved. □

Example 18.2 Let X be a ground set, $\mathcal{I}$ be an ideal of subsets of X, and let $\mathcal{F}$ denote the filter dual to $\mathcal{I}$. As we know, the family $\mathcal{A} = \mathcal{I} \cup \mathcal{F}$ is an algebra of subsets of X. Let us define a function

$$\mu : \mathcal{A} \to \{0, 1\}$$

by putting $\mu(Z) = 0$ if $Z \in \mathcal{I}$, and $\mu(Z) = 1$ if $Z \in \mathcal{F}$. Clearly, this μ is a two-valued finitely additive measure on X. Moreover, if $\mathcal{I}$ is a σ-ideal, then μ turns out to be a countably additive measure on X. According to Theorem 18.4, μ admits an extension to a finitely additive measure v defined on the power set $\mathcal{P}(X)$. But some values of the extended finitely additive measure v may differ from 0 and 1. However, in this special case Theorem 18.4 can be substantially strengthened. Namely, v can be chosen so that the range ran(v) of v remains the same, i.e., is equal to $\{0, 1\}$. To see why, observe that if one has a nonempty family of ideals (filters) in X, which is linearly ordered with respect to the inclusion relation, then the union of the family is again an ideal (filter) in X. This implies (by the Kuratowski–Zorn lemma) that every ideal (filter) in X is contained in a maximal ideal (filter) in X.

The maximal ideals (filters) are called *ultraideals* (*ultrafilters*).

Notice that an ideal $\mathcal{I}'$ (respectively, a filter $\mathcal{F}'$) is an ultraideal (respectively, an ultrafilter) in X if and only if for any set $Z \subset X$ either $Z \in \mathcal{I}'$ or $X \setminus Z \in \mathcal{I}'$ (respectively, either $Z \in \mathcal{F}'$ or $X \setminus Z \in \mathcal{F}'$). So, if $\mathcal{I}'$ is an ultraideal in X containing $\mathcal{I}$ and $\mathcal{F}'$ is the ultrafilter in X dual to $\mathcal{I}'$, then we may define a finitely additive measure v on the algebra

$$\mathcal{A}' = \mathcal{I}' \cup \mathcal{F}'$$

by the formula

$$(\forall Z \in \mathcal{I}')(\nu(Z) = 0) \ \& \ (\forall Z \in \mathcal{F}')(\nu(Z) = 1).$$

Evidently, ν extends μ, the domain $\mathcal{A}'$ of ν coincides with the power set $\mathcal{P}(X)$, and the range of ν is $\{0, 1\}$.

In some exercises presented below we suppose that the reader is more or less familiar with elements of the general theory of topological vector spaces over the field $\mathbb{R}$ (see, e.g., [38, 104]). For the sake of simplicity, we also assume that all topological vector spaces considered in exercises are Hausdorff (which is not necessary in certain situations).

Exercises

1. Let E be a topological vector space (over $\mathbb{R}$) and let $f : E \to \mathbb{R}$ be a linear functional not identically equal to zero.

    Demonstrate that the following three assertions are equivalent:

    (a) f is continuous on E;
    (b) the set $\ker(f) = f^{-1}(0)$ is a closed vector hyperplane in E;
    (c) f is bounded from above on some nonempty open subset of E.

2. Let E be a topological vector space (over $\mathbb{R}$), let G be a vector subspace of E, and let U be a nonempty open convex subset of E such that $G \cap U = \emptyset$.

    Prove that there exists a closed vector hyperplane H in E containing G and satisfying the equality $H \cap U = \emptyset$.

    For this purpose, consider the family $\mathcal{H}$ of all vector subspaces of E which contain G and have no common points with U. Observe that this $\mathcal{H}$ is partially ordered by the standard inclusion relation and apply to $\mathcal{H}$ the Kuratowski–Zorn lemma. Denoting by H a maximal element of $\mathcal{H}$, verify that H is a closed vector hyperplane in E.

3. Let E be a topological vector space (over $\mathbb{R}$) and let U and V be any two nonempty disjoint open convex subsets of E.

    Show that there exists a closed affine hyperplane H in E which strictly separates U and V, i.e., U and V lie in the distinct open half-spaces determined by H.

    For this purpose, consider in E the nonempty open convex set

    $$U - V = \{u - v : u \in U, \ v \in V\}$$

    and apply to $U - V$ and $\{0\}$ the result of Exercise 2.

18 Positive Linear Functionals

4. Let E be a topological vector space (over $\mathbb{R}$), let U be a nonempty open convex subset of E, and let K be a nonempty compact convex subset of E such that $U \cap K = \emptyset$.

   Prove that there exists a closed affine hyperplane H in E which separates U and K, i.e., U and K lie in the distinct closed half-spaces determined by H.

   For this purpose, apply to $U - K$ and $\{0\}$ the result of Exercise 2.

   Give an example which illustrates the situation where there is no affine hyperplane in E strictly separating U and K.

5. Let E be a topological vector space over $\mathbb{R}$ such that there exists a fundamental system of open convex neighborhoods of the neutral element 0 of E (the latter means the local convexity of E).

   Let X be a nonempty closed convex set in E and let Y be a nonempty compact convex set in E such that $X \cap Y = \emptyset$.

   Show that there exists a closed affine hyperplane H in E which strictly separates X and Y.

   For this purpose, first establish that there exists an open convex neighborhood $V(0)$ of 0 satisfying the relation

   $$(V(0) + X) \cap (V(0) + Y) = \emptyset.$$

   Then use for $(V(0) + X) - (V(0) + Y)$ and $\{0\}$ the result of Exercise 2.

6. Let E be a finite-dimensional topological vector space (over $\mathbb{R}$) and let F_1 and F_2 be two disjoint nonempty closed convex sets in E.

   Demonstrate that there exists an affine hyperplane in E separating F_1 and F_2.

   Give an example which illustrates the situation where there is no affine hyperplane in E strictly separating F_1 and F_2.

   Starting with the above result, infer that if A and B are any two nonempty convex sets in E without common points, then there exists an affine hyperplane in E separating A and B.

7. Let E be an infinite-dimensional separable Hilbert space over $\mathbb{R}$.

   Verify that there are two nonempty disjoint closed convex sets X and Y in E such that the set

   $$X - Y = \{x - y : x \in X, \ y \in Y\}$$

   is everywhere dense in E.

   Conclude that there exists no closed affine hyperplane in E separating X and Y.

8. Deduce the standard analytic form of the Hahn–Banach theorem (i.e., Theorem 18.3 of this chapter) from its geometric form (i.e., from the result of Exercise 2).

9. Let (E, C) be a real partially pre-ordered topological vector space, and suppose that the convex cone C has an interior point and $C \neq E$.

Show that there exists a nonzero linear continuous functional on E which is positive on C.

For this purpose, use the geometric form of the Hahn–Banach theorem.

10. Let $(E, \|\cdot\|)$ be an infinite-dimensional normed vector space over $\mathbb{R}$ and let L be a maximal (by inclusion) linearly independent subset of E (L is usually called a *linear basis* of E or an *algebraic basis* of E).

    Prove that there exists a base of the topological space E whose cardinality does not exceed card(L).

    For this purpose, first observe that there is an everywhere dense subset of E whose cardinality does not exceed card(L).

11. Let $(E, \|\cdot\|)$ be an infinite-dimensional normed vector space over the field $\mathbb{R}$.

    Demonstrate that there exists a partition $\{A, B\}$ of E into two everywhere dense convex subsets of E.

    For this purpose, keep in mind Exercise 10 and construct the required partition $\{A, B\}$ of E by using the method of transfinite recursion.

12. Consider the topological product vector space $E = \mathbb{R}^\mathbb{N}$ of all real-valued infinite sequences and show that:

    (a) E is a locally convex space;
    (b) E is separable and metrizable by a complete metric (i.e., E is a *Polish space*);
    (c) E does not admit any norm compatible with its topology.

    To establish (c), suppose that E admits a norm $\|\cdot\|$ producing the same topology. Then take in E any open ball

    $$B(x, r_1) = \{z \in E : \|z - x\| < r_1\},$$

    where $x \in E$ and $r_1 > 0$, and find another ball $B(y, r_2)$ in E such that, for every $t > 0$, the ball $B(y, tr_2)$ does not entirely contain $B(x, r_1)$. However, this circumstance contradicts the well-known properties of a norm.

    Remark 18.1 In connection with (c) of Exercise 12, see also [104], where the general theorem (due to Kolmogorov) is presented which gives a simple necessary and sufficient condition for a topological vector space to be isomorphic to a normed vector space.

13. Let $(E, \|\cdot\|)$ and $(F, \|\cdot\|)$ be two normed vector spaces over $\mathbb{R}$ and let A be a linear operator acting from E into F.

    Prove that the following two assertions are equivalent:

    (i) A is a continuous operator;
    (ii) for any linear continuous functional $g : F \to \mathbb{R}$, the composition $g \circ A$ is continuous.

    The implication (i) $\Rightarrow$ (ii) is trivial. To demonstrate the validity of the converse implication, use the following consequence of the Banach–Steinhaus theorem:

18 Positive Linear Functionals

If $\{y_n : n \in \mathbb{N}\}$ is an unbounded sequence of elements of F, then there exists a linear continuous functional $h : F \to \mathbb{R}$ such that $\{h(y_n) : n \in \mathbb{N}\}$ is an unbounded sequence in $\mathbb{R}$.

Remark 18.2 Let $(E, || \cdot ||)$ be a real normed vector space, F be a Polish topological vector space (over $\mathbb{R}$), and let A be a linear operator from E into F. Then, in general, the equivalence of the assertions (i) and (ii) of Exercise 13 fails to be true (cf., for instance, Exercise 6 of Appendix A).

Chapter 19
The Nonexistence of Universal Countably Additive Measures

We have mentioned several times Cantor's classical theorem which states that the strict inequality

$$\text{card}(E) < \text{card}(\mathcal{P}(E))$$

holds true for an arbitrary set E (see Chap. 3). In particular, taking $E = \mathbb{N}$, we obtain the existence of uncountable sets. The most important of them is the real line $\mathbb{R}$, whose cardinality is denoted by $\mathbf{c}$, and we have

$$\mathbf{c} = \text{card}(\mathcal{P}(\mathbb{N})) = \text{card}(\mathbb{N}^{\mathbb{N}}).$$

The least infinite cardinality ($= \text{card}(\mathbb{N})$) is usually denoted by ω and the least uncountable cardinality is denoted by ω_1. According to the definition of von Neumann's ordinal numbers (see, e.g., [76, 108, 111]), ω_1 is the set of all at most countable ordinals.

Actually, ω_1 can be identified with any well-ordered set $(E, \leq)$ having the following two properties:

(i) E is uncountable;
(ii) for each element $x \in E$, the initial interval $\{y \in E : y < x\}$ is at most countable.

Both cardinalities $\mathbf{c}$ and ω_1 are objects of **ZF** theory, i.e., are defined effectively—without the aid of the Axiom of Choice. Using **AC**, it is not difficult to demonstrate that ω_1 is indeed the smallest uncountable cardinal number, which means that if X is an arbitrary uncountable set, then the inequality $\omega_1 \leq \text{card}(X)$ holds and, in particular, we obviously get $\omega_1 \leq \mathbf{c}$.

On the other hand, it should be underlined that it is impossible to compare $\mathbf{c}$ and ω_1 within the framework of **ZF** theory. More precisely, the disjunction

$$\omega_1 \leq \mathbf{c} \lor \mathbf{c} \leq \omega_1$$

cannot be established in **ZF** theory (note that the same disjunction cannot be deduced even in the stronger **ZF** & **DC** theory).

It is only possible to prove in **ZF** that there exists a surjection from $\mathbf{c}$ onto ω_1 (in this connection, see Exercise 7).

Let E be a ground set with $\mathrm{card}(E) = \omega_1$. Our goal is to demonstrate, following Ulam's classical work [180], that one cannot define a nonzero finite (or σ-finite) countably additive measure μ on E such that $\mathrm{dom}(\mu) = \mathcal{P}(E)$ and $\mu(\{x\}) = 0$ for all elements $x \in E$ (cf. also [9]).

This circumstance is in some contrast with the measure extension theorems presented in Chaps. 16, 17, and 18.

In order to show the validity of Ulam's result, we need two lemmas.

Lemma 19.1 *Let μ be a finite finitely additive measure on a base set E and let $\{X_i : i \in I\}$ be a disjoint uncountable family of μ-measurable subsets of E.*

Then the set $\{i \in I : \mu(X_i) > 0\}$ is at most countable.

Proof Suppose otherwise, i.e., the set $J = \{i \in I : \mu(X_i) > 0\}$ is uncountable. For each natural number $k > 0$, let

$$J_k = \{i \in J : \mu(X_i) \geq 1/k\}.$$

Evidently, there exists a number $k_0 \in \mathbb{N} \setminus \{0\}$ such that the set J_{k_0} is also uncountable. Therefore, for every finite subfamily $\{X_{i_1}, X_{i_2}, \ldots, X_{i_n}\}$ of $\{X_i : i \in J_{k_0}\}$, we may write

$$\mu(X_{i_1} \cup X_{i_2} \cup \cdots \cup X_{i_n}) \geq n/k_0,$$

because the sets $X_{i_1}, X_{i_2}, \ldots, X_{i_n}$ are pairwise disjoint. Taking n sufficiently large and remembering that $\mu(E) < +\infty$, we come to a contradiction.

The obtained contradiction completes the proof. $\square$

Lemma 19.2 *If E is a base set of cardinality ω_1, then there exists a family*

$$\{E_{\xi,\zeta} : \xi < \omega, \ \zeta < \omega_1\}$$

of subsets of E, satisfying the following two conditions:

(a) *for each $\xi < \omega$, the partial family $\{E_{\xi,\zeta} : \zeta < \omega_1\}$ is disjoint;*
(b) *for each $\zeta < \omega_1$, the set $E \setminus \cup \{E_{\xi,\zeta} : \xi < \omega\}$ is at most countable.*

Proof First of all, we identify E with ω_1 by some bijection between these two sets, i.e., we put $E = \omega_1$. Then, for every $\eta < \omega_1$, we pick an injective mapping $f_\eta : [0, \eta] \to \omega$. The existence of an injection f_η is without any doubt, because the interval $[0, \eta]$ is at most countable.

Further, let us define the sets $E_{\xi,\zeta}$ ($\xi < \omega$, $\zeta < \omega_1$) by the formula

$$E_{\xi,\zeta} = \{\eta < \omega_1 : \zeta \leq \eta, \ f_\eta(\zeta) = \xi\}$$

and let us check that the family $\{E_{\xi,\zeta} : \xi < \omega, \ \zeta < \omega_1\}$ is as required.

To show (a), fix an ordinal $\xi < \omega$. Let $\zeta < \omega_1$ and $\zeta' < \omega_1$ be two distinct ordinals. Suppose for a moment that there is an ordinal η belonging to $E_{\xi,\zeta} \cap E_{\xi,\zeta'}$. Then we must have

$$f_\eta(\zeta) = f_\eta(\zeta') = \xi,$$

which contradicts the injectivity of f_η. So (a) holds.

To show (b), it suffices to observe that, for $\zeta < \omega_1$, the inclusion

$$\omega_1 \setminus \cup\{E_{\xi,\zeta} : \xi < \omega\} \subset [0, \zeta]$$

holds true, whence it follows that the set $\omega_1 \setminus \cup\{E_{\xi,\zeta} : \xi < \omega\}$ is at most countable, which proves (b).

Lemma 19.2 has thus been established. □

Remark 19.1 The family of subsets of E indicated in Lemma 19.2 is usually called *Ulam's transfinite matrix* corresponding to E. Ulam proved that there are certain analogs of his matrix for many other uncountable cardinal numbers (see [180]). Actually, his result stimulated the development of combinatorial set theory and of the theory of large cardinals (for further more detailed information, see [76, 80, 108, 111]).

Now, we are ready to prove the following fundamental result, also due to Ulam.

Theorem 19.1 *If* $\mathrm{card}(E) = \omega_1$, *then there exists no nonzero finite (or σ-finite) countably additive measure μ on E such that* $\mathrm{dom}(\mu) = \mathcal{P}(E)$ *and* $\mu(\{x\}) = 0$ *for all elements $x \in E$.*

Proof Let a family $\{E_{\xi,\zeta} : \xi < \omega, \ \zeta < \omega_1\}$ be as in Lemma 19.2 and let μ be a nonzero finite countably additive measure on E such that $\mu(\{x\}) = 0$ for each element $x \in E$.

We are going to show that $\mathrm{dom}(\mu) \neq \mathcal{P}(E)$.

Suppose to the contrary that all sets $E_{\xi,\zeta}$ are μ-measurable. Then condition (b) of Lemma 19.2 gives us that, for an arbitrary ordinal number $\zeta < \omega_1$, there is an ordinal $\xi(\zeta) < \omega$ such that $\mu(E_{\xi(\zeta),\zeta}) > 0$. Since the trivial inequality $\omega < \omega_1$ holds, the latter fact implies that there exists an ordinal $\xi < \omega$ satisfying the relation

$$\mathrm{card}(\{\zeta < \omega_1 : \xi = \xi(\zeta)\}) > \omega.$$

Consequently, defining

$$\Xi = \{\zeta < \omega_1 : \xi = \xi(\zeta)\},$$

we come to the uncountable disjoint family $\{E_{\xi,\zeta} : \zeta \in \Xi\}$ of μ-measurable sets, all of which have strictly positive μ-measure. But the existence of such a family contradicts Lemma 19.1.

The obtained contradiction ends the proof of Theorem 19.1. □

The principal result formulated in the above theorem implies the following.

Theorem 19.2 *Let E be a ground set of cardinality ω_1 and let μ be a nonzero finite (σ-finite) countably additive measure defined on some σ-algebra of subsets of E and vanishing at all singletons in E.*

Then the following two assertions hold true:

(1) $\mathrm{dom}(\mu) \neq \mathcal{P}(E)$;
(2) *there exists a countably additive measure ν on E properly extending the measure μ.*

Proof The validity of (1) is a direct consequence of Theorem 19.1.

To demonstrate the validity of (2), take any subset X of E which does not belong to $\mathrm{dom}(\mu)$. According to Theorem 17.4 from Chap. 17, there exists a countably additive measure ν on E extending μ and such that $X \in \mathrm{dom}(\nu)$. Now, it is clear that ν properly extends μ, so (2) is fulfilled. □

Remark 19.2 We thus see that, for a set E with $\mathrm{card}(E) = \omega_1$, the problem of defining a nonzero finite (σ-finite) countably additive measure on $\mathcal{P}(E)$ vanishing at all singletons in E has a negative solution. In other words, there exists no nonzero σ-finite universal countably additive measure on E vanishing at all singletons of E. At the same time, any nonzero σ-finite countably additive measure μ on E such that $\mu(\{x\}) = 0$ for each $x \in E$ admits a countably additive extension which is defined on some σ-algebra of subsets of E properly containing $\mathrm{dom}(\mu)$. Obviously, this process of extending μ can be continued finitely many times. But, in general, the same process cannot be continued countably many times, because the countable additivity of the final measure may be lost (however, the finite additivity is always preserved).

Developing naive (intuitive) set theory, Cantor came to his famous Continuum Hypothesis, according to which one has the equality $\mathrm{card}(\mathbb{R}) = \omega_1$, or, equivalently, the equality $\mathrm{card}([0, 1]) = \omega_1$. Many works were then devoted to this hypothesis (usually denoted by **CH**). In those works their authors intended to infer **CH** from the axioms of set theory, but all attempts in this direction turned out to be unsuccessful. At last, in 1939 K. Gödel proved that **CH** does not contradict the axioms of **ZFC** theory, so all the results obtained by using **CH** are consistent with **ZFC**. Finally, in 1963 P. Cohen proved that the negation of **CH** also does not contradict **ZFC** and, therefore, **CH** is independent of **ZFC** theory (for more details, see [76] or [108]).

Remark 19.3 There are various versions of **CH** and we would like to mention some of them below. For example, let us present four such versions.

(1) $2^\omega = \omega_1$.

19 The Nonexistence of Universal Countably Additive Measures   191

(2) For every infinite (in the standard sense) subset X of $\mathbb{R}$, either X is equinumerous with $\mathbb{N}$ or X is equinumerous with $\mathbb{R}$.
(3) For every infinite (in the sense of Bolzano and Dedekind) subset Y of $\mathbb{R}$, either Y is equinumerous with $\mathbb{N}$ or Y is equinumerous with $\mathbb{R}$.
(4) There exists no subset Z of $\mathbb{R}$ such that $\omega < \text{card}(Z) < \mathbf{c}$.

It is not difficult to show in **ZF** theory that the assertions (2), (3), (4) are mutually equivalent and that (1) implies each of (2), (3), and (4).

Similarly to (4), one can formulate the so-called Generalized Continuum Hypothesis (**GCH**):

For any cardinal $\mathbf{a} \geq \omega$, there is no cardinal $\mathbf{b}$ satisfying the inequalities

$$\mathbf{a} < \mathbf{b} < 2^{\mathbf{a}}.$$

It should be mentioned that, according to a theorem of Sierpiński, in **ZF** theory the above form of **GCH** implies **AC** (see, e.g., [167], where a stronger result is also presented). In **ZFC** theory the usual formulation of **GCH** is as follows:

For every cardinal $\mathbf{a} \geq \omega$, the equality $2^{\mathbf{a}} = \mathbf{a}^{+}$ holds, where $\mathbf{a}^{+}$ is the least cardinal strictly greater than $\mathbf{a}$.

Combining version (1) of **CH** with Ulam's Theorem 19.1, we get the next theorem.

Theorem 19.3 *Under the Continuum Hypothesis, there exists no nonzero finite countably additive measure defined on the algebra $\mathcal{P}([0, 1])$ and vanishing at all singletons in $[0, 1]$.*

In other words, under certain set-theoretical assumptions about the cardinality of the continuum $\mathbf{c}$, there is no nonzero finite countably additive universal measure on $[0, 1]$ vanishing at all singletons in $[0, 1]$.

Remark 19.4 On the other hand, the following important fact should be mentioned. If one assumes the existence of very large cardinal numbers, then a nonzero finite countably additive universal measure on $[0, 1]$, vanishing at all one-element subsets of $[0, 1]$, can exist. Moreover, it may happen that such a measure extends the standard Lebesgue measure λ on $[0, 1]$ (see [76]).

Exercises

1. A subset L of $\mathbb{R}$ is called a *Luzin set* if L is uncountable and, for every first category subset X of $\mathbb{R}$, one has $\text{card}(L \cap X) \leq \omega$ (see, e.g., [109, 118, 132, 141, 166]).
   Verify that:
   (a) any uncountable subset of a Luzin set is a Luzin set;
   (b) the union of a nonempty countable family of Luzin sets is a Luzin set;

(c) if L is a Luzin set and Z is a countable subset of $\mathbb{R}$, then $L \cup Z$ is also a Luzin set;
(d) if $f : \mathbb{R} \to \mathbb{R}$ is a homeomorphism and L is a Luzin set, then $f(L)$ is also a Luzin set;
(e) if L is a Luzin set in $\mathbb{R}$, then any first category subset of the space L (endowed with the induced topology) is at most countable.

2. Demonstrate that, under the Continuum Hypothesis $\mathbf{c} = \omega_1$, there exists a Luzin set on $\mathbb{R}$ (Luzin's theorem).

For this purpose, argue as follows. From the beginning, take into account that, for every first category set $X \subset \mathbb{R}$, there is a first category set $Y \subset \mathbb{R}$ containing X and representable as the union of countably many closed nowhere dense subsets of $\mathbb{R}$.

Keeping in mind the assumption $\mathbf{c} = \omega_1$, let $\{Y_\xi : \xi < \omega_1\}$ be the family of all closed nowhere dense sets in $\mathbb{R}$.

Using the method of transfinite recursion and remembering the fact that $\mathbb{R}$ is not of first category in itself, define a family $\{z_\xi : \xi < \omega_1\}$ of points of $\mathbb{R}$ such that

$$(\forall \xi < \omega_1)(z_\xi \notin (\cup\{Y_\zeta : \zeta < \xi\} \cup \{z_\zeta : \zeta < \xi\})).$$

Then put $L = \{z_\xi : \xi < \omega_1\}$ and check that L is a Luzin set in $\mathbb{R}$.

Remark 19.5 The reader can see that, using an argument completely analogous to the one presented in Exercise 2, it becomes possible to construct (under **CH**) a Luzin set entirely contained in the unit segment $[0, 1]$ or, more generally, entirely contained in a given subset of $\mathbb{R}$ of second Baire category.

3. Let E be a topological space and let X be a subset of E having the following two properties:

   (i) X is at most countable and everywhere dense in E;
   (ii) for each point $x \in X$, there exists a countable family $\{U_k(x) : k \in \mathbb{N}\}$ of neighborhoods of x such that $\{x\} = \cap\{U_k(x) : n \in \mathbb{N}\}$.

Let μ be a finite countably additive measure on E such that all open subsets of E (equivalently, all closed subsets of E) belong to the domain of μ, and μ vanishes at all points of X.

Prove that there exists a subset Z of E satisfying the following conditions:

(a) Z is of first category in E and is representable as the union of countably many closed sets in E;
(b) $\mu(E \setminus Z) = 0$ (in other words, Z is a first category support of μ).

For this purpose, define

$$X = \{x_n : n \in \mathbb{N}\}.$$

Take a natural number k. For each point x_n, find an open neighborhood $V_k(x_n)$ of x_n such that $\mu(V_k(x_n)) < 1/2^{k+n}$. Then define

$$V_k = \cup\{V_k(x_n) : n \in \mathbb{N}\}.$$

The set V_k is open and everywhere dense in E and $\mu(V_k) \leq 1/2^{k-1}$.
Further, putting

$$Y = \cap\{V_k : k \in \mathbb{N}\},$$

observe that the subset Y of E is of μ-measure zero. According to the definition of Y, the set $Z = E \setminus Y$ is a first category subset of E representable as the union of countably many nowhere dense closed sets, and the relation

$$\mu(E \setminus Z) = \mu(Y) = 0$$

is fulfilled. Thus, Z is a required first category support of the measure μ.
4. Let L be an arbitrary Luzin subset of $\mathbb{R}$.

Show that the following two assertions hold:

(a) $\mu^*(L) = 0$ for any finite (σ-finite) countably additive measure μ on $\mathbb{R}$ such that all open subsets of $\mathbb{R}$ belong to the domain of μ and $\mu(\{x\}) = 0$ whenever $x \in \mathbb{R}$;
(b) L is *universal measure zero*, i.e., if ν is any finite (σ-finite) countably additive measure on the topological space L such that all open subsets of L belong to dom(ν) and $\nu(\{z\}) = 0$ whenever $z \in L$, then ν is identically equal to zero.

Argue as follows. First observe that assertions (a) and (b) are equivalent. Hence it suffices to prove only the second assertion. Let ν be an arbitrary finite measure given on L and satisfying the assumptions of (b). Because L is a separable metric space, the result of Exercise 3 is applicable to L and ν. So there exists a first category subset Z of L on which ν is concentrated, i.e., $\nu(L \setminus Z) = 0$. Now, keep in mind that each first category subset of L is at most countable (see Exercise 1). In particular, card(Z) $\leq \omega$ and, consequently, $\nu(Z) = 0$.

Therefore, $\nu(L) = \nu(L \setminus Z) + \nu(Z) = 0$, as required.
5. Using the properties of Luzin sets indicated in Exercise 4, infer that under **CH** there exists no nonzero σ-finite countably additive universal measure on $[0, 1]$ vanishing at all one-element subsets of $[0, 1]$.

For this purpose, consider any bijection of a Luzin set L onto $[0, 1]$.

Remark 19.6 We thus see that Luzin sets on the unit segment $[0, 1]$ enable one to solve negatively (under **CH**) the problem of the existence of a nonzero finite countably additive universal measure on $[0, 1]$ vanishing at all one-element subsets of $[0, 1]$. Clearly, Luzin's method differs essentially from Ulam's purely set-theoretical method described in this chapter. Actually, Luzin's approach shows us that, assuming **CH**, there exists a countable family $\{X_i : i \in I\}$ of sets in $[0, 1]$ such that, for every nonzero finite (σ-finite) countably additive measure μ on $[0, 1]$ which vanishes at all singletons of $[0, 1]$, the inclusion

$$\{X_i : i \in I\} \subset \text{dom}(\mu)$$

is impossible. Indeed, if L is a Luzin set on $[0, 1]$ and $f : L \to [0, 1]$ is a bijection, then the role of $\{X_i : i \in I\}$ can be successfully carried out by the family $\{f(U_i) : i \in I\}$, where $\{U_i : i \in I\}$ is a countable base of the topological space L.

6. Work in **ZF** theory and prove Cantor's classical theorem stating that every at most countable linearly ordered set $(E, \preceq)$ has an isomorphic copy in $(\mathbb{Q}, \leq)$, where $\leq$ denotes the standard linear ordering of the set $\mathbb{Q}$ of all rational numbers.

   For this purpose, use the density of $\leq$, keep in mind the fact that there is neither a least nor a greatest element in $\mathbb{Q}$, and construct by induction a monomorphism of $(E, \preceq)$ into $(\mathbb{Q}, \leq)$.

7. Work in **ZF** theory and show that there exists a surjection from $\mathbb{R}$ onto ω_1.

   For this purpose, apply the result of Exercise 6 to the family of all at most countable ordinal numbers.

8. Let $\mathcal{M}$ denote the family of all those nonzero σ-finite countably additive measures on $\mathbb{R}$ which vanish at all singletons in $\mathbb{R}$.

   Assuming **CH**, demonstrate that there exists a function $f : \mathbb{R} \to \mathbb{R}$ which is absolutely nonmeasurable with respect to $\mathcal{M}$.

   For this purpose, consider some Luzin set L on $\mathbb{R}$ and take as f any injection of $\mathbb{R}$ into $\mathbb{R}$ such that $\text{ran}(f) = L$.

Remark 19.7 A Hausdorff topological space E is called a *Luzin space* if E is uncountable and the family of all first category subsets of E coincides with the family of all at most countable subsets of E. Obviously, a standard Luzin set on $\mathbb{R}$ which is everywhere dense in $\mathbb{R}$ can be considered as an example of a Luzin space. Any metrizable Luzin space without isolated points is separable and of universal measure zero (cf. Exercise 3 of this chapter). At the same time, there exists an uncountable everywhere dense subset of $\mathbb{R}$ which is universal measure zero, but is not a Luzin space (with respect to the induced topology).

9. Let E be an infinite set of cardinality $\mathbf{a}^+$, where $\mathbf{a}^+$ denotes the least cardinal number strictly greater than a cardinal number $\mathbf{a}$. Both $\mathbf{a}$ and $\mathbf{a}^+$ are identified with the corresponding initial ordinal numbers.

   Show that there exists a family $\{E_{\xi,\zeta} : \xi < \mathbf{a}, \zeta < \mathbf{a}^+\}$ of subsets of E, satisfying the following two conditions:

   (a) for each $\xi < \mathbf{a}$, the partial family $\{E_{\xi,\zeta} : \zeta < \mathbf{a}^+\}$ is disjoint;
   (b) for each $\zeta < \mathbf{a}^+$, the set $E \setminus \cup\{E_{\xi,\zeta} : \xi < \mathbf{a}\}$ has cardinality not exceeding $\mathbf{a}$.

19 The Nonexistence of Universal Countably Additive Measures

To establish the existence of the required family, argue similarly to the proof of Lemma 19.2.

10. A cardinal number **a** is called *measurable in Ulam's sense* (or *real-valued measurable*) if there exists a nonzero σ-finite countably additive measure μ on the algebra $\mathcal{P}(\mathbf{a})$ such that $\mu(\{x\}) = 0$ for every $x \in \mathbf{a}$.

    Demonstrate that:

    (i) if **a** is not measurable in Ulam's sense and $\mathbf{b} < \mathbf{a}$, then **b** also is not measurable in Ulam's sense;

    (ii) if **a** is not measurable in Ulam's sense, then $\mathbf{a}^+$ is not measurable in Ulam's sense;

    (iii) if $\{\mathbf{a}_i : i \in I\}$ is a family of cardinals such that all $\mathbf{a}_i$ ($i \in I$) and card(I) are not measurable in Ulam's sense, then the cardinal sum

    $$\mathbf{a} = \sum \{\mathbf{a}_i : i \in I\}$$

    is not measurable in Ulam's sense.

    For this purpose, use Exercise 9 and the method of transfinite induction.

11. Recall that an uncountable ordinal (cardinal) number ω_α is *weakly inaccessible* if its index α is a limit ordinal (i.e., $\alpha \neq \beta + 1$ for each ordinal $\beta < \alpha$) and ω_α itself is regular, i.e., there exists no representation

    $$\omega_\alpha = \sum \{\mathbf{a}_i : i \in I\},$$

    where all cardinals $\mathbf{a}_i$ ($i \in I$) and card(I) are strictly less than ω_α.

    Keeping in mind Exercise 10, verify that all those cardinal numbers which are strictly less than the first weakly inaccessible cardinal, are not measurable in Ulam's sense.

Remark 19.8 A cardinal number **a** is called *strongly inaccessible* if **a** is weakly inaccessible and, for any cardinal $\mathbf{b} < \mathbf{a}$, one has $2^\mathbf{b} < \mathbf{a}$. It is easy to check that, under **GCH**, every weakly inaccessible cardinal turns out to be strongly inaccessible. Since **GCH** does not contradict **ZFC** theory, it is consistent to assume that all those cardinals which are strictly less than the first strongly inaccessible cardinal are not measurable in Ulam's sense.

Taking into account that the existence of strongly inaccessible cardinal numbers cannot be established within **ZFC** theory (for more details, see [76] or [108]), one can conclude that the theory

ZFC & there exists no cardinal measurable in Ulam's sense

is equiconsistent with **ZFC** theory.

12. Show in **ZF** theory that the assertions (2), (3), (4) of Remark 19.3 are mutually equivalent.

For this purpose, use Exercise 14 from Chap. 4.
13. Let $(P, \preceq)$ be an arbitrary nonempty partially ordered set. A nonempty set $G \subset P$ is called a *filter* in $(P, \preceq)$ if

$$(\forall p \in G)(\forall q \in P)(p \preceq q \Rightarrow q \in G),$$

$$(\forall p \in G)(\forall q \in G)(\exists r \in G)(r \preceq p \ \& \ r \preceq q).$$

Work in **ZF** & **DC** theory and prove the Rasiowa–Sikorski lemma stating that if $\mathcal{D}$ is a countable family of coinitial subsets of P and $p \in P$, then there exists a filter G intersecting all members of $\mathcal{D}$ and such that $p \in G$.

For this purpose, let $\mathcal{D} = \{D_n : n \in \mathbb{N}\}$ and define by recursion a sequence $\{p_n : n \in \mathbb{N}\}$ of elements of P. First, take as p_0 any element of the nonempty set

$$\{q \in D_0 : q \preceq p\}.$$

Further, supposing that p_n has already been defined, take as p_{n+1} any element of the nonempty set

$$\{q \in D_{n+1} : q \preceq p_n\}.$$

Finally, introduce the set $G = \{q \in P : (\exists n \in \mathbb{N})(p_n \preceq q)\}$ and check that G is as required.

14. Demonstrate that the following two sentences are equivalent:
    (i) the Continuum Hypothesis ($2^\omega = \omega_1$);
    (ii) for any nonempty partially ordered set $(P, \preceq)$ and for every family $\mathcal{D}$ of coinitial subsets of P with card($\mathcal{D}$) < **c**, there exists a filter $G \subset P$ which intersects each member of $\mathcal{D}$.

Argue as follows. First, observe that the implication (i) $\Rightarrow$ (ii) is an immediate consequence of Exercise 13.

To establish the converse implication (ii) $\Rightarrow$ (i), suppose that (ii) holds, but $\omega_1 < $ **c**. Consider the partially ordered set $(P, \preceq)$, where P is the family of all those functions whose domains are finite subsets of ω and whose ranges are subsets of ω_1, and $p \preceq q$ means that p extends q. For any $\xi < \omega_1$, define

$$D_\xi = \{p \in P : \xi \in \mathrm{ran}(p)\}.$$

Verify that all sets D_ξ ($\xi < \omega_1$) are coinitial in P. By virtue of (ii), there exists a filter G in P which has common elements with any D_ξ. Infer from this that $\cup G$ is a surjection from ω onto ω_1. Since the latter is impossible, the obtained contradiction yields the required result.

19 The Nonexistence of Universal Countably Additive Measures

Remark 19.9 Let $(P, \preceq)$ be a partially ordered set.

Two elements p and q of P are called *inconsistent* (*incompatible*) if there is no $r \in P$ such that $r \preceq p$ and $r \preceq q$.

By definition, a set $Q \subset P$ is *totally inconsistent* (*totally incompatible*) if any two distinct elements of Q are inconsistent (incompatible).

One says that $(P, \preceq)$ satisfies the *countable chain condition* (*ccc*) if every totally inconsistent subset of P is at most countable.

It makes sense to recall here that at present Martin's Axiom (abbreviation **MA**) is successfully used, instead of **CH**, in many topics of set-theoretical analysis, general topology, and universal algebra.

The standard formulation of **MA** is as follows:

If $(P, \preceq)$ is a nonempty partially ordered set satisfying the countable chain condition and $\mathcal{D}$ is a family of coinitial subsets of P with card($\mathcal{D}$) < **c**, then there exists a filter $G \subset P$ which intersects all members of $\mathcal{D}$.

The next statement (analogous to Baire's classical theorem on category) is a purely topological equivalent of **MA**:

If E is an arbitrary nonempty compact topological space satisfying the Suslin condition, then E cannot be covered by a family of nowhere dense subsets whose cardinality is strictly less than **c**.

Obviously, the Continuum Hypothesis implies Martin's Axiom (cf. Exercise 14). It has been proved that the assertion **MA** & ($\neg$**CH**) is consistent with **ZFC** set theory (see, for instance, [76, 108]). Furthermore, it has been shown that the size of **c** is not precisely determined by **MA** and can be arbitrarily large. So, **CH** is substantially stronger than **MA**. At the same time, **MA** enables one to establish that the cardinal **c** is regular and is not real-valued measurable (see [46], where numerous interesting and important consequences of **MA** are thoroughly discussed). Notice also that, under **MA**, the behavior of those sets whose cardinalities are strictly less than **c** is rather similar to the behavior of those sets which are at most countable. For example, let I be a set with card(I) < **c** and let $\{X_i : i \in I\}$ be a family of subsets of $\mathbb{R}$. Then, assuming **MA**, the following two assertions hold:

(a) if all X_i ($i \in I$) are of Lebesgue measure zero, then $\cup \{X_i : i \in I\}$ is of Lebesgue measure zero;
(b) if all X_i ($i \in I$) are of first category in $\mathbb{R}$, then $\cup \{X_i : i \in I\}$ is of first category in $\mathbb{R}$.

Chapter 20
Radon Measures

In this chapter we will be concerned with measures which are given on topological spaces and possess certain regularity properties.

Let E be a Hausdorff topological space and let μ be a finite finitely additive measure defined on some algebra $\mathcal{A}$ of subsets of E.

As usual, the members of $\mathcal{A}$ are called μ-measurable sets in E.

We say that μ is a *Radon measure* if, for any μ-measurable set X and for any real $\varepsilon > 0$, there exists a compact set $C \subset X$ such that

$$C \in \mathrm{dom}(\mu), \qquad \mu(X \setminus C) = \mu(X) - \mu(C) < \varepsilon.$$

Radon measures play an important role in different topics of mathematical analysis and probability theory (cf. [17, 20, 38, 48, 52]). In most cases, these measures are defined on the Borel σ-algebra $\mathcal{B}(E)$ of E.

Recall that $\mathcal{B}(E)$ is the σ-algebra of subsets of E generated by the family of all open (equivalently, by the family of all closed) sets in E.

A Hausdorff topological space is called *Radon* if every finite countably additive Borel measure on E is Radon.

The following theorem is due to A.D. Alexandrov (see, e.g., [17, 38]).

Theorem 20.1 *Let E be a Hausdorff topological space and let μ be a finite finitely additive Radon measure defined on some algebra $\mathcal{A}$ of subsets of E.*

Then μ is a countably additive measure on $\mathcal{A}$.

Proof Let $\{X_n : n \in \mathbb{N}\}$ be a countable disjoint family of μ-measurable sets such that $\cup \{X_n : n \in \mathbb{N}\}$ is also μ-measurable. By virtue of the finite additivity and monotonicity of μ, we always have the inequality

$$\Sigma\{\mu(X_n) : n \in \mathbb{N}\} \le \mu(\cup\{X_n : n \in \mathbb{N}\}).$$

So, it suffices to establish the reverse inequality

$$\mu(\cup\{X_n : n \in \mathbb{N}\}) \leq \Sigma\{\mu(X_n) : n \in \mathbb{N}\}.$$

For this purpose, let

$$X = \cup\{X_n : n \in \mathbb{N}\}$$

and take any real $\varepsilon > 0$. Applying the Radon property of μ to each set $E \setminus X_n$, we can find an open set $U_n \in \operatorname{dom}(\mu)$ such that

$$X_n \subset U_n, \qquad \mu(U_n) - \mu(X_n) < \varepsilon/2^{n+1}.$$

Let $C \in \operatorname{dom}(\mu)$ be a compact set in E which satisfies the relations

$$C \subset X, \qquad \mu(X \setminus C) < \varepsilon.$$

Since the family $\{U_n : n \in \mathbb{N}\}$ is an open covering of C, there is a finite subfamily $\{U_n : 0 \leq n \leq k\}$ of $\{U_n : n \in \mathbb{N}\}$ which also covers C. Then we may write

$$\mu(X) < \mu(C) + \varepsilon \leq \Sigma\{\mu(U_n) : 0 \leq n \leq k\}) + \varepsilon$$
$$\leq \Sigma\{\mu(X_n) : n \in \mathbb{N}\} + 2\varepsilon,$$

whence it follows, in view of the arbitrary smallness of ε, that

$$\mu(X) \leq \Sigma\{\mu(X_n) : n \in \mathbb{N}\},$$

which completes the proof of Alexandrov's theorem. □

Remark 20.1 The argument used in the proof of Theorem 20.1 essentially relies on corresponding topological properties of E, namely, on the notion of compactness for subsets of E. However, we do not need here the compactness concept in the standard sense, i.e., the existence of a finite subcovering of any open covering. For our purposes, it suffices to require that all countable open coverings of certain subsets of E contain finite subcoverings. Under this much weaker assumption, the assertion of Theorem 20.1 still remains valid in an appropriate reformulation (cf. Exercise 1).

Theorem 20.2 *Let E be a Hausdorff topological space and let μ be a finite finitely additive Radon measure defined on the Borel σ-algebra $\mathcal{B}(E)$ of E.*

Then μ is a countably additive Radon measure on $\mathcal{B}(E)$.

Proof Applying Theorem 20.1 to the given measure μ, we obtain at once the desired result. □

20 Radon Measures

Let E be a topological space and let μ be a finite finitely additive Borel measure on E (i.e., the domain of μ coincides with the Borel σ-algebra $\mathcal{B}(E)$ of E).

We shall say that μ is an *inner regular measure* if, for any set $X \in \mathcal{B}(E)$ and for each real $\varepsilon > 0$, there exists a closed set $Y \subset X$ such that

$$\mu(X) - \mu(Y) = \mu(X \setminus Y) < \varepsilon.$$

Analogously, we shall say that μ is an *outer regular measure* if, for any set $X \in \mathcal{B}(E)$ and for each real $\varepsilon > 0$, there exists an open set $Z \supset X$ such that

$$\mu(Z) - \mu(X) = \mu(Z \setminus X) < \varepsilon.$$

Naturally, we shall say that μ is a *regular measure* if μ is simultaneously inner regular and outer regular.

Lemma 20.1 *Let E be a topological space and let μ be a finite finitely additive Borel measure on E.*

Then the following three assertions are equivalent:

(1) *μ is an inner regular measure;*
(2) *μ is an outer regular measure;*
(3) *μ is a regular measure.*

Proof Obviously, it suffices to establish the equivalence (1) $\Leftrightarrow$ (2).

Suppose that μ is inner regular, X is a Borel set in E, and $\varepsilon > 0$. Then for the Borel set $E \setminus X$ there exists a closed set $Y \subset E \setminus X$ such that

$$\mu(E \setminus X) - \mu(Y) = \mu((E \setminus X) \setminus Y) < \varepsilon.$$

Clearly, the set $E \setminus Y$ is open in E and

$$X \subset E \setminus Y, \qquad \mu((E \setminus Y) \setminus X) < \varepsilon,$$

which shows us that μ is also an outer regular measure.

A quite analogous argument yields that if the given measure μ is outer regular, then the same μ is inner regular.

Lemma 20.1 has thus been proved. $\square$

Lemma 20.2 *Let E be a topological space in which every closed set is representable as the intersection of countably many open sets in E, and let μ be a finite countably additive Borel measure on E.*

Then μ is a regular measure on E.

Proof First note that, according to the assumption of this lemma, every open subset of E is representable as the union of countably many closed sets in E.

Consider the family $\mathcal{F}$ of all those Borel sets $X \subset E$ which satisfy the following three conditions:

(a) there are two sets $Y \subset E$ and $Z \subset E$ such that

$$Y \subset X \subset Z;$$

(b) Y is the union of countably many closed sets in E and Z is the intersection of countably many open sets in E;
(c) $\mu(X) = \mu(Y) = \mu(Z)$.

A straightforward verification shows that if U is an open set in E, then $U \in \mathcal{F}$, and if $X \in \mathcal{F}$, then $E \setminus X \in \mathcal{F}$.

Let us demonstrate that if $\{X_n : n \in \mathbb{N}\} \subset \mathcal{F}$, then $\cup \{X_n : n \in \mathbb{N}\} \in \mathcal{F}$. For this purpose, define

$$X = \cup\{X_n : n \in \mathbb{N}\}$$

and observe at once that there exists a countable family $\{Y_i : i \in I\}$ of closed subsets of E such that

$$\cup\{Y_i : i \in I\} \subset X, \qquad \mu(X) = \mu(\cup\{Y_i : i \in I\}).$$

At the same time, for every real $\varepsilon > 0$, there exists an open set U_n in E such that

$$X_n \subset U_n, \qquad \mu(U_n \setminus X_n) < \varepsilon/2^{n+1}.$$

Consequently, we may write $\cup\{X_n : n \in \mathbb{N}\} \subset \cup\{U_n : n \in \mathbb{N}\}$ and

$$\mu(\cup\{U_n : n \in \mathbb{N}\} \setminus \cup\{X_n : n \in \mathbb{N}\}) \leq \sum\{\mu(U_n \setminus X_n) : n \in \mathbb{N}\}$$
$$< \varepsilon(1/2 + 1/4 + 1/8 + \cdots) = \varepsilon.$$

Taking into account the arbitrary smallness of ε, we readily infer that there exists a countable family $\{Z_j : j \in J\}$ of open subsets of E such that

$$X \subset \cap\{Z_j : j \in J\}, \qquad \mu(X) = \mu(\cap\{Z_j : j \in J\}).$$

Therefore, we get $X = \cup\{X_n : n \in \mathbb{N}\} \in \mathcal{F}$.

From the above it immediately follows that $\mathcal{F}$ is a certain σ-algebra of subsets of E containing all open sets in E, so we finally come to the equality

$$\mathcal{F} = \mathcal{B}(E),$$

and μ is a regular measure in view of the conditions (a), (b), and (c).

This ends the proof of Lemma 20.2. □

Theorem 20.3 *Every finite countably additive Borel measure on a metric space (M, d) is regular.*

Proof The assertion of this theorem is a direct consequence of Lemma 20.2, because any closed subset of M is effectively representable as the intersection of countably many open sets in M. Indeed, if F is a nonempty closed set in M, then we have

$$F = \cap\{U_n : n \in \mathbb{N} \setminus \{0\}\},$$

where

$$U_n = \{x \in M : d(x, F) < 1/n\}$$

for each natural number $n > 0$. Here $d(x, F)$ denotes, as usual, the distance between x and F.

Theorem 20.3 is thus proved. □

For Polish topological spaces (i.e., for complete separable metric spaces), Theorem 20.3 can be substantially strengthened. Namely, we have the next theorem, due to Ulam.

Theorem 20.4 *Every finite countably additive Borel measure on a Polish space (M, d) is Radon.*

Proof Let μ be a finite countably additive Borel measure on (M, d). By virtue of Theorem 20.3, μ is regular. So it suffices to show that, for any closed set $F \subset M$ and for each real $\varepsilon > 0$, there exists a compact set $K \subset F$ satisfying the inequality

$$\mu(F \setminus K) < \varepsilon.$$

Consider the subspace (F, d_F) of (M, d). Since F is closed in M, the subspace (F, d_F) is Polish. For every nonzero natural number n, there exists a countable family $\{B_i : i \in I_n\}$ of closed balls in M such that

$$(\forall i \in I_n)(\operatorname{diam}(B_i) < 1/n),$$

$$F \subset \cup\{B_i : i \in I_n\}.$$

Further, there exists a finite set $J_n \subset I_n$ satisfying the inequality

$$\mu(F) - \mu(F \cap (\cup\{B_j : j \in J_n\})) < \varepsilon/2^n.$$

Now, we put $Y_n = F \cap (\cup\{B_j : j \in J_n\})$ and let

$$Y = \cap\{Y_n : n \in \mathbb{N} \setminus \{0\}\}.$$

Clearly, the set Y is compact, because it is totally bounded and closed in the complete metric space (M, d) (see, e.g., [35, 38, 40, 84, 104, 109]).

Also, $Y \subset F$ and

$$\mu(F - Y) \leq \sum \{\mu(F) - \mu(Y_n) : n \in \mathbb{N} \setminus \{0\}\}$$
$$< \varepsilon(1/2 + 1/4 + \cdots + 1/2^n + \cdots) = \varepsilon.$$

This completes the proof of Ulam's theorem. □

Remark 20.2 The natural question arises whether any finite countably additive Borel measure given on a complete metric space is Radon. It turns out that the answer to this question depends on additional axioms of set theory (in this connection, see [27, 95, 141]). In fact, the assumption that every complete metric space is Radon does not contradict **ZFC** set theory. More precisely, it has been established that the following two statements are equivalent in this theory:

(i) there exists no cardinal number measurable in Ulam's sense;
(ii) all complete metric spaces are Radon.

Remark 20.3 There are many examples of compact topological spaces which are not Radon. One of them is presented in Exercise 7 below.

Exercises

1. Let E be a topological space and let μ be a finite finitely additive measure defined on some algebra of subsets of E. Suppose that, for any μ-measurable set X and for any real $\varepsilon > 0$, there exists a closed countably quasi-compact set $C \subset X$ such that

$$C \in \text{dom}(\mu), \qquad \mu(X \setminus C) = \mu(X) - \mu(C) < \varepsilon.$$

Show that the measure μ is countably additive.

For this purpose, argue similarly to the proof of Theorem 20.1 of the present chapter.

2. Let E be a topological space in which every closed set is an intersection of countably many open sets (or, equivalently, in which every open set is a union of countably many closed sets).

Prove that the Borel σ-algebra $\mathcal{B}(E)$ coincides with the smallest (by inclusion) family $\mathcal{F}$ of subsets of E such that:

(a) all open sets belong to $\mathcal{F}$;
(b) if $\{X_n : n \in \mathbb{N}\}$ is any monotone (by inclusion) sequence of sets belonging to $\mathcal{F}$, then the set $\lim\{X_n : n \in \mathbb{N}\}$ also belongs to $\mathcal{F}$.

3. Let ω_1 denote, as usual, the least uncountable ordinal (cardinal) number, let the set

$$E = \omega_1 = [0, \omega_1[$$

of all at most countable ordinal numbers be equipped with its natural order topology, and let $\{F_n : n \in \mathbb{N}\}$ be an arbitrary countable family of unbounded from above (equivalently, uncountable) closed subsets of E.

Verify that the set $F = \cap\{F_n : n \in \mathbb{N}\}$ is also closed and unbounded from above in E.

Argue as follows. Without loss of generality, one may assume that, for each $n \in \mathbb{N}$, the equality

$$\text{card}(\{k \in \mathbb{N} : F_k = F_n\}) = \omega$$

holds true. Clearly, F is closed. To show the unboundedness from above of F, take any ordinal $\xi < \omega_1$ and define by recursion a strictly increasing sequence $\{\xi_n : n \in \mathbb{N}\}$ of countable ordinals so that

$$\xi_0 = \xi, \qquad (\forall n \in \mathbb{N})(\xi_n \in F_n).$$

Then check that the ordinal $\sup\{\xi_n : n \in \mathbb{N}\}$ belongs to F.

4. Show that the topological space $E = [0, \omega_1[$ has the following properties:
   (a) for each $\xi < \omega_1$, the subspace $[0, \xi]$ of E is compact and metrizable;
   (b) the space E is not metrizable;
   (c) there exists an uncountable disjoint family of nonempty open intervals in E, so E is not separable;
   (d) any compact subset of E is bounded in E (in particular, E itself is not compact);
   (e) the space E is locally compact and countably compact.

5. Consider the set $E' = \omega_1 + 1 = [0, \omega_1]$ equipped with its natural order topology.
   Demonstrate that E' can be treated as Alexandrov's one-point compactification of the space $E = [0, \omega_1[$ (for necessary information about Alexandrov's one-point compactification of a locally compact topological space, see e.g. [19, 40, 84]).

6. For every closed subset F of the space $E = [0, \omega_1[$, define

$$v(F) = \begin{cases} 0 & \text{if } F \text{ is bounded,} \\ 1 & \text{if } F \text{ is unbounded from above.} \end{cases}$$

Prove that the function v can be uniquely extended to a countably additive Borel measure v' on the space E and that this v' is not a Radon measure.

Remark 20.4 The measure v' of Exercise 6 is called the *Dieudonné measure* on E (see, e.g., [52, 84]).

7. Verify that the compact space $E' = [0, \omega_1]$ is not Radon.

   For this purpose, define a countably additive Borel measure μ on E' such that:

   (a) $\mu(\{\omega_1\}) = 0$;
   (b) μ induces the Dieudonné measure ν' on $E = [0, \omega_1[$.

   Conclude that this μ is not a Radon measure.

8. Let X be a ground set of cardinality ω_1, equipped with the discrete topology, and let X' denote Alexandrov's one-point compactification of X.

   Show that X' is a non-metrizable Radon topological space.

   For this purpose, take into account Ulam's theorem stating that there exists no nonzero finite countably additive measure defined on the power set $\mathcal{P}(X)$ of X and vanishing at all singletons in X (see Chap. 19).

9. Let E be a Hausdorff topological space and let μ be a finite countably additive Borel measure on E which satisfies the following condition:

   For every open set $U \subset E$ and for every real $\varepsilon > 0$, there exists a compact set $C \subset U$ such that

   $$\mu(U) - \mu(C) = \mu(U \setminus C) < \varepsilon.$$

   Prove that μ is a Radon measure on E.

   Argue step by step as follows.

   First, applying the above condition to $U = E$, reduce this exercise to the special case when E is a compact space. So assume below that E is compact.

   Then introduce several auxiliary notions.

   A set $X \in \text{dom}(\mu)$ is μ-*approximable from above* if, for any real $\varepsilon > 0$, there exists an open set $U \subset E$ such that

   $$X \subset U, \qquad \mu(U) - \mu(X) = \mu(U \setminus X) < \varepsilon.$$

   A set $Y \in \text{dom}(\mu)$ is μ-*approximable from below* if, for any real $\varepsilon > 0$, there exists a compact set $C \subset E$ such that

   $$C \subset Y, \qquad \mu(Y) - \mu(C) = \mu(Y \setminus C) < \varepsilon.$$

   A set $Z \in \text{dom}(\mu)$ is μ-*approximable* if Z is μ-approximable from above and from below.

   Starting with these definitions, check that:

   (a) every open set in E is μ-approximable (in view of the assumption on μ);
   (b) a set X is μ-approximable from above if and only if the set $E \setminus X$ is μ-approximable from below;
   (c) X is μ-approximable if and only if $E \setminus X$ is μ-approximable.

Let $\{X_n : n \in \mathbb{N}\}$ be an arbitrary countable family of μ-approximable sets in E. Let

$$X = \cup\{X_n : n \in \mathbb{N}\}$$

and take any real $\varepsilon > 0$. For each $n \in \mathbb{N}$, there exist an open subset U_n of E and a compact subset C_n of E, which fulfil the relations:

(d) $C_n \subset X_n \subset U_n$;
(e) $\mu(U_n \setminus C_n) < \varepsilon/2^{n+1}$.

Now, introduce the notation $U = \cup\{U_n : n \in \mathbb{N}\}$. Obviously, U is an open set in E and $X \subset U$. We have the inclusion

$$U \setminus X \subset \cup\{U_n \setminus C_n : n \in \mathbb{N}\},$$

from which it immediately follows that $\mu(U \setminus X) < \varepsilon$. This means that X is μ-approximable from above.

Further, we have the inclusions

$$X \setminus \cup\{C_n : n \in \mathbb{N}\} \subset U \setminus \cup\{C_n : n \in \mathbb{N}\} \subset \cup\{U_n \setminus C_n : n \in \mathbb{N}\},$$

which imply

$$\mu(X \setminus \cup\{C_n : n \in \mathbb{N}\}) < \varepsilon.$$

Then keep in mind the equality

$$\mu(X \setminus \cup\{C_n : n \in \mathbb{N}\}) = \lim_{n \to \infty} \mu(\{X \setminus \cup\{C_m : 0 \leq m \leq n\})$$

and infer that there exists a natural number n_0 such that

$$\mu(\{X \setminus \cup\{C_m : 0 \leq m \leq n_0\}) \leq 2\varepsilon.$$

Since $\cup\{C_m : 0 \leq m \leq n_0\}$ is a compact subset of E, the latter means that X is μ-approximable from below.

Summarizing all the above, conclude that the family of all μ-approximable sets is a σ-algebra containing all open sets in E, whence it follows that this family coincides with $\mathcal{B}(E)$. In particular, μ is a Radon measure.

Argue in an analogous manner and establish the following similar result:

Let E be a topological space and let μ be a finite countably additive Borel measure on E such that, for any open subset U of E and for any real $\varepsilon > 0$, there exists a closed subset Z of E satisfying the relations

$$Z \subset U, \qquad \mu(U \setminus Z) = \mu(U) - \mu(Z) < \varepsilon.$$

Then μ is a regular Borel measure on E.

10. Prove Henry's theorem stating that every finite finitely additive Radon measure ν on a Hausdorff topological space E can be extended to a countably additive measure ν' on E which also is Radon and contains in its domain the Borel σ-algebra $\mathcal{B}(E)$ of E.

Argue as follows. Consider the family of all those finitely additive measures μ which are defined on algebras of subsets of E, possess the Radon property, and extend ν. Verify that:

(a) this family has a maximal element ν' (with respect to the inclusion relation);
(b) ν' is a countably additive Radon measure on a σ-algebra of subsets of E;
(c) all closed subsets of E belong to dom(ν'), so all Borel subsets of E also belong to dom(ν').

For (a), use the Kuratowski–Zorn lemma.

For (b), use Theorem 20.1 of this chapter, which states that every finite finitely additive Radon measure is countably additive.

For (c), assume on the contrary that some closed subset X of E does not belong to dom(ν') and, using Marczewski's method described in Chap. 17, extend ν' to a Radon measure ν'' such that $X \in$ dom(ν''). But this contradicts the maximality of ν', so gives the required result.

11. Let x and y be any two distinct elements, let $E = \{x, y\}$, and let

$$\mathcal{T} = \{\emptyset, \{x\}, E\}.$$

For the two-element topological space $(E, \mathcal{T})$, define

$$\mu(\emptyset) = 0, \qquad \mu(\{x\}) = \mu(\{y\}) = 1, \qquad \mu(E) = 2.$$

Verify that the finite countably additive Borel measure μ on $(E, \mathcal{T})$ is not inner regular and, consequently, is not outer regular.

Chapter 21
Invariant and Quasi-Invariant Measures

In this chapter we consider some basic notions and general facts from the theory of invariant and quasi-invariant measures. More concrete examples of such measures and important results about them are discussed, with various applications, in many surveys, monographs and text-books (see, e.g., [17, 20, 63, 69, 95, 183, 188]).

Let E be a ground set and let G be a group of bijections of E onto itself (in other words, G is a subgroup of $(Sym(E), \circ)$).

In the sequel, we shall say that the pair (E, G) is a *space equipped with a transformation group*.

Two subsets X and Y of E are called G-*congruent* if there exists at least one $g \in G$ such that $g(X) = Y$.

Clearly, G-congruence for the subsets of E is an equivalence relation on the power set $\mathcal{P}(E)$.

In particular, we have an equivalence relation S_G on E canonically associated with G and defined as follows:

$$S_G(x, y) \Leftrightarrow (x \in E \ \& \ y \in E \ \& \ (\exists g \in G)(g(x) = y)).$$

The corresponding S_G-equivalence classes are usually called G-*orbits* of points of E.

We say that a group G is *transitive* on E or *acts transitively* on E if there is only one G-orbit in E (which necessarily coincides with E).

Obviously, we can say that G acts transitively on E if and only if all singletons in E are pairwise G-congruent.

A group G *acts freely* on E if, for each $g \in G \setminus \{Id_E\}$ and for each $x \in E$, one has $g(x) \neq x$. This means that, for any two distinct $g \in G$ and $h \in G$ and for any $x \in E$, the relation $g(x) \neq h(x)$ holds true.

Example 21.1 If G acts freely on E, then all G-orbits have the same cardinality (equal to $\operatorname{card}(G)$) and the formula

$$\operatorname{card}(E) = \operatorname{card}(G) \cdot \operatorname{card}(Orb_G)$$

is fulfilled, where Orb_G denotes the set of all G-orbits in E.

Example 21.2 The group of all translations of the Euclidean space $\mathbb{R}^n$ acts transitively and freely on $\mathbb{R}^n$. If $n \geq 1$, then the group of all isometric transformations of $\mathbb{R}^n$ is transitive, but does not act freely on $\mathbb{R}^n$. Also, if $n \geq 2$, then the group of all rotations of $\mathbb{R}^n$ about its origin is not transitive.

For a pair (E, G), the concepts of G-invariant and G-quasi-invariant measures are naturally introduced.

Let μ be a measure defined on some G-invariant algebra of subsets of E. It makes sense to recall (cf. Exercises 1 and 2 from Chap. 17) that the G-invariance of $\operatorname{dom}(\mu)$ is expressed by the formula

$$(\forall g \in G)(\forall X \in \operatorname{dom}(\mu))(g(X) \in \operatorname{dom}(\mu)).$$

We say that μ is a *G-quasi-invariant measure* if

$$(\forall g \in G)(\forall X \in \operatorname{dom}(\mu))(\mu(X) = 0 \Leftrightarrow \mu(g(X)) = 0).$$

We say that μ is a *G-invariant measure* if

$$(\forall g \in G)(\forall X \in \operatorname{dom}(\mu))(\mu(X) = \mu(g(X))).$$

Clearly, every G-invariant measure is also G-quasi-invariant, but the converse assertion is not true, in general.

The G-quasi-invariant measures considered below are usually assumed to be countably additive and defined on G-invariant σ-algebras of subsets of E.

Example 21.3 For a natural number $n > 0$, the standard Lebesgue measure λ_n on the Euclidean space $\mathbb{R}^n$ is invariant with respect to the group of all isometric transformations of $\mathbb{R}^n$. The same measure is quasi-invariant (but not invariant) with respect to the group of all affine transformations of $\mathbb{R}^n$.

Example 21.4 Two measures μ_1 and μ_2 on E are called *equivalent* if

$$\operatorname{dom}(\mu_1) = \operatorname{dom}(\mu_2), \qquad (\forall X \in \operatorname{dom}(\mu_1))(\mu_1(X) = 0 \Leftrightarrow \mu_2(X) = 0).$$

It directly follows from the definition of quasi-invariant measures that, for any two equivalent measures μ_1 and μ_2, the measure μ_1 is G-quasi-invariant if and only if the measure μ_2 is G-quasi-invariant.

21 Invariant and Quasi-Invariant Measures

Remark 21.1 Suppose that in (E, G) the group G consists only of the identity transformation of E. Then the concepts of G-quasi-invariant and G-invariant measures are reduced to the ordinary measures. Consequently, the general theory of quasi-invariant (invariant) measures includes the theory of ordinary measures.

Recall that a G-quasi-invariant measure μ on a ground set E is σ-*finite* if there exists a countable covering $\{X_i : i \in I\}$ of E such that

$$(\forall i \in I)(X_i \in \text{dom}(\mu) \ \& \ \mu(X_i) < +\infty).$$

Recall also that a measure μ on E is a *probability measure* (in short, *probability*) if $\mu(E) = 1$.

The following simple lemma shows us that nonzero σ-finite countably additive quasi-invariant measures can be replaced by equivalent probability countably additive quasi-invariant measures.

Lemma 21.1 *Let μ be any nonzero σ-finite countably additive G-quasi-invariant measure on a space (E, G).*

Then there exists a probability countably additive G-quasi-invariant measure ν on (E, G) equivalent to μ.

Proof The case when $\mu(E) < +\infty$ is trivial. So suppose that the equality $\mu(E) = +\infty$ holds. Since μ is σ-finite, there exists a countable family $\{X_n : n \in \mathbb{N}\}$ of μ-measurable sets which are of finite μ-measure and collectively cover E.

Without loss of generality, we may assume that this family is disjoint and $\mu(X_n) > 0$ for each $n \in \mathbb{N}$. Now, for every set $X \in \text{dom}(\mu)$, let us put

$$\nu(X) = \sum \{\mu(X \cap X_n)/(2^{n+1}\mu(X_n)) : n \in \mathbb{N}\}.$$

A straightforward verification gives us that:

(a) ν is a probability countably additive measure on E equivalent to μ;
(b) ν is G-quasi-invariant (in view of Example 21.4).

This completes the proof of Lemma 21.1. □

Lemma 21.2 *Let μ be an arbitrary σ-finite countably additive measure and let $\{Y_j : j \in J\}$ be a disjoint family of μ-measurable sets.*

Then the set $\{j \in J : \mu(Y_j) > 0\}$ is at most countable.

The proof of this lemma is not difficult and is left to the reader (cf. the proof of Lemma 19.1 in Chap. 19).

A set $X \subset E$ is called *almost G-invariant* with respect to a measure μ given on (E, G) if

$$(\forall g \in G)(\mu(g(X) \triangle X) = 0).$$

Let us stress that in the above definition we do not assume that μ is G-quasi-invariant or that X is a μ-measurable set.

Lemma 21.3 *Let μ be a σ-finite countably additive G-quasi-invariant measure on (E, G) and let X be an arbitrary μ-measurable set.*

Then there exists a μ-measurable set Y which satisfies the following three conditions:

(1) $X \subset Y$;
(2) $Y = \cup \{g_i(X) : i \in I\}$ *for some countable family* $\{g_i : i \in I\}$ *of elements from G;*
(3) Y *is almost G-invariant with respect to μ.*

Proof We will construct the required Y by using the method of transfinite recursion.

First, take $Y_0 = X$. Suppose now that, for an ordinal $\xi < \omega_1$, the increasing (by inclusion) family $\{Y_\zeta : \zeta < \xi\}$ of μ-measurable sets has already been defined such that each set Y_ζ is representable in the form $\cup \{g_i(X) : i \in I_\zeta\}$, where $\{g_i : i \in I_\zeta\} \subset G$ and I_ζ is at most countable. Consider the set

$$Z = \cup\{Y_\zeta : \zeta < \xi\}.$$

If Z is almost G-invariant with respect to μ, then we are done, putting $Y = Z$.

In the other case, there exists a $g \in G$ such that

$$\mu(g(Z) \setminus Z) > 0 \vee \mu(Z \setminus g(Z)) > 0.$$

We may assume, without loss of generality, that $\mu(g(Z) \setminus Z) > 0$. Indeed, if $\mu(Z \setminus g(Z)) > 0$, then in view of the G-quasi-invariance of μ, we have $\mu(g^{-1}(Z) \setminus Z) > 0$ and the transformation g can be replaced by its inverse g^{-1}.

Further, we put $Y_\xi = g(Z) \cup Z$ and observe that the set Y_ξ satisfies the analog of (2) and $\mu(Y_\xi \setminus Y_\zeta) > 0$ for any ordinal number $\zeta < \xi$.

Proceeding in this manner, we come to the increasing (by inclusion) ω_1-sequence $\{Y_\xi : \xi < \omega_1\}$ of μ-measurable sets. Finally, taking the disjoint uncountable family $\{Y_{\xi+1} \setminus Y_\xi : \xi < \omega_1\}$ of μ-measurable sets and remembering that

$$\mu(Y_{\xi+1} \setminus Y_\xi) > 0 \qquad (\xi < \omega_1),$$

we get a contradiction to Lemma 21.2. The obtained contradiction ends the proof.
□

Remark 21.2 The above argument yields a much stronger result than Lemma 21.3. Actually, by using this argument we have established that the assertion of Lemma 21.3 remains true for any countably additive G-quasi-invariant measure μ on E, which satisfies the countable chain condition (or Suslin's condition). This condition means that there exists no uncountable disjoint family of μ-measurable sets all of which are of strictly positive μ-measure.

The following definition is very important for the general theory of invariant and quasi-invariant measures.

Let μ be a countably additive G-quasi-invariant measure on a space (E, G).

We say that μ is *metrically transitive* with respect to G (or μ is *G-ergodic*) if, for every μ-measurable set X, the relation

$$(\forall g \in G)(\mu(g(X) \triangle X) = 0)$$

implies the disjunction $\mu(X) = 0 \lor \mu(E \setminus X) = 0$.

In other words, μ is G-ergodic if and only if every μ-measurable almost G-invariant set (with respect to μ) either is of μ-measure zero or is the complement of a μ-measure zero set.

Theorem 21.1 *For a σ-finite countably additive G-quasi-invariant measure μ on E, the following two assertions are equivalent:*

(1) *μ is G-ergodic;*
(2) *for any μ-measurable set X with $\mu(X) > 0$, there exists a countable family $\{g_i : i \in I\} \subset G$ such that $\mu(E \setminus \cup\{g_i(X) : i \in I\}) = 0$.*

Proof It can easily be seen that (2) always implies (1) even without assuming the σ-finiteness of μ.

Let us show that (1) implies (2). For this purpose, suppose (1) and take any μ-measurable set X with $\mu(X) > 0$. According to Lemma 21.3, there exists a countable family $\{g_i : i \in I\} \subset G$ such that the set

$$Y = \cup\{g_i(X) : i \in I\}$$

contains X and is almost G-invariant with respect to μ. By virtue of (1), we have either $\mu(Y) = 0$ or $\mu(E \setminus Y) = 0$. Since $X \subset Y$ and $\mu(X) > 0$, the equality $\mu(Y) = 0$ is impossible. Therefore,

$$\mu(E \setminus \cup\{g_i(X) : i \in I\}) = 0,$$

which completes the proof of the theorem. □

Theorem 21.2 *Let E be a ground set endowed with a transformation group G such that G is uncountable and acts freely in E.*

Then, for every nonzero σ-finite countably additive G-quasi-invariant measure μ on E, there exists a subset of E nonmeasurable with respect to μ.

Proof We may assume, without loss of generality, that μ is a probability measure (see Lemma 21.1) and that $\text{card}(G) = \omega_1$.

Suppose to the contrary that all subsets of E are μ-measurable. Denote by X some selector of the family of all G-orbits in E. Since G acts freely in E, the family $\{g(X) : g \in G\}$ is a partition of E into uncountably many sets, all of which are G-congruent. In view of the finiteness and G-quasi-invariance of μ, we can apply

Lemma 21.2. So we get

$$\mu(g(X)) = 0 \qquad (g \in G).$$

Remembering that card$(G) = \omega_1$, we may rewrite the family $\{g(X) : g \in G\}$ in the form $\{X_\xi : \xi < \omega_1\}$, where:

(a) $\mu(X_\xi) = 0$ for each ordinal $\xi < \omega_1$;
(b) $X_\xi \cap X_\zeta = \emptyset$ for any two distinct ordinals $\xi < \omega_1$ and $\zeta < \omega_1$;
(c) $\mu(\cup\{X_\xi : \xi < \omega_1\}) = \mu(E) = 1$.

Now, for each subset Ξ of ω_1, we can put

$$\nu(\Xi) = \mu(\cup\{X_\xi : \xi \in \Xi\}).$$

A straightforward verification gives us that ν is a countably additive probability measure defined on the power set $\mathcal{P}(\omega_1)$ of ω_1, and this ν vanishes at all singletons of ω_1. But the existence of such a measure ν contradicts Ulam's well-known theorem (see Chap. 19).

The obtained contradiction ends the proof of Theorem 21.2. □

Now, let us briefly consider the question of the existence of a countably additive invariant probability measure for the special case of (E, G).

Let E be a nonempty compact metric space and let $u : E \to E$ be some homeomorphism of E onto itself. Since u is a certain transformation of E, the natural question arises whether there exists a Borel countably additive probability measure on E invariant with respect to the group of transformations of E, generated by u. Evidently, this group is cyclic and coincides with the family

$$[u] = \{u^m : m \in \mathbb{Z}\},$$

where $\mathbb{Z}$ denotes the set of all integers. In general, $[u]$ is infinite but, sometimes, may be finite.

It turns out that a measure with the required properties does always exist for the space $(E, [u])$. This celebrated result is due to Krylov and Bogoliubov (see [106]).

Before proving their theorem, let us make several preliminary remarks.

As usual, we denote by the symbol $C(E)$ $(= C(E, \mathbb{R}))$ the Banach space of all real-valued continuous functions on E (equipped with the standard norm of uniform convergence, i.e., the sup-norm).

By virtue of the classical Riesz theorem (see Chap. 28), there is a one-to-one correspondence between the Radon probability measures on E and those linear positive functionals on $C(E)$ which take value 1 on the characteristic function χ_E. In other words, any given Radon probability measure ν on E can be identified with some uniquely determined linear normalized positive (hence continuous) functional on $C(E)$. For the sake of convenience and brevity, this functional will be denoted by the same symbol ν.

21 Invariant and Quasi-Invariant Measures

Obviously, the set M of such functionals ν is convex in the dual space $(C(E))^*$ (recall that $(C(E))^*$ consists of all real-valued continuous linear functionals defined on the Banach space $C(E)$). We equip $(C(E))^*$ with the weakest topology with respect to which all $f \in C(E)$ considered as linear functionals on $(C(E))^*$ become continuous, i.e., all linear mappings of the form

$$\phi \to \phi(f) \qquad (\phi \in (C(E))^*),$$

where $f \in C(E)$, are continuous.

As usual, this topology is denoted by the symbol $\sigma((C(E))^*, C(E))$.

The set M turns out to be compact with respect to $\sigma((C(E))^*, C(E))$. The latter fact immediately follows from the fundamental theorem on products of quasi-compact spaces, by taking into account that M is a closed subset of the topological product

$$\prod_{f \in C(E)} [-\|f\|, \|f\|].$$

In addition, since E is a compact metric space, $C(E)$ is separable (see Chap. 32) and hence M is metrizable as well, because M can also be regarded as a closed subset of the topological product

$$\prod_{f \in D} [-\|f\|, \|f\|],$$

where D is a countable everywhere dense subset of $C(E)$.

So we derive that M is a compact metrizable space, and this fact implies that any sequence of elements of M contains at least one convergent subsequence.

Theorem 21.3 *Under the above assumptions on E and u, there exists a Radon probability measure μ on E such that, for every Borel subset B of E, the relation*

$$\mu(B) = \mu(u(B)) = \mu(u^{-1}(B))$$

holds true.

Proof For an arbitrary functional $\nu \in M$ and for each natural number n, define a functional ν_n by the formula

$$\nu_n(f) = \frac{\nu(f) + \nu(f \circ u) + \cdots + \nu(f \circ u^n)}{n+1} \qquad (f \in C(E)).$$

Since M is convex, we have $\nu_n \in M$. So we obtain a certain sequence $\{\nu_n : n \in \mathbb{N}\}$ of points from M. Let $\{n(k) : k \in \mathbb{N}\}$ be a strictly increasing family of natural numbers, such that the corresponding partial sequence $\{v_{n(k)} : k \in \mathbb{N}\}$ converges to some point $\mu \in M$. Let us verify that

$$\mu(f) = \mu(f \circ u) \qquad (f \in C(E)).$$

Indeed, we may write

$$\mu(f) = \lim_{k \to \infty} v_{n(k)}(f), \qquad \mu(f \circ u) = \lim_{k \to \infty} v_{n(k)}(f \circ u).$$

A straightforward calculation shows that

$$v_{n(k)}(f) - v_{n(k)}(f \circ u) = \frac{v(f) - v(f \circ u^{n(k)+1})}{n(k) + 1},$$

which implies

$$|v_{n(k)}(f) - v_{n(k)}(f \circ u)| \leq \frac{2\|v\| \cdot \|f\|}{n(k) + 1}.$$

Letting k tend to infinity and keeping in mind the fact that $n(k) \geq k$, we come to the equality $\mu(f) = \mu(f \circ u)$ for every function $f \in C(E)$. In addition, since u is a homeomorphism of E onto itself, we also get

$$\mu(f \circ u^{-1}) = \mu(f \circ u^{-1} \circ u) = \mu(f)$$

for each $f \in C(E)$.

Now, let A be any open subset of E and let χ_A denote, as usual, the characteristic function of A. As has been shown in Chap. 13, there exists an increasing sequence $\{f_n : n \in \mathbb{N}\}$ of real-valued non-negative continuous functions on E such that

$$\chi_A(x) = \lim_{n \to \infty} f_n(x) \qquad (x \in E).$$

Therefore,

$$\chi_{u(A)}(x) = \lim_{n \to \infty} (f_n \circ u^{-1})(x) \qquad (x \in E).$$

These two formulas together with the $[u]$-invariance of the functional μ imply

$$\mu(A) = \mu(u(A)) = \mu(u^{-1}(A)).$$

Since μ is a Radon probability measure, we finally obtain

$$\mu(B) = \mu(u(B)) = \mu(u^{-1}(B))$$

for every Borel subset B of E. Thus, the measure μ is invariant under the cyclic group $[u]$. □

Remark 21.3 It should be mentioned that the group $[u]$ is small in the sense that it is at most countable. Kakutani [79] and Markov [124] extended Theorem 21.3 to a commutative (and, more generally, solvable) group of homeomorphisms of E. Another well-known and important result was obtained by Haar who established that, for every locally compact group $(\Gamma, \cdot)$, there exists a nonzero Radon measure on Γ which is invariant under the group of all left translations of Γ. The theory of Haar measure is thoroughly presented in [17, 20, 63, 69] with many applications in harmonic analysis on topological groups. Some aspects of the general theory of invariant and quasi-invariant measures are discussed in the extensive survey [188].

Exercises

1. Let (E, G) be a space equipped with a transformation group and let μ be a finite countably additive G-quasi-invariant measure defined on some G-invariant algebra of subsets of E.

    Denote by μ' the unique measure on E which extends μ and is defined on the σ-algebra generated by $\text{dom}(\mu)$.

    Can one assert that μ' is also a G-quasi–invariant measure?

2. Give a detailed proof of Lemma 21.2.

3. Let $\{\mu_i : i \in I\}$ ba a family of σ-finite countably additive measures, all of which are defined on some σ-algebra of subsets of a ground set E.

    Verify that:

    (a) if the set I is finite, then the measure $\Sigma_{i \in I} \mu_i$ is σ-finite;
    (b) if the set I is countably infinite, then it may happen that the measure $\Sigma_{i \in I} \mu_i$ is not σ-finite.

4. Let $\{\mu_i : i \in I\}$ be a countable family of σ-finite countably additive measures, all of which are defined on some σ-algebra $\mathcal{S}$ of subsets of a ground set E. Suppose, in addition, that there exists a family $\{X_i : i \in I\} \subset \mathcal{S}$ such that:

    (a) $\mu_i(X_j) = 0$ for any two distinct indices i and j from I;
    (b) $E = \cup\{X_i : i \in I\}$.

    Show that the measure $\Sigma_{i \in I} \mu_i$ is σ-finite.

5. Let μ_1 and μ_2 be two equivalent G-quasi-invariant measures on a space (E, G).

    Check that if the measure μ_1 is G-ergodic, then the measure μ_2 is G-ergodic too.

6. Present an example of a countably additive measure which is not σ-finite, but satisfies the countable chain condition.

7. Give a proof of Lemma 21.3 without using ordinal numbers and the method of transfinite induction.

For this purpose, take an arbitrary nonzero σ-finite countably additive G-quasi-invariant measure μ on E and denote by ν a probability measure equivalent to μ. For any μ-measurable set X with $\mu(X) > 0$, consider the supremum r of all real numbers of the form $\nu(\cup\{g_i(X) : i \in I\})$, where

$$\{g_i : i \in I\} \subset G, \qquad \mathrm{card}(I) \leq \omega, \qquad X \subset \cup\{g_i(X) : i \in I\}.$$

Verify that this r can be attained by using some at most countable family $\{g_j : j \in J\}$ of transformations from G. Finally, show that the set

$$Y = \cup\{g_j(X) : j \in J\}$$

is as required.

Explain why the above argument does not work for nonzero countably additive quasi-invariant measures satisfying Suslin's condition.

8. Prove a more general result than Theorem 21.2. Namely, under the assumptions of the theorem, demonstrate that every μ-measurable set X with $\mu(X) > 0$ contains a subset nonmeasurable with respect to μ.

To obtain this result, argue similarly to the proof of Theorem 21.2.

9. Let μ be a σ-finite countably additive G-quasi-invariant measure defined on a σ-algebra of subsets of a space (E, G), and let Y be an almost G-invariant subset of E with respect to this μ.

Show that:

(a) any μ-measurable hull of Y is almost G-invariant with respect to μ;
(b) any μ-measurable kernel of Y is also almost G-invariant with respect to μ.

10. Give an example of a σ-finite countably additive G-quasi-invariant measure μ on a space (E, G) such that there exists a set $Z \subset E$ which is not almost G-invariant with respect to μ, but a μ-measurable hull and a μ-measurable kernel of Z are almost G-invariant with respect to μ.

Observe that any such Z is not a μ-measurable subset of E.

11. Let μ be a σ-finite countably additive G-invariant G-ergodic measure on a space (E, G) and let X and Y be any two μ-measurable sets in E such that $\mu(X) = \mu(Y) < +\infty$.

Prove that there exist two disjoint countable families of μ-measurable sets $\{X_i : i \in I\}$ and $\{Y_i : i \in I\}$ satisfying the following three relations:

(a) $\cup\{X_i : i \in I\} \subset X$ and $\cup\{Y_i : i \in I\} \subset Y$;
(b) $\mu(X \setminus \cup\{X_i : i \in I\}) = \mu(Y \setminus \cup\{Y_i : i \in I\}) = 0$;
(c) for each index $i \in I$, the set X_i is G-congruent to Y_i.

In other words, one may say that the given sets X and Y are almost countably G-equidecomposable with respect to the measure μ.

21 Invariant and Quasi-Invariant Measures

12. Consider the product topological space

$$E = \mathbb{R}^{\mathbb{N}} = \prod \{R_k \ : \ k \in \mathbb{N}\},$$

where R_k coincides with $\mathbb{R}$ for any natural number k.

For each $k \in \mathbb{N}$, denote by λ_k the standard Borel R_k-invariant measure on R_k and let ν_k be the Borel probability measure on R_k whose restriction to the segment $[-1/2, 1/2] \subset R_k$ coincides with the standard Borel probability measure on $[-1/2, 1/2]$. Further, consider the product measure

$$\mu_n = (\otimes \{\lambda_k \ : \ k \leq n\}) \otimes (\otimes \{\nu_k \ : \ k > n\})$$

(see Appendix B) and define

$$G_n = R_0 \times \cdots \times R_n \times \{0\} \times \{0\} \times \{0\} \times \cdots,$$
$$Y_n = R_0 \times \cdots \times R_n \times [-1/2, 1/2] \times [-1/2, 1/2] \times [-1/2, 1/2] \times \cdots$$

for every natural number n.

Now, for each Borel subset Z of E, put

$$\mu(Z) = \lim_{n \to \infty} \mu_n(Z \cap Y_n)$$

and verify that:

(a) μ is well-defined, i.e., the above limit (possibly, equal to $+\infty$) exists;
(b) μ is a nonzero σ-finite countably additive Borel measure on E;
(c) μ is invariant with respect to the group $G = \cup \{G_k \ : \ k \in \mathbb{N}\}$ which is an everywhere dense vector subspace of the topological vector space E;
(d) μ is a G-ergodic measure;
(e) μ is invariant under the symmetry of E whose fixed point coincides with the neutral element of E.

13. Let H be a real infinite-dimensional separable Hilbert space.

Show that there exists a nonzero σ-finite countably additive Borel measure ν on H such that:

(a) ν is invariant with respect to some everywhere dense vector subspace V of H;
(b) ν is V-ergodic;
(c) ν is invariant under the symmetry of H whose fixed point coincides with zero of H;
(d) $\nu(U) = +\infty$ for any nonempty open set U in H.

For this purpose, identify H with the standard space ℓ_2 of all real-valued square summable sequences and keep in mind the result of Exercise 12.

14. Let ℓ_1 be the Banach space of all real-valued absolutely summable sequences and let W be the everywhere dense vector subspace of ℓ_1 consisting of all those elements of ℓ_1 which differ from zero in ℓ_1 only at finitely many natural indices.

Demonstrate that there exists a nonzero σ-finite countably additive Borel measure on ℓ_1 which is W-invariant, W-ergodic, and is also invariant under the symmetry of ℓ_1 whose fixed point coincides with zero of ℓ_1.

15. Let $(E, ||\cdot||)$ be an arbitrary nonseparable normed vector space over $\mathbb{R}$ and let μ be a nonzero σ-finite countably additive measure on E quasi-invariant with respect to some everywhere dense subgroup of E.

Check that there exists a ball in E which is nonmeasurable with respect to μ.

Remark 21.4 If the space $(E, ||\cdot||)$ of Exercise 15 has an everywhere dense subset of cardinality ω_1, then $(E, ||\cdot||)$ and $\mathbb{R}$ are isomorphic as vector spaces over the field $\mathbb{Q}$ of rational numbers. It readily follows from this fact that there exists a measure ν on E which is isomorphic (only in the measure-theoretic sense) to the Lebesgue measure λ on $\mathbb{R}$ and, in addition, ν is invariant under the group of all translations of E.

16. For any real separable Banach space $(E, ||\cdot||)$, prove that there exists a continuous linear surjective mapping $\phi : \ell_1 \to E$.

Argue as follows. Denote by B the closed unit ball in E, i.e., put

$$B = \{e \in E : ||e|| \leq 1\}.$$

Since E is separable, one can choose a sequence of points

$$\{e_n : n \in \mathbb{N}\} \subset B$$

which is everywhere dense in B. Take an arbitrary element

$$x = \{x_n : n \in \mathbb{N}\} \in \ell_1$$

and define $\phi(x) = \sum\{x_n e_n : n \in \mathbb{N}\}$.

Note that ϕ is well-defined, because of the inequalities

$$\sum_{n \in \mathbb{N}} ||x_n e_n|| \leq \sum_{n \in \mathbb{N}} |x_n| < +\infty.$$

Also, it can easily be seen that ϕ is a linear continuous mapping acting from ℓ_1 into E.

It remains to check that ϕ is a surjection. In view of the linearity of ϕ, it is sufficient to establish that $B \subset \phi(\ell_1)$.

Let e be any point of B. Construct, by ordinary recursion, an injective infinite subfamily $\{e_{n(k)} : k \in \mathbb{N}\}$ of the sequence $\{e_n : n \in \mathbb{N}\}$ in such a manner that the relation

21 Invariant and Quasi-Invariant Measures

$$\|e - e_{n(0)}/2^0 - \cdots - e_{n(k)}/2^k\| \leq 1/2^{k+1}$$

is fulfilled for each $k \in \mathbb{N}$. At the first step, one can find an element $e_{n(0)}$ satisfying the inequality

$$\|e - e_{n(0)}\| \leq 1/2.$$

Suppose now that for a natural number k the partial family

$$\{e_{n(0)}, e_{n(1)}, \ldots, e_{n(k)}\}$$

has already been defined. Then one may write

$$\|2^{k+1}(e - e_{n(0)}/2^0 - \cdots - e_{n(k)}/2^k)\| \leq 1.$$

Since the sequence $\{e_n : n \in \mathbb{N}\}$ is everywhere dense in B, one can choose a natural index $n(k+1)$ such that

$$n(k+1) > \max\{n(0), n(1), \ldots, n(k)\}$$

and

$$\|2^{k+1}(e - e_{n(0)}/2^0 - \cdots - e_{n(k)}/2^k) - e_{n(k+1)}\| \leq 1/2.$$

From the latter inequality one immediately obtains

$$\|e - e_{n(0)}/2^0 - \cdots - e_{n(k)}/2^k - e_{n(k+1)}/2^{k+1}\| \leq 1/2^{k+2}.$$

Proceeding in this manner, one comes to the desired family $\{e_{n(k)} : k \in \mathbb{N}\}$.
Finally, put

$$y = \{y_n : n \in \mathbb{N}\} \in \mathbb{R}^\mathbb{N},$$

where $y_n = 0$ if n differs from all $n(k)$ ($k \in \mathbb{N}$), and $y_n = 1/2^k$ if $n = n(k)$ for some $k \in \mathbb{N}$.

Then it is easy to deduce that $y \in \ell_1$ and $\phi(y) = e$.

Consequently, the inclusion $B \subset \phi(\ell_1)$ holds true, which yields the required result.

17. For a real Banach space $(E, \|\cdot\|)$, prove that the following two assertions are equivalent:

(a) E is separable;
(b) there exists a Borel countably additive probability measure on E quasi-invariant with respect to some everywhere dense vector subspace of E.

For this purpose, use the results of Exercises 15 and 16.

18. Let $(G, \cdot)$ be a topological group and let H and K be subsets of G. Suppose that the following two conditions are fulfilled:

(1) H cannot be covered by a countable family of quasi-compact subsets of G;
(2) K can be covered by countably many quasi-compacts in G.

Show that there exists an uncountable disjoint family consisting of left H-translates of K.

For this purpose, construct the required family by using the method of transfinite recursion over countable ordinals.

19. Let $(G, \cdot)$ be a Polish topological group and let H be a subgroup of G which cannot be covered by countably many compact subsets of G.

Demonstrate that there exists no nonzero σ-finite countably additive left H-quasi-invariant Borel measure on G.

In particular, if $(G, \cdot)$ is a non-locally compact Polish topological group, then there exists no nonzero σ-finite countably additive left G-quasi-invariant Borel measure on G.

To establish this result, use Exercise 18.

20. Prove that the group of all isometric transformations of the n-dimensional Euclidean space $\mathbb{R}^n$, where $n \geq 1$, is generated by the family of all mirror symmetries (i.e., orthogonal reflections with respect to affine hyperplanes in $\mathbb{R}^n$).

Also, check that for any affine hyperplane Γ in $\mathbb{R}^n$ any nonempty open subset U of $\mathbb{R}^n$ can be represented in the form

$$U = \cup\{K_i : i \in I\},$$

where $\{K_i : i \in I\}$ is a countable family of n-dimensional closed cubes whose interiors are pairwise disjoint and, in addition, each cube K_i of the family has a facet parallel to Γ.

Infer from these two geometric facts that the standard Lebesgue measure λ_n on $\mathbb{R}^n$ is invariant under the group of all isometric transformations of $\mathbb{R}^n$.

Chapter 22
Pointwise Limits of Finite Sums of Periodic Functions

In this chapter we will be dealing with some representations of real-valued functions on $\mathbb{R}$ in the form of limits of finite sums of real-valued periodic functions on $\mathbb{R}$.

To begin, let us prove the following simple theorem.

Theorem 22.1 *For every function $f : \mathbb{R} \to \mathbb{R}$, there exists a sequence of real-valued functions $\{f_n : n \in \mathbb{N}\}$ such that:*

(1) *each f_n ($n \in \mathbb{N}$) is a periodic function on $\mathbb{R}$;*
(2) $f = \sum\{f_n : n \in \mathbb{N}\}.$

Proof Take any function $f : \mathbb{R} \to \mathbb{R}$. We shall construct the required sequence $\{f_n : n \in \mathbb{N}\}$ by using ordinary induction (recursion).

Suppose that, for a natural number n, the partial sequence $\{f_k : k < n\}$ of real-valued periodic functions on $\mathbb{R}$ has already been defined and consider the function

$$\phi_n = f - \sum\{f_k : k < n\}.$$

Let f_n be the real-valued function on $\mathbb{R}$ whose period is $2(n+1)$ and which satisfies the relation

$$f_n|[-(n+1), n+1[\ = \ \phi_n|[-(n+1), n+1[.$$

Obviously, the described recursive construction uniquely determines the sequence of functions $\{f_n : n \in \mathbb{N}\}$.

Now, it is easy to see that, for each natural number n, the restriction of $f - \sum\{f_k : k \leq n\}$ to the interval $[-(n+1), n+1[$ is identically equal to zero. This directly implies that the series of functions $\sum\{f_n : n \in \mathbb{N}\}$ converges pointwise on $\mathbb{R}$ to the given function f.

Theorem 22.1 has thus been proved. □

Remark 22.1 The preceding argument is effective, i.e., it does not rely on any form of the Axiom of Choice (**AC**), and actually establishes that if an initial function $f : \mathbb{R} \to \mathbb{R}$ is of a good descriptive structure, then the corresponding periodic functions f_n ($n \in \mathbb{N}$) can be supposed to be of the same structure. For instance, if f is Borel (respectively, Lebesgue measurable or possessing the Baire property), then all f_n can also be taken to be Borel (respectively, Lebesgue measurable or possessing the Baire property).

Remark 22.2 In the proof of Theorem 22.1 the periods of the functions f_n ($n \in \mathbb{N}$) increase and tend to infinity as n tends to infinity. It would be interesting to describe those cases which enable one to have in Theorem 22.1 some uniformly bounded periods for all functions f_n.

Briefly speaking, Theorem 22.1 yields that every function $f : \mathbb{R} \to \mathbb{R}$ is expressible as a pointwise limit of finite sums of real-valued periodic functions on $\mathbb{R}$. In this connection, it makes sense to consider an example of a real-valued analytic function on $\mathbb{R}$ which is not representable as a uniform limit of finite sums of real-valued periodic functions on $\mathbb{R}$.

For this purpose, we need several auxiliary facts and propositions concerning periodic functions.

Let $\{T_1, T_2, \ldots, T_m\}$ be a nonempty finite set of strictly positive real numbers.

For any function $f : \mathbb{R} \to \mathbb{R}$ and for any natural number $k \leq m$, we introduce the notation

$$f(k, x) = \sum \{f(x + T_{i_1} + T_{i_2} + \cdots + T_{i_k}) : \{i_1, i_2, \ldots, i_k\} \subset \{1, 2, \ldots, m\}\},$$

where x ranges over $\mathbb{R}$ and $\{i_1, i_2, \ldots, i_k\}$ stands for a k-element subset of $\{1, 2, \ldots, m\}$.

Using this notation, one can formulate the following lemma.

Lemma 22.1 *If a function $g : \mathbb{R} \to \mathbb{R}$ is representable as a sum of $m \geq 1$ periodic functions on $\mathbb{R}$ whose periods are respectively $T_1, T_2, \ldots, T_m$, then the formula*

$$g(m, x) - g(m - 1, x) + g(m - 2, x) - \cdots + (-1)^m g(0, x) = 0$$

holds for each point x of $\mathbb{R}$.

The above formula can readily be established by using induction on m. We omit the details and leave the proof to the reader.

Lemma 22.2 *Let $f : \mathbb{R} \to \mathbb{R}$ be a function and let $g : \mathbb{R} \to \mathbb{R}$ be a function representable as a sum of m periodic functions on $\mathbb{R}$, where m is a nonzero natural number. Suppose also that*

$$|f(x) - g(x)| \leq L \qquad (x \in \mathbb{R}),$$

where L is some positive real constant (not depending on x). Then the inequality

$$|f(m, x) - f(m - 1, x) + f(m - 2, x) - \cdots + (-1)^m f(0, x)| \leq 2^m L$$

is satisfied for each $x \in \mathbb{R}$.

Proof Let k be an arbitrary natural number not exceeding m. According to our assumption, we may write

$$|f(x + T_{i_1} + T_{i_2} + \cdots + T_{i_k}) - g(x + T_{i_1} + T_{i_2} + \cdots + T_{i_k})| \leq L$$

for any k-element subset $\{i_1, i_2, \ldots, i_k\}$ of $\{1, 2, \ldots, m\}$ and for each $x \in \mathbb{R}$. These inequalities imply that

$$|f(m, x) - g(m, x)| \leq L \qquad (x \in \mathbb{R}),$$
$$|f(m - 1, x) - g(m - 1, x)| \leq mL \qquad (x \in \mathbb{R}),$$
$$|f(m - 2, x) - g(m - 2, x)| \leq \frac{m(m - 1)}{2} L \qquad (x \in \mathbb{R}),$$

....

$$|f(2, x) - g(2, x)| \leq \frac{m(m - 1)}{2} L \qquad (x \in \mathbb{R}),$$
$$|f(1, x) - g(1, x)| \leq mL \qquad (x \in \mathbb{R}),$$
$$|f(0, x) - g(0, x)| \leq L \qquad (x \in \mathbb{R}),$$

whence it follows that

$$|(f(m, x) - f(m - 1, x) + f(m - 2, x) - \cdots + (-1)^m f(0, x))$$
$$- (g(m, x) - g(m - 1, x) + g(m - 2, x) - \cdots + (-1)^m g(0, x))| \leq 2^m L$$

for every $x \in \mathbb{R}$. Finally, taking into account Lemma 22.1, i.e., the equality

$$g(m, x) - g(m - 1, x) + g(m - 2, x) - \cdots + (-1)^m g(0, x) = 0,$$

we obtain that

$$|f(m, x) - f(m - 1, x) + f(m - 2, x) - \cdots + (-1)^m f(0, x)| \leq 2^m L$$

for any $x \in \mathbb{R}$, which completes the proof. $\square$

Lemma 22.3 *Let a function* $f : \mathbb{R} \to \mathbb{R}$ *be given by the formula*

$$f(x) = \exp(x^2) \qquad (x \in \mathbb{R})$$

and let m be an arbitrary nonzero natural number. Then one has

$$\lim_{x \to +\infty} |f(m, x) - f(m-1, x) + f(m-2, x) - \cdots + (-1)^m f(0, x)| = +\infty$$

for any strictly positive real numbers $T_1, T_2, \ldots, T_m$.

Proof It suffices to observe that if a real variable x tends to $+\infty$, then the growth of the function

$$x \to \exp((x + T_1 + T_2 + \cdots + T_m)^2) \qquad (x \in [0, +\infty[)$$

substantially exceeds the growth of all functions of the form

$$x \to \exp((x + T_{i_1} + T_{i_2} + \cdots + T_{i_k})^2) \qquad (x \in [0, +\infty[),$$

where $k < m$.

This observation yields at once the desired result. $\square$

Theorem 22.2 *Let $f : \mathbb{R} \to \mathbb{R}$ be as in Lemma 22.3.*

Then this f cannot be represented as a uniform limit of finite sums of periodic functions on $\mathbb{R}$.

Proof Suppose to the contrary that f is representable as a uniform limit of some sequence $\{f_n : n \in \mathbb{N}\}$ of real-valued functions, where each f_n ($n \in \mathbb{N}$) is a finite sum of periodic functions on $\mathbb{R}$. Clearly, we can find a natural number n such that

$$|f(x) - f_n(x)| \leq 1 \qquad (x \in \mathbb{R}).$$

In accordance with our assumption, f_n is a sum of $m = m(n)$ periodic functions on $\mathbb{R}$ with certain periods

$$T_1 > 0, \quad T_2 > 0, \ldots, \quad T_m > 0,$$

where m is a nonzero natural number. Keeping in mind Lemma 22.2, we must have

$$|f(m, x) - f(m-1, x) + f(m-2, x) - \cdots + (-1)^m f(0, x)| \leq 2^m$$

for all $x \in \mathbb{R}$. However, the latter relation contradicts Lemma 22.3.

The contradiction obtained ends the proof of Theorem 22.2. $\square$

On the other hand, in the sequel we will show that the function f of Theorem 22.2 can be represented as the pointwise limit of a sequence of finite sums of real-valued periodic functions on $\mathbb{R}$, which uniformly converges on every bounded subinterval of $\mathbb{R}$.

To get this result, we will apply the techniques of Hamel bases (see, e.g., [65, 87, 92, 107, 158]).

Recall that, by definition, a *Hamel basis* of $\mathbb{R}$ is any maximal (with respect to inclusion) rationally independent set in $\mathbb{R}$. So our further argument will be essentially based on an uncountable form of the Axiom of Choice (**AC**), because the existence of a Hamel basis cannot be established by using only countable forms of this axiom. Actually, in **ZF** & **DC** theory the existence of a Hamel basis implies the existence of a subset of $\mathbb{R}$ nonmeasurable in the Lebesgue sense.

First of all, we need the amazing fact that any real polynomial on $\mathbb{R}$ is a finite sum of periodic functions on $\mathbb{R}$ (cf. [105]). For the sake of completeness, we will give its proof here. Before proving it, let us introduce a useful auxiliary notion from combinatorial set theory.

Let E be an arbitrary ground set and let $\{E_j : j \in J\}$ be a family of subsets of E.

This family is called *independent* (in the set-theoretical sense) if, for every finite subset J_0 of J, any intersection of the form $\cap \{E'_j : j \in J_0\}$, where

$$(\forall j \in J_0)(E'_j = E_j \vee E'_j = E \setminus E_j),$$

is nonempty.

Notice that independent families of sets play an important role in universal algebra, general topology, measure theory, and probability theory (cf. [69, 87, 111]).

According to one of Tarski's results, if a set E is infinite, then there exists an independent family $\{E_j : j \in J\}$ of subsets of E such that

$$\mathrm{card}(J) = 2^{\mathrm{card}(E)}.$$

In connection with the above equality, see, for instance, [111] or Exercise 4 of this chapter.

Lemma 22.4 *Any real polynomial on $\mathbb{R}$ can be represented as a finite sum of real-valued periodic functions on $\mathbb{R}$.*

Proof Obviously, it suffices to establish that, for every natural number $n > 0$, the function

$$x \to x^n \quad (x \in \mathbb{R})$$

is expressible in the form of a finite sum of periodic functions on $\mathbb{R}$.

Consider the real line $\mathbb{R}$ as a vector space over the field $\mathbb{Q}$ of all rational numbers and let $\{e_i : i \in I\}$ be a Hamel basis of this space. We may assume, without loss of generality, that

$$(\forall i \in I)(|e_i| \leq 1).$$

The set I is uncountable and, actually, we have $\mathrm{card}(I) = \mathbf{c}$, where $\mathbf{c}$ denotes the cardinality of the continuum. So, in view of Tarski's above-mentioned result, there exists an independent family $\{I_j : j \in J\}$ of subsets of I such that $\mathrm{card}(J) = 2^{\mathbf{c}}$.

In our further considerations we need only the fact that J is infinite.

For any set I_j define the function $g_j : \mathbb{R} \to \mathbb{R}$ by the formula

$$g_j(x) = \mathrm{pr}_{U_j}(x) \qquad (x \in \mathbb{R}),$$

where pr_{U_j} stands for the canonical projection of $\mathbb{R}$ onto the vector space U_j (over $\mathbb{Q}$) generated by the family $\{e_i : i \in I_j\}$.

Analogously, define the function $h_j : \mathbb{R} \to \mathbb{R}$ by the formula

$$h_j(x) = \mathrm{pr}_{V_j}(x) \qquad (x \in \mathbb{R}),$$

where pr_{V_j} stands for the canonical projection of $\mathbb{R}$ onto the vector space V_j (over $\mathbb{Q}$) generated by the family $\{e_i : i \in I \setminus I_j\}$.

Both g_j and h_j are additive functions acting from $\mathbb{R}$ into itself and, since $\mathbb{R}$ is a direct algebraic sum of U_j and V_j, we have the relation

$$x = g_j(x) + h_j(x) \qquad (x \in \mathbb{R}).$$

In addition, it is easy to see that any nonzero element of U_j is a period of the function h_j, and any nonzero element of V_j is a period of the function g_j.

Now, take a subset J_0 of J with $\mathrm{card}(J_0) = n$. Then we may write

$$J_0 = \{j_1, j_2, \ldots, j_n\}.$$

Observe that the real polynomial

$$x \to x^n \qquad (x \in \mathbb{R})$$

is a finite sum of products of the form $f_{j_1} f_{j_2} \cdots f_{j_n}$, where each f_j ($j \in J_0$) is either g_j or h_j. Keeping in mind that $\{I_j : j \in J_0\}$ is an independent family of sets, we readily obtain that, for some index i from I, the element e_i of our Hamel basis is a period of the function $f_{j_1} f_{j_2} \cdots f_{j_n}$.

This completes the proof of Lemma 22.4. □

Theorem 22.3 *The function $f : \mathbb{R} \to \mathbb{R}$ defined by*

$$f(x) = \exp(x^2) \qquad (x \in \mathbb{R})$$

is representable as the limit of a certain sequence of finite sums of real-valued periodic functions on $\mathbb{R}$, and this sequence converges uniformly on any bounded subinterval of $\mathbb{R}$. Moreover, all the above-mentioned periodic functions have some uniformly bounded periods.

Proof For every natural number n, let us introduce the notation

$$f_n(x) = 1 + x^2/1! + x^4/2! + \cdots + x^{2n}/n! \qquad (x \in \mathbb{R}).$$

Clearly, the sequence of functions $\{f_n : n \in \mathbb{N}\}$ converges pointwise to f and this convergence is uniform on any bounded subinterval of $\mathbb{R}$. According to Lemma 22.4, each function f_n ($n \in \mathbb{N}$) is a finite sum of periodic functions, and at least one of the positive periods of any of those summands does not exceed 1 (see the proof of Lemma 22.4).

We thus come to the required result. □

Remark 22.3 It would be interesting to find a characterization of those functions $f : \mathbb{R} \to \mathbb{R}$ which are representable in the form

$$f = \sum \{f_n : n \in \mathbb{N}\},$$

where the series $\sum \{f_n : n \in \mathbb{N}\}$ converges uniformly on $\mathbb{R}$ and all the terms f_n ($n \in \mathbb{N}$) are finite sums of periodic functions.

The proof of Theorem 22.3 given above substantially uses the Hamel basis $\{e_i : i \in I\}$ of $\mathbb{R}$ and certain additive functions associated with $\{e_i : i \in I\}$.

Applying an analogous technique, it becomes possible to examine some strengthened versions of the well-known Darboux property for additive functions acting from $\mathbb{R}$ into $\mathbb{R}$.

Recall that a function $g : \mathbb{R} \to \mathbb{R}$ has the *Darboux property* if, for any two real numbers a and b, where $a < b$, the line segment with the endpoints $g(a)$ and $g(b)$ is entirely contained in $g([a, b])$.

We shall say that a function $g : \mathbb{R} \to \mathbb{R}$ has the *strong Darboux property* if $g(\Delta) = \mathbb{R}$ for every non-degenerate subinterval Δ of $\mathbb{R}$.

We shall say that a function $g : \mathbb{R} \to \mathbb{R}$ has the *super-strong Darboux property* if $g(P) = \mathbb{R}$ for every nonempty perfect subset P of $\mathbb{R}$.

Remark 22.4 It is not difficult to verify that:

(a) there are additive functions acting from $\mathbb{R}$ into itself which do not have the Darboux property;
(b) there are additive functions acting from $\mathbb{R}$ into itself which have the Darboux property, but do not have the strong Darboux property;
(c) there are additive functions acting from $\mathbb{R}$ into itself which have the strong Darboux property, but do not have the super-strong Darboux property.

To show the existence of additive functions which possess the super-strong Darboux property, a more delicate approach is necessary.

Theorem 22.4 *There exist real-valued additive functions on $\mathbb{R}$ possessing the super-strong Darboux property.*

Any function $g : \mathbb{R} \to \mathbb{R}$ with the super-strong Darboux property satisfies the following two relations:

(1) *g is nonmeasurable in the Lebesgue sense;*
(2) *for every Lebesgue nonmeasurable set $X \subset \mathbb{R}$, the set $g^{-1}(X)$ is also Lebesgue nonmeasurable.*

Proof Let J be a set of cardinality continuum $\mathbf{c}$ and let $\{P_j : j \in J\}$ denote the family of all nonempty perfect subsets of $\mathbb{R}$.

By using a fairly standard argument based on the method of transfinite induction, it can be demonstrated that there is a disjoint family $\{E_j : j \in J\}$ of subsets of $\mathbb{R}$ such that:

(i) $(\forall j \in J)(E_j \subset P_j)$;
(ii) $(\forall j \in J)(\text{card}(E_j) = \mathbf{c})$;
(iii) the set $\cup \{E_j : j \in J\}$ is linearly independent over the field $\mathbb{Q}$.

We omit the proof of the existence of $\{E_j : j \in J\}$ and leave it to the reader (in this connection, see also [101, 102]).

Further, taking into account (ii) and (iii), it is easy to define an additive function

$$h : \mathbb{R} \to \mathbb{R}$$

such that $h(E_j) = \mathbb{R}$ for each index $j \in J$. Consequently, in view of (i), we get $h(P) = \mathbb{R}$ for every nonempty perfect set P in $\mathbb{R}$. This implies that h has the super-strong Darboux property.

Now, let $g : \mathbb{R} \to \mathbb{R}$ be an arbitrary function possessing the super-strong Darboux property and let λ denote the standard Lebesgue measure on $\mathbb{R}$.

It is clear that g cannot be λ-measurable (because g is unbounded on each uncountable compact subset of $\mathbb{R}$).

Take any Lebesgue nonmeasurable set $X \subset \mathbb{R}$ and suppose to the contrary that the set $g^{-1}(X)$ is λ-measurable. Consider the two possible cases.

1. $\lambda(g^{-1}(X)) > 0$.

    In this case, the set $g^{-1}(X)$ contains a nonempty perfect subset. So we get the relations

$$\mathbb{R} = g(g^{-1}(X)) \subset X, \qquad X = \mathbb{R},$$

which is impossible because of the Lebesgue nonmeasurability of X.

2. $\lambda(g^{-1}(X)) = 0$.

In this case, $\lambda(\mathbb{R} \setminus g^{-1}(X)) > 0$ and the set $\mathbb{R} \setminus g^{-1}(X)$ contains a nonempty perfect subset. So we come to the equality

$$g(\mathbb{R} \setminus g^{-1}(X)) = \mathbb{R},$$

which contradicts the relation $X \cap g(\mathbb{R} \setminus g^{-1}(X)) = \emptyset$.

The contradiction obtained in both above cases completes the proof. □

Remark 22.5 In the formulation of Theorem 22.4 the classical Lebesgue measure λ can be replaced by the completion of any nonzero σ-finite Borel measure on $\mathbb{R}$ vanishing at all singletons in $\mathbb{R}$. Note that the analog of Theorem 22.4 in terms of the Baire property is also valid (see Chap. 30 about the Baire property of sets in general topological spaces). Namely, every function $g : \mathbb{R} \to \mathbb{R}$ possessing the super-strong Darboux property does not have the Baire property, and if X is an arbitrary subset of $\mathbb{R}$ lacking the Baire property, then the set $g^{-1}(X)$ also does not have the Baire property.

These facts can be established by using an argument similar to the previous one.

Exercises

1. Give a proof of Lemma 22.1 applying induction on m.
2. Give a detailed proof of Lemma 22.3.
3. Work in **ZF** theory and show that, for every nonzero natural number n, there exists a family $\{X_i : 1 \leq i \leq n\}$ of subsets of $\mathbb{R}$ such that, for any function

$$f : \{1, 2, \ldots, n\} \to \{0, 1\},$$

the relation

$$\mathrm{card}(X_1^{f(1)} \cap X_2^{f(2)} \cap \cdots \cap X_n^{f(n)}) = \mathfrak{c}$$

holds, where $X_i^{f(i)} = X_i$ if $f(i) = 0$, and $X_i^{f(i)} = \mathbb{R} \setminus X_i$ if $f(i) = 1$.

For this purpose, replace $\mathbb{R}$ by the unit cube $[0, 1]^n$ of the Euclidean space $\mathbb{R}^n$ and consider n affine hyperplanes in $\mathbb{R}^n$ such that each of those hyperplanes passes through the center of $[0, 1]^n$ and is parallel to an appropriate facet of $[0, 1]^n$.

4. Work in **ZFC** set theory and prove Tarski's theorem stating that if a set E is infinite, then there exists an independent family $\{E_j : j \in J\}$ of subsets of E such that

$$\mathrm{card}(J) = 2^{\mathrm{card}(E)}.$$

For this purpose, argue step by step as follows.

(a) Firstly, show that there exists a family $\{X_j : j \in J\}$ of subsets of E such that $\mathrm{card}(J) = 2^{\mathrm{card}(E)}$ and $X_i \setminus X_j \neq \emptyset$ for any two distinct indices i and j from J.

(b) Secondly, show that there exists a family $\{Y_j : j \in J\}$ of subsets of E such that $\mathrm{card}(J) = 2^{\mathrm{card}(E)}$ and, for any finite injective sequence

$(j_0, j_1, j_2, \ldots, j_k)$ of indices from J, the relation

$$Y_{j_0} \setminus (Y_{j_1} \cup Y_{j_2} \cup \cdots \cup Y_{j_k}) \neq \emptyset$$

holds true (here k ranges over $\mathbb{N}$).

(c) Finally, show the existence of an independent family $\{E_j : j \in J\}$ of subsets of E such that $\mathrm{card}(J) = 2^{\mathrm{card}(E)}$.

Infer from the above result that, for the family Φ_E of all ultrafilters in E, the equality

$$\mathrm{card}(\Phi_E) = 2^{2^{\mathrm{card}(E)}}$$

is fulfilled.

5. Verify that:

   (a) there are nonzero additive functions acting from $\mathbb{R}$ into $\mathbb{Q}$ (so they do not have the Darboux property);
   (b) there are additive functions acting from $\mathbb{R}$ into $\mathbb{R}$ having the Darboux property, but lacking the strong Darboux property;
   (c) there are additive functions acting from $\mathbb{R}$ into $\mathbb{R}$ having the strong Darboux property, but lacking the super-strong Darboux property.

6. Prove the two assertions formulated in Remark 22.5.

7. Show that if an additive function $f : \mathbb{R} \to \mathbb{R}$ is not linear over $\mathbb{R}$, then:

   (a) f is everywhere discontinuous on $\mathbb{R}$;
   (b) f is not measurable in the Lebesgue sense;
   (c) f does not have the Baire property;
   (d) the graph of f is everywhere dense in the plane $\mathbb{R}^2$.

   Infer from (b) and (c) that any real-valued additive function on $\mathbb{R}$ with the strong Darboux property is not measurable in the Lebesgue sense and does not possess the Baire property.

8. Give an example of a nonzero function $g : \mathbb{R} \to \mathbb{R}$ such that:

   (a) g is additive;
   (b) all values of g are rational numbers (consequently, g is everywhere discontinuous on $\mathbb{R}$);
   (c) g is measurable with respect to the completion of some nonzero σ-finite Borel measure on $\mathbb{R}$ vanishing at all singletons in $\mathbb{R}$.

   For this purpose, use the fact that there exists a nonempty perfect subset of $\mathbb{R}$ which is linearly independent over the field $\mathbb{Q}$ (cf. [107]).

9. Let $f : \mathbb{R} \to \mathbb{R}$ be any periodic function.
   Check that:

   (a) the set $P(f)$ of all periods of f (including 0) is a subgroup of the additive group $(\mathbb{R}, +)$;

(b) if f is continuous, then $P(f)$ is a closed subgroup of $(\mathbb{R}, +)$;
(c) if f is continuous and the group $P(f)$ is non-discrete, then f is constant;
(d) if f is the Dirichlet function, i.e., $f = \chi_{\mathbb{Q}}$, then $P(f) = \mathbb{Q}$.

10. Let E be a ground set equipped with a transformation group G, let μ be a σ-finite countably additive G-invariant G-ergodic measure on E, and let f be a real-valued μ-measurable function on E.

    Show that the following two assertions are equivalent:

    (1) f is *almost G-invariant*, i.e., for each transformation $g \in G$, the functions f and $f \circ g$ are μ-almost identical;
    (2) f is constant μ-almost everywhere.

    Argue step by step as follows.

    First, observe that the implication (2) $\Rightarrow$ (1) trivially holds true.

    Now, suppose (1) and, for any two real numbers t_1 and t_2, where $t_1 < t_2$, introduce the notation

    $$X([t_1, t_2]) = f^{-1}([t_1, t_2]).$$

    Then verify that

    $$(\forall g \in G)(\mu(g(X([t_1, t_2])) \triangle X([t_1, t_2])) = 0),$$

    which means that the set $X([t_1, t_2])$ is almost G-invariant with respect to μ (see Chap. 21).

    By virtue of the G-ergodicity of μ, deduce that

    $$\mu(X([t_1, t_2])) = 0 \quad \vee \quad \mu(E \setminus X([t_1, t_2])) = 0.$$

    Then, starting with the above fact, construct by recursion two families

    $$\{\Delta_n : n \in \mathbb{N}\}, \qquad \{E_n : n \in \mathbb{N}\}$$

    satisfying the following five conditions:

    (a) Δ_n is a non-degenerate segment in $\mathbb{R}$ for any $n \in \mathbb{R}$;
    (b) $\Delta_{n+1} \subset \Delta_n$ for any $n \in \mathbb{N}$;
    (c) $\lim_{n\to\infty} \lambda(\Delta_n) = 0$;
    (d) $E_n = X(\Delta_n)$ for each $n \in \mathbb{N}$;
    (e) $\mu(E \setminus E_n) = 0$ for any $n \in \mathbb{N}$.

    Finally, denoting $\cap \{E_n : n \in \mathbb{N}\}$ by E', check that:

    (i) $\mu(E \setminus E') = 0$;
    (ii) the restriction of f to the set E' is constant (actually, the range of $f|E'$ coincides with $\cap \{\Delta_n : n \in \mathbb{N}\}$).

11. Let $f : \mathbb{R} \to \mathbb{R}$ be a function and let T_1 and T_2 be two periods of f such that the fraction T_1/T_2 is an irrational real number.

Prove that the additive group $P(f)$ of all periods of f (including 0) is everywhere dense in $\mathbb{R}$.

For this purpose, first demonstrate that, for any real $\varepsilon > 0$, there exist two integers m and n which fulfil the inequalities

$$0 < |mT_1 - nT_2| < \varepsilon,$$

and then deduce the required result.

12. Let $f : \mathbb{R} \to \mathbb{R}$ be an arbitrary λ-measurable function.

Show that the disjunction of the following two assertions holds true:

(a) for any two periods T_1 and T_2 of f, the fraction T_1/T_2 is rational;
(b) f is λ-equivalent to a constant function.

To obtain the required result, take into account that the measure λ is G-ergodic for each everywhere dense subgroup G of the group $(\mathbb{R}, +)$, and use Exercises 10 and 11.

Chapter 23
Absolutely Nonmeasurable Sets in Commutative Groups

In Chap. 17 the concepts of a real-valued absolutely nonmeasurable function and of an absolutely nonmeasurable set were introduced with respect to a given class $\mathcal{M}$ of σ-finite countably additive measures on a ground set E. In accordance with these concepts, we will examine below those subsets of uncountable commutative (i.e., abelian) groups which are extremely bad from the viewpoint of the general theory of nonzero σ-finite countably additive quasi-invariant measures on such groups.

Let $(G, +)$ be an arbitrary commutative group identified with the group of all its translations. All measures on $(G, +)$ considered in the sequel will be assumed to be countably additive.

A subset Z of G is called *G-absolutely nonmeasurable* (in G) if there is no probability G-quasi-invariant measure μ on G such that $Z \in \text{dom}(\mu)$.

In other words, the G-absolute nonmeasurability of a set $Z \subset G$ means that Z is absolutely nonmeasurable with respect to the family (class) of all nonzero σ-finite G-quasi-invariant measures on $(G, +)$ (cf. Chap. 17).

It is known that if a commutative group $(G, +)$ is uncountable, then there exist G-absolutely nonmeasurable subsets of G (see [92]).

It should be noted that the analogous problem on the existence of absolutely nonmeasurable subsets still remains open for an arbitrary uncountable group (not necessarily commutative). Namely, we do not know whether any uncountable group $(\Gamma, \cdot)$ contains a subset absolutely nonmeasurable with respect to the class of all those nonzero σ-finite measures on Γ which are quasi-invariant under the group of left translations of Γ.

The main goal of the present chapter is to obtain a characterization of those sets in an uncountable commutative group $(G, +)$ which contain at least one G-absolutely nonmeasurable subset.

In this context, it is useful to compare the result given below with Exercise 8 from Chap. 21.

To begin, we need some preliminary notions and auxiliary statements.

We shall say that a set X in a commutative group $(G, +)$ is G-*thick* (in G) if there exists a countable family $\{g_j : j \in J\}$ of elements of G such that

$$\cup \{g_j + X : j \in J\} = G.$$

A set Y in a commutative group $(G, +)$ is called G-*negligible* (in G) if the following two conditions are fulfilled:

(a) there exists a nonzero σ-finite G-quasi-invariant measure μ on G such that $Y \in \text{dom}(\mu)$;
(b) for every σ-finite G-quasi-invariant measure ν on G, the relation $Y \in \text{dom}(\nu)$ implies the equality $\nu(Y) = 0$.

Various nontrivial examples of negligible sets in uncountable commutative groups can be found in [92] and [102] (see also Exercise 2 of this chapter).

It can easily be deduced from the definition of G-negligible sets that no such set is G-thick in G.

Notice, by the way, that if $(G, +)$ is an uncountable commutative group, then any at most countable subset of G is trivially G-negligible in G.

In our further considerations, we shall use the following simple lemma.

Lemma 23.1 *Let $(G, +)$ and $(H, +)$ be two commutative groups and let*

$$\phi : (G, +) \to (H, +)$$

be a surjective homomorphism (epimorphism) from $(G, +)$ onto $(H, +)$.
Then the following four assertions hold:

(1) *if a set $A \subset H$ is H-negligible in H, then the set $\phi^{-1}(A)$ is G-negligible in G;*
(2) *if a set $B \subset H$ is H-absolutely nonmeasurable in H, then the set $\phi^{-1}(B)$ is G-absolutely nonmeasurable in G;*
(3) *if a set $C \subset H$ admits uncountably many pairwise disjoint translates in H, then the set $\phi^{-1}(C)$ admits uncountably many pairwise disjoint translates in G;*
(4) *if a set $D \subset H$ is H-thick in H, then the set $\phi^{-1}(D)$ is G-thick in G.*

We omit the straightforward verification of the validity of assertions (1)–(4) and leave it to the reader.

Remark 23.1 It is not difficult to demonstrate that if a subset X of a commutative group $(G, +)$ admits uncountably many pairwise disjoint translates in G, then X cannot be G-thick in G. It follows from this fact that such X is necessarily a G-negligible subset of G (see Exercise 2).

Lemma 23.2 *For any uncountable commutative group $(H, +)$, there exist two subgroups H_0 and H_1 of H such that:*

(1) $\text{card}(H_0) = \omega$;
(2) $\text{card}(H_1) = \omega_1$;
(3) $H_0 \cap H_1 = \{0\}$, *where 0 denotes the neutral element of $(H, +)$.*

Proof For each natural number $k > 0$, let

$$T_k = \{h \in H : kh = 0\}$$

and consider the two possible cases.

1. There exists a natural number $k > 0$ such that the subgroup T_k of H is uncountable.

   In this case, suppose (without loss of generality) that this k is the least natural number for which $\operatorname{card}(T_k) > \omega$. Take an arbitrary subgroup H_0 of H with $\operatorname{card}(H_0) = \omega$ and let H' be a maximal (with respect to the inclusion relation) subgroup of T_k satisfying $H_0 \cap H' = \{0\}$. In view of the definition of k, the group H' is necessarily uncountable. Then we may take a subgroup H_1 of H' with $\operatorname{card}(H_1) = \omega_1$ and so we obtain the required result.

2. All subgroups T_k ($0 < k < \omega$) are at most countable.

   In this case, denote by H_0 a subgroup of H containing $\cup \{T_k : 0 < k < \omega\}$ and such that $\operatorname{card}(H_0) = \omega$. Let H' be a maximal (with respect to the inclusion relation) subgroup of H satisfying $H_0 \cap H' = \{0\}$. Again, it is not hard to verify that H' is necessarily uncountable. Then, similarly to the previous case, we take a subgroup H_1 of H' with $\operatorname{card}(H_1) = \omega_1$ and come to the desired result. □

Remark 23.2 Let us mention that the purely algebraic fact formulated in Lemma 23.2 can also be deduced from some general theorems concerning the structure of infinite commutative groups. For instance, this fact readily follows from a profound theorem of Kulikov (see [50, 114]).

Lemma 23.3 *Let $(H, +)$ be a commutative group of cardinality ω_1. Then there exists a subset X of H such that:*

(1) *X is H-thick in H;*
(2) *some ω_1-sequence $\{h_\xi : \xi < \omega_1\}$ of elements of H has the property that, for any two distinct ordinals $\xi < \omega_1$ and $\zeta < \omega_1$, the set*

$$X_{\xi,\zeta} = (h_\xi + X) \cap (h_\zeta + X)$$

is at most countable.

In particular, the set X is H-absolutely nonmeasurable and, for any two distinct ordinals $\xi < \omega_1$ and $\zeta < \omega_1$, the set $X_{\xi,\zeta}$ is H-negligible.

Proof According to Lemma 23.2, there exist two subgroups H_0 and H_1 of H satisfying the relations

$$\operatorname{card}(H_0) = \omega,$$
$$\operatorname{card}(H_1) = \omega_1,$$
$$H_0 \cap H_1 = \{0\}.$$

Clearly, we may take an ω_1-sequence $\{\Gamma_\xi : \xi < \omega_1\}$ of subgroups of H, such that:

(a) $\Gamma_0 = H_0$;
(b) for each ordinal $\xi < \omega_1$, the equality $\mathrm{card}(\Gamma_\xi) = \omega$ is fulfilled;
(c) for each ordinal $\xi < \omega_1$, the set $\cup\{\Gamma_\zeta : \zeta < \xi\}$ is a proper subset of Γ_ξ (in particular, this ω_1-sequence of subgroups of H is strictly increasing by inclusion);
(d) $\cup\{\Gamma_\xi : \xi < \omega_1\} = H$.

Further, for any ordinal number $\xi < \omega_1$, we put

$$Y_\xi = \Gamma_\xi \setminus \cup\{\Gamma_\zeta : \zeta < \xi\}.$$

A straightforward verification shows that the family of sets $\{Y_\xi : \xi < \omega_1\}$ forms a partition of H and each Y_ξ is a Γ'_ξ-invariant subset of H, where the group Γ'_ξ is defined by the formula

$$\Gamma'_\xi = \cup\{\Gamma_\zeta : \zeta < \xi\} \cup \{0\}.$$

According to relation (c), the group Γ'_ξ is a proper subgroup of Γ_ξ. Also, we have

$$\mathrm{card}(Y_\xi) = \omega, \qquad \Gamma'_\xi + Y_\xi = Y_\xi \qquad (\xi < \omega_1).$$

Now, for each ordinal number $\xi < \omega_1$, we introduce the group

$$H_{1,\xi} = H_1 \cap \Gamma'_\xi.$$

Obviously, the ω_1-sequence $\{H_{1,\xi} : \xi < \omega_1\}$ of groups is increasing by inclusion and

$$\cup\{H_{1,\xi} : \xi < \omega_1\} = H_1.$$

Fix for a while an ordinal $\xi < \omega_1$ and consider the two partitions of Y_ξ canonically associated with the groups H_0 and $H_{1,\xi}$, respectively.

In other words, we will be dealing with the partition of Y_ξ into H_0-orbits and with the partition of Y_ξ into $H_{1,\xi}$-orbits. Taking into account the relation $H_0 \cap H_1 = \{0\}$, we infer that the above-mentioned two partitions of Y_ξ are transversal to each other. The latter phrase means that every equivalence class of the first partition has at most one common element with every equivalence class of the second partition. Starting with this fact, we will define by recursion a finite or countably infinite sequence

$$\{x_{\xi,k} : k = 0, 1, \ldots\} \subset Y_\xi$$

such that:

(i) $H_0 + \{x_{\xi,k} : k = 0, 1, \ldots\} = Y_\xi$;

(ii) for any two distinct natural numbers k and m, the point $x_{\xi,k}$ does not belong to the orbit $H_{1,\xi} + x_{\xi,m}$.

For this purpose, let $\{Z_{\xi,k} : k = 0, 1, \ldots\}$ denote the injective family of all those H_0-orbits which are contained in Y_ξ.

Take $Z_{\xi,0}$ and choose arbitrarily a point $x_{\xi,0}$ in $Z_{\xi,0}$.

Suppose that, for a natural number $k \geq 1$, the elements

$$x_{\xi,0} \in Z_{\xi,0},\ x_{\xi,1} \in Z_{\xi,1},\ \ldots,\ x_{\xi,k-1} \in Z_{\xi,k-1}$$

have already been defined and that they lie in pairwise distinct $H_{1,\xi}$-orbits. Consider the set

$$P_k = (H_{1,\xi} + x_{\xi,0}) \cup (H_{1,\xi} + x_{\xi,1}) \cup \cdots \cup (H_{1,\xi} + x_{\xi,k-1}).$$

Evidently, we have the relations

$$\operatorname{card}(P_k \cap Z_{\xi,k}) \leq k, \qquad \operatorname{card}(Z_{\xi,k}) = \omega.$$

Consequently, there exists an element $x \in Z_{\xi,k} \setminus P_k$. So, we can put

$$x_{\xi,k} = x.$$

Therefore, for each ordinal $\xi < \omega_1$, we get the corresponding sequence $\{x_{\xi,k} : k = 0, 1, \ldots\}$ of elements from Y_ξ, fulfilling conditions (i) and (ii).

Now, let us introduce the notation

$$X = \{x_{\xi,k} : \xi < \omega_1,\ k = 0, 1, \ldots\}.$$

We are going to verify that the set X is H-absolutely nonmeasurable in H.

Indeed, on the one hand, we may write

$$H_0 + X = \cup\{H_0 + \{x_{\xi,k} : k = 0, 1, \ldots\} : \xi < \omega_1\} = \cup\{Y_\xi : \xi < \omega_1\} = H,$$

and the above relation implies that if the set X is measurable with respect to a nonzero σ-finite H-quasi-invariant measure μ on H, then the inequality $\mu(X) > 0$ must hold.

On the other hand, let us take an arbitrary element $g \in H_1 \setminus \{0\}$.

Then there exists an ordinal $\xi_0 < \omega_1$ for which $g \in H_{1,\xi_0}$. Further, for any $\xi < \omega_1$, let

$$X_\xi = \{x_{\xi,k} : k = 0, 1, \ldots\}.$$

Clearly, we have the relation

$$(\forall \xi < \omega_1)(\operatorname{card}(X_\xi) \leq \omega).$$

Also, the equality $X = \cup\{X_\xi : \xi < \omega_1\}$ holds true, whence it follows that

$$(g + X) \cap X = \cup\{(g + X_\zeta) \cap X_\eta : \zeta < \omega_1, \eta < \omega_1\}.$$

If $\zeta < \omega_1$ and $\eta < \omega_1$ satisfy the relations $\xi_0 < \zeta$ and $\xi_0 < \eta$, then a straightforward verification gives us

$$(g + X_\zeta) \cap X_\eta = \emptyset.$$

We thus get the inclusion

$$(g + X) \cap X \subset (\cup\{(g + X_\zeta) : \zeta \leq \xi_0\}) \cup (\cup\{X_\eta : \eta \leq \xi_0\})$$

and, therefore,

$$\operatorname{card}((g + X) \cap X) \leq \omega.$$

Now, suppose that g and h are any two distinct elements of H_1. Then

$$g - h \neq 0, \qquad g - h \in H_1,$$

and, according to the fact established above, we may write

$$\operatorname{card}((g - h + X) \cap X) \leq \omega,$$

which implies at once that

$$\operatorname{card}((g + X) \cap (h + X)) \leq \omega.$$

The last relation shows us that if the set X is measurable with respect to a σ-finite H-quasi-invariant measure μ on H, then the equality $\mu(X) = 0$ must be fulfilled.

So, we get from the presented argument that simultaneously $\mu(X) > 0$ and $\mu(X) = 0$. Obviously, this yields a contradiction and hence X is an H-absolutely nonmeasurable subset of H.

Lemma 23.3 has thus been proved. □

Remark 23.3 Assuming the Continuum Hypothesis (**CH**), it immediately follows from Lemma 23.3 that there exists a subset X of the standard commutative group $(\mathbb{R}, +)$, satisfying the following two conditions:

(a) $\cup\{g_j + X : j \in J\} = \mathbb{R}$ for some countable family $\{g_j : j \in J\} \subset \mathbb{R}$;
(b) there exists an uncountable family $\{h_i : i \in I\} \subset \mathbb{R}$ such that

$$\operatorname{card}((h_i + X) \cap (h_k + X)) \leq \omega$$

for any two distinct indices $i \in I$ and $k \in I$.

It makes sense to compare the existence of such X (of course, under **CH**) with the fact mentioned in Remark 23.1.

We also need the next purely algebraic lemma.

Lemma 23.4 *For every uncountable commutative group $(G, +)$, there exist a commutative group $(H, +)$ of cardinality ω_1 and a surjective homomorphism (epimorphism) $\phi : (G, +) \to (H, +)$.*

For a proof, see Exercise 7 of this chapter.

Lemma 23.5 *Let $(G, +)$ be an arbitrary uncountable commutative group. Then there exists a subset T of G such that:*

(1) *T is G-thick in G;*
(2) *some ω_1-sequence $\{g_\xi : \xi < \omega_1\}$ of elements of G has the property that, for any two distinct ordinals $\xi < \omega_1$ and $\zeta < \omega_1$, the set*

$$T_{\xi,\zeta} = (g_\xi + T) \cap (g_\zeta + T)$$

admits uncountably many pairwise disjoint translates in G.

In particular, the set T is G-absolutely nonmeasurable in G and the set $T_{\xi,\zeta}$ is G-negligible in G.

Proof By virtue of Lemma 23.4, there exist a commutative group $(H, +)$ with $\text{card}(H) = \omega_1$ and a surjective homomorphism

$$\phi : (G, +) \to (H, +).$$

Further, according to Lemma 23.3, there exists a subset X of H such that:

(i) X is H-thick in H;
(ii) there is an ω_1-sequence $\{h_\xi : \xi < \omega_1\}$ of elements of H having the property that, for any two distinct ordinals $\xi < \omega_1$ and $\zeta < \omega_1$, the set

$$X_{\xi,\zeta} = (h_\xi + X) \cap (h_\zeta + X)$$

is at most countable.

Let us stipulate $T = \phi^{-1}(X)$ and let us verify that the set T is as required.

Assertion (4) of Lemma 23.1 directly implies that $\phi^{-1}(X)$ is G-thick in G. Indeed, in view of (i), there exists a countable family $\{h_j : j \in J\}$ of elements from H satisfying the equality

$$\cup \{h_j + X : j \in J\} = H.$$

For each index $j \in J$, denote by g_j an element of G such that $\phi(g_j) = h_j$ (the existence of g_j is evident, because ϕ is a surjective homomorphism of $(G, +)$ onto

$(H, +))$. Now, it is clear that

$$\cup \{g_j + T : j \in J\} = \phi^{-1}(\cup \{h_j + X : j \in J\}) = \phi^{-1}(H) = G,$$

so (1) of Lemma 23.5 holds true.

Further, for each ordinal number $\xi < \omega_1$, denote by g_ξ an element of G such that $\phi(g_\xi) = h_\xi$. Again, the existence of g_ξ ($\xi < \omega_1$) follows from the fact that ϕ is an epimorphism of $(G, +)$ onto $(H, +)$.

For any two distinct ordinals $\xi < \omega_1$ and $\zeta < \omega_1$, consider the set

$$T_{\xi,\zeta} = (g_\xi + T) \cap (g_\zeta + T).$$

Obviously, we may write

$$T_{\xi,\zeta} = \phi^{-1}((h_\xi + X) \cap (h_\zeta + X)) = \phi^{-1}(X_{\xi,\zeta}).$$

By virtue of (ii), the set $X_{\xi,\zeta}$ is at most countable. Consequently, the set $T_{\xi,\zeta}$ admits uncountably many pairwise disjoint translates in G and so this $T_{\xi,\zeta}$ is G-negligible in G.

The above properties of the set T also imply that T is G-absolutely nonmeasurable in the group $(G, +)$.

This completes the proof of Lemma 23.5. □

Now, we are able to formulate and prove the main result of the present chapter.

Theorem 23.1 *Let A be an arbitrary subset of an uncountable commutative group $(G, +)$.*

The following two assertions are equivalent:

(1) *A contains a G-absolutely nonmeasurable subset of $(G, +)$;*
(2) *A is G-thick in $(G, +)$.*

Proof First, let us demonstrate the validity of the implication

$$(1) \Rightarrow (2).$$

Suppose (1), and let T be some G-absolutely nonmeasurable subset of A. It is easy to check that every G-absolutely nonmeasurable set in G is simultaneously G-thick in G. So, there exists a countable family $\{g_j : j \in J\}$ of elements from G such that

$$\cup \{g_j + T : j \in J\} = G.$$

Since $T \subset A$, we also have

$$G = \cup\{g_j + T : j \in J\} \subset \cup\{g_j + A : j \in J\} \subset G$$

and, consequently,

$$G = \cup \{g_j + A : j \in J\},$$

which shows us that A is G-thick in G, i.e., assertion (2) holds true.

To prove the validity of the converse implication (2) $\Rightarrow$ (1), suppose that (2) is fulfilled. It is not hard to see that (2) implies the existence of a countable subgroup G_0 of G such that

$$G_0 + A = \cup \{g + A : g \in G_0\} = G.$$

Consider the quotient group $H = G/G_0$. Since the initial group G is uncountable and the group G_0 is countable, the group H is uncountable. Let

$$\phi : (G, +) \to (H, +)$$

stand for the canonical surjective homomorphism of G onto H. Obviously, we get the equality $\phi(A) = H$.

Further, according to Lemma 23.5, there exists an H-absolutely nonmeasurable subset T of H such that:

(a) T is H-thick in H;
(b) some ω_1-sequence $\{h_\xi : \xi < \omega_1\}$ of elements of H has the property that, for any two distinct ordinals $\xi < \omega_1$ and $\zeta < \omega_1$, the set

$$T_{\xi,\zeta} = (h_\xi + T) \cap (h_\zeta + T)$$

admits uncountably many pairwise disjoint translates in H.

Consider the set $\phi^{-1}(T)$. By Lemma 23.1, $\phi^{-1}(T)$ is a G-absolutely nonmeasurable subset of G. At the same time, the set $\phi^{-1}(T)$ satisfies the following two conditions:

(c) there is a disjoint family $\{Z_i : i \in I\}$ of G_0-orbits in G such that

$$\phi^{-1}(T) = \cup \{Z_i : i \in I\};$$

(d) every G_0-orbit Z_i ($i \in I$) has nonempty intersection with the set A.

In view of (d), there exists a selector Z of $\{Z_i : i \in I\}$ entirely contained in A. We assert that the set Z is G-absolutely nonmeasurable in G.
Indeed, on the one hand, the relation

$$G_0 + Z = \cup\{Z_i : i \in I\} = \phi^{-1}(T)$$

and the G-absolute nonmeasurability of $\phi^{-1}(T)$ imply that Z is G-thick in the group G.

On the other hand, the inclusion $Z \subset \phi^{-1}(T)$ and relation (b) imply (in view of (3) of Lemma 23.1) that there exists an ω_1-sequence $\{g_\xi : \xi < \omega_1\}$ of elements of G such that, for any two distinct ordinals $\xi < \omega_1$ and $\zeta < \omega_1$, the set

$$Z_{\xi,\zeta} = (g_\xi + Z) \cap (g_\zeta + Z)$$

admits uncountably many pairwise disjoint translates in G.

These two properties of Z enable us to conclude that Z is a G-absolutely nonmeasurable subset of A.

Theorem 23.1 has thus been proved. □

Let $(\mathbb{R}, +)$ denote, as usual, the additive group of all real numbers and let $(\mathbb{Q}, +)$ denote the subgroup of $\mathbb{R}$ consisting of all rational numbers.

Recall that a *Vitali set* in $\mathbb{R}$ is any selector of the quotient set $\mathbb{R}/\mathbb{Q}$.

According to Vitali's celebrated theorem, every Vitali subset of $\mathbb{R}$ is absolutely nonmeasurable with respect to the class of all those $\mathbb{Q}$-invariant measures on $\mathbb{R}$ which extend the Lebesgue measure λ (see, e.g., [92, 134, 183]).

As a consequence of Theorem 23.1, we get the following.

Theorem 23.2 *Every Vitali set contains an $\mathbb{R}$-absolutely nonmeasurable subset of $\mathbb{R}$.*

Proof Let V be an arbitrary Vitali set on the real line $\mathbb{R}$. The definition of Vitali sets implies at once that all of them are $\mathbb{R}$-thick in $\mathbb{R}$ (and even $\mathbb{Q}$-thick in $\mathbb{R}$). Therefore, we may directly apply Theorem 23.1 to V. In this way, we obtain that there exists an $\mathbb{R}$-absolutely nonmeasurable set entirely contained in V. □

Remark 23.4 In connection with Theorem 23.2, it should be noted that there are many Vitali sets in $\mathbb{R}$ which are not $\mathbb{R}$-absolutely nonmeasurable. Moreover, it can be shown that there exist a Vitali set W in $\mathbb{R}$ and an $\mathbb{R}$-quasi-invariant measure μ on $\mathbb{R}$ such that μ is an extension of the measure λ and $W \in \text{dom}(\mu)$. For further details, see Exercise 8 of this chapter.

Remark 23.5 Let E be a real infinite-dimensional separable Hilbert space and let B denote the closed unit ball in E. It turns out that B is an E-absolutely nonmeasurable subset of E (see [86]). Moreover, this fact is provable within **ZF & DC** theory.

Remark 23.6 Let $(G, +)$ be an uncountable commutative group. The following natural question arises:

Does there exist a subset of G absolutely nonmeasurable with respect to the class of all nonzero σ-finite finitely additive G-quasi-invariant measures on G?

The answer to this question is negative and immediately follows from the well-known result of Banach stating that there exists a nonzero finite finitely additive G-invariant measure μ on G such that $\text{dom}(\mu) = \mathcal{P}(G)$ (see, e.g., [183]). In fact, the analogous result was obtained by Banach for any uncountable *solvable group* $(H, \cdot)$.

However, if a group $(\Gamma, \cdot)$ contains an uncountable free subgroup (so this $(\Gamma, \cdot)$ is a *strongly paradoxical group*), then there exist subsets of Γ which are absolutely nonmeasurable with respect to the class of all those nonzero finitely additive left Γ-quasi-invariant measures on Γ which satisfy Suslin's condition (for more details, see [133] and [183]).

Exercises

1. Check that, for a subset Z of a commutative group $(G, +)$, the following two assertions are equivalent:

   (a) Z is G-thick in $(G, +)$;
   (b) Z does not belong to any G-invariant σ-ideal of subsets of G.

   Conclude from the above-mentioned equivalence that no G-negligible set in a commutative group $(G, +)$ can be G-thick in G.
   Formulate and prove the analogous result for a group $(\Gamma, \cdot)$ (not necessarily commutative).

2. Let a subset X of a commutative group $(G, +)$ be such that there are uncountably many pairwise disjoint translates of X in G.
   Verify that X cannot be G-thick in G and infer from this fact that X is a G-negligible subset of G.
   For this purpose, suppose on the contrary that X is G-thick and obtain that any uncountable family of G-translates of X is not disjoint.

3. Give a detailed proof of Lemma 23.1 of this chapter.
   The argument is similar for all assertions (1)–(4) of the lemma. For instance, consider (2) and argue as follows. Let a set $B \subset H$ be H-absolutely nonmeasurable in H and suppose to the contrary that the set $\phi^{-1}(B)$ is not G-absolutely nonmeasurable in G. Then there exists a probability countably additive G-quasi-invariant measure μ on G such that

   $$\phi^{-1}(B) \in \text{dom}(\mu).$$

   Let $\mathcal{A}$ denote the family of all those sets $Z \subset H$ for which $\phi^{-1}(Z) \in \text{dom}(\mu)$. Verify that $\mathcal{A}$ is an H-invariant σ-algebra of subsets of H and that the formula

   $$\nu(Z) = \mu(\phi^{-1}(Z)) \qquad (Z \in \mathcal{A})$$

   defines a probability countably additive H-quasi-invariant measure ν on H such that the set B is ν-measurable. But the latter contradicts the H-absolute nonmeasurability of B.

4. Recall that a commutative group $(H, +)$ is *divisible* if, for every nonzero natural number k and for every $h \in H$, the equation $kx = h$ has at least one solution x_0 in H (i.e., $kx_0 = h$).
   Check that the product group of any family of commutative divisible groups is also divisible.
   Prove that any partial homomorphism ϕ from a commutative group $(G, +)$ into a divisible commutative group $(H, +)$ can be extended to a homomorphism ϕ^* from $(G, +)$ into $(H, +)$.
   For this purpose, argue as follows.

First, consider any partial homomorphism $\phi' : G \to H$ extending ϕ. Let g be an arbitrary element of $G \setminus \mathrm{dom}(\phi')$.

Show that ϕ' can be extended to a partial homomorphism $\phi'' : G \to H$ such that $g \in \mathrm{dom}(\phi'')$.

Here only two cases are possible.

(a) $kg \in \mathrm{dom}(\phi')$ for some natural number $k > 0$.

In this case, one may assume, without loss of generality, that k is the least nonzero natural number with property (a). Since H is divisible, there exists an element $h \in H$ satisfying the relation

$$kh = \phi'(kg).$$

Put $\phi''(g) = h$ and prove that this equality produces a well-defined partial homomorphism $\phi'' : G \to H$ which extends ϕ' and whose domain coincides with the group generated by $\mathrm{dom}(\phi') \cup \{g\}$.

(b) $kg \notin \mathrm{dom}(\phi')$ for all natural numbers $k > 0$.

In this case, pick an arbitrary element $h \in H$ and put $\phi''(g) = h$. Then verify that this relation also produces a well-defined partial homomorphism $\phi'' : G \to H$ which extends ϕ' and whose domain coincides with the group generated by $\mathrm{dom}(\phi') \cup \{g\}$.

Finally, apply the Kuratowski–Zorn lemma to obtain a homomorphism $\phi^* : G \to H$ which extends the initial partial homomorphism ϕ.

5. Demonstrate that, for every commutative group $(G, +)$, there exists a divisible commutative group $(G^*, +)$ containing an isomorphic copy of $(G, +)$.

For this purpose, consider the family $\{\phi_i : i \in I\}$ of all homomorphisms of G into the commutative divisible circle group

$$\mathbf{S}_1 = \{(x, y) \in \mathbb{R}^2 : x^2 + y^2 = 1\}$$

and, using the result of Exercise 4, check that this family separates the elements of G, i.e., for any two distinct elements $g \in G$ and $h \in G$, there is an index $i \in I$ such that $\phi_i(g) \neq \phi_i(h)$.

Then infer that the mapping $\phi = \{\phi_i : i \in I\}$ defined by

$$\phi(g) = \{\phi_i(g) : i \in I\} \qquad (g \in G)$$

is a monomorphism of G into the product group $(\mathbf{S}_1)^I$ and, taking into account that $(\mathbf{S}_1)^I$ is a divisible group, obtain the required result.

6. Show that, for every commutative group $(G, +)$, there exists a divisible commutative group $(G^*, +)$ satisfying the following two conditions:

(a) $(G^*, +)$ contains an isomorphic copy of $(G, +)$;
(b) $\mathrm{card}(G^*) \leq \mathrm{card}(G) + \omega$.

Argue as follows. According to the result of Exercise 5, there exists a divisible commutative group $(H, +)$ containing $(G, +)$ as a subgroup. Construct by recursion an increasing (with respect to the standard inclusion relation) countable family $\{G_n : n < \omega\}$ of subgroups of H.
First of all, put $G_0 = G$.
Suppose that, for a natural number n, the group $G_n \subset H$ has already been defined. For any nonzero natural number k and for each element $g \in G_n$, consider the equation $kx = g$ in the group H. Since H is divisible, one has a solution $x(k, g)$ of this equation in H. Take as G_{n+1} the subgroup of H generated by the set

$$G_n \cup \{x(k, g) : k \in \mathbb{N} \setminus \{0\}, \ g \in G_n\}.$$

Verify that $\text{card}(G_{n+1}) \leq \text{card}(G_n) + \omega$. Finally, define

$$G^* = \cup \{G_n : n < \omega\}$$

and establish that $(G^*, +)$ is a desired divisible commutative group.

7. Give a proof of Lemma 23.4, i.e., demonstrate that, for every uncountable commutative group $(G, +)$, there exist a commutative group $(H, +)$ of cardinality ω_1 and a surjective homomorphism $\phi : (G, +) \to (H, +)$.

    For this purpose, argue as follows. Since G is uncountable, it contains a subgroup G_1 of cardinality ω_1. In its turn, G_1 is contained in some divisible commutative group G_2 of the same cardinality ω_1 (see Exercise 6). Let ψ denote the identical embedding of G_1 into G_2. This ψ can also be treated as a partial homomorphism of G into G_2. According to Exercise 4, for ψ there exists an extension ψ^* which is a homomorphism of G into G_2. Conclude that the group $H = \psi^*(G)$ and the homomorphism $\phi = \psi^*$ are as required.

8. Show that there exist a Vitali set W in $\mathbb{R}$ and an $\mathbb{R}$-quasi-invariant measure μ on $\mathbb{R}$ such that μ is an extension of the standard Lebesgue measure λ on $\mathbb{R}$ and $W \in \text{dom}(\mu)$.

    For this purpose, take a Hamel basis B of $\mathbb{R}$ such that $1 \in B$ and denote by W the vector space (over $\mathbb{Q}$) generated by $B \setminus \{1\}$. Observe that W is a Vitali set in $\mathbb{R}$ and

    $$\lambda_*(W) = \lambda_*(\mathbb{R} \setminus W) = 0,$$

    where λ_* is, as usual, the inner Lebesgue measure on $\mathbb{R}$.
    Let $\{W + q : q \in \mathbb{Q}\}$ denote the injective countable family of all translates of W in $\mathbb{R}$. Consider the family $\mathcal{A}$ of all those subsets X of $\mathbb{R}$ which are representable in the form

    $$X = \cup\{(W + q) \cap X_q : q \in \mathbb{Q}, \ X_q \in \text{dom}(\lambda)\},$$

    and define on $\mathcal{A}$ an appropriate countably additive $\mathbb{R}$-quasi-invariant measure μ extending λ. Finally, keep in mind that $W + 0 = W \in \mathcal{A}$.

9. Prove that there exists a Vitali set in $\mathbb{R}$ which is $\mathbb{R}$-absolutely nonmeasurable. For this purpose, consider $\mathbb{R}$ as a vector space over the field $\mathbb{Q}$ and represent this space in the form of a direct sum:

$$\mathbb{R} = \mathbb{Q} + U_1 + U_2,$$

where both U_1 and U_2 are also vector spaces over $\mathbb{Q}$ and $\operatorname{card}(U_1) = \omega_1$.
Using Lemma 23.3, infer that there is a subset X of $\mathbb{Q} + U_1$ satisfying the following two conditions:

(a) X is $(\mathbb{Q} + U_1)$-absolutely nonmeasurable in $\mathbb{Q} + U_1$;
(b) for each $z \in \mathbb{Q} + U_1$, the set $(z + \mathbb{Q}) \cap X$ is a singleton.

Deduce from (a) and (b) that $X + U_2$ is a required Vitali set in $\mathbb{R}$.

10. Verify that if $(G, +)$ is a countable commutative group, then there exists a nonzero σ-finite G-invariant measure whose domain coincides with the power set $\mathcal{P}(G)$ of G.

Conclude that there is no G-absolutely nonmeasurable subset of G.

11. Let Z be an $\mathbb{R}$-thick subset of $\mathbb{R}$ such that there exists some family $\{h_\xi : \xi < \omega_1\} \subset \mathbb{R}$ having the property that

$$\operatorname{card}((h_\xi + Z) \cap (h_\zeta + Z)) \leq \omega \qquad (\xi < \omega, \ \zeta < \omega, \ \xi \neq \zeta).$$

Prove that in this case the Continuum Hypothesis (**CH**) holds true.
Argue as follows. Suppose on the contrary that $\omega_1 < \mathbf{c}$ and observe that the set Z is of cardinality $\mathbf{c}$. Using the method of transfinite recursion, construct a decreasing (by inclusion) ω_1-sequence $\{Z_\xi : \xi < \omega_1\}$ of subsets of $\mathbb{R}$ such that:

(a) for each $\xi < \omega_1$, the set Z_ξ is contained in Z and $\operatorname{card}(Z \setminus Z_\xi) \leq \omega$;
(b) $(h_\xi + Z_\xi) \cap (h_\zeta + Z_\zeta) = \emptyset$ for any two distinct countable ordinals ξ and ζ.

Then define $X = \cap \{Z_\xi : \xi < \omega_1\}$ and check that:

(c) $\operatorname{card}(Z \setminus X) \leq \omega_1$, so X is also an $\mathbb{R}$-thick set in $\mathbb{R}$;
(d) $(h_\xi + X) \cap (h_\zeta + X) = \emptyset$ for any two distinct countable ordinals ξ and ζ.

Taking into account Remark 23.1 of this chapter, obtain a contradiction which yields the required result.

Chapter 24
Radon Spaces

All measures considered throughout this chapter are assumed to be countably additive.

We would like to begin by recalling a general concept from measure theory (see Chap. 17).

Let E be a ground set and let M be a family of σ-finite measures on E (note that the domains of measures from M may be various σ-algebras of subsets of E).

We say that a set $X \subset E$ is *absolutely* (*universally*) *measurable* with respect to M if X is measurable with respect to each measure from M.

Accordingly, it makes sense to call a function $f : E \to \mathbb{R}$ *absolutely* (*universally*) *measurable* with respect to M if the f-pre-images of all open subsets of $\mathbb{R}$ are absolutely (universally) measurable with respect to M.

It is not hard to see that:

(a) the family of all those sets in E which are absolutely measurable with respect to M is a σ-algebra of subsets of E;
(b) the family of all those real-valued functions on E which are absolutely measurable with respect to M is closed under natural algebraic operations (such as addition, multiplication, etc.) and is also closed under taking the pointwise limits of sequences of functions.

In Chap. 20 of this lecture course we were concerned with some basic properties of Radon measures on Hausdorff topological spaces. Here we would like to discuss the important concept of a Radon topological space, which has found a lot of applications in various branches of functional analysis, convex analysis, optimization, probability theory and stochastic processes, etc.

First, we recall the precise definition of a Radon measure and the associated notion of a Radon space. In the sequel, we will restrict our consideration to those measures which are either finite or σ-finite.

Let E be a Hausdorff topological space and let μ be a σ-finite measure defined on some σ-algebra of subsets of E.

We recall that μ is a *Radon measure* if, for every μ-measurable set X, the equality

$$\mu(X) = \sup\{\mu(K) : K \in \mathrm{dom}(\mu),\ K \subset X,\ K \text{ is compact}\}$$

holds true.

In other words, μ is a Radon measure if, for every μ-measurable set X, there exists a sequence $\{K_n : n \in \mathbb{N}\}$ of compact subsets of X such that

$$(\forall n \in \mathbb{N})(K_n \in \mathrm{dom}(\mu)), \qquad \mu(X \setminus \cup\{K_n : n \in \mathbb{N}\}) = 0.$$

For a finite measure defined on some σ-algebra of subsets of E, the above definition can be replaced by the following equivalent definition (cf. Chap. 20).

A finite measure ν on a Hausdorff space E is Radon if, for any ν-measurable set X and for each real $\varepsilon > 0$, there exists a compact set $K \subset X$ such that $K \in \mathrm{dom}(\nu)$ and $\nu(X) < \nu(K) + \varepsilon$.

Obviously, given any two σ-finite equivalent measures μ and ν, one can assert that μ is a Radon measure if and only if ν is a Radon measure.

We say that a Hausdorff topological space E is a *Radon space* if every σ-finite measure defined on the Borel σ-algebra $\mathcal{B}(E)$ of E is a Radon measure.

As we already know, any nonzero σ-finite measure is equivalent to a probability measure (see Lemma 21.1 from Chap. 21).

This implies that a Hausdorff topological space E is a Radon space if and only if every probability measure defined on the Borel σ-algebra $\mathcal{B}(E)$ is a Radon measure.

Example 24.1 The Lebesgue measure λ_n on the Euclidean n-dimensional space $\mathbb{R}^n$ provides a standard example of a Radon measure. More generally, let X be a Polish topological space and let μ be the completion of a σ-finite Borel measure on X. Then, according to a result of Ulam, μ turns out to be a Radon measure (see Theorem 20.4 in Chap. 20). Therefore, any Polish topological space is a Radon space.

Furthermore, let (Y, d) be a complete metric space whose topological weight $w(Y)$ is not measurable in Ulam's sense, which means that no set E of cardinality $w(Y)$ admits a probability measure defined on the σ-algebra $\mathcal{P}(E)$ and vanishing at all singletons in E. Let ν be the completion of a σ-finite Borel measure on Y. Then, analogously to Ulam's theorem mentioned above, ν turns out to be a Radon measure, because ν has a separable support (see, e.g., [27, 95, 141]). Therefore, (Y, d) is a Radon space.

To give some other nontrivial examples of Radon spaces, recall that a topological space E is *analytic* (*Suslin*) if either $E = \emptyset$ or E is a metrizable continuous image of the canonical Baire product space $\mathbb{N}^\mathbb{N}$, where $\mathbb{N}$ is equipped with the discrete topology. Analytic spaces play an important role in many topics of real analysis, measure theory, and general topology (see, for instance, [33, 82, 109, 119]). Any Borel subset of a Polish space X is simultaneously an analytic subspace of X. On the other hand, if X is uncountable, then there are analytic subspaces of X which are not Borel subsets of X (see [26, 76, 82, 109, 119]).

24 Radon Spaces

Example 24.2 Let X be a Polish space and let E be an analytic subset of X, endowed with the induced topology. According to a well-known result of Luzin, E is a Radon space (for a proof, see e.g. [17, 26, 82, 109, 119]). It follows from this fact that all members of the σ-algebra $\mathcal{S}(X)$ generated by the analytic subsets of X are absolutely measurable sets with respect to the class of the completions of all σ-finite Borel measures on X. Notice that the σ-algebra of all absolutely measurable sets with respect to the same class of measures may be substantially wider than $\mathcal{S}(X)$. For instance, if $X = \mathbb{R}$, then, under the Continuum Hypothesis, the σ-algebra of absolutely measurable sets with respect to the class of the completions of all σ-finite Borel measures on $\mathbb{R}$ has among its members any Luzin subset of $\mathbb{R}$ which is universal measure zero (see Exercise 4 from Chap. 19) and trivially is a Radon space. However, no Luzin set possesses the Baire property, while all members of $\mathcal{S}(\mathbb{R})$ have this property (cf. [109]). Moreover, the cardinality of $\mathcal{S}(\mathbb{R})$ is equal to $\mathbf{c}$, while (assuming **CH**) the cardinality of the family of all Luzin subsets of $\mathbb{R}$ is $2^\mathbf{c}$.

The following simple result shows that certain uncountable families of open sets behave nicely with respect to Radon measures.

Theorem 24.1 *Let E be a Hausdorff topological space and let μ be a σ-finite Radon measure on E such that $\mathcal{B}(E) \subset \text{dom}(\mu)$. Suppose that $\{U_i : i \in I\}$ is a family of open subsets of E which is filtered (i.e., directed) with respect to the standard inclusion relation $\subset$.*

Then the equality $\mu(\cup\{U_i : i \in I\}) = \sup\{\mu(U_i) : i \in I\}$ holds true.

Proof Since μ is a Radon measure and $\mathcal{B}(E) \subset \text{dom}(\mu)$, we may write

$$\mu(\cup\{U_i : i \in I\}) = \sup\{\mu(K) : K \subset \cup\{U_i : i \in I\}, \; K \text{ is compact}\}.$$

Let K be any compact subset of $\cup\{U_i : i \in I\}$. Then the family $\{U_i : i \in I\}$ is an open covering of K, so there exist finitely many sets

$$U_{i_1}, U_{i_2}, \ldots, U_{i_m}$$

from this family such that $K \subset U_{i_1} \cup U_{i_2} \cup \cdots \cup U_{i_m}$.

Since $\{U_i : i \in I\}$ is a filtered family of sets, we can find an index $i \in I$ satisfying the inclusion $U_{i_1} \cup U_{i_2} \cup \cdots \cup U_{i_m} \subset U_i$ and, consequently, satisfying the relations

$$K \subset U_i, \qquad \mu(K) \leq \mu(U_i).$$

This implies at once the inequality

$$\mu(\cup\{U_i : i \in I\}) \leq \sup\{\mu(U_i) : i \in I\}.$$

The reverse inequality

$$\sup\{\mu(U_i) : i \in I\} \leq \mu(\cup\{U_i : i \in I\})$$

is obvious, and so we obtain the desired equality

$$\mu(\cup\{U_i : i \in I\}) = \sup\{\mu(U_i) : i \in I\}.$$

Theorem 24.1 has thus been proved. □

Remark 24.1 In the literature, the fact expressed by Theorem 24.1 is sometimes called the τ-*smoothness* (or τ-*additivity*) of Radon measures.

For our further purposes, it will be convenient to denote by $\mathcal{CBM}(E)$ the class of the completions of all σ-finite Borel measures on a topological space E.

Let us note that if some subset Y of E is absolutely (universally) measurable with respect to the class $\mathcal{CBM}(E)$ and $Z \in \mathcal{B}(Y)$, then Z is also absolutely measurable with respect to $\mathcal{CBM}(E)$. This almost trivial fact will be essentially used below.

Theorem 24.2 *Let X be a Hausdorff topological space and let Y be a Radon subspace of X.*

Then Y is absolutely measurable with respect to the class $\mathcal{CBM}(X)$.

Proof Let μ' be an arbitrary nonzero measure from $\mathcal{CBM}(X)$. By definition, this means that there exists a nonzero σ-finite Borel measure μ on X such that μ' coincides with the completion of μ. Without loss of generality, we may assume that μ is a probability measure, i.e., $\mu(X) = 1$. We may also suppose that Y is μ-thick in X, i.e., $\mu^*(Y) = 1$, where μ^* denotes the outer measure associated with μ (if $\mu^*(Y) < 1$, then we can replace X by any μ-measurable hull of Y).

Now, we define a probability Borel measure ν on Y by the following formula:

$$\nu(B \cap Y) = \mu(B) \qquad (B \in \mathcal{B}(X)).$$

The μ^*-thickness of Y implies that ν is well-defined. Further, since Y is a Radon space, the measure ν must be Radon. So there exists a σ-compact set $Z \subset Y$ such that

$$1 = \mu^*(Y) = \nu(Y) = \nu(Z).$$

This Z is a Borel set in the space X (being a countable union of compact subsets of X). Consequently, Z is μ-measurable and, by the definition of ν, we have

$$1 = \nu(Y) = \nu(Z) = \mu(Z) \leq \mu_*(Y) \leq 1,$$

where μ_* denotes the inner measure associated with μ. Therefore, we finally obtain the relation

$$1 = \mu^*(Y) = \mu_*(Y),$$

which immediately implies the μ'-measurability of Y and hence the universal measurability of Y with respect to the class $\mathcal{CBM}(X)$.

This ends the proof of Theorem 24.2. □

Under natural additional assumptions on a topological space X, the converse statement to Theorem 24.2 can be established.

Theorem 24.3 *Let X be a Radon topological space and let Y be a subset of X absolutely measurable with respect to the class $CBM(X)$.*
Then Y is a Radon space with respect to the induced topology.

Proof Let μ be an arbitrary Borel probability measure on Y. We will demonstrate that μ is a Radon measure. In order to do this, take any set $Z \in \mathcal{B}(Y)$. Since Y is absolutely measurable with respect to the class $CBM(X)$, the set Z is also absolutely measurable with respect to the same class.

Let us define a Borel probability measure ν on X by the formula

$$\nu(A) = \mu(A \cap Y) \qquad (A \in \mathcal{B}(X)).$$

Since X is a Radon space, ν must be a Radon measure on X. Obviously, for ν we have the relations

$$\nu^*(Y) = \nu_*(Y) = 1,$$

$$\nu^*(Z) = \nu_*(Z), \qquad \mu(Z) \le \nu^*(Z).$$

There exists a set $B \in \mathcal{B}(X)$ such that

$$B \subset Z, \qquad \nu(B) = \nu_*(Z) = \nu^*(Z).$$

At the same time, the definition of ν and the relation $B \subset Z \subset Y$ yield at once

$$\nu(B) = \mu(B) \le \mu(Z).$$

Remembering that ν is a Radon measure on X, we may write $\nu(B) = \nu(P)$, where P is a σ-compact subset of X entirely contained in B (hence in Z). Evidently, P is also a σ-compact subset of Z and $\mu(P) = \nu(P)$. We finally obtain

$$\mu(Z) \le \nu^*(Z) = \nu_*(Z) = \nu(B) = \nu(P) = \mu(P) \le \mu(Z),$$

$$\mu(Z) = \mu(P),$$

which implies that μ is a Radon probability measure on Y.

Thus, Y is a Radon topological space and the proof of Theorem 24.3 is complete. □

Various canonical set-theoretical operations over Radon spaces can be considered (such as topological sums, topological products, inductive and projective limits, etc.). Some of them preserve the class of Radon spaces, while others do not (for more details, see the exercises of the present chapter).

Example 24.3 One of the most natural topological operations is taking a continuous image of a given space. But it turns out that even in a class of rather good topological spaces a continuous image of a Radon space can be a non-Radon space. The standard example of this sort is provided by co-analytic (co-Suslin) subsets of the real line $\mathbb{R}$. Indeed, according to the above-mentioned result of Luzin, all analytic and co-analytic subsets of $\mathbb{R}$ are absolutely measurable with respect to the class $\mathcal{CBM}(\mathbb{R})$, so they are Radon spaces. On the other hand, under the Constructibility Axiom $\mathbf{V} = \mathbf{L}$ of Gödel (see, for instance, [76, 108]), there exists a set $X \subset \mathbb{R}$ having the following properties:

(*) X is a co-analytic set (i.e., $\mathbb{R} \setminus X$ is an analytic set);
(**) some continuous image of X is a Lebesgue nonmeasurable subset of $\mathbb{R}$.

This profound result of Gödel indicates that, in general, one cannot assert that any continuous image of a Radon space is also a Radon space. Under Martin's Axiom and the negation of the Continuum Hypothesis, i.e. in the theory

$$\mathbf{ZFC} \ \& \ \mathbf{MA} \ \& \ (\neg \mathbf{CH}),$$

the situation is much better, at least for continuous images of co-analytic sets. Namely, in this theory, if Y is a co-analytic subset of $\mathbb{R}$ and $f : Y \to \mathbb{R}$ is a continuous mapping, then $f(Y)$ is a Radon subspace of $\mathbb{R}$ (see [76]).

Now, we would like to recall an important notion of measurability of real-valued functions which are defined on a Hausdorff topological space E equipped with a σ-finite measure μ.

We shall say that a function $f : E \to \mathbb{R}$ is *measurable in the Luzin sense* (with respect to μ) if there exists a sequence $\{K_n : n \in \mathbb{N}\}$ of compact subsets of E such that:

(1) $(\forall n \in \mathbb{N})(K_n \in \mathrm{dom}(\mu))$ and $\mu(E \setminus \cup\{K_n : n \in \mathbb{N}\}) = 0$;
(2) for each natural index n, the restriction $f|K_n$ is a continuous function.

Example 24.4 By virtue of Luzin's classical C-property (see [104, 121, 134, 141]), every λ-measurable function $f : \mathbb{R} \to \mathbb{R}$ is measurable in the Luzin sense with respect to λ, where λ denotes, as usual, the standard Lebesgue measure on $\mathbb{R}$.

The next theorem is a substantial generalization of the above widely known example.

Theorem 24.4 *Let E be a Hausdorff topological space and let μ be a σ-finite Radon measure on E. Denote by μ' the completion of μ.*

Then every μ'-measurable function $f : E \to \mathbb{R}$ is measurable in the Luzin sense with respect to μ'.

Proof Without loss of generality, we may suppose that μ is a probability measure and that the set $\mathrm{ran}(f)$ is entirely contained in the half-open unit interval $[0, 1[$.

Actually, it suffices to prove the following assertion:

For any real $\varepsilon > 0$, there exists a compact set $K_\varepsilon \in \text{dom}(\mu)$ such that $\mu(K_\varepsilon) \geq 1 - \varepsilon$ and the restriction $f|K_\varepsilon$ is continuous.

The argument presented below imitates the proof of Luzin's C-property of all λ-measurable real-valued functions and, in fact, is fairly standard.

Take a nonzero natural number n and consider the sets

$$X_{i,n} = \{x \in E : i/n \leq f(x) < (i+1)/n\} \quad (i = 0, 1, \ldots, n-1),$$

all of which are pairwise disjoint and μ'-measurable. These sets collectively cover the whole space E. Since μ is a Radon measure, there are compact sets

$$K_{i,n} \in \text{dom}(\mu) \quad (i = 0, 1, \ldots, n-1)$$

such that

$$K_{i,n} \subset X_{i,n}, \quad \mu(X_{i,n} \setminus K_{i,n}) < \varepsilon/n2^n \quad (i = 0, 1, \ldots, n-1).$$

For each natural index $i \in \{0, 1, \ldots, n-1\}$, let $f_{i,n}$ denote the constant function on $K_{i,n}$ whose range coincides with the singleton $\{i/n\}$, and let f_n be the least (by inclusion) extension of all these $f_{i,n}$ ($i = 0, 1, \ldots, n-1$).

One can easily verify the validity of the following five assertions:

(a) the function f_n is defined on the compact μ-measurable set

$$K_n = \cup\{K_{i,n} : 0 \leq i < n\}$$

and is continuous on K_n;
(b) $\mu(K_n) \geq 1 - \varepsilon/2^n$;
(c) the set $K_\varepsilon = \cap\{K_n : n = 1, 2, \ldots\}$ is compact and μ-measurable, and the inequality $\mu(K_\varepsilon) \geq 1 - \varepsilon$ holds;
(d) the sequence of functions $\{f_n|K_\varepsilon : n = 1, 2, \ldots\}$ converges uniformly to the function $f|K_\varepsilon$;
(e) the function $f|K_\varepsilon$ is continuous (as the limit of a uniformly convergent sequence of continuous functions).

Theorem 24.4 has thus been proved. □

Exercises

1. Let E be an arbitrary Hausdorff topological space.
Check that any nonzero σ-finite Radon measure on the Borel σ-algebra of E is equivalent to a Radon probability measure on the same σ-algebra. Infer from this fact that the following two assertions are equivalent:

(a) every Borel probability measure on E is Radon;
(b) E is a Radon topological space.

Also, verify that any finite Radon measure μ on E is outer regular in the sense that, for every μ-measurable set X, the equality

$$\mu(X) = \inf\{\mu(U) : U \in \text{dom}(\mu) \ \& \ X \subset U \ \& \ U \text{ is open in } E\}$$

holds true.

Keeping in mind Theorem 24.1 of this chapter, prove that if μ is a finite Radon measure with $\text{dom}(\mu) = \mathcal{B}(E)$ and a family $\{K_i : i \in I\}$ of compact subsets of E is filtered by the reverse inclusion relation $\supset$, then the relation

$$\mu(\cap\{K_i : i \in I\}) = \inf\{\mu(K_i) : i \in I\}$$

is fulfilled.

2. Show that if some subset Y of a topological space E is absolutely (universally) measurable with respect to the class $\mathcal{CBM}(E)$ and $Z \in \mathcal{B}(Y)$, then Z is also absolutely measurable with respect to $\mathcal{CBM}(E)$.

3. Let E be a Hausdorff topological space, S be an algebra of subsets of E, and let μ be a σ-finite measure on S satisfying the following condition:

(a) for each set $X \in S$ with $\mu(X) < +\infty$ and for any real $\varepsilon > 0$, there exists a compact set $K \subset X$ such that $K \in S$ and $\mu(X \setminus K) < \varepsilon$.

Let $\sigma(S)$ denote the σ-algebra generated by S and let μ' be the measure on $\sigma(S)$ extending μ by Carathéodory's classical method (see Chap. 17).
Prove that the analogous condition holds true for $\sigma(S)$ and μ', i.e.,

(b) for each set $X \in \sigma(S)$ with $\mu'(X) < +\infty$ and for any real $\varepsilon > 0$, there exists a compact set $K \subset X$ such that $K \in \sigma(S)$ and $\mu'(X \setminus K) < \varepsilon$.

In order to demonstrate this fact, keep in mind that $\sigma(S)$ coincides with the monotone class generated by S.

4. Let I be a nonempty finite set of indices and let J be a countably infinite set of indices. Suppose that $\{\mu_i : i \in I\}$ is a family of σ-finite Radon measures and $\{\nu_j : j \in J\}$ is a family of probability Radon measures.
Show that the two associated product measures

$$\mu = \otimes\{\mu_i : i \in I\}, \qquad \nu = \otimes\{\nu_j : j \in J\}$$

are Radon.

For this purpose, take into account the fact formulated in Exercise 3 (also, keep in mind some general properties of product measures, which are discussed in Appendix B).

Remark 24.2 Assuming the Continuum Hypothesis, it can be proved that there exist two compact Radon spaces X and Y such that the compact product space $X \times Y$ is not Radon (see [182]).

5. Let I be a nonempty set of indices, $\{E_i : i \in I\}$ be a family of compact topological spaces and let, for each index $i \in I$, a Radon probability measure μ_i be given on the space E_i.
   Check that the product probability measure $\mu = \otimes\{\mu_i : i \in I\}$ is Radon on the compact space $\prod\{E_i : i \in I\}$.
   For this purpose, use again the fact formulated in Exercise 3.

   Remark 24.3 If $\{\mu_i : i \in I\}$ is an uncountable family of Radon probability measures on Polish topological spaces, then one cannot assert that the product measure $\mu = \otimes\{\mu_i : i \in I\}$ is Radon.

6. Let E be a Hausdorff topological space and let μ be a σ-finite Radon measure defined on the Borel σ-algebra $\mathcal{B}(E)$ of E.
   Show that, for an arbitrary function $f : E \to \mathbb{R}$, the following three assertions are equivalent:
   (a) f is measurable in the Luzin sense (with respect to μ);
   (b) there exists a disjoint countable family $\{K_i : i \in I\}$ of compact subsets of E such that

   $$\mu(E \setminus \cup\{K_i : i \in I\}) = 0$$

   and the restriction $f|K_i$ is continuous for each index $i \in I$;
   (c) for any Borel set $X \subset E$ with $\mu(X) > 0$, there exists a compact set $K \subset X$ such that $\mu(K) > 0$ and the restriction $f|K$ is continuous.

7. Let E be a base set, μ be a nonzero σ-finite measure defined on some σ-algebra of subsets of E, and let $\{X_i : i \in I\}$ be a partition of E into μ-measure zero sets.
   Supposing that $\text{card}(I)$ is not measurable in the Ulam sense, demonstrate that there exists a set $J \subset I$ such that the corresponding union

   $$X_J = \cup\{X_j : j \in J\}$$

   is not μ-measurable.

8. Let ν be a σ-finite measure on a ground set E and let Y be a ν-measurable subset of E with $\nu(Y) > 0$.
   This Y is called an *atom* of ν if, for every ν-measurable set $Y_0 \subset Y$, one has the disjunction

   $$\nu(Y_0) = 0 \,\vee\, \nu(Y \setminus Y_0) = 0.$$

   A σ-finite measure ν is called *non-atomic* if there are no atoms of ν.

Let now μ be a nonzero non-atomic σ-finite measure on a base set E.
Show that there exists a partition $\{Z_i : i \in I\}$ of E satisfying the following two relations:

(a) $\text{card}(I) \leq \mathbf{c}$;
(b) each set Z_i ($i \in I$) is of μ-measure zero.

Argue as follows. First, establish the auxiliary fact:
For any real $\varepsilon > 0$ and for every μ-measurable subset X of E, there exists a disjoint countable family $\{X_j : j \in J\}$ of μ-measurable subsets of X such that

$$X = \cup\{X_j : j \in J\}, \qquad (\forall j \in J)(\mu(X_j) < \varepsilon).$$

Then, for every nonzero natural number k, construct by recursion a countable family of pairwise disjoint μ-measurable sets

$$\{X_{j_1, j_2, \ldots, j_k} : j_1 \in \mathbb{N}, j_2 \in \mathbb{N}, \ldots, j_k \in \mathbb{N}\}$$

so that the next three conditions are fulfilled:

$$X_{j_1} \supset X_{j_1, j_2} \supset \cdots \supset X_{j_1, j_2, \ldots, j_k} \supset \cdots,$$

$$\cup \{X_{j_1, j_2, \ldots, j_k} : j_1 \in \mathbb{N}, j_2 \in \mathbb{N}, \ldots, j_k \in \mathbb{N}\} = E,$$

$$\mu(X_{j_1, j_2, \ldots, j_k}) < 1/k.$$

Finally, take the nonempty intersections of the form

$$X_{j_1} \cap X_{j_1, j_2} \cap \cdots \cap X_{j_1, j_2, \ldots, j_k} \cap \cdots$$

as members Z_i ($i \in I$) of the required partition of E.

9. Let I be a set of indices with cardinality nonmeasurable in the Ulam sense, let $\{E_i : i \in I\}$ be a family of Radon topological spaces, and let E denote the topological sum of $\{E_i : i \in I\}$.
Show that E is also a Radon topological space.
Deduce from this fact that if X is an infinite discrete topological space and X^* is Alexandrov's compactification of X, then the following two assertions are equivalent:

(a) $\text{card}(X)$ is not measurable in Ulam's sense;
(b) X^* is a Radon space.

Conclude from (a) and (b) that the existence of a compact Radon space whose cardinality is arbitrarily large and all subsets of which are Borel is consistent with **ZFC** set theory.

10. Observe that every Hausdorff universal measure zero space is a Radon space.
Let I be a set of indices whose cardinality is not measurable in Ulam's sense and let $\{E_i : i \in I\}$ be a family of universal measure zero spaces.
Verify that the topological sum of $\{E_i : i \in I\}$ is also a universal measure zero space.

11. Let $\{E_n : n \in \{1, 2, \ldots, k\}\}$ be a finite family of universal measure zero topological spaces.
Check that the product topological space $\prod\{E_n : n \in \{1, 2, \ldots, k\}\}$ is also a universal measure zero space.
Give an example of a countable family $\{X_n : n \in \mathbb{N}\}$ of universal measure zero topological spaces such that the product topological space $\prod\{X_n : n \in \mathbb{N}\}$ is not a universal measure zero space.

12. Let I be a nonempty countable set of indices, let $\{E_i : i \in I\}$ be a family of topological spaces, and let μ_i be a Borel probability measure on each E_i such that there is a separable support Z_i of μ_i (i.e., $\mu_i(Z_i) = 1$ and Z_i has a countable everywhere dense subset).
Prove that the product probability measure $\mu = \otimes\{\mu_i : i \in I\}$ on the topological product space $\prod\{E_i : i \in I\}$ also possesses a separable support.
Present an example which shows that the assumption of the countability of I is essential for obtaining the above result.

13. Recall that a set $X \subset \mathbb{R}$ is *totally imperfect* in $\mathbb{R}$ if no nonempty perfect subset of $\mathbb{R}$ is contained in X.
A set $B \subset \mathbb{R}$ is called a *Bernstein subset* of $\mathbb{R}$ if both sets B and $\mathbb{R} \setminus B$ are *totally imperfect* in $\mathbb{R}$ (see [12, 53, 109, 132, 141]).
Demonstrate that:

(a) there exist Bernstein subsets of $\mathbb{R}$;
(b) no Bernstein set (equipped with the induced topology) is a Radon space.

For (a), use some well-ordering of $\mathbb{R}$ and the method of transfinite induction.

Chapter 25
Nonmeasurable Sets with Respect to Radon Measures

Let us recall an important notion due to Gnedenko and Kolmogorov [56], which plays a significant role in modern probability theory and the theory of stochastic processes.

Let E be a base (ground) set, S be a σ-algebra of subsets of E, and let μ be a probability countably additive measure whose domain coincides with S.

The triplet (E, S, μ) is usually called a *probability measure space* (or, briefly, a *probability space*).

By definition, a *random variable* on (E, S, μ) is any real-valued function on E measurable with respect to μ.

A probability space (E, S, μ) is said to be *perfect* if, for any random variable $f : E \to \mathbb{R}$, there exists a Borel subset T of $\mathbb{R}$ such that

$$T \subset \operatorname{ran}(f), \qquad \mu(f^{-1}(T)) = 1.$$

From Theorem 24.4 of Chap. 24 one can infer the following theorem.

Theorem 25.1 *Suppose that E is a Hausdorff topological space and μ is a Radon probability measure on a σ-algebra of subsets of E.*

Then $(E, \operatorname{dom}(\mu), \mu)$ is a perfect probability space.

Consequently, if E is a Radon topological space, then, for every Borel probability measure ν on E, the probability space $(E, \mathcal{B}(E), \nu)$ is a perfect space.

Proof Let $f : E \to \mathbb{R}$ be a random variable on the probability measure space $(E, \operatorname{dom}(\mu), \mu)$.

According to Theorem 24.4 of Chap. 24, this f is measurable in the Luzin sense with respect to the completion of μ, i.e., there exists some countable family $\{K_n : n \in \mathbb{N}\}$ of compact subsets of E such that:

(1) $(\forall n \in \mathbb{N})(K_n \in \operatorname{dom}(\mu))$ and $\mu(E \setminus \cup\{K_n : n \in \mathbb{N}\}) = 0$;
(2) for each $n \in \mathbb{N}$, the restriction $f|K_n$ is continuous.

Let us define

$$T = \cup\{f(K_n) : n \in \mathbb{N}\}.$$

Since all $f(K_n)$ are compact subsets of $\mathbb{R}$ (as continuous images of compact sets), T is a σ-compact set and hence is of type F_σ in $\mathbb{R}$. In addition, one can easily see that

$$1 \geq \mu(f^{-1}(T)) \geq \mu(\cup\{K_n : n \in \mathbb{N}\}) = 1,$$

so $\mu(f^{-1}(T)) = 1$, which yields the required result and completes the proof of the theorem. □

Theorem 25.2 *Let (E, S, μ) be a probability space and let $f : E \to \mathbb{R}$ be a random variable on E such that:*

(1) $\mu(f^{-1}(t)) = 0$ *for each point $t \in \mathbb{R}$;*
(2) *the range of f contains no uncountable closed subset of $\mathbb{R}$.*

Then the space (E, S, μ) is not perfect.

Proof Suppose to the contrary that (E, S, μ) is a perfect probability space. Then, by virtue of the definition, there exists a Borel subset T of $\mathbb{R}$ such that

$$T \subset \mathrm{ran}(f), \qquad \mu(f^{-1}(T)) = 1.$$

In view of condition (1), T cannot be countable, so is uncountable. But then, according to the well-known theorem of Alexandrov and Hausdorff (see, e.g., [26, 76, 82, 109, 119]), T must contain an uncountable closed subset of $\mathbb{R}$, which contradicts condition (2).

The obtained contradiction ends the proof of Theorem 25.2. □

Before formulating the next result, let us recall the notion of an atom of a given σ-finite measure μ (cf. Exercise 8 from Chap. 24).

A μ-measurable set X is an *atom* of μ if $\mu(X) > 0$ and, for each μ-measurable set $Y \subset X$, one has

$$\mu(Y) = 0 \vee \mu(X \setminus Y) = 0.$$

A measure μ is called *non-atomic* if there are no atoms of μ.

Theorem 25.3 *Let (E, S, μ) be a perfect probability space with a non-atomic measure μ.*

Then the σ-algebra S differs from the power set $\mathcal{P}(E)$, i.e., there exists a subset of E nonmeasurable with respect to μ.

Proof Suppose to the contrary that $S = \mathcal{P}(E)$. Then every real-valued function on E may be treated as a random variable on the probability space (E, S, μ). Since our μ does not have atoms, there exists a partition $\{X_i : i \in I\}$ of E such that:

(a) $\operatorname{card}(I) \leq \mathbf{c}$;
(b) $\mu(X_i) = 0$ for each index $i \in I$.

The existence of a partition $\{X_i : i \in I\}$ of E with the above-mentioned properties (a) and (b) can be justified by a fairly standard argument (see again Exercise 8 from Chap. 24).

Let T be a totally imperfect subset of $\mathbb{R}$ with $\operatorname{card}(T) = \operatorname{card}(I)$. Obviously, this T can be realized as a certain subset of a Bernstein set in $\mathbb{R}$ (cf. Exercise 13 from Chap. 24).

Let ϕ be a bijection acting from I onto T. We define a real-valued function f on E as follows:

$$f(x) = \phi(i) \text{ if and only if } x \in X_i.$$

By virtue of our assumption, f is a random variable on the space $(E, \mathcal{S}, \mu)$ and $\mu(f^{-1}(t)) = 0$ for each point $t \in \mathbb{R}$. On the other hand, it is clear that the range of f coincides with T and does not contain any uncountable closed subset of $\mathbb{R}$. This contradicts Theorem 25.2.

The obtained contradiction completes the proof. □

As a corollary of Theorem 25.3, one can deduce the following highly nontrivial result (see [126]).

Theorem 25.4 *Let E be a Hausdorff topological space and let μ be a nonzero σ-finite Radon measure on E vanishing at all points of E.*

Then $\operatorname{dom}(\mu)$ differs from $\mathcal{P}(E)$, i.e., there exists a subset of E nonmeasurable with respect to μ.

Proof We may assume, without loss of generality, that μ is a Radon probability measure on E. Taking into account Theorems 25.1, 25.2 and 25.3, it suffices to demonstrate that μ does not possess atoms, i.e., there exists no μ-measurable set X such that $\mu(X) > 0$ and, for any μ-measurable set $Y \subset X$, the disjunction

$$\mu(Y) = 0 \quad \vee \quad \mu(X \setminus Y) = 0$$

holds.

Suppose to the contrary that such an atom X does exist. Since μ is a Radon measure, we may assume that X is compact. Consider the family

$$\mathcal{F} = \{Y \subset X : Y \text{ is compact and } \mu(Y) = \mu(X)\}.$$

Clearly, this family is centered. Consequently, the set $X_0 = \cap\{Y : Y \in \mathcal{F}\}$ is nonempty and compact as well. Moreover, using again the fact that μ is a Radon measure, it is not difficult to check that

$$\mu(X_0) = \mu(X) > 0.$$

(In this connection, see also Exercise 1 from Chap. 24.) So, X_0 must be uncountable in view of the assumption that μ vanishes at all singletons in E. Let y and z be any two distinct points from X_0. According to Urysohn's classical theorem (see, e.g., [40] or [84]), there exists a continuous function

$$\phi : X_0 \to [0, 1]$$

such that $\phi(y) = 1$ and $\phi(z) = 0$. Let us put

$$Y = \phi^{-1}([0, 1/2]), \qquad Z = \phi^{-1}([1/2, 1]).$$

Both Y and Z are compact subsets of X_0 and

$$X_0 = Y \cup Z, \qquad y \notin Y, \qquad z \notin Z.$$

Therefore, we get

$$0 < \mu(X) = \mu(X_0) \leq \mu(Y) + \mu(Z),$$
$$\mu(Y) = \mu(X) \ \vee \ \mu(Z) = \mu(X).$$

Without loss of generality, we may suppose that $\mu(Y) = \mu(X)$. So $Y \in \mathcal{F}$ and, according to the definition of X_0, we must have $X_0 \subset Y$, which contradicts the fact that Y is a proper subset of X_0 (because $y \in X_0 \setminus Y$).

The obtained contradiction ends the proof of Theorem 25.4. □

Remark 25.1 For further extensions and generalizations of Theorem 25.4 and some related results, see [25, 29, 47].

Exercises

1. Show that every finite Radon measure μ defined on the Borel σ-algebra of a compact space E has a smallest closed support, i.e., there exists a least (with respect to the inclusion relation) closed subset $K = K(\mu)$ of E such that $\mu(E \setminus K) = 0$.
   Observe that this K is unique and possesses the property that $\mu(U(x)) > 0$ for each point $x \in K$ and for any open neighborhood $U(x)$ of x.
   Check that the Dieudonné measure ν' on the compact space $[0, \omega_1]$ does not have a minimal closed support (with respect to the inclusion relation).
2. Let E be an arbitrary Hausdorff topological space.
   Prove that the following two assertions are equivalent:

   (a) there exists a nonzero σ-finite Radon measure μ defined on the Borel σ-algebra of E and vanishing at all points of E;

(b) there exists a compact subset of E which can be continuously mapped onto the closed unit interval $[0, 1]$.

Argue as follows. Let (a) be satisfied and let K be a compact subset of E such that $0 < \mu(K) < +\infty$. According to Exercise 1, there exists a smallest closed support K_0 of the restriction of μ to K. Infer that all elements of K_0 are its condensation points and, by using ordinary recursion, construct a dyadic system of uncountable closed subsets of K_0 (cf. Exercise 14 from Chap. 4). Then deduce that K_0 contains a compact subset which can be continuously mapped onto the Cantor space $\{0, 1\}^\omega$ and, therefore, can be continuously mapped onto $[0, 1]$. This yields (b) and the validity of the implication (a) $\Rightarrow$ (b).

To demonstrate the validity of the converse implication (b) $\Rightarrow$ (a), suppose that (b) is satisfied. Let F be a compact subset of E such that some function

$$h : F \to [0, 1]$$

is a continuous surjection. Consider the σ-algebra of sets

$$S = \{h^{-1}(B) : B \in \mathcal{B}([0, 1])\}$$

and put

$$\nu(h^{-1}(B)) = \lambda(B) \qquad (B \in \mathcal{B}([0, 1])).$$

Check that ν is a Radon non-atomic probability measure on S. By using Exercise 10 from Chap. 20, extend ν to a Radon probability measure defined on the Borel σ-algebra $\mathcal{B}(F)$ and vanishing at all singletons in F. This yields (a) and so proves the implication (b) $\Rightarrow$ (a).

3. Let E be a locally compact topological space and let μ be a finite countably additive τ-smooth Borel measure on E (see Remark 24.1 from Chap. 24). Prove that μ is a Radon measure.

For this purpose, first verify that if U is any open set in E, then there exists a σ-compact subset X of U such that $\mu(U) = \mu(X)$. Finally, use Exercise 9 of Chap. 20.

4. Recall that a cardinal number **a** is *two-valued measurable* if, for a set E of cardinality **a**, there exists a two-valued countably additive probability measure μ such that

$$\mathrm{dom}(\mu) = \mathcal{P}(E), \qquad (\forall x \in E)(\mu(\{x\}) = 0).$$

According to Ulam's classical result [180], the existence of two-valued measurable cardinals cannot be proved within **ZFC** set theory.

Observe that every two-valued probability measure space is a perfect space and infer from this fact that if $\mathrm{card}(E)$ is a two-valued measurable cardinal number, then there exists a perfect space of the form $(E, \mathcal{P}(E), \mu)$, where μ

is a two-valued probability measure whose domain coincides with $\mathcal{P}(E)$ and which vanishes at all singletons in E.

Conclude that Theorem 25.4 of this chapter does not admit a generalization (within **ZFC** set theory) to the class of all perfect probability spaces.

5. Let E be a Hausdorff topological space and let μ be a σ-finite Radon measure defined on the Borel σ-algebra $\mathcal{B}(E)$ of E. Suppose, in addition, that an ω_1-sequence $\{F_\xi : \xi < \omega_1\}$ of closed subsets of E is given such that

$$\mu(F_\xi) = 0 \quad (\xi < \omega_1).$$

Assuming Martin's Axiom with the negation of the Continuum Hypothesis, show that

$$\mu_*(\cup\{F_\xi : \xi < \omega_1\}) = 0,$$

where μ_* denotes the inner measure produced by μ.

For this purpose, use the fact that every finite Radon measure defined on the Borel σ-algebra of a compact space has a smallest closed support (see Exercise 1). Also, keep in mind that the above-mentioned support is compact and satisfies Suslin's condition, so a topological version of Martin's Axiom can be applied in this case (see [46] or [108] about various forms of **MA**).

6. Let (E, d) be a complete metric space with card$(E) = \mathbf{c}$ and without isolated points.

Assuming that $\mathbf{c}$ is not measurable in Ulam's sense, show that there exists a subset B of E which is absolutely nonmeasurable with respect to the class $\mathcal{CBM}_0(E)$ of the completions of all those nonzero σ-finite Borel measures on E which vanish at all points of E.

For this purpose, take as B an appropriate analog of a Bernstein subset of $\mathbb{R}$ (cf. Exercise 13 of Chap. 24).

7. Equip the two-element set $\{0, 1\}$ with the discrete topology and consider the topological product space $E = \{0, 1\}^\mathbf{c}$ (this is the *generalized Cantor space*, whose topological weight is equal to $\mathbf{c}$).

Assuming that the cardinal $\mathbf{c}$ is measurable in Ulam's sense, verify that no subset of E is absolutely nonmeasurable with respect to the class $\mathcal{CBM}_0(E)$ of the completions of all those nonzero σ-finite Borel measures on E which vanish at every singleton in E.

In order to obtain the required result, identify $\mathbf{c}$ with the least ordinal number α such that card$(\alpha) = \mathbf{c}$ and consider in E the family of characteristic functions

$$D = \{\chi_{[0,\xi]} : \xi < \alpha\}.$$

Check that D is a discrete Borel subset of E such that card$(D) = \mathbf{c}$. Further, using D, define a Borel probability measure μ on E which vanishes at all singletons of E and whose completion μ' coincides with the power set of E. This trivially implies the non-existence of absolutely nonmeasurable subsets of E with respect to the class $\mathcal{CBM}_0(E)$.

In addition, taking into account Theorem 25.4 of the present chapter, conclude that E is not a Radon space under the above assumption on **c**.

8. Let E be a Hilbert space (over $\mathbb{R}$) whose orthonormal basis is of cardinality **c**, and suppose that **c** is measurable in Ulam's sense.
   Demonstrate that:

   (a) there exists no subset of E which is absolutely nonmeasurable with respect to the class $\mathcal{CBM}_0(E)$;

   (b) there exists a set $X \subset E$ such that X is absolutely nonmeasurable with respect to the class of the completions of all nonzero σ-finite Radon measures on E vanishing at all singletons in E.

9. Assume Martin's Axiom with the negation of the Continuum Hypothesis. Let E be a Polish space and let $X \subset E$ be a continuous image of some co-analytic subset of a Polish space.
   Prove that X is a Radon topological space.
   For this purpose, use the classical result of Luzin and Sierpiński, according to which X is representable as the union of an ω_1-sequence of Borel subsets of E (see, e.g., [26, 76, 109, 119]).

10. Let E be a compact universal measure zero space, E' be a topological space, and let $f : E \to E'$ be a surjective function such that the f-pre-images of all open sets in E' are G_δ-subsets of E.
    Show that either E' is universal measure zero or E' is not a Radon space.
    For this purpose, keep in mind Exercise 10 from Chap. 20.
    Let X be a discrete topological space with card$(X) = \omega_1$, let $x \notin X$, and let $X^* = X \cup \{x\}$ denote Alexandrov's compactification of X. Consider any bijection $h : X^* \to [0, \omega_1]$ such that $h(x) = \omega_1$.
    Verify that the h-pre-images of all open sets in $[0, \omega_1]$ are G_δ-subsets of X^* and observe that the space X^* is universal measure zero, while the space $[0, \omega_1]$ is not Radon.

11. Let Y be a Hausdorff universal measure zero space and let Z be a Radon space. Check that the topological product $Y \times Z$ is a Radon space.

12. Let X be a subset of the real line $\mathbb{R}$.
    Prove that the following two assertions are equivalent:

    (a) X is a Bernstein set in $\mathbb{R}$, i.e., both sets X and $\mathbb{R} \setminus X$ are totally imperfect in $\mathbb{R}$;

    (b) X is absolutely nonmeasurable with respect to the class $\mathcal{CBM}_0(\mathbb{R})$.

    Replacing $\mathbb{R}$ by an uncountable Polish space E, formulate and prove the analogous equivalence for a subset Z of E.

Remark 25.2 The reader can see that the standard definition of Bernstein sets is purely topological. Exercise 12 shows that it is possible to define Bernstein sets in terms of the absolute nonmeasurability with respect to a concrete class of measures on $\mathbb{R}$.

13. Give an example of a Bernstein subset B of $\mathbb{R}$ measurable with respect to some $\mathbb{R}$-invariant measure μ on $(\mathbb{R}, +)$ which extends the Lebesgue measure λ.
    For this purpose, use the method of transfinite recursion and construct a Bernstein set B so that

    $$(\forall t \in \mathbb{R})(\mathrm{card}(B \triangle (t + B)) < \mathbf{c}).$$

    Then, applying Marczewski's method, define the required measure μ.
    Conclude from the above result that there exist Bernstein sets in $\mathbb{R}$ which are not absolutely nonmeasurable with respect to the class of all $\mathbb{R}$-invariant measures on $\mathbb{R}$ extending λ.

14. Consider the topological space $(\mathbb{R}, \mathcal{S})$, where $\mathcal{S}$ denotes the *Sorgenfrey topology* on $\mathbb{R}$ (see [40, 84] or Exercise 1 from Appendix E).
    Demonstrate that:

    (a) the Borel σ-algebra of this space coincides with the standard Borel σ-algebra of $\mathbb{R}$;
    (b) each compact subset of this space is at most countable.

    For (a), use the fact that $(\mathbb{R}, \mathcal{S})$ is hereditarily Lindelöf.
    For (b), first verify that no compact subset of $(\mathbb{R}, \mathcal{S})$ contains a strictly increasing infinite sequence of points. Then take into account the fact that any uncountable subset of $\mathbb{R}$ contains many strictly increasing infinite sequences of points.
    Conclude from (a) and (b) that the Lebesgue measure λ is not a Radon measure on $(\mathbb{R}, \mathcal{S})$ and hence $(\mathbb{R}, \mathcal{S})$ is not a Radon space.

15. Let $\{X_i : i \in I\}$ be an arbitrary partition of $\mathbb{R}$ into λ-measure zero sets.
    Prove that there exists a subset J of I such that the set $\cup \{X_i : i \in J\}$ is not λ-measurable.
    Argue as follows. Since $\mathrm{card}(I) \leq \mathbf{c}$, one may suppose that I is a totally imperfect set in $\mathbb{R}$ (e.g., I can be realized as an appropriate subset of a Bernstein set in $\mathbb{R}$). Define a function $f : \mathbb{R} \to \mathbb{R}$ by the formula

    $$f(x) = i \Leftrightarrow x \in X_i.$$

    Then show that f is not λ-measurable. For this purpose, suppose on the contrary that f is measurable in the Lebesgue sense and, using Luzin's classical theorem, take a compact subset K of $\mathbb{R}$ with $\lambda(K) > 0$ such that the restricted function $f|K$ is continuous. Observe that $f(K)$ is a compact set contained in I. So, by definition of I, one has the inequality

    $$\mathrm{card}(f(K)) \leq \omega.$$

    Consequently, K is covered by some countable subfamily of $\{X_i : i \in I\}$, which is impossible in view of the relation $\lambda(K) > 0$. The obtained contradiction yields that f is not λ-measurable, i.e, there exists a Borel set

B in $\mathbb{R}$ for which the set $f^{-1}(B)$ is not λ-measurable. Finally, let $J = B \cap I$ and, keeping in mind that

$$f^{-1}(B) = f^{-1}(J) = \cup\{X_i : i \in J\},$$

obtain the required result.

16. Let $\{Y_i : i \in I\}$ be an arbitrary partition of $\mathbb{R}$ into sets of first Baire category. Prove that there exists a subset J of I such that the set $\cup \{Y_i : i \in J\}$ does not possess the Baire property.

    Argue similarly to the scheme presented in Exercise 15. Instead of Luzin's theorem, use the following purely topological fact:

    If g is any function from $\mathbb{R}$ into $\mathbb{R}$ having the Baire property, then there exists a subset Z of $\mathbb{R}$ which is of first category and for which the restricted function $g|(\mathbb{R} \setminus Z)$ is continuous.

17. Work in **ZF** & **DC** theory and verify that if there exists a totally imperfect subset X of $\mathbb{R}$ with $\text{card}(X) = \mathbf{c}$, then there exists a subset of $\mathbb{R}$ nonmeasurable in the Lebesgue sense.

    For this purpose, apply again Luzin's theorem as in Exercise 15.

18. Supposing that there exists a partition $\{X_\xi : \xi < \omega_1\}$ of $\mathbb{R}$ into λ-measure zero sets, deduce that no measure on $\mathbb{R}$ extending λ is defined on the family of all subsets of $\mathbb{R}$.

    For this purpose, use Ulam's result stating that the cardinal ω_1 is not real-valued measurable.

Remark 25.3 Working within **ZFC** theory, it is impossible to establish the existence of a partition $\{X_\xi : \xi < \omega_1\}$ of $\mathbb{R}$ into λ-measure zero sets. Indeed, the conjunction **MA** & $\neg$**CH** directly implies that such a partition of $\mathbb{R}$ does not exist. At the same time, under **CH**, one trivially has

$$\mathbb{R} = \cup\{X_\xi : \xi < \omega_1\},$$

where X_ξ ($\xi < \omega_1$) are all pairwise distinct singletons in $\mathbb{R}$.

19. Work in **ZF** & **DC** theory and, assuming that there exists a partition $\{X_\xi : \xi < \omega_1\}$ of $\mathbb{R}$ into λ-measure zero sets, infer that there exists a λ-nonmeasurable subset of $\mathbb{R}$.

    Argue as follows. For every point $x \in \mathbb{R}$, denote by $\xi(x)$ the unique ordinal ξ such that x belongs to the set X_ξ. Then in the Euclidean plane $\mathbb{R}^2$ consider the set

    $$Z = \{(x, y) : \xi(x) \leq \xi(y)\}$$

    and, using Fubini's theorem, establish that Z is nonmeasurable with respect to the two-dimensional Lebesgue measure λ_2 on $\mathbb{R}^2$. Conclude from this fact that there is a subset of $\mathbb{R}$ nonmeasurable with respect to λ.

Chapter 26
The Radon–Nikodym Theorem

One of the most important results in classical measure theory is the Radon–Nikodym theorem (see [17, 63, 104, 154]). In the present chapter we give the standard proof of this remarkable theorem. Also, in the sequel we will touch upon nontrivial logical interrelations between the Radon–Nikodym theorem and some other statements of mathematical analysis.

Let E be a ground set and let S be a σ-algebra of subsets of E.

A functional $\nu : S \to \mathbb{R}$ is called a *signed measure* (or, sometimes, *charge*) if $\nu(\emptyset) = 0$ and ν is countably additive on S, i.e.,

$$\nu(\cup\{X_n : n \in \mathbb{N}\}) = \sum\{\nu(X_n) : n \in \mathbb{N}\}$$

for every disjoint countable family $\{X_n : n \in \mathbb{N}\}$ of members of S.

The above formula trivially implies that the series

$$\sum\{\nu(X_n) : n \in \mathbb{N}\}$$

commutatively converges to the real number $\nu(\cup\{X_n : n \in \mathbb{N}\})$, i.e., the value of the series does not change after any permutation of its terms. This is equivalent to the fact that the series is absolutely convergent; in other words, we have

$$\sum\{|\nu(X_n)| : n \in \mathbb{N}\} < +\infty.$$

The following helpful result is known as the Hahn–Jordan decomposition theorem.

Theorem 26.1 *Let $\nu : S \to \mathbb{R}$ be a signed measure defined on a σ-algebra S of subsets of a ground set E.*

Then there are two sets $A \in S$ and $B \in S$ such that:

(1) $A \cap B = \emptyset$ and $A \cup B = E$;
(2) *A is positive with respect to ν, i.e., for any set $X \in S$, one has the inequality* $\nu(X \cap A) \geq 0$;
(3) *B is negative with respect to ν, i.e., for any set $X \in S$, one has the inequality* $\nu(X \cap B) \leq 0$.

Proof Consider the family $\mathcal{A} \subset S$ of all positive subsets of E with respect to ν. Clearly, this family is countably additive and is closed under taking those subsets of positive sets which belong to S. Let us denote

$$t = \sup\{\nu(A) : A \in \mathcal{A}\}.$$

In view of the countable additivity of $\mathcal{A}$, there exists an $A \in \mathcal{A}$ such that $\nu(A) = t < +\infty$. Let us define

$$B = E \setminus A$$

and let us check that B is negative with respect to ν.

Suppose for a moment otherwise, i.e., there exists a set $X \in S$ such that $X \subset B$ and $\nu(X) > 0$. Obviously, the set X cannot be positive (because of $\nu(A) = t$), so there is an $X_0 \in S$ such that $X_0 \subset X$ and $\nu(X_0) < 0$.

Now, assume that for a nonzero ordinal number $\xi < \omega_1$, we have already defined a disjoint family $\{X_\zeta : \zeta < \xi\}$ of members of S such that

$$(\forall \zeta < \xi)(X_\zeta \subset X \ \& \ \nu(X_\zeta) < 0).$$

Consider the set $X \setminus \cup\{X_\zeta : \zeta < \xi\}$, which obviously belongs to S. Since $\nu(X) > 0$ and $\nu(X_\zeta) < 0$ for all ordinals $\zeta < \xi$, we must have

$$\nu(X \setminus \cup\{X_\zeta : \zeta < \xi\}) > 0.$$

The set $X \setminus \cup\{X_\zeta : \zeta < \xi\}$ cannot be positive, so there exists a set $Y \in S$ satisfying the relations

$$Y \subset X \setminus \cup\{X_\zeta : \zeta < \xi\}, \qquad \nu(Y) < 0.$$

We then put $X_\xi = Y$. Proceeding in this manner, we come to the disjoint uncountable family $\{X_\xi : \xi < \omega_1\} \subset S$, all members of which are of strictly negative ν-measure. But it is easy to see that, in this case, there is a countable ordinal η such that

$$\nu(\cup\{X_\zeta : \zeta < \eta\}) = -\infty,$$

which is impossible.

The obtained contradiction completes the proof. □

In general, a decomposition $\{A, B\}$ of Theorem 26.1 is not uniquely determined. But it is easy to see that if $\{A', B'\}$ is another such decomposition, then

$$\nu(A \setminus A') = \nu(A' \setminus A) = 0,$$
$$\nu(B \setminus B') = \nu(B' \setminus B) = 0.$$

Keeping in mind these equalities, we can say that the above decomposition $\{A, B\}$ of E is unique in some natural sense.

So, we may associate with the initial signed measure ν the two finite countably additive measures

$$\nu_1 : S \to \mathbb{R}, \qquad \nu_2 : S \to \mathbb{R}$$

defined by the following formulas:

$$\nu_1(X) = \nu(X \cap A), \qquad \nu_2(X) = -\nu(X \cap B),$$

where X is any member of S. According to these definitions of ν_1 and ν_2, we get

$$\nu(X) = \nu_1(X) - \nu_2(X) \qquad (X \in S)$$

or, equivalently, $\nu = \nu_1 - \nu_2$.

The last equality is also called a *Hahn–Jordan representation* of the signed measure ν in the form of the difference of two associated finite measures.

Clearly, the sum $\nu_1 + \nu_2$ is a measure on S which is called the *variation* of ν. It is usually denoted by the symbol $|\nu|$.

Now, let us introduce the important notion of absolute continuity of a signed measure with respect to a countably additive measure.

Let E be a base set and let S be a σ-algebra of subsets of E.

A signed measure $\nu : S \to \mathbb{R}$ is called *absolutely continuous* with respect to a countably additive measure $\mu : S \to \mathbb{R}$ if, for any set $X \in S$, the relation $\mu(X) = 0$ implies the relation $\nu(X) = 0$.

Theorem 26.2 *The following two statements hold:*

(1) *a signed measure ν is absolutely continuous with respect to a measure μ if and only if the variation $|\nu|$ is absolutely continuous with respect to μ;*
(2) *a signed measure ν is absolutely continuous with respect to a measure μ if and only if, for any real $\varepsilon > 0$, there exists a real $\delta > 0$ such that*

$$\mu(X) < \delta \Rightarrow |\nu(X)| < \varepsilon$$

whenever $X \in S$.

Proof The statement (1) is easily verified, so we leave it to the reader.

In order to demonstrate the validity of (2), let us first remark that if, for any real $\varepsilon > 0$, there exists a real $\delta > 0$ satisfying

$$(\forall X \in \mathcal{S})(\mu(X) < \delta \Rightarrow |\nu(X)| < \varepsilon),$$

then ν turns out to be absolutely continuous with respect to μ. Indeed, it suffices to consider any set $X \in \mathcal{S}$ with $\mu(X) = 0 < \delta$. For such X we must have $|\nu(X)| < \varepsilon$, where $\varepsilon > 0$ may be arbitrarily small. The latter immediately implies $|\nu(X)| = 0$.

Suppose now that ν is absolutely continuous with respect to μ, but the formula

$$(\forall \varepsilon > 0)(\exists \delta > 0)(\forall X \in \mathcal{S})(\mu(X) < \delta \Rightarrow |\nu(X)| < \varepsilon)$$

is not true, i.e., there exists a real $\varepsilon_0 > 0$ for which no suitable real $\delta > 0$ can be found.

The latter means that, for any natural number n, there is a set $X_n \in \mathcal{S}$ such that $\mu(X_n) < 1/2^n$, but $|\nu|(X_n) \geq \varepsilon_0$. Let

$$X = \limsup\{X_n : n \in \mathbb{N}\}.$$

Obviously, we may write

$$\mu(X) = 0, \qquad |\nu|(X) \geq \varepsilon_0,$$

which contradicts the absolute continuity of ν with respect to μ.

Theorem 26.2 has thus been proved. $\square$

Before formulating the main result of this chapter, we would like to recall the notion of the equivalence of real-valued functions with respect to a given countably additive measure μ on a base (ground) set E. Let

$$f : E \to \mathbb{R}, \qquad g : E \to \mathbb{R}$$

be two functions (not necessarily μ-measurable).

We say that these two functions are μ-*equivalent* if the set

$$\{x \in E : f(x) \neq g(x)\}$$

is of μ-measure zero.

It is not hard to verify that if the measure μ is complete, then the μ-equivalence just introduced is indeed an equivalence relation on the family of all real-valued functions on E.

As a rule, real-valued μ-equivalent functions are identified.

Now, using the notion of real-valued μ-integrable functions (see Appendix A) and following a fairly standard argument, we will deduce the fundamental Radon–Nikodym theorem from the Hahn–Jordan theorem on a decomposition of a signed measure.

26 The Radon–Nikodym Theorem

Theorem 26.3 *Let E be a ground set, let S be a σ-algebra of subsets of E, and let a signed measure $\nu : S \to \mathbb{R}$ be absolutely continuous with respect to a finite countably additive measure μ on S.*

Then there exists a μ-integrable function $f^* : E \to \mathbb{R}$ such that the equality

$$\nu(X) = \int_X f^*(x) \, d\mu(x)$$

holds true for every set $X \in S$.

Moreover, a function f^* is unique up to the μ-equivalence of real-valued functions on E.

Proof Since any signed measure is representable as the difference of two measures, it suffices to restrict further considerations to the case when ν is a measure, i.e., $\nu(X) \geq 0$ for each $X \in S$.

We shall say that a non-negative μ-measurable function $f : E \to \mathbb{R}$ is *admissible* if

$$\int_X f(x) \, d\mu(x) \leq \nu(X)$$

for all sets $X \in S$.

Denote by $\mathcal{F}$ the family of all admissible functions. Since $f = 0$ belongs to $\mathcal{F}$, we have $\mathcal{F} \neq \emptyset$.

Take any two functions $g \in \mathcal{F}$ and $h \in \mathcal{F}$, let $\phi = \sup(g, h)$, and put

$$C = \{x \in E : g(x) \geq h(x)\}, \qquad D = \{x \in E : g(x) < h(x)\}.$$

Obviously, the sets C and D are μ-measurable and

$$C \cap D = \emptyset, \qquad C \cup D = E.$$

Then, for every $X \in S$, we may write

$$\int_X \phi(x) \, d\mu(x) = \int_{X \cap C} g(x) \, d\mu(x) + \int_{X \cap D} h(x) \, d\mu(x)$$
$$\leq \nu(X \cap C) + \nu(X \cap D) = \nu(X),$$

which shows us that

$$\phi = \sup(g, h) \in \mathcal{F}.$$

Observe also that if $\{\psi_n : n \in \mathbb{N}\} \subset \mathcal{F}$ is an increasing sequence of functions and

$$\psi(x) = \lim_{n \to \infty} \psi_n(x) \qquad (x \in E),$$

then a well-known property of the Lebesgue integral (see Appendix A) gives us that $\psi \in \mathcal{F}$.

Further, let us introduce the notation

$$t = \sup\{\int_E f(x)\,d\mu(x) : f \in \mathcal{F}\}.$$

Obviously, we have

$$t \leq \nu(E) < +\infty$$

and, according to the above, there exists a function $f^* \in \mathcal{F}$ such that

$$t = \int_E f^*(x)\,d\mu(x).$$

It remains to show for this f^* that

$$\nu(X) = \int_X f^*(x)\,d\mu(x) \qquad (X \in \mathcal{S}).$$

Suppose otherwise, i.e., the measure

$$\theta : \mathcal{S} \to [0, +\infty[$$

defined by

$$\theta(X) = \nu(X) - \int_X f^*(x)\,d\mu(x) \qquad (X \in \mathcal{S})$$

is not identically zero or, equivalently, $\theta(E) > 0$. Since $\mu(E) < +\infty$, we can find a real $\varepsilon > 0$ such that $\theta(E) > \varepsilon\mu(E)$. So we come to the signed measure $\theta - \varepsilon\mu$. Let $\{A, B\}$ be a Hahn–Jordan decomposition of E for this signed measure, where A is a positive component in $\{A, B\}$ and B is a negative component in $\{A, B\}$. In particular, we have

$$\theta(X \cap A) \geq \varepsilon\mu(X \cap A) \qquad (X \in \mathcal{S}).$$

We claim that $\mu(A) > 0$. Indeed, assuming for a moment that $\mu(A) = 0$ and remembering the absolute continuity of ν with respect to μ, we get $\nu(A) = 0$ and, consequently,

$$\theta(A) = -\int_A f^*(x)\,d\mu(x) \leq 0, \qquad (\theta - \varepsilon\mu)(A) \leq 0,$$

$$(\theta - \varepsilon\mu)(E) = (\theta - \varepsilon\mu)(A) + (\theta - \varepsilon\mu)(B) \leq 0,$$

which contradicts the inequality $\theta(E) > \varepsilon\mu(E)$. Therefore, $\mu(A) > 0$.

26 The Radon–Nikodym Theorem

Further, for each set $X \in \mathcal{S}$, we may write

$$\begin{aligned}
v(X) &= \int_X f^*(x)\,d\mu(x) + \theta(X) \\
&\geq \int_X f^*(x)\,d\mu(x) + \theta(X \cap A) \\
&\geq \int_X f^*(x)\,d\mu(x) + \varepsilon\mu(X \cap A) \\
&= \int_X (f^*(x) + \varepsilon\chi_A(x))\,d\mu(x),
\end{aligned}$$

where χ_A denotes, as usual, the characteristic function of A.

We thus infer that the function $f^* + \varepsilon\chi_A$ belongs to $\mathcal{F}$. But, simultaneously, we have

$$t = \int_E f^*(x)\,d\mu(x) < \int_E (f^* + \varepsilon\chi_A)(x)\,d\mu(x),$$

which contradicts the definition of f^*. The obtained contradiction enables us to conclude that θ is identically equal to zero and

$$v(X) = \int_X f^*(x)\,d\mu(x) \qquad (X \in \mathcal{S}).$$

The uniqueness of f^* (up to μ-equivalence) is more or less trivial and its detailed checking is left to the reader.

The Radon–Nikodym theorem has thus been proved. □

The function f^* of Theorem 26.3 is called the *Radon–Nikodym derivative* of v with respect to μ and is usually denoted by $dv/d\mu$.

Some easy properties of the Radon–Nikodym derivatives are described in Exercise 3. The importance of such derivatives is especially seen in probability theory, where the concept of conditional expectation is introduced and extensively studied (see, for instance, [17, 27, 177]).

The assertion of Theorem 26.3 remains valid for any σ-finite measure μ. This more general situation can readily be reduced to the case of finite measures (see Exercise 4 below).

It will be shown in the sequel that the Radon–Nikodym theorem is closely connected with the Riesz theorem on representation of linear continuous functionals defined on a real Hilbert space (see Chap. 29).

Exercises

1. Let $a_0 + a_1 + \cdots + a_n + \cdots$ be a series of real numbers such that the series

$$a_{\phi(0)} + a_{\phi(1)} + \cdots + a_{\phi(n)} + \cdots$$

converges for every bijective mapping $\phi : \mathbb{N} \to \mathbb{N}$.
Demonstrate that $a_0 + a_1 + \cdots + a_n + \cdots$ is absolutely convergent, i.e.,

$$|a_0| + |a_1| + \cdots + |a_n| + \cdots < +\infty.$$

Conclude from this result that in the definition of a signed measure (charge) one deals only with absolutely convergent series.

2. Show that the Radon–Nikodym derivative is unique up to the μ-equivalence of real-valued functions on E.

3. Verify the validity of the following formulas for the Radon–Nikodym derivatives:

   (a) $d(\nu_1 + \nu_2)/d\mu = d\nu_1/d\mu + d\nu_2/d\mu$;
   (b) $d(t\nu)/d\mu = t(d\nu/d\mu)$ for each $t \in \mathbb{R}$;
   (c) $(d\nu/d\theta) \cdot (d\theta/d\mu) = d\nu/d\mu$.

4. Check that the assertion of the Radon–Nikodym theorem remains valid in the more general case where a countably additive measure μ on a ground set E is assumed to be σ-finite.
   For this purpose, represent E in the form

$$E = \cup\{E_i : i \in I\},$$

where the set I is at most countable, all sets E_i are pairwise disjoint and μ-measurable, and $\mu(E_i) < +\infty$. This decomposition of E canonically produces the family $\{\mu_i : i \in I\}$ of finite measures on E such that

$$(\forall i \in I)(\mathrm{dom}(\mu_i) = \mathrm{dom}(\mu)),$$

$$\mu = \Sigma\{\mu_i : i \in I\}.$$

Then apply Theorem 26.3 to each measure μ_i and obtain the required result.

5. Let μ and ν be two nonzero σ-finite countably additive measures on a σ-algebra S of subsets of a base set E.
   Prove that the following two assertions are equivalent:

   (a) $(\forall X \in S)(\mu(X) = 0 \Leftrightarrow \nu(X) = 0)$;
   (b) the Radon–Nikodym derivative $d\nu/d\mu$ can be chosen to be strictly positive on E.

Moreover, if the condition (a) is fulfilled (i.e., if μ and ν are equivalent measures), then the formula

$$(d\nu/d\mu) \cdot (d\mu/d\nu) = 1$$

holds.

6. Let E be a ground set and let S be an algebra of subsets of E. Suppose that ν is a finite finitely additive measure on S and μ is a finite countably additive measure on S. Suppose also that ν is absolutely continuous with respect to μ, i.e., the relation

$$(\forall \varepsilon > 0)(\exists \delta > 0)(\forall X \in S)(\mu(X) < \delta \Rightarrow \nu(X) < \varepsilon)$$

holds true.

Demonstrate that ν is countably additive on S.

For this purpose, verify that ν is upper semicontinuous at $\emptyset$.

Chapter 27
Decompositions of Linear Functionals

Let E be a nonempty ground set and let $\mathcal{F}$ be some vector space of real-valued functions on E.

Suppose that θ is a linear functional on the vector space $\mathcal{F}$.

We shall say that θ satisfies condition (*) if

$$(\forall f \in \mathcal{F})(0 \leq f \Rightarrow \sup\{\theta(g) : g \in \mathcal{F},\ 0 \leq g \leq f\} < +\infty).$$

Recall that, by definition, a linear functional $\phi : \mathcal{F} \to \mathbb{R}$ is *positive* (on $\mathcal{F}$) if, for any function $f \geq 0$ from $\mathcal{F}$, one has $\phi(f) \geq 0$.

Under a natural additional assumption on $\mathcal{F}$, the following statement yields a representation of every linear functional satisfying condition (*) in the form of the difference of two positive linear functionals.

Theorem 27.1 *Suppose that a vector space $\mathcal{F} \subset \mathbb{R}^E$ is a vector lattice, i.e., $\mathcal{F}$ is closed with respect to the two standard operations:*

$$(f, g) \to \inf(f, g) \qquad (f \in \mathcal{F},\ g \in \mathcal{F}),$$
$$(f, g) \to \sup(f, g) \qquad (f \in \mathcal{F},\ g \in \mathcal{F}).$$

Then, for any linear functional $\theta : \mathcal{F} \to \mathbb{R}$ satisfying condition (*), there exist two positive linear functionals

$$\phi : \mathcal{F} \to \mathbb{R}, \qquad \psi : \mathcal{F} \to \mathbb{R}$$

such that the equality $\theta = \phi - \psi$ holds true.

Proof Consider an arbitrary linear functional $\theta : \mathcal{F} \to \mathbb{R}$ for which condition (*) is fulfilled.

For every function $f \geq 0$ belonging to $\mathcal{F}$, define

$$\phi_0(f) = \sup\{\theta(g) : g \in \mathcal{F},\ 0 \leq g \leq f\}.$$

It is obvious from this definition that

$$\phi_0(f) \geq \theta(0) = 0, \qquad \phi_0(f) \geq \theta(f).$$

Further, take any two functions $f_1 \geq 0$ and $f_2 \geq 0$ belonging to $\mathcal{F}$. We may write

$$\begin{aligned}
\phi_0(f_1 + f_2) &= \sup\{\theta(g) : g \in \mathcal{F},\ 0 \leq g \leq f_1 + f_2\} \\
&\geq \sup\{\theta(g_1 + g_2) : \{g_1, g_2\} \subset \mathcal{F},\ 0 \leq g_1 \leq f_1,\ 0 \leq g_2 \leq f_2\} \\
&= \sup\{\theta(g_1) : g_1 \in \mathcal{F},\ 0 \leq g_1 \leq f_1\} + \sup\{\theta(g_2) : g_2 \in \mathcal{F},\ 0 \leq g_2 \leq f_2\} \\
&= \phi_0(f_1) + \phi_0(f_2).
\end{aligned}$$

So we get the inequality

$$\phi_0(f_1 + f_2) \geq \phi_0(f_1) + \phi_0(f_2).$$

On the other hand, if g is any function from $\mathcal{F}$ such that $0 \leq g \leq f_1 + f_2$, then defining

$$g_1 = \inf(g, f_1), \qquad g_2 = g - g_1,$$

it can readily be seen that

$$0 \leq g_1 \leq f_1, \qquad 0 \leq g_2 \leq f_2,$$

$$g = g_1 + g_2.$$

This immediately implies

$$\begin{aligned}
\phi_0(f_1 + f_2) &= \sup\{\theta(g) : g \in \mathcal{F},\ 0 \leq g \leq f_1 + f_2\} \\
&\leq \sup\{\theta(g_1 + g_2) : \{g_1, g_2\} \subset \mathcal{F},\ 0 \leq g_1 \leq f_1,\ 0 \leq g_2 \leq f_2\} \\
&= \phi_0(f_1) + \phi_0(f_2)
\end{aligned}$$

and, therefore,

$$\phi_0(f_1 + f_2) \leq \phi_0(f_1) + \phi_0(f_2).$$

Thus, we have the equality

$$\phi_0(f_1+f_2)=\phi_0(f_1)+\phi_0(f_2)$$

and it is also clear that

$$\phi_0(tf)=t\phi_0(f)$$

for all positive $t\in\mathbb{R}$ and for all functions $f\geq 0$ from $\mathcal{F}$.

It turns out that the obtained functional ϕ_0 admits an extension to a positive linear functional on $\mathcal{F}$. Indeed, every function $f\in\mathcal{F}$ can be expressed in the form

$$f=\sup(f,0)-\sup(-f,0),$$

where both functions $\sup(f,0)$ and $\sup(-f,0)$ belong to $\mathcal{F}$ (in view of condition of the theorem) and are non-negative. Putting

$$\phi(f)=\phi_0(\sup(f,0))-\phi_0(\sup(-f,0)),$$

we get a linear mapping

$$\phi:\mathcal{F}\to\mathbb{R}.$$

To show that ϕ is well-defined, it suffices to consider any two representations

$$f=f_1-f_2,\qquad f=h_1-h_2$$

of f in the form of the difference of two non-negative functions from $\mathcal{F}$. Since

$$f_1+h_2=f_2+h_1,$$

we may write

$$\phi_0(f_1)+\phi_0(h_2)=\phi_0(f_2)+\phi_0(h_1),$$
$$\phi_0(f_1)-\phi_0(f_2)=\phi_0(h_1)-\phi_0(h_2),$$

and we obtain

$$\phi(f)=\phi(f_1-f_2)=\phi(h_1-h_2).$$

So the value $\phi(f)$ does not depend on the above-mentioned representations of f.

Evidently, the functional ϕ extends ϕ_0 and is a linear functional on $\mathcal{F}$.

Finally, if $f\in\mathcal{F}$ and $f\geq 0$, then by definition of ϕ_0, we have

$$\phi(f)=\phi_0(f)\geq\theta(f),$$
$$(\phi-\theta)(f)\geq 0,$$

which implies that the linear functional

$$\psi = \phi - \theta$$

is positive on the space $\mathcal{F}$. Therefore, we come to the required representation of θ in the form $\theta = \phi - \psi$.

The proof of Theorem 27.1 is thus complete. □

Remark 27.1 The reader can verify that the argument used in the proof of Theorem 27.1 belongs to **ZF** theory, because it does not rely on the Axiom of Choice (or on any nontrivial version of **AC**). Actually, the additional condition (*) enables one to avoid **AC** in the process of proving Theorem 27.1. This is important in various applications of this theorem. Indeed, if the original linear functional θ admits a representation in the form $\phi - \psi$, where ϕ and ψ are positive linear functionals, then it is quite desirable to have the descriptive structure of ϕ and ψ as nice as the descriptive structure of θ. For instance, if θ is continuous in some sense, then both ϕ and ψ must also be continuous in the same sense (cf. Theorem 27.4 below).

Remark 27.2 In view of the trivial equality

$$\sup(f, g) + \inf(f, g) = f + g,$$

it suffices to suppose in the formulation of Theorem 27.1 that the vector space $\mathcal{F}$ is closed only under one of the standard operations sup and inf.

Let us also note that a more abstract version of Theorem 27.1 has been established, by using the terminology of the general theory of vector lattices. The argument for proving such a generalized version remains essentially the same (cf. [20, 38]).

Let $\mathcal{F}$ be a vector space of real-valued bounded functions on a nonempty ground set E.

Clearly, one can equip $\mathcal{F}$ with the standard sup-norm (i.e., the norm of uniform convergence) by putting

$$||f|| = \sup\{|f(x)| : x \in E\} \qquad (f \in \mathcal{F}).$$

So we obtain the normed vector space $(\mathcal{F}, ||\cdot||)$.

In general, $(\mathcal{F}, ||\cdot||)$ is not complete with respect to $||\cdot||$.

Theorem 27.2 *Let θ be an arbitrary continuous linear functional on the normed vector space $(\mathcal{F}, ||\cdot||)$.*

Then θ satisfies condition (*).

Proof We only have to check that if $f \in \mathcal{F}$ is any non-negative function, then

$$\sup\{\theta(g) : g \in \mathcal{F}, \ 0 \leq g \leq f\} < +\infty.$$

By virtue of the continuity of θ, there exists a natural number n such that

$$|\theta(h)| \leq n \cdot \sup\{|h(x)| : x \in E\}$$

for all functions $h \in \mathcal{F}$. Now, take arbitrarily a non-negative function $g \in \mathcal{F}$ satisfying $g \leq f$. Then we may write

$$\theta(g) \leq |\theta(g)| \leq n \cdot \sup\{|g(x)| : x \in E\} = n \cdot \sup\{g(x) : x \in E\}.$$

Keeping in mind that

$$\sup\{g(x) : x \in E\} \leq \sup\{f(x) : x \in E\},$$

we infer that

$$\theta(g) \leq n \cdot \sup\{f(x) : x \in E\} = n \cdot \sup\{|f(x)| : x \in E\}.$$

Since the value

$$n \cdot \sup\{|f(x)| : x \in E\}$$

is a fixed real number not depending on the choice of g, condition (*) is trivially fulfilled.

Theorem 27.2 has thus been proved. □

Theorem 27.3 *With the previous notation, suppose that all real-valued constant functions on E belong to $(\mathcal{F}, ||\cdot||)$.*

If θ is a positive linear functional on $(\mathcal{F}, ||\cdot||)$, then θ is continuous on $(\mathcal{F}, ||\cdot||)$.

Proof Take any function $f \in \mathcal{F}$. Obviously,

$$-||f|| \leq f(x) \leq ||f|| \quad (x \in E)$$

or, equivalently,

$$-f \leq ||f||, \quad f \leq ||f||.$$

Since the functional θ is positive, we deduce

$$-\theta(f) = \theta(-f) \leq \theta(||f||),$$
$$-\theta(||f||) \leq \theta(f) \leq \theta(||f||).$$

Therefore, we may write

$$|\theta(f)| \leq \theta(||f||) = \theta(\chi_E)||f||,$$

where χ_E denotes the function on E identically equal to 1 (i.e., the characteristic function of E). This gives us the continuity of θ and ends the proof. □

The next important theorem immediately follows from the above results.

Theorem 27.4 *Suppose that $\mathcal{F}$ is a vector space of real-valued bounded functions on a nonempty ground set E and suppose that $\mathcal{F}$ has the following two properties:*

(1) *$\mathcal{F}$ is a vector lattice;*
(2) *all real-valued constant functions on E belong to $\mathcal{F}$.*

Let θ be a continuous linear functional on $(\mathcal{F}, \|\cdot\|)$, where $\|\cdot\|$ is the standard sup-norm on $\mathcal{F}$.

Then θ admits a representation in the form

$$\theta = \phi - \psi,$$

where both ϕ and ψ are positive continuous linear functionals on $(\mathcal{F}, \|\cdot\|)$.

Remark 27.3 Theorem 27.4 is due to F. Riesz and has many applications in functional analysis and measure theory (see, e.g., [17, 20, 38], and Chap. 28 of this book).

Exercises

1. Let E be an arbitrary vector space over $\mathbb{R}$.
   Show that it is possible to introduce an inner (scalar) product $\langle \cdot, \cdot \rangle$ for E, so E becomes a real pre-Hilbert space.
   For this purpose, consider some algebraic basis $\{e_i : i \in I\}$ of E. Any vector $x \in E$ admits a unique representation in the form

$$x = \Sigma\{t_i e_i : i \in I\},$$

where $\{t_i : i \in I\}$ is a family of real numbers and the set $\{i \in I : t_i \neq 0\}$ is finite. If $y \in E$ is another vector (not necessarily distinct from x), then y has an analogous unique representation

$$y = \Sigma\{r_i e_i : i \in I\}.$$

Now, define

$$\langle x, y \rangle = \Sigma\{t_i r_i : i \in I\}$$

and check that $\langle \cdot, \cdot \rangle$ is an inner product of elements of E.

Remark 27.4 The result of Exercise 1 essentially uses the existence of an algebraic basis of E, so it heavily relies on the Axiom of Choice. Moreover, as was proved in [14], within **ZF** set theory the existence of an algebraic basis of any vector space (over an arbitrary field) is equivalent to **AC**.

2. Deduce from the previous exercise that every vector space E over $\mathbb{R}$ can be considered as a real normed space.
3. Let $\{e_i : i \in I\}$ be a fixed algebraic basis of a real vector space E.
   By definition, a linear functional $g : E \to \mathbb{R}$ possesses a *finite support* (with respect to $\{e_i : i \in I\}$) if the set

$$\{i \in I : g(e_i) \neq 0\}$$

is finite.
Show that, after introducing in E the inner product $\langle \cdot, \cdot \rangle$ as is described in Exercise 1, for any such linear functional g, one has a representation

$$g(x) = \langle x, e \rangle \quad (x \in E),$$

where e is some vector in E uniquely determined by g.

4. Let $\{e_i : i \in I\}$ be again an algebraic basis of a real vector space E. For any two vectors

$$x = \Sigma\{t_i e_i : i \in I\}, \quad y = \Sigma\{r_i e_i : i \in I\},$$

write $x \preceq y$ if and only if

$$(\forall i \in I)(t_i \leq r_i).$$

Verify that the binary relation $\preceq$ is a partial ordering of E (canonically associated with the basis $\{e_i : i \in I\}$). Moreover, verify that this partial ordering is invariant under all translations of E and under any homothety of E with center 0 and with a positive coefficient. Describe the convex cone

$$C = \{x \in E : 0 \preceq x\}$$

determined by $\preceq$ (cf. Chap. 18).
A linear functional $f : E \to \mathbb{R}$ will be called *positive* (with respect to $\{e_i : i \in I\}$) if the relation $x \preceq y$ implies $f(x) \leq f(y)$.
Equivalently, one may say that the same f is positive if the relation $0 \preceq x$ implies $0 \leq f(x)$.
Show that the following two assertions are also equivalent:

(a) f is a positive linear functional on E;
(b) $f(e_i) \geq 0$ for each index $i \in I$.

Prove that every linear functional $\theta : E \to \mathbb{R}$ admits a representation in the form

$$\theta = \phi - \psi,$$

where both ϕ and ψ are positive linear functionals on $(E, \preceq)$.
For this purpose, denote $\theta(e_i)$ by a_i, where $i \in I$. Choose, for each $i \in I$, two positive real numbers b_i and c_i so that $a_i = b_i - c_i$. Then define

$$\phi(e_i) = b_i, \qquad \psi(e_i) = c_i \qquad (i \in I).$$

Extend ϕ and ψ to some linear functionals on E and, finally, check that the linear functionals obtained in this manner are as required.

Remark 27.5 Let us emphasize that the above representation of θ in the form $\theta = \phi - \psi$ depends on the choice of a basis $\{e_i : i \in I\}$, so one can conclude that this representation substantially depends on **AC**.

5. Let E be a real topological vector space such that its algebraic (linear) dimension (i.e., the cardinality of an algebraic basis of E) is greater than or equal to the local weight of zero in E.

   In other words, suppose that there exists a fundamental system $\mathcal{U}(0)$ of neighborhoods of 0 in E such that $\text{card}(\mathcal{U}(0))$ does not exceed the algebraic dimension of E.

   Demonstrate that there exists a linear functional on E discontinuous at each point of E.

   For this purpose, define the required everywhere discontinuous linear functional on E by using the method of transfinite recursion.

6. Let $(E, ||\cdot||)$ be an arbitrary infinite-dimensional normed vector space over $\mathbb{R}$. Infer from Exercise 5 that there exists an everywhere discontinuous linear functional on E.

Chapter 28
Linear Continuous Functionals and Radon Measures

Let E be a Hausdorff topological space and let $f : E \to \mathbb{R}$ be a function.

Recall that, by definition, a subset X of E is a *compact support* of f (in E) if X is compact and $f|(E \setminus X) = 0$.

If $f : E \to \mathbb{R}$ is continuous, then the set $\{x \in E : f(x) \neq 0\}$ is open and the set $\{x \in E : f(x) = 0\}$ is closed. If, in addition, f has a compact support X, then necessarily $\{x \in E : f(x) \neq 0\} \subset X$.

More concretely, in the sequel we will use the standard notation

$$\mathrm{supp}(f) = \text{the closure of the set } \{x \in E : f(x) \neq 0\},$$

so the inclusion $\mathrm{supp}(f) \subset Y$ trivially implies $f|(E \setminus Y) = 0$.

Let E be an arbitrary nonempty compact topological space. As usual, we denote by $C(E)$ ($= C(E, \mathbb{R})$) the space of all real-valued continuous functions on E, which is equipped with the standard sup-norm, i.e.,

$$||f|| = \sup |f(x)| \qquad (x \in E)$$

for every $f \in C(E)$. Thus, $C(E)$ is a real Banach space.

Consider any positive linear functional $\theta : C(E) \to \mathbb{R}$. Our goal is to associate with this θ an appropriate finite countably additive Radon measure μ on E with $\mathrm{dom}(\mu) = \mathcal{B}(E)$.

We begin by defining μ on the family of all open subsets of E.

Take any open set $U \subset E$ and put

$$\mu(U) = \sup\{\theta(g) : g \in C(E),\ 0 \leq g \leq 1,\ \mathrm{supp}(g) \subset U\}.$$

Note that if $g \in C(E)$ and $0 \leq g \leq 1$, then $\theta(g) \leq \theta(\chi_E)$, where χ_E denotes, as usual, the characteristic function of E which is identically equal to 1. So we always have $0 \leq \mu(U) \leq \theta(\chi_E)$.

Moreover, the above definition of μ on the family of all open subsets of E trivially implies the monotonicity of μ, i.e., $\mu(U) \leq \mu(V)$ whenever $V \subset E$ is also an open set in E and $U \subset V$.

Lemma 28.1 *If $\{U_n : n \in \mathbb{N}\}$ is an arbitrary countable family of open sets in E, then the inequality*

$$\mu(\cup\{U_n : n \in \mathbb{N}\}) \leq \Sigma\{\mu(U_n) : n \in \mathbb{N}\}$$

holds true.

Proof Let $U = \cup\{U_n : n \in \mathbb{N}\}$ and pick a real $\varepsilon > 0$. By definition of $\mu(U)$, there exists a function $g \in C(E)$ such that

$$0 \leq g \leq 1, \qquad \operatorname{supp}(g) \subset U, \qquad \mu(U) \leq \theta(g) + \varepsilon.$$

Since $\{U_n : n \in \mathbb{N}\}$ is an open covering of the compact set $\operatorname{supp}(g)$, there exists a finite subcovering $\{U_n : 0 \leq n \leq k\}$ of $\operatorname{supp}(g)$. In view of Exercise 2 of this chapter, one can represent g in the form

$$g - \Sigma\{g_n : 0 \leq n \leq k\},$$

where the functions g_n ($0 \leq n \leq k$) satisfy the relations

$$g_n \in C(E), \qquad 0 \leq g_n \leq 1, \qquad \operatorname{supp}(g_n) \subset U_n.$$

This implies at once

$$\mu(U) \leq \theta(g) + \varepsilon = \Sigma\{\theta(g_n) : 0 \leq n \leq k\} + \varepsilon \leq \Sigma\{\mu(U_n) : n \in \mathbb{N}\} + \varepsilon,$$

which gives us the desired result, because $\varepsilon > 0$ may be arbitrarily small.

Lemma 28.1 has thus been proved. □

Now, for every set $X \subset E$, we define

$$\mu(X) = \inf\{\mu(U) : U \text{ is open in } E \text{ and } X \subset U\}.$$

Let us underline that the definition just introduced is compatible with the definition of μ introduced earlier for the open subsets of E. Indeed, in view of the monotonicity of μ on the family of all open sets in E, the value $\mu(U)$ remains the same for any open $U \subset E$.

It is also clear that the extended functional μ preserves the monotonicity property, i.e., if $X \subset Y \subset E$, then $\mu(X) \leq \mu(Y)$.

Moreover, similarly to the outer measures, μ is countably subadditive, i.e.,

$$\mu(\cup\{X_n : n \in \mathbb{N}\}) \leq \Sigma\{\mu(X_n) : n \in \mathbb{N}\}$$

for any countable family $\{X_n : n \in \mathbb{N}\}$ of subsets of E. To see the latter fact, it suffices to argue by the standard scheme which we have already used after introducing the outer measure associated with a given finite finitely additive measure (cf. the proof of Theorem 15.1 from Chap. 15).

Indeed, take a real $\varepsilon > 0$ and, for each natural number n, find an open set $U_n \subset E$ such that

$$X_n \subset U_n, \quad \mu(X_n) \geq \mu(U_n) - \varepsilon/2^{n+1}.$$

Then apply Lemma 28.1 and write

$$\mu(\cup\{X_n : n \in \mathbb{N}\}) \leq \mu(\cup\{U_n : n \in \mathbb{N}\})$$
$$\leq \Sigma\{\mu(U_n) : n \in \mathbb{N}\}$$
$$\leq \Sigma\{\mu(X_n) : n \in \mathbb{N}\} + \varepsilon,$$

whence the countable subadditivity (and, consequently, the finite subadditivity) of μ immediately follows.

So we have defined the finite non-negative countably subadditive functional μ on the family of all subsets of a compact space E.

Lemma 28.2 *The following four assertions hold:*

(1) *if U is an open set in E and $X \subset U$ is a compact set in E, then*

$$\mu(U) = \mu(X) + \mu(U \setminus X);$$

(2) *if X and Y are compact subsets of E and $X \subset Y$, then*

$$\mu(Y) = \mu(X) + \mu(Y \setminus X)$$

and, in particular, if Z_1 and Z_2 are any two disjoint compact sets in E, then

$$\mu(Z_1 \cup Z_2) = \mu(Z_1) + \mu(Z_2);$$

(3) *if $\{X_n : n \in \mathbb{N}\}$ is an increasing (by inclusion) countable family of compact sets in E, then*

$$\mu(\cup\{X_n : n \in \mathbb{N}\}) = \lim_{n \to \infty} \mu(X_n);$$

(4) *if $\{Y_n : n \in \mathbb{N}\}$ is a decreasing (by inclusion) countable family of compact sets in E, then*

$$\mu(\cap\{Y_n : n \in \mathbb{N}\}) = \lim_{n \to \infty} \mu(Y_n).$$

Proof To establish (1), first observe that by virtue of the finite subadditivity of μ, we can write the inequality

$$\mu(U) \leq \mu(X) + \mu(U \setminus X).$$

At the same time, for each real $\varepsilon > 0$, there exists a function $g \in C(E)$ such that

$$0 \leq g \leq 1, \qquad \mathrm{supp}(g) \subset U \setminus X, \qquad \mu(U \setminus X) \leq \theta(g) + \varepsilon.$$

Putting $W = U \setminus \mathrm{supp}(g)$, we see that

$$X \subset W \subset U, \qquad \mathrm{supp}(g) \cap W = \emptyset.$$

Let $h \in C(E)$ be a function satisfying the relations

$$0 \leq h \leq 1, \qquad \mathrm{supp}(h) \subset W, \qquad \theta(h) \geq \mu(W) - \varepsilon.$$

Then we can write

$$0 \leq g + h \leq 1, \qquad \mathrm{supp}(g+h) \subset U, \qquad \theta(g+h) \leq \mu(U).$$

Finally, we obtain

$$\mu(X) + \mu(U \setminus X) \leq \mu(W) + (\theta(g) + \varepsilon)$$
$$\leq \theta(h) + \varepsilon + \theta(g) + \varepsilon$$
$$= \theta(g+h) + 2\varepsilon$$
$$\leq \mu(U) + 2\varepsilon,$$

which shows us that

$$\mu(X) + \mu(U \setminus X) \leq \mu(U)$$

and, consequently,

$$\mu(U) = \mu(X) + \mu(U \setminus X).$$

To demonstrate (2), take again a real $\varepsilon > 0$ and find an open set $U \subset E$ such that

$$Y \subset U, \qquad \mu(U) \leq \mu(Y) + \varepsilon.$$

According to (1), we may write

$$\mu(Y) \geq \mu(U) - \varepsilon$$
$$= \mu(U \setminus X) + \mu(X) - \varepsilon$$
$$\geq \mu(Y \setminus X) + \mu(X) - \varepsilon,$$

whence it follows that

$$\mu(Y) \geq \mu(Y \setminus X) + \mu(X).$$

On the other hand, we know that the converse inequality

$$\mu(Y) \leq \mu(X) + \mu(Y \setminus X)$$

always holds true, which gives us the desired result.

To establish (3), let us represent the set $\cup \{X_n : n \in \mathbb{N}\}$ in the form

$$\cup \{X_n : n \in \mathbb{N}\} = X_0 \cup (X_1 \setminus X_0) \cup \cdots \cup (X_n \setminus X_{n-1}) \cup \cdots.$$

By virtue of the countable subadditivity of μ, this representation implies

$$\mu(\cup\{X_n : n \in \mathbb{N}\}) \leq \mu(X_0) + \Sigma\{\mu(X_n \setminus X_{n-1}) : n \in \mathbb{N} \setminus \{0\}\}$$

and, according to (2), we get

$$\mu(\cup\{X_n : n \in \mathbb{N}\}) \leq \lim_{n \to \infty} \mu(X_n).$$

The converse inequality is trivial, so (3) holds true.

In order to demonstrate (4), first observe that

$$\mu(\cap\{Y_n : n \in \mathbb{N}\}) \leq \lim_{n \to \infty} \mu(Y_n)$$

in view of the monotonicity of μ. Further, take an arbitrary real $\varepsilon > 0$ and consider an open set $U \subset E$ such that

$$\cap\{Y_n : n \in \mathbb{N}\} \subset U, \qquad \mu(U) \leq \mu(\cap\{Y_n : n \in \mathbb{N}\}) + \varepsilon.$$

Observe that there exists an $m \in \mathbb{N}$ for which $Y_m \subset U$. Consequently,

$$\lim_{n \to \infty} \mu(Y_n) \leq \mu(Y_m) \leq \mu(U) \leq \mu(\cap\{Y_n : n \in \mathbb{N}\}) + \varepsilon,$$

which implies the inequality

$$\lim_{n \to \infty} \mu(Y_n) \leq \mu(\cap\{Y_n : n \in \mathbb{N}\})$$

and yields the required result.

This ends the proof of Lemma 28.2. □

Taking in (1) of Lemma 28.2 the entire compact space E as U, we get

$$\theta(\chi_E) = \mu(E) = \mu(Z) + \mu(E \setminus Z)$$

for each compact (or open) set Z in E.

The next lemma provides a slightly more general result.

Lemma 28.3 *The following two assertions hold true for any monotone (with respect to inclusion) sequence $\{U_n : n \in \mathbb{N}\}$ of open sets in E:*

(1) $\mu(\lim\{U_n : n \in \mathbb{N}\}) = \lim_{n \to \infty} \mu(U_n)$;
(2) $\mu(E) = \mu(\lim\{U_n : n \in \mathbb{N}\}) + \mu(E \setminus \lim\{U_n : n \in \mathbb{N}\})$.

Proof Let $\{U_n : n \in \mathbb{N}\}$ be an increasing sequence of open sets in E. Putting $Y_n = E \setminus U_n$ for each $n \in \mathbb{N}$, we come to the decreasing sequence $\{Y_n : n \in \mathbb{N}\}$ of compact subsets of E. Applying (4) of Lemma 28.2, we may write

$$\mu(\cap\{Y_n : n \in \mathbb{N}\}) = \lim_{n \to \infty} \mu(Y_n)$$
$$= \lim_{n \to \infty} (\mu(E) - \mu(U_n))$$
$$= \mu(E) - \lim_{n \to \infty} \mu(U_n),$$
$$\mu(E) - \mu(\cap\{Y_n : n \in \mathbb{N}\}) = \lim_{n \to \infty} \mu(U_n).$$

At the same time,

$$\mu(E) - \mu(\cap\{Y_n : n \in \mathbb{N}\}) = \mu(E \setminus \cap\{Y_n : n \in \mathbb{N}\})$$
$$= \mu(\cup\{U_n : n \in \mathbb{N}\}),$$

which yields the required result for all increasing sequences of open subsets of E.

Let now $\{U_n : n \in \mathbb{N}\}$ be a decreasing sequence of open subsets of E. Take any real $\varepsilon > 0$ and find an open subset V of E such that

$$\cap\{U_n : n \in \mathbb{N}\} \subset V, \qquad \mu(V) \leq \mu(\cap\{U_n : n \in \mathbb{N}\}) + \varepsilon.$$

Using the relation

$$E \setminus V \subset E \setminus \cap\{U_n : n \in \mathbb{N}\} = \cup\{E \setminus U_n : n \in \mathbb{N}\}$$

and Lemma 28.2, we have

$$\mu(E) - \mu(V) \leq \mu(E) - \lim_{n \to \infty} \mu(U_n)$$

or, equivalently, $\mu(V) \geq \lim_{n \to \infty} \mu(U_n)$. Therefore,

$$\lim_{n \to \infty} \mu(U_n) \leq \mu(V) \leq \mu(\cap\{U_n : n \in \mathbb{N}\}) + \varepsilon,$$

whence it follows that

$$\lim_{n \to \infty} \mu(U_n) \leq \mu(\cap\{U_n : n \in \mathbb{N}\}).$$

28 Linear Continuous Functionals and Radon Measures

The reverse inequality is trivial, so assertion (1) holds.

Assertion (2) can readily be deduced from (1).

Lemma 28.3 has thus been proved. □

Let now X be an arbitrary subset of our nonempty compact space E.

We shall say that X is μ-*approximable from above* (by using open sets in E) if, for any real $\varepsilon > 0$, there exists an open set $U \subset E$ such that

$$X \subset U, \qquad \mu(U \setminus X) < \varepsilon.$$

We shall say that X is μ-*approximable from below* (by using compact sets in E) if, for any real $\varepsilon > 0$, there exists a compact set $K \subset E$ such that

$$K \subset X, \qquad \mu(X \setminus K) < \varepsilon.$$

Accordingly, we shall say that X is μ-*approximable* if X is simultaneously μ-approximable from above and from below.

Recall that the analogous concepts have already been considered in Exercise 9 of Chap. 20.

Lemma 28.4 *The following three assertions hold:*

(1) *every open set in E is μ-approximable;*
(2) *$X \subset E$ is μ-approximable from above (from below) if and only if $E \setminus X$ is μ-approximable from below (from above);*
(3) *$X \subset E$ is μ-approximable if and only if $E \setminus X$ is μ-approximable.*

Proof The assertions (2) and (3) are trivial, so it suffices to demonstrate (1) or, equivalently, to establish that any open set U in E is μ-approximable from below.

For this purpose, take a real $\varepsilon > 0$ and consider the compact set $E \setminus U$. There exists an open set $V \subset E$ such that

$$E \setminus U \subset V, \qquad \mu(V) \leq \mu(E \setminus U) + \varepsilon.$$

Then we have $E \setminus V \subset U$ and $E \setminus V$ is a compact set in E. Further, applying Lemma 28.2, we may write

$$\begin{aligned}
\mu(U \setminus (E \setminus V)) &= \mu(U) - (\mu(E) - \mu(V)) \\
&= \mu(U) - \mu(E) + \mu(V) \\
&\leq \mu(U) - \mu(E) + (\mu(E \setminus U) + \varepsilon) \\
&= \varepsilon,
\end{aligned}$$

which completes the proof. □

Lemma 28.5 *All Borel sets in E are μ-approximable and the restriction of μ to the Borel σ-algebra $\mathcal{B}(E)$ is a finite Radon measure on E.*

Proof First, observe that the family of all μ-approximable sets is closed under taking the complements of its members (see (3) of Lemma 28.4).

Now, let $\{X_n : n \in \mathbb{N}\}$ be a countable family of μ-approximable sets in E. Take a real $\varepsilon > 0$. For each $n \in \mathbb{N}$, there exist an open set $U_n \subset E$ and a compact set $P_n \subset E$ such that

$$P_n \subset X_n \subset U_n,$$

$$\mu(U_n \setminus X_n) < \varepsilon/2^{n+2}, \qquad \mu(X_n \setminus P_n) < \varepsilon/2^{n+2}.$$

Since we have

$$\mu(U_n \setminus P_n) \leq \mu(U_n \setminus X_n) + \mu(X_n \setminus P_n),$$

the inequality

$$\mu(U_n \setminus P_n) < \varepsilon/2^{n+1}$$

is fulfilled. Let

$$U = \cup\{U_n : n \in \mathbb{N}\}, \qquad X = \cup\{X_n : n \in \mathbb{N}\}, \qquad P = \cup\{P_n : n \in \mathbb{N}\}.$$

Obviously, we may write $P \subset X \subset U$ and

$$U \setminus X \subset U \setminus P \subset \cup\{U_n \setminus P_n : n \in \mathbb{N}\},$$

$$\mu(U \setminus X) \leq \mu(U \setminus P) \leq \Sigma\{\mu(U_n \setminus P_n) : n \in \mathbb{N}\} < \varepsilon.$$

Consequently, the set X is μ-approximable from above.

In view of (1) of Lemma 28.3, there exists a natural number m such that

$$\mu(U \setminus \cup\{P_n : 0 \leq n \leq m\}) \leq 2\varepsilon.$$

Evidently, $\cup\{P_n : 0 \leq n \leq m\}$ is a compact set and

$$\mu(X \setminus \cup\{P_n : 0 \leq n \leq m\}) \leq \mu(U \setminus \cup\{P_n : 0 \leq n \leq m\}) \leq 2\varepsilon,$$

whence it follows that the set X is μ-approximable from below.

Keeping in mind (1) of Lemma 28.4, we deduce that the family of all μ-approximable sets in E is a σ-algebra containing $\mathcal{B}(E)$.

Now, take any disjoint countable family $\{Z_n : n \in \mathbb{N}\}$ of μ-approximable sets. The countable subadditivity of μ implies the inequality

$$\mu(\cup\{Z_n : n \in \mathbb{N}\}) \leq \Sigma\{\mu(Z_n) : n \in \mathbb{N}\}.$$

28 Linear Continuous Functionals and Radon Measures

In order to establish the reverse inequality it suffices to demonstrate the finite additivity of μ on the family of all μ-approximable sets.

For this purpose, consider any two disjoint μ-approximable sets X and Y and take a real $\varepsilon > 0$. There exist two compact sets P and Q in E such that

$$P \subset X, \quad Q \subset Y, \quad \mu(X \setminus P) < \varepsilon/2, \quad \mu(Y \setminus Q) < \varepsilon/2.$$

Clearly, we can write

$$\mu(X) + \mu(Y) \leq \mu(P) + \mu(X \setminus P) + \mu(Q) + \mu(Y \setminus Q)$$
$$\leq \mu(P \cup Q) + \varepsilon$$
$$\leq \mu(X \cup Y) + \varepsilon,$$

which gives us $\mu(X) + \mu(Y) \leq \mu(X \cup Y)$ and so

$$\mu(X \cup Y) = \mu(X) + \mu(Y).$$

Therefore, μ is countably additive on the family of all μ-approximable sets and the restriction of μ to this family is a Radon measure (according to the definition of μ-approximable sets).

Lemma 28.5 has thus been proved. □

Remark 28.1 The described measure μ on $\mathcal{B}(E)$ is called the *Radon measure associated with a given positive linear functional* θ. We would like to especially underline that the construction of μ uses only countable forms of the Axiom of Choice (e.g., the axiom **DC** is completely sufficient). An important special case is when E coincides with the unit segment $[0, 1]$ of $\mathbb{R}$ and θ is the classical Riemann integral on $C([0, 1])$. In this case, the associated Radon measure is the standard Borel probability measure on $[0, 1]$ and its completion turns out to be the standard Lebesgue probability measure μ_0 on $[0, 1]$. Obviously, we have the natural analog μ_n of μ_0 for any segment $[n, n+1]$ of $\mathbb{R}$, where n ranges over the set $\mathbb{Z}$ of all integers. Taking the direct sum of the family of measures $\{\mu_n : n \in \mathbb{Z}\}$, we obtain the ordinary Lebesgue measure λ on $\mathbb{R}$.

Before formulating the next lemma, let us recall once more that every positive linear functional θ on the space $C(E)$ is automatically continuous. Indeed, if $f \in C(E)$, then the relation

$$-\|f\|\chi_E \leq f \leq \|f\|\chi_E$$

holds true, whence it follows, by virtue of the positivity of θ, that

$$-\|f\|\theta(\chi_E) \leq \theta(f) \leq \|f\|\theta(\chi_E)$$

or, equivalently,

$$|\theta(f)| \leq \theta(\chi_E)\|f\|,$$

which means the continuity of θ.

Lemma 28.6 *For any positive linear functional θ on $C(E)$ and for the associated with θ Radon measure μ, the following three assertions hold:*

(1) *if a set $U \subset E$ is open in E and $f \in C(E)$ is such that $f \leq \chi_U$, then $\theta(f) \leq \mu(U)$;*
(2) *if a set $P \subset E$ is compact in E and $g \in C(E)$ is such that $g \geq \chi_P$, then $\theta(g) \geq \mu(P)$;*
(3) *if a set $P \subset E$ is compact in E and $h \in C(E)$ is such that $h \leq \chi_P$, then $\theta(h) \leq \mu(P)$.*

Proof In order to show (1), first observe that

$$f \leq \chi_U \Leftrightarrow \sup(0, f) \leq \chi_U,$$
$$\theta(f) \leq \theta(\sup(0, f)).$$

Consequently, without loss of generality one may assume in (1) that $0 \leq f$.

Notice also that $f(x) > 0$ implies $x \in U$. Further, by definition,

$$\mu(U) = \sup\{\theta(\phi) : \phi \in C(E),\ 0 \leq \phi \leq 1,\ \mathrm{supp}(\phi) \subset U\}.$$

Take any $\phi \in C(E)$ which satisfies $0 \leq \phi \leq 1$ and $\mathrm{supp}(\phi) \subset U$. For every natural number $n > 0$, let

$$P_n = \{x \in E : f(x) \geq 1/n\}.$$

Obviously, P_n is contained in U and is compact in E. Since E is a normal topological space, there exists a $\phi_n \in C(E)$ such that

$$0 \leq \phi_n \leq 1, \qquad \phi_n|P_n = f|P_n, \qquad \mathrm{supp}(\phi_n) \subset U.$$

It can readily be checked that

$$0 \leq f(x) - \min(f(x), \phi_n(x)) < 1/n \qquad (x \in E),$$

so the sequence of continuous functions $\{\inf(f, \phi_n) : n \in \mathbb{N} \setminus \{0\}\}$ converges uniformly to f and all members of this sequence have their supports in U. Therefore, using the continuity of θ, we get

$$\theta(f) = \lim_{n \to \infty} \theta(\inf(f, \phi_n)) \leq \mu(U).$$

To demonstrate (2), consider the function $f = 1 - g$. We have $f \leq \chi_{(E \setminus P)}$ and, according to (1), we may write

$$\theta(1 - g) \leq \mu(E \setminus P), \qquad \mu(E) - \theta(g) \leq \mu(E) - \mu(P),$$

which gives us $\theta(g) \geq \mu(P)$.

Finally, to establish (3), take any real $\varepsilon > 0$ and find an open set $U \subset E$ such that $P \subset U$ and $\mu(U) \leq \mu(P) + \varepsilon$. Since $h \leq \chi_P$ and $P \subset U$, we have $h \leq \chi_U$. By virtue of (1), we may write

$$\theta(h) \leq \mu(U) \leq \mu(P) + \varepsilon,$$

which implies $\theta(h) \leq \mu(P)$ and ends the proof of Lemma 28.6. □

Now, we are able to formulate and prove a fundamental theorem of mathematical analysis, which is due to F. Riesz (cf. [17, 20, 38, 104]).

Theorem 28.1 *Let E be a nonempty compact topological space, let θ be a positive linear functional on $C(E)$, and let μ be the Radon measure on $\mathcal{B}(E)$ associated with θ.*

Then, for every function $f \in C(E)$, one has $\theta(f) = \int_E f(x) \, d\mu(x)$.

Proof First, consider the situation when $f \geq 0$. To show the above equality for such f, it suffices to additionally assume that $0 \leq f \leq 1$. For any nonzero $n \in \mathbb{N}$, let us introduce the notation:

$$P_k = \{x \in E : f(x) \geq k/n\} \qquad (k = 0, 1, \ldots, n),$$
$$\phi_k = \inf(f, k/n) \qquad (k = 0, 1, \ldots, n),$$
$$f_k = \phi_k - \phi_{k-1} \qquad (k = 1, 2, \ldots, n).$$

With this notation, we may write

$$f = \Sigma\{f_k : 1 \leq k \leq n\},$$

and it is not hard to verify that

$$(1/n)\chi_{P_k} \leq f_k \leq (1/n)\chi_{P_{k-1}} \qquad (k = 1, 2, \ldots, n).$$

By virtue of (2) and (3) of Lemma 28.6, we have

$$(1/n)\mu(P_k) \leq \theta(f_k) \leq (1/n)\mu(P_{k-1}) \qquad (k = 1, 2, \ldots n),$$

whence it follows that

$$(1/n)\Sigma\{\mu(P_k) : 1 \leq k \leq n\} \leq \theta(f) \leq (1/n)\Sigma\{\mu(P_k) : 0 \leq k \leq n\}.$$

At the same time, it is clear that if $k \in \{1, 2, \ldots, n\}$, then

$$(1/n)\mu(P_k) \leq \int_E f_k(x)\,d\mu(x) \leq (1/n)\mu(P_{k-1}),$$

which implies the inequalities

$$(1/n)\Sigma\{\mu(P_k) : 1 \leq k \leq n\} \leq \int_E f(x)\,d\mu(x) \leq (1/n)\Sigma\{\mu(P_k) : 0 \leq k \leq n\}.$$

Consequently, we get

$$|\theta(f) - \int_E f(x)\,d\mu(x)| \leq (1/n)\mu(E) \qquad (n \in \mathbb{N} \setminus \{0\}),$$

which yields the equality $\theta(f) = \int_E f(x)\,d\mu(x)$.

Finally, if f is an arbitrary function from $C(E)$, then f admits a representation $f = g - h$, where

$$g \in C(E), \qquad h \in C(E), \qquad g \geq 0, \qquad h \geq 0,$$

and we can write

$$\theta(f) = \theta(g) - \theta(h) = \int_E g(x)\,d\mu(x) - \int_E h(x)\,d\mu(x) = \int_E f(x)\,d\mu(x).$$

Theorem 28.1 has thus been proved. □

A functional $\nu : \mathcal{B}(E) \to \mathbb{R}$ is called a *Radon signed measure* (or a *Radon charge*) if ν can be represented in the form $\nu = \mu_1 - \mu_2$, where both μ_1 and μ_2 are finite Radon measures defined on $\mathcal{B}(E)$.

Theorem 28.1 directly leads to the following result (also due to F. Riesz).

Theorem 28.2 *Let E be a nonempty compact topological space.*

If θ is a continuous linear functional on the space $C(E)$, then θ admits a representation

$$\theta(f) = \int_E f(x)\,d\nu(x) \qquad (f \in C(E)),$$

where ν is a uniquely determined Radon signed measure defined on $\mathcal{B}(E)$.

Proof The conditions of Theorem 27.4 from Chap. 27 are fulfilled for the given functional θ, so θ is representable in the form $\theta = \theta_1 - \theta_2$, where both θ_1 and θ_2 are positive continuous linear functionals on $C(E)$.

By virtue of Theorem 28.1 of this chapter, we may write

28 Linear Continuous Functionals and Radon Measures

$$\theta_1(f) = \int_E f(x)\,d\mu_1(x) \qquad (f \in C(E)),$$

$$\theta_2(f) = \int_E f(x)\,d\mu_2(x) \qquad (f \in C(E)).$$

Here μ_1 and μ_2 are some finite Radon measures defined on $\mathcal{B}(E)$. Now, putting $\nu = \mu_1 - \mu_2$, we get

$$\theta(f) = \theta_1(f) - \theta_2(f) = \int_E f(x)\,d(\mu_1 - \mu_2)(x) = \int_E f(x)\,d\nu(x).$$

The uniqueness of such ν is discussed in Exercise 6 of the present chapter. □

Remark 28.2 The reader can verify that the proof of Theorem 28.2 uses only a countable form of the Axiom of Choice (cf. Remark 28.1). There are known profound generalizations of Theorem 28.2 to sufficiently large classes of non-compact topological spaces E. Naturally, any E from such a class is considered with the associated Banach space $C_b(E)$ of all real-valued bounded continuous functions on E (for more details, see e.g. [38]). As a rule, generalizations of this type are based on uncountable forms of **AC**.

Exercises

1. Let E be a compact topological space, K be a compact (equivalently, closed) subset of E, and let

$$K \subset U_1 \cup U_2 \cup \cdots \cup U_n,$$

where $U_1, U_2, \ldots, U_n$ are some open sets in E.
Show that there exist compact sets $K_1, K_2, \ldots, K_n$ in E, for which the relations

$$K_1 \subset U_1, \quad K_2 \subset U_2, \quad \ldots, \quad K_n \subset U_n,$$
$$K = K_1 \cup K_2 \cup \cdots \cup K_n$$

are fulfilled.
For this purpose, argue as follows. For each natural number m from $\{1, 2, \ldots, n\}$, denote $E \setminus U_m$ by P_m and observe that

$$P_1 \cap P_2 \cap \cdots \cap P_n \cap K = \emptyset.$$

Then verify that in E there are open sets $V_1, V_2, \ldots, V_n, V$ such that

$$P_1 \subset V_1, \quad P_2 \subset V_2, \quad \ldots, \quad P_n \subset V_n, \quad K \subset V,$$

$$V_1 \cap V_2 \cap \cdots \cap V_n \cap V = \emptyset.$$

Finally, let

$$K_1 = (E \setminus V_1) \cap K, \quad K_2 = (E \setminus V_2) \cap K, \quad \ldots, \quad K_n = (E \setminus V_n) \cap K$$

and check that the sets $K_1, K_2, \ldots, K_n$ are as required.

2. Let E be a nonempty compact topological space, let $g : E \to [0, 1]$ be a continuous function, and let $\{U_m : m = 1, 2, \ldots, n\}$ be a finite open covering of $\mathrm{supp}(g)$.

Prove that g can be represented in the form

$$g = \Sigma\{g_m : m = 1, 2, \ldots, n\},$$

where the functions g_m satisfy the relations

$$g_m \in C(E), \qquad 0 \le g_m \le 1, \qquad \mathrm{supp}(g_m) \subset U_m.$$

Argue step by step. First, using the result of Exercise 1, find in E compact sets $K_1, K_2, \ldots, K_n$ such that

$$K_1 \subset U_1, \quad K_2 \subset U_2, \quad \ldots, \quad K_n \subset U_n,$$

$$\mathrm{supp}(g) = K_1 \cup K_2 \cup \cdots \cup K_n.$$

Further, keeping in mind the fact that E is a normal space, for each m from $\{1, 2, \ldots, n\}$ take an open set V_m in E such that $K_m \subset V_m$ and the closure of V_m is contained in U_m. Once again, since E is a normal space, for each $m \in \{1, 2, \ldots, n\}$ there exists a continuous function ϕ_m acting from E into $[0, 1]$ and satisfying the following two conditions:

$$\phi_m|(E \setminus V_m) = 0, \qquad \phi_m|K_m = 1.$$

Now, define a function $g_m : E \to [0, 1]$ as follows:

$$g_m(x) = \begin{cases} 0 & \text{if } g(x) = 0, \\ g(x)\phi_m(x)/(\phi_1(x) + \phi_2(x) + \cdots + \phi_n(x)) & \text{if } g(x) > 0. \end{cases}$$

Finally, demonstrate that the obtained functions g_m ($m = 1, 2, \ldots, n$) are as required.

3. Give a detailed argument for proving assertion (2) of Lemma 28.3.

4. Returning to the proof of Theorem 28.1, verify in detail the inequalities

$$(1/n)\chi_{P_k} \leq f_k \leq (1/n)\chi_{P_{k-1}} \qquad (k = 1, 2, \ldots, n).$$

5. Analyze thoroughly all steps of the proof of Theorem 28.1 and check that the **DC** axiom suffices to establish the validity of this theorem.

6. Let E be a nonempty normal topological space and let μ_1 and μ_2 be two finite Radon measures on the Borel σ-algebra $\mathcal{B}(E)$ of E such that, for every real-valued continuous function f on E, the equality

$$\int_E f(x)\,\mathrm{d}\mu_1(x) = \int_E f(x)\,\mathrm{d}\mu_2(x)$$

is fulfilled.

Show that the measures μ_1 and μ_2 are identical to each other.

For this purpose, argue as follows. It suffices to verify that, for any open subset U of E, the equality $\mu_1(U) = \mu_2(U)$ holds. Take an arbitrary real $\varepsilon > 0$ and find a compact set $K \subset U$ satisfying the relations

$$\mu_1(U) - \varepsilon \leq \mu_1(K) \leq \mu_1(U), \qquad \mu_2(U) - \varepsilon \leq \mu_2(K) \leq \mu_2(U).$$

Consider a continuous function $g : E \to [0, 1]$ such that $g|K$ is 1 and $g|(E\setminus U)$ is 0. Then, keeping in mind the relations

$$\mu_1(K) \leq \int_E g(x)\,\mathrm{d}\mu_1(x) \leq \mu_1(U), \qquad \mu_2(K) \leq \int_E g(x)\,\mathrm{d}\mu_2(x) \leq \mu_2(U),$$

$$\int_E g(x)\,\mathrm{d}\mu_1(x) = \int_E g(x)\,\mathrm{d}\mu_2(x),$$

infer that $|\mu_1(U) - \mu_2(U)| \leq \varepsilon$, which implies the required result.

Deduce from this result the uniqueness of the signed measure ν in Theorem 28.2.

7. Formulate and prove a certain analog of Theorem 28.2 for the case of a nonempty non-compact locally compact topological space E.

In this case, instead of the family $C(E)$, consider the family $C_0(E)$ of all those real-valued continuous functions on E which vanish at infinity.

Recall that, by definition, a real-valued function f on E belongs to $C_0(E)$ if f is continuous and, for every real $\varepsilon > 0$, there exists a compact set K_ε in E such that $|f(x)| \leq \varepsilon$ whenever $x \in E \setminus K_\varepsilon$.

Check that $C_0(E)$ is a Banach space with respect to the standard sup-norm, consider Alexandrov's compactification E^* of E, and demonstrate that any linear continuous functional θ on $C_0(E)$ can be extended to a linear continuous functional θ^* on $C(E^*)$. Then apply to θ^* the Riesz representation theorem and, finally, obtain for θ a suitable Radon signed measure on the Borel σ-algebra $\mathcal{B}(E)$.

8. Let E be a nonempty ground set, $\mathcal{J}$ be a σ-ideal of subsets of E, and let $\mathcal{F}$ be the dual filter for $\mathcal{J}$. Denote by $\mathcal{A}$ the σ-algebra generated by $\mathcal{J}$ (in fact, one has the equality $\mathcal{A} = \mathcal{J} \cup \mathcal{F}$).
Define a probability countably additive measure μ on $\mathcal{A}$ by putting:

$$\mu(X) = \begin{cases} 0 & \text{if } X \in \mathcal{J}, \\ 1 & \text{if } X \in \mathcal{F}. \end{cases}$$

Let $g : E \to \mathbb{R}$ be an arbitrary μ-measurable function.
Show that g is constant on some member of $\mathcal{F}$, so g is μ-equivalent to a constant function.

9. Let $E = [0, \omega_1[$ be equipped with its standard order topology and let $h : E \to \mathbb{R}$ be any function continuous with respect to this topology.
Demonstrate that there exists an ordinal $\xi < \omega_1$ such that the restriction of h to the interval $]\xi, \omega_1[$ is a constant function.
For this purpose, argue as follows.
First of all, consider on E the Dieudonné probability measure ν' (see Exercise 6 from Chap. 20) and infer that the given function h is measurable with respect to ν'. Consequently, by virtue of Exercise 8, there exists a closed unbounded subset X of E such that $h|X$ is a constant function, all values of which coincide with some real number t.
Further, for every natural number n, introduce the set

$$Y_n = \{\zeta \in E : |h(\zeta) - t| \geq 1/(n+1)\}.$$

Verify that Y_n is a closed subset of E and $Y_n \cap X = \emptyset$. Deduce from this fact that Y_n is necessarily bounded in E.
Finally, let $Y = \cup\{Y_n : n \in \mathbb{N}\}$ and conclude that Y is also bounded in E, i.e., there exists $\xi < \omega_1$ such that $Y \subset [0, \xi]$. Obviously, for each countable ordinal ζ strictly greater than ξ, one has $h(\zeta) = t$, i.e., ξ is as required.

10. Using the notation of Exercise 9, let E' be Alexandrov's compactification of E. Actually, $E' = [0, \omega_1] = \omega_1 + 1$.
Define a Borel probability measure μ_1 on E' as follows.
For each Borel subset Z of E', put:

$$\mu_1(Z) = \begin{cases} 1 & \text{if } Z \text{ contains the point } \omega_1, \\ 0 & \text{if } Z \text{ does not contain } \omega_1. \end{cases}$$

Also, define a Borel probability measure μ_2 on E' so that the restriction of μ_2 to the Borel σ-algebra of E is identical to the Dieudonné measure ν'.
Observe that μ_1 is a Radon measure and μ_2 is not a Radon measure.
Further, introduce a continuous linear functional $\theta : C(E') \to \mathbb{R}$ by the formula

$$\theta(f) = f(\omega_1) \qquad (f \in C(E')),$$

28 Linear Continuous Functionals and Radon Measures

and show that, for any function $f \in C(E')$, the equalities

$$\theta(f) = \int_{E'} f(x)\,\mathrm{d}\mu_1(x) = \int_{E'} f(x)\,\mathrm{d}\mu_2(x)$$

hold true.

Conclude from this result that the Radon property of the signed measure ν in Theorem 28.2 is essential for the uniqueness of ν.

Chapter 29
Linear Continuous Functionals on a Real Hilbert Space

In Chap. 28 we were concerned with the Riesz theorem on the representation of a continuous linear functional $\theta : C(E) \to \mathbb{R}$, where $C(E)$ is the Banach space of all real-valued continuous functions defined on a nonempty compact topological space E.

As widely known, another celebrated representation theorem, also due to Riesz, describes the structure of all continuous linear functionals on a real Hilbert space. This second theorem is based on the following important projection lemma.

Lemma 29.1 *Let $(E, ||\cdot||)$ be a real pre-Hilbert space (i.e., the norm $||\cdot||$ is given by some inner product $\langle \cdot, \cdot \rangle$ in E) and let $Z \subset E$ be a nonempty convex set complete with respect to the metric induced by $||\cdot||$.*

Then, for every point $x \in E$, there exists a unique point $y \in Z$ such that:

(1) $||x - y|| = \inf\{||x - z|| : z \in Z\}$;
(2) $\langle x - y, z - y \rangle \le 0$ for any $z \in Z$.

Proof For each point $x \in E$, let

$$d = \inf\{||x - z|| : z \in Z\}$$

and, for any nonzero natural number n, choose a point $z_n \in Z$ satisfying

$$||x - z_n||^2 \le d^2 + 1/n.$$

So we come to the sequence $\{z_n : n \in \mathbb{N} \setminus \{0\}\}$ of points of Z.

We are going to demonstrate that $\{z_n : n \in \mathbb{N} \setminus \{0\}\}$ is a Cauchy sequence in Z. Indeed, according to the *parallelogram formula* for a pre-Hilbert space (see Exercise 1), we may write

$$2||x - z_n||^2 + 2||x - z_m||^2 = ||z_n - z_m||^2 + ||z_n + z_m - 2x||^2$$

whenever $n > 0$ and $m > 0$. Equivalently, we have

$$||x - z_n||^2/2 + ||x - z_m||^2/2 = ||z_n - z_m||^2/4 + ||x - (z_n + z_m)/2||^2,$$

whence it follows, in view of $(z_n + z_m)/2 \in Z$, that

$$(d^2 + 1/n)/2 + (d^2 + 1/m)/2 \geq ||z_n - z_m||^2/4 + d^2.$$

After easy calculations, we get the inequality

$$||z_n - z_m||^2 \leq 2(1/n + 1/m),$$

which shows us that $\{z_n : n \in \mathbb{N} \setminus \{0\}\}$ is a fundamental sequence in Z.

Therefore, this sequence converges to some point $y \in Z$, and it is clear that $||x - y|| = d$. Further, assuming for a moment that there are at least two distinct points $z' \in Z$ and $z'' \in Z$ such that

$$||x - z'|| = ||x - z''|| = d$$

and keeping in mind that $\langle (z' - x)/2, (z'' - x)/2 \rangle < d^2/4$, we readily infer that

$$||x - (z' + z'')/2||^2 =$$

$$||(x - z')/2||^2 + ||(x - z'')/2||^2 + 2\langle (z' - x)/2, (z'' - x)/2 \rangle < d^2,$$

whence it immediately follows that

$$||x - (z' + z'')/2|| < d.$$

However, the last inequality contradicts the definition of d. So, (1) holds.

Now, let us deduce (2) from (1). Suppose to the contrary that there exists a $z \in Z$ such that

$$\langle x - y, z - y \rangle > 0.$$

In view of the convexity of Z, the line segment $[z, y]$ is entirely contained in Z. Formally, for any $t \in [0, 1]$, the point $y + t(z - y)$ belongs to Z. The above implies that if $t > 0$ is sufficiently small, then

$$||x - (y + t(z - y))||^2 < ||x - y||^2 = d^2,$$

which contradicts (1).

Lemma 29.1 has thus been proved. □

Under the assumptions of Lemma 29.1, for each point $x \in E$, the unique point $y \in Z$ satisfying the equality

$$||x - y|| = \inf\{||x - z|| : z \in Z\}$$

is called the *projection* of x to Z and is denoted $\mathrm{pr}_Z(x)$.

Lemma 29.2 *Let $(E, ||\cdot||)$ be again a real pre-Hilbert space, let $Z \subset E$ be an affine linear manifold in E complete with respect to the metric induced by $||\cdot||$, and let x be any point of E.*

Then, for every point $z \in Z$, the equality

$$\langle x - \mathrm{pr}_Z(x), z - \mathrm{pr}_Z(x)\rangle = 0$$

holds true.

Proof Since any affine linear manifold in E is trivially a convex subset of E, we may apply Lemma 29.1 proved above. According to (2) of Lemma 29.1, we have

$$\langle x - \mathrm{pr}_Z(x), z - \mathrm{pr}_Z(x)\rangle \leq 0$$

whenever $z \in Z$. Consider the point z' in E defined by

$$z' = 2\mathrm{pr}_Z(x) - z.$$

It is easy to see that z' also belongs to Z. So we must have

$$\langle x - \mathrm{pr}_Z(x), z' - \mathrm{pr}_Z(x)\rangle \leq 0,$$

which immediately implies that

$$\langle x - \mathrm{pr}_Z(x), z - \mathrm{pr}_Z(x)\rangle \geq 0$$

and, consequently, the required equality

$$\langle x - \mathrm{pr}_Z(x), z - \mathrm{pr}_Z(x)\rangle = 0$$

is fulfilled, which ends the proof. □

Actually, Lemma 29.2 states that if Z is an affine linear manifold in $(E, ||\cdot||)$ complete with respect to the metric induced by $||\cdot||$ and x is any point of E, then the vector $x - \mathrm{pr}_Z(x)$ is orthogonal to Z.

In addition, it directly follows from Lemma 29.2 that if Z is a proper affine linear manifold in $(E, ||\cdot||)$ complete with respect to the metric induced by $||\cdot||$, then there exists a nonzero vector in E which is orthogonal to Z.

Now, we can formulate and prove the Riesz theorem on the structure of continuous linear functionals defined on a real Hilbert space (see, e.g., [35, 38, 104]). Recall, by the way, that a *real Hilbert space* is any real pre-Hilbert space complete with respect to its natural metric.

Theorem 29.1 *Let $(H, \langle \cdot, \cdot \rangle)$ be a real Hilbert space and let $\theta : H \to \mathbb{R}$ be a continuous linear functional on H.*

Then there exists an element $g \in H$ such that

$$\theta(h) = \langle g, h \rangle \qquad (h \in H).$$

Moreover, an element $g \in H$ with this property is unique.

Proof To begin, let us define

$$H(\theta) = \{h \in H : \theta(h) = 0\}.$$

It can readily be verified that $H(\theta)$ is a closed vector subspace of H. If $H(\theta) = H$, then θ is identically equal to zero, and we may take $g = 0$.

Suppose that $H(\theta) \neq H$. Then $H(\theta)$ is a closed vector hyperplane in H and there is a nonzero vector $g \in H$ orthogonal to $H(\theta)$ (hence $\theta(g) \neq 0$). In other words, using Lemma 29.2, we come to a representation of H in the form of a direct sum

$$H = H(\theta) + \mathbb{R}g,$$

where the vector g is orthogonal to all vectors from $H(\theta)$. We may assume, without loss of generality, that

$$\theta(g) = \|g\|^2.$$

Now, for each vector $h \in H$, we can write

$$h = h' + tg,$$

where $h' \in H(\theta)$ and $t \in \mathbb{R}$. So we have

$$\theta(h) = \theta(h') + t\theta(g) = t\theta(g) = t\|g\|^2,$$
$$\langle g, h \rangle = \langle g, h' \rangle + \langle g, tg \rangle = t\|g\|^2.$$

Consequently, we get the equality $\theta(h) = \langle g, h \rangle$ for any $h \in H$.

The uniqueness of g is trivial, because of the following. If

$$\langle g_1, h \rangle = \langle g_2, h \rangle = 0 \qquad (h \in H),$$

then, putting $h = g_1 - g_2$, we immediately obtain from the above equalities that

$$||g_1 - g_2||^2 = 0,$$

i.e., g_1 is identical with g_2.

This completes the proof of Theorem 29.1. □

Example 29.1 Let us mention one special case of Theorem 29.1:

If θ is a continuous linear functional on a real infinite-dimensional separable Hilbert space H (which can be identified with the space of all square-summable real sequences), then one has the formula

$$\theta(x) = x_0 a_0 + x_1 a_1 + \cdots + x_n a_n + \cdots,$$

where $x = (x_0, x_1, \ldots, x_n, \ldots)$ is any vector from H and $a = (a_0, a_1, \ldots, a_n, \ldots)$ is a fixed vector of H uniquely determined by θ.

Example 29.2 Consider the family $C([0, 1])$ of all real-valued continuous functions on the segment $[0, 1]$ and equip this family with the standard sup-norm

$$||f|| = \sup\{|f(x)| : x \in [0, 1]\} \qquad (f \in C([0, 1])).$$

So we come to the classical separable Banach space $(C([0, 1]), ||\cdot||)$ which possesses many remarkable properties. One of them states that every separable metric space can be isometrically embedded into $C([0, 1])$ (in this connection, see also [150] where a much stronger result is presented).

It is not difficult to find two functions f and g from $C([0, 1])$ such that

$$||f|| = ||g|| = ||f + g|| = ||f - g|| = 1.$$

For these functions, we trivially have

$$2(||f||^2 + ||g||^2) = 4 \neq 2 = ||f + g||^2 + ||f - g||^2.$$

Consequently, the parallelogram formula is false for the sup-norm $||\cdot||$, which immediately implies that $||\cdot||$ is not produced by an inner product in $C([0, 1])$.

Now, we would like to discuss a concrete situation closely connected with both theorems of Riesz on representations of continuous linear functionals.

Consider the two classical function spaces $C([0, 1])$ and $L_2([0, 1])$ associated with the unit segment $[0, 1]$. As above, $C([0, 1])$ is the space of all real-valued continuous functions on $[0, 1]$ equipped with the topology of uniform convergence, and $L_2([0, 1])$ is the space of all real-valued Lebesgue square-integrable functions on $[0, 1]$ (as usual, equivalent Lebesgue measurable functions are identified). The second space $L_2([0, 1])$ is equipped with the canonical inner product

$$\langle f, g \rangle = \int_0^1 f(x)g(x)\,d\lambda(x),$$

where λ stands for the standard Lebesgue probability measure on $[0, 1]$.

So, $L_2([0, 1])$ is a real Hilbert space and $C([0, 1])$ is a certain vector subspace of $L_2([0, 1])$.

Suppose that a continuous linear functional $\theta : C([0, 1]) \to \mathbb{R}$ is given. Then the following question arises quite naturally:

Under which necessary and sufficient conditions can this θ be extended to a continuous linear functional

$$\theta' : L_2([0, 1]) \to \mathbb{R}?$$

To answer this question, first recall that according to the Riesz theorem from Chap. 28, there exists a unique Radon signed measure μ on $[0, 1]$ such that $\text{dom}(\mu) = \mathcal{B}([0, 1])$ and

$$\theta(f) = \int_0^1 f(x)\,d\mu(x) \qquad (f \in C([0, 1])).$$

Our goal is to prove the next theorem.

Theorem 29.2 *With the above notation, the following two conditions are equivalent:*

(1) *the functional θ admits an extension to a continuous linear functional θ' on $L_2([0, 1])$;*
(2) *μ is absolutely continuous with respect to λ and its Radon–Nikodym derivative $d\mu/d\lambda$ belongs to the space $L_2([0, 1])$.*

Proof First, suppose that condition (1) is fulfilled. Then we have

$$\theta'(f) = \int_0^1 f(x)g(x)\,d\lambda(x) \qquad (f \in L_2([0, 1])),$$

where θ' extends θ and g is some fixed function from $L_2([0, 1])$. In particular, the preceding formula holds for any $f \in C([0, 1])$, so

$$\theta'(f) = \theta(f) = \int_0^1 f(x)g(x)\,d\lambda(x) \qquad (f \in C([0, 1])).$$

Note, by the way, that in view of the inequality

$$|g| \leq (1 + g^2)/2,$$

the function g is λ-integrable, i.e., $g \in L_1([0, 1])$. Putting

29 Linear Continuous Functionals on a Real Hilbert Space

$$\nu(X) = \int_X g(x)\,d\lambda(x) \qquad (X \in \mathcal{B}([0,1])),$$

we infer that the Radon signed measure ν is absolutely continuous with respect to λ and the corresponding Radon–Nikodym derivative $d\nu/d\lambda$ is equal to $g \in L_2([0,1])$. Taking into account that the relation

$$\theta(f) = \int_0^1 f(x)\,d\mu(x) = \int_0^1 f(x)g(x)\,d\lambda(x) = \int_0^1 f(x)\,d\nu(x)$$

holds true for every function $f \in C([0,1])$, we conclude that $\mu = \nu$ (see Theorem 28.2 of Chap. 28) and so condition (2) holds for μ.

Conversely, suppose that condition (2) is fulfilled, i.e.,

$$\theta(f) = \int_0^1 f(x)\,d\mu(x) \qquad (f \in C([0,1])),$$

where μ is absolutely continuous with respect to λ and the Radon–Nikodym derivative $d\mu/d\lambda$ belongs to $L_2([0,1])$. Denoting $d\mu/d\lambda$ by g, we can write

$$\theta(f) = \int_0^1 f(x)\,d\mu(x) = \int_0^1 f(x)g(x)\,d\lambda(x) \qquad (f \in C([0,1])).$$

It is clear now that we are able to define the continuous linear functional

$$\theta' : L_2([0,1]) \to \mathbb{R}$$

by the analogous formula

$$\theta'(f) = \int_0^1 f(x)g(x)\,d\lambda(x) \qquad (f \in L_2([0,1])).$$

This θ' trivially extends θ, so condition (1) is fulfilled.

Theorem 29.2 has thus been proved. $\square$

Exercises

1. Let $(E, \|\cdot\|)$ be a real normed vector space.
   Show that the following two assertions are equivalent:

   (a) the norm of E is induced by some inner product (i.e., E is a real pre-Hilbert space);

   (b) for any two vectors x and y from E, the parallelogram formula

$$2(||x||^2 + ||y||^2) = ||x+y||^2 + ||x-y||^2$$

holds.

2. Let (E, d) be a metric space and let Z be a nonempty subset of E. Verify that, for any two points x_1 and x_2 from E, the inequality

$$|d(x_1, Z) - d(x_2, Z)| \leq d(x_1, x_2)$$

is fulfilled and infer from this fact that the real-valued function

$$x \to d(x, Z) \qquad (x \in E)$$

is uniformly continuous on E (moreover, this function satisfies the Lipschitz condition with constant 1).

3. Let Z be a nonempty compact convex subset of a real pre-Hilbert space E. Prove that the projection mapping

$$x \to \mathrm{pr}_Z(x) \qquad (x \in E)$$

is continuous on E.

4. Let Z be a nonempty closed convex subset of the Euclidean space $\mathbb{R}^n$. Show that the projection mapping

$$x \to \mathrm{pr}_Z(x) \qquad (x \in \mathbb{R}^n)$$

is continuous on $\mathbb{R}^n$.

5. Give an example of two functions f and g from $C([0, 1])$ such that

$$||f|| = |g|| = ||f + g|| = ||f - g|| = 1$$

and conclude that the sup-norm on $C[0, 1]$ is not produced by an inner (scalar) product in $C[0, 1]$.

For this purpose, take the required f and g so that the relations

$$\mathrm{supp}(f) = [0, 1/2], \qquad \mathrm{supp}(g) = [1/2, 1]$$

are fulfilled.

6. Let $(H, \langle \cdot, \cdot \rangle)$ be an arbitrary real Hilbert space.
Using Lemma 29.2 of this chapter and the Kuratowski–Zorn lemma, prove that there exists a family $\{h_i : i \in I\}$ of elements of H satisfying the following relations:

(a) $||h_i|| = 1$ for each index $i \in I$;
(b) $\langle h_i, h_j \rangle = 0$ for any two distinct indices i and j from I;

(c) $\{h_i : i \in I\}$ is maximal with respect to inclusion.

Such a family $\{h_i : i \in I\}$ is called an *orthonormal Hilbert basis* of H.
In addition, show that all orthonormal bases of H have the same cardinality (which is called the *Hilbert dimension* of H).
For this purpose, argue as follows. If H is finite-dimensional (as a real vector space), then the situation is more or less clear. It remains to consider the case when H is infinite-dimensional.
Take any orthonormal basis $\{h'_j : j \in J\}$ of H different from the orthonormal basis $\{h_i : i \in I\}$ mentioned above. Observe that $\mathrm{card}(J) \geq \omega$ and verify the validity of these two facts:

(d) the family $\{B(h_i, 1/2) : i \in I\}$ of open balls in H is disjoint;
(e) the set of all rational linear combinations of vectors from $\{h'_j : j \in J\}$ is everywhere dense in H.

Infer from (d) and (e) that $\mathrm{card}(I) \leq \mathrm{card}(J)$. By a similar argument, $\mathrm{card}(J) \leq \mathrm{card}(I)$, which yields the required equality $\mathrm{card}(I) = \mathrm{card}(J)$.

7. Let $(H, \langle \cdot, \cdot \rangle)$ be again a real Hilbert space and let $\{h_i : i \in I\}$ be an orthonormal basis of H.
Demonstrate that, for each vector $x \in H$, one has a unique representation of x in the form

$$x = \Sigma\{a_i h_i : i \in I\},$$

where all coefficients $a_i = a_i(x)$ are some real numbers and only countably many of them differ from zero.
Moreover, check that $a_i = \langle x, h_i \rangle$ for any $i \in I$, and the so-called *Parseval equality*

$$||x||^2 = \Sigma\{a_i^2 : i \in I\}$$

holds true.
Also, if $y = \Sigma\{b_i h_i : i \in I\}$ is any other vector from H, then

$$\langle x, y \rangle = \Sigma\{a_i b_i : i \in I\},$$

which can be treated as a generalized Parseval equality.

Remark 29.1 It readily follows from Exercises 6 and 7 that two real Hilbert spaces H_1 and H_2 are isomorphic to each other if and only if the Hilbert dimension of H_1 coincides with the Hilbert dimension of H_2.

8. Let E be a finite-dimensional vector space over $\mathbb{R}$ and let $n = \dim(E)$. Suppose that a family $\{f_1, f_2, \ldots, f_n\}$ of real-valued linear functionals on E is given. Demonstrate that the following two assertions are equivalent:

(a) the functionals $f_1, f_2, \ldots, f_n$ are linearly independent (as elements of the dual vector space of all linear functionals on E);
(b) there exist vectors $e_1, e_2, \ldots, e_n$ in E such that $f_i(e_i) = 1$ whenever i is a natural number from $[1, n]$ and $f_i(e_j) = 0$ whenever i and j are distinct natural numbers from $[1, n]$.

For this purpose, equip E with the structure of a real Hilbert space and then apply to this space the Riesz representation theorem.

9. Let H be a real Hilbert space and let $f : H \to \mathbb{R}$ be a continuous linear functional.
Show that there exists a separable closed vector subspace $G = G_f$ of H such that f is completely determined by its restriction to G.
In other words, if x and y are any two vectors from H satisfying the equality $\mathrm{pr}_G(x) = \mathrm{pr}_G(y)$, then $f(x) = f(y)$.
For this purpose, use again the Riesz representation theorem and Exercise 7.

10. Let $(H, \langle \cdot, \cdot \rangle)$ be a real Hilbert space and let B be any orthonormal Hilbert basis of H.
Prove that if $\mathrm{card}(B) > 1$, then $\mathrm{card}(H) = (\mathrm{card}(B))^\omega$.
For this purpose, argue as follows. The case when H is finite-dimensional is almost trivial, so one may reduce the argument to the case when H is infinite-dimensional (i.e., $\mathrm{card}(B) \geq \omega$).
The inequality $\mathrm{card}(H) \leq (\mathrm{card}(B))^\omega$ easily follows from the fact that the set of all rational linear combinations of vectors from B is everywhere dense in H.
Further, denote by $Inj(\omega, B)$ the family of all injective functions from ω into B and let ϕ be any function from $Inj(\omega, B)$. In H define the vector

$$h(\phi) = \sum \{(1/3)^n \phi(n) : n \in \mathbb{N}\}$$

and consider a mapping $\Phi : Inj(\omega, B) \to H$ given by the formula

$$\Phi(\phi) = h(\phi) \qquad (\phi \in Inj(\omega, B)).$$

Verify that Φ is an injective mapping. Finally, taking into account the relation

$$\mathrm{card}(Inj(\omega, B)) = (\mathrm{card}(B))^\omega,$$

deduce $(\mathrm{card}(B))^\omega \leq \mathrm{card}(H)$ and conclude that the required equality

$$(\mathrm{card}(B))^\omega = \mathrm{card}(H)$$

holds true.

Remark 29.2 Let X be a nonempty set, Y be an infinite set, and let $\mathrm{card}(X) \leq \mathrm{card}(Y)$. Then the family $Inj(X, Y)$ of all injective functions from X into Y is equinumerous with the family Y^X of all mappings from X to Y. Indeed, it suffices to observe that if $f \in Y^X$, then the function f^* defined by

$$f^*(x) = (x, f(x)) \qquad (x \in X)$$

is an injection from X into $X \times Y$. Moreover, $\operatorname{card}(X \times Y) = \operatorname{card}(Y)$ and if $g \in Y^X$ differs from f, then $f^* \neq g^*$.

In Exercise 10 the special case of the above result has been used when $X = \omega$ and $Y = B$.

11. Work within **ZF** & **DC** theory and demonstrate that, for a real Hilbert space $(H, \langle \cdot, \cdot \rangle)$, the following two assertions are equivalent:

    (a) H is separable (as a metric space);
    (b) there exists a countable Hilbert basis of H.

    Deduce from this fact that any two real infinite-dimensional separable Hilbert spaces are isomorphic to each other.

12. Following an argument due to von Neumann, show that the Riesz representation theorem of this chapter implies the Radon–Nikodym theorem (see Chap. 26).
    For this purpose, first of all reduce the question to the case of a measurable space $(E, \mathcal{S})$ equipped with two nonzero finite countably additive measures μ and ν on the σ-algebra $\mathcal{S}$ such that ν is absolutely continuous with respect to μ.
    Denote $\mu + \nu$ by ρ and observe that, for each set $X \in \mathcal{S}$, the equality $\rho(X) = \nu(X)$ is equivalent to the relation

    $$\mu(X) = \nu(X) = \rho(X) = 0.$$

    Then argue step by step.

    (a) For every function f from the real Hilbert space $L_2(E, \rho)$, define

    $$\phi(f) = \int_E f(x) \, d\nu(x)$$

    and check that ϕ is a positive continuous linear functional on $L_2(E, \rho)$.
    In order to establish (a), use the Cauchy–Schwarz inequality (in its integral form) and get

    $$|\phi(f)| \leq \nu(E)^{1/2} \left(\int_E f^2(x) \, d\nu(x) \right)^{1/2} \leq \nu(E)^{1/2} \left(\int_E f^2(x) \, d\rho(x) \right)^{1/2},$$

    which yields the continuity of ϕ on $L_2(E, \rho)$.

    (b) According to the Riesz theorem of this chapter, for ϕ there exists a nonnegative function $g \in L_2(E, \rho)$ such that

    $$\phi(f) = \int_E f(x) g(x) \, d\rho(x) \qquad (f \in L_2(E, \rho)).$$

In particular, taking as f the characteristic function of any set $X \in \mathcal{S}$, one has

$$\nu(X) = \int_X g(x)\,d\rho(x) \qquad (X \in \mathcal{S}).$$

Infer from the above formula that for ρ-almost all $x \in E$ (equivalently, for μ-almost all $x \in E$), the inequalities $0 \le g(x) \le 1$ are satisfied.

(c) Keeping in mind the equality $\rho = \mu + \nu$, deduce

$$\rho(\{x \in E : g(x) = 1\}) = \mu(\{x \in E : g(x) = 1\}) = 0.$$

Replacing (if necessary) the function g by some μ-equivalent $\mathcal{S}$-measurable function, one may assume that $0 \le g(x) < 1$ whenever $x \in E$.

(d) Introduce the non-negative function $g_0 = g/(1-g)$, take into account the formula

$$\int_E f(x)(1-g(x))\,d\nu(x) = \int_E f(x)g(x)\,d\mu(x) \qquad (f \in L_2(E,\rho))$$

and infer that this formula remains valid for any real-valued non-negative μ-measurable function f on E.

Finally, putting $f = \chi_X/(1-g)$ for $X \in \mathcal{S}$, conclude that

$$\nu(X) = \int_X g_0(x)\,d\mu(x) \qquad (X \in \mathcal{S}),$$

which is precisely the Radon–Nikodym theorem for the given measures μ and ν on $\mathcal{S}$.

13. Let m be a nonzero natural number, $(E, \langle \cdot, \cdot \rangle)$ be a real pre-Hilbert space, and let $e_1, e_2, \ldots, e_m$ be some elements of E. Consider a real-valued function ϕ on E defined by

$$\phi(x) = ||x - e_1||^2 + ||x - e_2||^2 + \cdots + ||x - e_m||^2 \qquad (x \in E).$$

Prove that $x_0 = (e_1 + e_2 + \cdots + e_m)/m$ is the unique point at which ϕ attains its infimum.

For this purpose, keep in mind that the formula

$$\sum \{||x - e_i||^2 : 1 \le i \le m\} - m||x - (e_1 + e_2 + \cdots + e_m)/m||^2$$
$$= \sum \{||e_i||^2 : 1 \le i \le m\} + ||(e_1 + e_2 + \cdots + e_m)/m||^2$$

holds whenever $x \in E$.

29 Linear Continuous Functionals on a Real Hilbert Space

14. Under the assumptions of Exercise 13, suppose additionally that

$$\|e_i - e_j\| \geq 2$$

for any two distinct natural numbers i and j from $\{1, 2, \ldots, m\}$. Demonstrate that $\sum\{\|x - e_i\|^2 : 1 \leq i \leq m\} \geq 2(m - 1)$ whenever $x \in E$. For this purpose, assume without loss of generality that

$$e_1 + e_2 + \cdots + e_m = 0$$

and check that $m \sum\{\|e_i\|^2 : 1 \leq i \leq m\} \geq 2m(m - 1)$. Then use the result given in Exercise 13.

15. Preserve all the assumptions of Exercise 14 and consider any ball B in E containing the finite set $\{e_1, e_2, \ldots, e_m\}$.
Show that $r(B) \geq (2(m - 1)/m)^{1/2}$, where $r(B)$ denotes the radius of B.

Chapter 30
The Baire Property in Topological Spaces

Without any doubt, the concept of measurability of sets and functions is very important for modern mathematical analysis and its numerous applications. In general topology there is an interesting analog of the concept of measurability, which is also important and applicable in many fields of mathematics. This second concept will be briefly discussed in the present chapter.

Let $(E, \mathcal{T})$ be a topological space. Recall that the symbol $\mathcal{B}(E)$ denotes the Borel σ-algebra of E, i.e., the σ-algebra generated by the family $\mathcal{T}$ of all open sets in E.

In a nonempty space E we always have the ideal $\mathcal{N}(E)$ of all nowhere dense subsets of E and if E is not of first category in itself, then we have the σ-ideal $\mathcal{K}(E)$ of all first category subsets of E.

Let $\mathcal{B}a(E)$ stand for the σ-algebra generated by $\mathcal{T} \cup \mathcal{N}(E)$.

We shall say that a set $X \subset E$ has (possesses) the *Baire property* in E if the relation $X \in \mathcal{B}a(E)$ holds true (see, e.g., [109, 141]).

The above definition immediately implies that $\mathcal{B}(E) \subset \mathcal{B}a(E)$ and, in general, this inclusion is proper.

For instance, $\mathcal{B}a(\mathbb{R})$ properly contains $\mathcal{B}(\mathbb{R})$, because

$$\text{card}(\mathcal{B}a(\mathbb{R})) = 2^{\mathbf{c}}, \quad \text{card}(\mathcal{B}(\mathbb{R})) = \mathbf{c}.$$

Indeed, the first equality follows from the fact that all subsets of the nowhere dense Cantor set on $\mathbb{R}$ belong to $\mathcal{B}a(\mathbb{R})$ and the second equality follows from the fact that the standard topology of $\mathbb{R}$ has a countable base.

It is not difficult to check that a set $X \subset E$ has the Baire property in E if and only if X admits a representation in the form

$$X = (U \cup Y) \setminus Z,$$

where U is an open subset of E and Y and Z are some first category subsets of E.

It directly follows from the above representation that the Baire property of subsets of E is preserved (actually, is invariant) under the group of all homeomorphisms of E onto itself.

Let E and F be two topological spaces. As known, a given mapping $g : E \to F$ is *measurable in the Borel sense* (in short, is a *Borel mapping*) if, for any open set $V \subset F$, the set $g^{-1}(V)$ belongs to $\mathcal{B}(E)$.

Accordingly, we shall say that a mapping $h : E \to F$ has (possesses) the *Baire property* if, for any open set $V \subset F$, the set $h^{-1}(V)$ belongs to the σ-algebra $\mathcal{B}a(E)$.

Evidently, every Borel mapping acting from E into F has the Baire property.

We will denote by $\mathcal{B}a(E, F)$ the family of all those mapping from E into F which have the Baire property.

So $\mathcal{B}(E, F) \subset \mathcal{B}a(E, F)$, where $\mathcal{B}(E, F)$ denotes, as usual, the family of all Borel mappings from E into F.

Theorem 30.1 *Let $(E, \mathcal{T})$ be a topological space and let $\mathbb{R}$ be equipped with its standard topology.*

Then $\mathcal{B}a(E, \mathbb{R})$ is an algebra of functions and is a lattice with respect to the two canonical operations

$$(f, g) \to \inf(f, g), \qquad (f, g) \to \sup(f, g).$$

Moreover, if a sequence of functions $\{f_n : n \in \mathbb{N}\} \subset \mathcal{B}a(E, \mathbb{R})$ is such that

$$\lim_{n \to \infty} f_n(x) = f(x) \qquad (x \in E),$$

then $f \in \mathcal{B}a(E, \mathbb{R})$.

We omit the easy proof of Theorem 30.1.

Remark 30.1 The definition of sets (functions) possessing the Baire property is effective, i.e., does not appeal to any form of the Axiom of Choice. Most sets and functions studied in classical real analysis have this property. In particular, all analytic (and all co-analytic) sets in a Polish topological space have the Baire property (see, e.g., [82, 109]). But there are important examples of subsets of $\mathbb{R}$ which do not possess the Baire property. For instance, no Vitali set on $\mathbb{R}$ (and no Bernstein set on $\mathbb{R}$) possesses this property. It has been established that the existence of subsets of $\mathbb{R}$ lacking the Baire property cannot be proved within **ZF & DC** theory (see [76]).

The following statement is due to Banach and is useful in many situations (cf. [109, 141]).

Lemma 30.1 *Let E be a topological space and let $\{U_i : i \in I\}$ be a family of open subsets of E all of which are of first category in E.*

Then the open set $U = \cup\{U_i : i \in I\}$ is also of first category in E.

Proof Consider the family $\mathcal{V}$ of all those open sets $V \subset E$ which satisfy the relation

$$(\exists i \in I)(V \subset U_i).$$

Clearly, each set $V \in \mathcal{V}$ is of first category in E. Let $\{V_j : j \in J\}$ be a maximal (by inclusion) disjoint subfamily of $\mathcal{V}$. Note that the existence of $\{V_j : j \in J\}$ immediately follows from the Kuratowski–Zorn lemma.

It can readily be verified that the open set

$$W = \cup\{V_j : j \in J\}$$

is everywhere dense in U. So, it suffices to demonstrate that W is of first category in E (because $U \setminus W$ is contained in the boundary of W and the boundary of each open subset of E is nowhere dense in E).

For any $j \in J$, we may write

$$V_j = \cup\{X_{n,j} : n \in \mathbb{N}\},$$

where all sets $X_{n,j}$ are nowhere dense in E. Let us put

$$X_n = \cup\{X_{n,j} : j \in J\} \qquad (n \in \mathbb{N}).$$

It is easy to check that any set X_n, where $n \in \mathbb{N}$, is also nowhere dense in E. Finally, since

$$W = \cup\{V_j : j \in J\} = \cup\{X_n : n \in \mathbb{N}\},$$

we can conclude that W is of first category in E, which ends the proof. □

Let E be a topological space and let G be a group of homeomorphisms of E onto itself.

We shall say that G acts *almost transitively* in E (in the topological sense) if, for any two nonempty open subsets U and V of E, there exists a $g \in G$ such that $g(U) \cap V \neq \emptyset$ (cf. [110]).

Obviously, if a group G of homeomorphisms of a topological space E acts transitively on E, then G acts almost transitively on E. The converse assertion fails to be true in general.

Remark 30.2 It is useful to compare the above notion with the metrical transitivity (i.e., ergodicity) of a quasi-invariant measure given on a ground set E equipped with a transformation group G (see Chap. 21). Indeed, the G-ergodicity of a σ-finite G-quasi-invariant measure μ on E can be formulated in the following equivalent form:

For any two μ-measurable sets X and Y with $\mu(X) > 0$ and $\mu(Y) > 0$, there exists a transformation g from G such that $\mu(g(X) \cap Y) > 0$.

In this connection, see also Exercise 8 of the present chapter.

Theorem 30.2 *Let E be a topological space and let G be a group of homeomorphisms of E onto itself, acting almost transitively in E.*

Then at least one of the following two assertions holds true:

(1) *every nonempty open subset of E is of second category in E (i.e., E is a Baire topological space);*
(2) *E is of first category in itself.*

Proof Suppose that assertion (1) is false. Then there exists a nonempty open set $U \subset E$ of first category in E. Let us define

$$W = \cup\{g(U) : g \in G\}$$

and let us demonstrate that the open set W is everywhere dense in E. Indeed, supposing for a moment otherwise, we come to the existence of a nonempty open subset V of E such that $V \cap W = \emptyset$. But, since G acts almost transitively in E, there is an $h \in G$ for which $h(U) \cap V \neq \emptyset$. At the same time,

$$h(U) \subset \cup\{g(U) : g \in G\} = W,$$

which contradicts the equality $V \cap W = \emptyset$. The obtained contradiction shows us that W is everywhere dense in E. So we have

$$E = \operatorname{cl}(W) = W \cup \operatorname{bd}(W),$$

where $\operatorname{bd}(W)$ denotes, as usual, the boundary of W. According to Lemma 30.1, W is a first category subset of E. Taking into account that $\operatorname{bd}(W)$ is nowhere dense in E, the latter immediately implies that E itself is a first category space, i.e., assertion (2) holds true. □

The next nice result concerning the Baire property in topological groups is due to Banach, Kuratowski, and Pettis (cf. [38, 84]).

Lemma 30.2 *Let $(G, \cdot)$ be a topological group and let X be a subset of G having the Baire property and of second category in G.*
Then the set $X \cdot X^{-1}$ is a neighborhood of the neutral element of G.

Proof Since $X \subset G$ and X is of second category, G is of second category too. In view of Lemma 30.1 (applied to G and to the group of all left translations of G), all nonempty open subsets of G are of second category, i.e., G is a Baire topological space. Further, the given set X admits a representation in the form

$$X = (U \cup Y) \setminus Z,$$

where U is a nonempty open set in G and Y and Z are some first category subsets of G. Let

$$V = \{g \in G : gU \cap U \neq \emptyset\}.$$

A straightforward verification shows that V is an open set in G containing the neutral element of G. So, it suffices to check that $V \subset X \cdot X^{-1}$ or, equivalently, that

$$(\forall g \in V)(gX \cap X \neq \emptyset).$$

Obviously, for any $g \in V$, we may write

$$(gU \cap U) \setminus (gZ \cup Z) \subset gX \cap X.$$

Since the set $gU \cap U$ is nonempty and open, it is of second category, while the set $gZ \cup Z$ is of first category. Consequently,

$$(gU \cap U) \setminus (gZ \cup Z) \neq \emptyset, \qquad gX \cap X \neq \emptyset,$$

which completes the proof of Lemma 30.2. □

Remark 30.3 Consider the standard Lebesgue measure λ on the real line $\mathbb{R}$. For this measure, H. Steinhaus [171] proved the following useful result:

If X is a λ-measurable set on $\mathbb{R}$ with $\lambda(X) > 0$, then the associated difference set

$$X - X = \{x - y : x \in X, \ y \in X\}$$

is necessarily a neighborhood of zero in $\mathbb{R}$.

Clearly, this fact may be treated as a measure-theoretical analog of Lemma 30.2. Further extensions and generalizations of the above-mentioned result of Steinhaus can be found in [17, 63, 69, 101, 140].

Theorem 30.3 *Let $(G, +)$ be a commutative topological group of second category in itself and let*

$$f : (G, +) \to (\mathbb{R}, +)$$

be an additive functional having the Baire property.
Then f is continuous at all points of G.

Proof It suffices to show that the given f is continuous at the neutral element (i.e., zero) of G. Let us define

$$X_n = f^{-1}([-n, n]) \qquad (n \in \mathbb{N}).$$

Obviously, $G = \cup\{X_n : n \in \mathbb{N}\}$ and all sets X_n possess the Baire property in G. Since G is of second category, there exists an $m \in \mathbb{N}$ such that X_m is of second category. Using Lemma 30.2, we infer that there is a neighborhood V of zero of G satisfying the relations

$$V \subset X_m - X_m,$$
$$f(V) \subset [-2m, 2m].$$

In other words, f is bounded on V.

Now, a simple argument (left to the reader) shows that the boundedness of our additive function f on some nonempty open subset of G necessarily implies the continuity of f (cf. also [107]).

Theorem 30.3 has thus been proved. □

Exercises

1. Give several nontrivial examples of a topological space $(E, \mathcal{T})$ for which the Borel σ-algebra $\mathcal{B}(E)$ is identical with the σ-algebra $\mathcal{B}a(E)$.
   Moreover, show that there exists a topology $\mathcal{T}_d$ on $\mathbb{R}$ which properly contains the standard topology of $\mathbb{R}$ and for which
   $$\mathcal{B}(\mathbb{R}, \mathcal{T}_d) = \mathcal{B}a(\mathbb{R}, \mathcal{T}_d) = \mathcal{L}(\mathbb{R}),$$
   where $\mathcal{L}(\mathbb{R})$ denotes the σ-algebra of all Lebesgue measurable subsets of $\mathbb{R}$.

   Remark 30.4 It should be noted that the so-called density topology on $\mathbb{R}$ plays the role of the topology $\mathcal{T}_d$ mentioned in Exercise 1 (see, for instance, [141]).

2. Let E be a topological space and let X be a subset of E.
   Check that X has the Baire property in E if and only if X admits a representation in the form
   $$X = (U \cup Y) \setminus Z,$$
   where U is an open subset of E and Y and Z are first category subsets of E.
   Also, show that, in general, such a representation is not unique for X.
3. Give a detailed proof of Theorem 30.1.
4. Demonstrate that no Vitali set on $\mathbb{R}$ has the Baire property.
5. Prove that no Bernstein set on $\mathbb{R}$ has the Baire property.
6. Let E be a topological space and let G be a group of homeomorphisms of E onto itself, such that there exists a point $x \in E$ whose G-orbit is everywhere dense in E.
   Verify that G acts almost transitively in E.
   In addition, show that there are a topological space F of second category and a group H of homeomorphisms of F onto itself, satisfying the following two conditions:

   (a) H acts almost transitively in F;
   (b) the H-orbits of all points of F are closed and nowhere dense in F.

7. Let (E, d) be a metric space and let G be a group of d-isometric transformations of E onto itself.
   Prove that the following three assertions are equivalent:

   (a) G acts almost transitively in E;
   (b) there exists a point in E whose G-orbit is everywhere dense in E;
   (c) the G-orbit of any point of E is everywhere dense in E.

8. Let E be a topological space and let G be a group of homeomorphisms of E onto itself, acting almost transitively in E.
   Check that if X and Y are any two subsets of E, both having the Baire property and of second category, then there exists a $g \in G$ such that the set $g(X) \cap Y$ is also of second category.

9. Let $(G, \cdot)$ be a topological group, and let X_1 and X_2 be two subsets of G both having the Baire property and of second category.
   Show that the set $X_1 \cdot X_2$ contains a nonempty open subset of G.
   For this purpose, argue similarly to the proof of Lemma 30.2.

10. Recall that a topological space E is *resolvable* (in the sense of Hewitt) if there exists a partition of E into two everywhere dense subsets of E (see Chap. 6 and Appendix E).
    According to this definition, a topological space E is *irresolvable* if the above-mentioned partition does not exist for E.
    Moreover, E is called *totally irresolvable* if no open subset of E is resolvable (with respect to the induced topology).
    Verify that in a totally irresolvable topological space E any set with empty interior is nowhere dense in E.
    Conclude from this fact that each subset of a totally irresolvable topological space E has the Baire property in E.

11. Let E be a topological space such that there is a group of homeomorphisms of E onto itself, acting almost transitively in E.
    Show that the following two assertions are equivalent:

    (a) E is a resolvable space;
    (b) there exists an open set in E which is a resolvable subspace of E.

12. Let $(G, \cdot)$ be a topological group (not necessarily Hausdorff).
    Check that the following two assertions are equivalent:

    (a) G is an irresolvable topological space;
    (b) G is a totally irresolvable topological space.

    For this purpose, use the result presented in Exercise 11.

13. Let E be a set and let $\mathcal{I}$ be an ideal of subsets of E.
    By definition, a family $\mathcal{B} \subset \mathcal{I}$ is a base of $\mathcal{I}$ if, for any set $X \in \mathcal{I}$, there exists a set $Y \in \mathcal{B}$ such that $X \subset Y$.
    Let now $(E, \mathcal{T})$ be an arbitrary topological space.

Recall that a family $\mathcal{V}$ of nonempty open subsets of E is a *pseudo-base* (or a π-*base*) of E if, for any nonempty open set U in E, there exists a set $V \in \mathcal{V}$ such that $V \subset U$.

Let $(E, \mathcal{T})$ be an infinite topological space of second category and let G be a group of homeomorphisms of E onto itself, acting freely and transitively on E. Suppose also that there is a pseudo-base of E whose cardinality is strictly less than card(E).

Assuming the Generalized Continuum Hypothesis (**GCH**), prove that there exists a subset of E which does not possess the Baire property.

Argue as follows. First, denote card(E) by **a**. Without loss of generality, one may assume that $\mathbf{a} > \omega$. Let $\mathcal{V}$ be a pseudo-base of E with minimal cardinality **b**. According to the condition, one has $\mathbf{b} < \mathbf{a}$, so by virtue of **GCH** the inequality $2^{\mathbf{b}} \leq \mathbf{a}$ is fulfilled. Observe also that E is a Baire topological space (because the group G acts transitively in E).

It is not difficult to verify that there exists a base $\mathcal{B}$ of $\mathcal{K}(E)$ for which the relation

$$\text{card}(\mathcal{B}) \leq 2^{\mathbf{b}} \leq \mathbf{a}$$

holds true. Denote by $\mathcal{H}$ the family of all those sets $X \subset E$ which admit a representation in the form $X = V \setminus B$, where $V \in \mathcal{V}$ and $B \in \mathcal{B}$. Taking into account the above relation, one has

$$\text{card}(\mathcal{H}) \leq \mathbf{b} \cdot 2^{\mathbf{b}} \leq \mathbf{a}.$$

In addition to this, keeping in mind that G acts freely and transitively on E, one infers that

$$(\forall X \in \mathcal{H})(\text{card}(X) = \mathbf{a}).$$

Therefore, a Bernstein type transfinite construction can be applied to the family $\mathcal{H}$ (cf. Appendix E). So, one obtains a partition of E into two sets Z_1 and Z_2 such that

$$(\forall X \in \mathcal{H})(Z_1 \cap X \neq \emptyset \ \& \ Z_2 \cap X \neq \emptyset).$$

Finally, it is not hard to see that neither Z_1 nor Z_2 possesses the Baire property in E.

14. With the notation and under the conditions of the previous exercise, demonstrate that there are $2^{\mathbf{a}}$ many subsets of E, none of which has the Baire property. For this purpose, slightly modify the above argument.

15. Let C be the classical Cantor set on the real line $\mathbb{R}$.
    Check that the equalities

$$C - C = [-1, 1],$$
$$C + C = [0, 2]$$

hold true.

Compare the first equality with the Steinhaus result mentioned in Remark 30.3, and compare the second equality with the result formulated in Exercise 9.

16. Under appropriate assumptions on two topological groups $(G, \cdot)$ and $(H, \cdot)$, try to formulate and prove a generalization of Theorem 30.3 for a homomorphism $f : G \to H$ having the Baire property.

17. Consider the product space $E = \mathbb{R}^{\mathbb{N}}$ equipped with the standard topological and vector structures. Clearly, this E can be expressed as

$$E = \prod \{R_n : n \in \mathbb{N}\},$$

where all multipliers R_n coincide with $\mathbb{R}$.

Show that any real-valued linear continuous functional f on E is of the form

$$f = t_1 \mathrm{pr}_{n_1} + t_2 \mathrm{pr}_{n_2} + \cdots + t_m \mathrm{pr}_{n_m},$$

where m is a nonzero natural number, $t_1, t_2, \ldots, t_m$ are some real coefficients, $\{n_1, n_2, \ldots, n_m\}$ is a finite subset of $\mathbb{N}$, and the symbol pr_n denotes, as usual, the canonical projection of E onto R_n.

For this purpose, first establish that f depends only on finitely many arguments, i.e., there exists a finite subset M of $\mathbb{N}$ such that $f(x) = f(y)$ whenever $x \in E$ and $y \in E$ satisfy the equality $\mathrm{pr}_M(x) = \mathrm{pr}_M(y)$. Then apply to the finite-dimensional vector space

$$E_M = \prod \{R_n : n \in M\}$$

and to the linear continuous functional $f|E_M$ the Riesz representation theorem (see Chap. 29).

18. Denote by E_c the vector space of all real-valued convergent sequences and let E be the space of Exercise 17. Define a linear functional g on E_c by the formula

$$g(\{t_n : n \in \mathbb{N}\}) = \lim_{n \to \infty} t_n \qquad (\{t_n : n \in \mathbb{N}\} \in E_c).$$

Working in **ZF** & **DC** theory, prove that if there exists a linear functional

$$g^* : E \to \mathbb{R}$$

extending g, then there exists a subset of $\mathbb{R}$ lacking the Baire property in $\mathbb{R}$.

For this purpose, using the result of Exercise 17, first establish that g^* is necessarily discontinuous on E and infer that g^* does not possess the Baire property (in view of Theorem 30.3). Therefore, there exists a subset of E without the Baire property. Finally, keeping in mind the fact that there is a Borel isomorphism between E and $\mathbb{R}$ preserving the category of sets, conclude that $\mathbb{R}$ also contains a subset without the Baire property.

19. Show that any nonempty topological space E admits a decomposition in the form

$$E = X \cup Y,$$

where X is a first category subset of E and Y is an open Baire subspace of E. For this purpose, use Lemma 30.1 of this chapter.

20. Let E be a topological space and let $f : E \to \mathbb{R}$ be a lower semicontinuous function.

Work in **ZF** theory and demonstrate that the set $D(f)$ of all discontinuity points of f is of first category in E.

For this purpose, argue as follows. Keeping in mind Exercise 12 from Chap. 10, one may suppose that f is a bounded lower semicontinuous function on E. For each real number $r > 0$, define

$$D_r(f) = \{x \in E : O_f(x) \geq r\}$$

and show that the closed set $D_r(f)$ is nowhere dense in E. Supposing for a moment otherwise, let U be a nonempty open subset of $D_r(f)$. Then one can write

$$t = \sup\{f(x) : x \in U\} < +\infty.$$

Let a point $z \in U$ satisfy the relation $f(z) > t - r/2$. Check that there is some neighborhood $V(z) \subset U$ of z and there are two points y_1 and y_2 in $V(z)$ such that

$$f(y_2) - f(y_1) > r/2, \qquad f(y_1) > t - r/2.$$

These two inequalities imply $f(y_2) > t$, which contradicts the definition of t. The obtained contradiction yields the required result.

21. Work again within **ZF** theory and prove the classical theorem of Baire:

If E is an arbitrary Baire topological space and $f \in Ba_1(E, \mathbb{R})$, then the set $D(f)$ of all discontinuity points of f is of first category in E; consequently, the set $C(f)$ of all continuity points of f is everywhere dense in E.

For this purpose, argue as follows. As is well known, the relation

$$C(f) = E \setminus D(f) = \bigcap_{1 \leq n < \omega} \{x \in E : O_f(x) < 1/n\}$$

holds, where all sets $\{x \in E : O_f(x) < 1/n\}$ are open in E (see Chap. 6). So it suffices to demonstrate that all these sets are everywhere dense in E.

Actually, it suffices to show that, for any real $\varepsilon > 0$ and for each nonempty open set $U \subset E$, there exists a nonempty open set $W \subset U$ such that

$$(\forall x \in W)(\forall y \in W)(|f(x) - f(y)| < \varepsilon).$$

Keeping in mind that $f \in Ba_1(E, \mathbb{R})$, take a sequence $\{f_k : k < \omega\}$ of real-valued continuous functions on E such that

$$f(x) = \lim_{k \to \infty} f_k(x) \qquad (x \in E).$$

Further, for every natural number k, introduce the set

$$X_k = \{x \in E : (\forall i \geq k)(\forall j \geq k)(|f_i(x) - f_j(x)| \leq \varepsilon/3)\}.$$

All sets X_k ($k < \omega$) are closed in E and

$$(\forall k < \omega)(X_k \subset X_{k+1}), \qquad E = \cup\{X_k : k < \omega\}.$$

Consequently, one has the equality

$$U = (U \cap X_0) \cup (U \cap X_1) \cup \cdots \cup (U \cap X_k) \cup \cdots .$$

Since E is a Baire space, there exists an $n \in \mathbb{N}$ such that $\text{int}(U \cap X_n) \neq \emptyset$. Let V be a nonempty open subset of E entirely contained in $U \cap X_n$. If x is an arbitrary point of V, then

$$(\forall i \geq n)(\forall j \geq n)(|f_i(x) - f_j(x)| \leq \varepsilon/3).$$

Putting $j = n$ and letting i tend to $+\infty$, one gets

$$(\forall x \in V)(|f(x) - f_n(x)| \leq \varepsilon/3).$$

Therefore, one can write

$$|f(y) - f(x)| \leq |f(y) - f_n(y)| + |f_n(y) - f_n(x)| + |f_n(x) - f(x)|$$
$$\leq 2\varepsilon/3 + |f_n(y) - f_n(x)|$$

for any two points x and y from V. Finally, since f_n is a continuous function, there exists a nonempty open set $W \subset V$ such that

$$(\forall x \in W)(\forall y \in W)(|f_n(y) - f_n(x)| < \varepsilon/3).$$

This gives at once the relation

$$(\forall x \in W)(\forall y \in W)(|f(y) - f(x)| < \varepsilon),$$

which completes the proof of the Baire theorem.

22. Work within **ZFC** theory and show that the Baire theorem presented in the previous exercise remains valid for an arbitrary topological space E.
    For this purpose, use Exercise 19.

23. Let E be a topological space and let $f : E \to \mathbb{R}$ be a function having the Baire property.
    Verify that the graph of f is a first category subset of the topological product space $E \times \mathbb{R}$.
    Also, give an example of a function $g : \mathbb{R} \to \mathbb{R}$ such that:

    (a) the graph of g is a nowhere dense set in the plane $\mathbb{R}^2$;
    (b) g does not have the Baire property.

24. Denote by Q the set of all rational numbers in the unit segment $[0, 1]$ and consider the topological product space $\mathbb{R}^Q$. Define a mapping ϕ from the function space $C([0, 1])$ into $\mathbb{R}^Q$ by the formula

    $$\phi(f) = (f(q_0), f(q_1), \ldots, f(q_n), \ldots) \qquad (f \in C([0, 1])),$$

    where $\{q_0, q_1, \ldots, q_n, \ldots\}$ is a fixed bijective enumeration of Q.
    Demonstrate that:

    (a) ϕ is an injective continuous linear operator from $C([0, 1])$ into $\mathbb{R}^Q$ (as usual, $C([0, 1])$ is equipped with the standard sup-norm);
    (b) the inverse partial operator ϕ^{-1} is discontinuous at all points of $\phi(C([0, 1]))$;
    (c) $\phi(C([0, 1]))$ is a Borel subspace of $\mathbb{R}^Q$;
    (d) for any closed ball X in $C([0, 1])$, the set $\phi(X)$ is Borel in $\mathbb{R}^Q$;
    (e) for any Borel set Y in $C([0, 1])$, the set $\phi(Y)$ is Borel in $\mathbb{R}^Q$.

    Argue as follows. The assertions (a) and (b) are almost trivial.
    For (c), take into account that if $z \in \mathbb{R}^Q$, then the equivalence

    $$z \in \phi(C([0, 1])) \Leftrightarrow$$

    $$(\forall \varepsilon > 0)(\exists \delta > 0)(\forall q \in Q)(\forall r \in Q)(|q - r| \geq \delta \vee |z_q - z_r| < \varepsilon)$$

    holds true.
    Finally, using (c), deduce (d) and (e).

Chapter 31
The Stone–Weierstrass Theorem

In this chapter we will be dealing with algebras of real-valued functions on a given nonempty ground set.

More precisely, the main objects of our study in the present chapter are various subalgebras $\mathcal{A}$ of the algebra $C(X)$ (= $C(X, \mathbb{R})$) of all real-valued continuous functions on a nonempty compact topological space X (cf. Chap. 28). Recall that $C(X)$ is a real Banach space with respect to the natural norm of uniform convergence:

$$||f|| = \sup\{|f(x)| : x \in X\} \qquad (f \in C(X)).$$

Also, $C(X)$ is a lattice of real-valued functions on X, i.e., for any two functions f and g from $C(X)$, one has

$$\inf(f, g) \in C(X), \qquad \sup(f, g) \in C(X).$$

By definition, a family $\mathcal{A} \subset C(X)$ is an *algebra* in $C(X)$ (or a *subalgebra* of $C(X)$) if $\mathcal{A}$ is a vector subspace of $C(X)$ and

$$(\forall f \in \mathcal{A})(\forall g \in \mathcal{A})(f \cdot g \in \mathcal{A}).$$

Here we are primarily interested in those algebras $\mathcal{A}$ which are everywhere dense in $C(X)$. Our goal is to give a characterization of all such algebras.

In the sequel, we will need two important properties of subfamilies of $C(X)$.

Recall that a family $\mathcal{F} \subset C(X)$ *separates points* of X if, for any two distinct points x and y from X, there exists an $f \in \mathcal{F}$ satisfying $f(x) \neq f(y)$.

We shall say that $\mathcal{F} \subset C(X)$ *does not vanish on the points of* X if, for each point $x \in X$, there exists an $f \in \mathcal{F}$ such that $f(x) \neq 0$.

Remark 31.1 An easy argument shows that if $\mathcal{F} \subset C(X)$ does not separate points of X, then $\mathcal{F}$ cannot be everywhere dense in $C(X)$.

Also, if $\mathcal{F} \subset C(X)$ vanishes at some point x_0 of X, i.e., if

$$(\forall f \in \mathcal{F})(f(x_0) = 0),$$

then $\mathcal{F}$ cannot be everywhere dense in $C(X)$.

Thus, each of the two introduced properties of $\mathcal{F} \subset C(X)$ is necessary for $\mathcal{F}$ to be everywhere dense in $C(X)$.

The Stone–Weierstrass theorem states that, for a subalgebra $\mathcal{A}$ of $C(X)$, the conjunction of the above two properties guarantees the everywhere density of $\mathcal{A}$ in $C(X)$. This remarkable theorem has many applications in various fields of mathematics (see, e.g., [35, 38, 55, 84, 172]). Its proof is based on several lemmas.

Lemma 31.1 *There exists a sequence $\{p_n : n \in \mathbb{N}\}$ of real polynomials on the segment $[0, 1]$, satisfying the following conditions:*

(1) *this sequence is increasing on $[0, 1]$;*
(2) *for any $n \in \mathbb{N}$ and for any $t \in [0, 1]$, one has $0 \leq p_n^2(t) \leq t$;*
(3) *this sequence uniformly converges to the function $\phi : [0, 1] \to [0, 1]$, where $\phi(t) = t^{1/2}$ for all $t \in [0, 1]$.*

Proof Let us define by recursion the required sequence of polynomials. First, put $p_0(t) = 0$ for each t from $[0, 1]$. Then, assuming that p_n has already been defined, put

$$p_{n+1}(t) = p_n(t) + (1/2)(t - p_n^2(t)) \qquad (t \in [0, 1]).$$

To show the validity of (1) and (2), it suffices to check (2).

Supposing that $0 \leq p_n^2(t) \leq t$ whenever t belongs to $[0, 1]$, we may write

$$t - p_{n+1}^2(t) = (t^{1/2} + p_{n+1}(t))(t^{1/2} - p_{n+1}(t)),$$

so it remains to establish the inequality $t^{1/2} - p_{n+1}(t) \geq 0$, which is equivalent to

$$t^{1/2} - p_n(t) - (1/2)(t - p_n^2(t)) = (t^{1/2} - p_n(t))(1 - (1/2)(t^{1/2} + p_n(t))) \geq 0.$$

By the inductive assumption, the latter formula holds true for all $t \in [0, 1]$.

Since the constructed sequence of polynomials is increasing and bounded from above on $[0, 1]$, it pointwise converges to some function ϕ, and from the definition of these polynomials we readily obtain that

$$\phi(t) = \phi(t) + (1/2)(t - \phi^2(t)).$$

Thus, $\phi(t) = t^{1/2}$ whenever $t \in [0, 1]$.

Finally, the function ϕ is trivially continuous. Therefore, by Dini's theorem (see Chap. 10), the convergence of $\{p_n : n \in \mathbb{N}\}$ is uniform on $[0, 1]$, i.e., (3) is satisfied.

31 The Stone–Weierstrass Theorem

This completes the proof of Lemma 31.1. □

Let $\mathcal{A}$ be a subalgebra of $C(X)$. Obviously, we can consider the closure $\mathrm{cl}(\mathcal{A})$ of $\mathcal{A}$ in $C(X)$. It is not hard to verify that $\mathrm{cl}(\mathcal{A})$ is also a subalgebra of $C(X)$.

Lemma 31.2 *Let $\mathcal{A}$ be a subalgebra of $C(X)$ and let $f \in \mathcal{A}$. Then the function $|f|$ belongs to $\mathrm{cl}(\mathcal{A})$.*

Proof Since $\|f\| < +\infty$, we may take a real constant $c > 0$ such that $\|f\|/c \leq 1$. Consider the sequence of elements of $\mathcal{A}$:

$$p_0(f^2/c^2),\ p_1(f^2/c^2),\ \ldots,\ p_n(f^2/c^2),\ \ldots,$$

where the polynomials $p_0, p_1, \ldots, p_n, \ldots$ are as in Lemma 31.1. By virtue of this lemma, the above sequence converges uniformly on X to the function $|f|/c$. Hence, the function $|f|/c$ belongs to $\mathrm{cl}(\mathcal{A})$. So the function $|f| = c \cdot |f|/c$ also belongs to $\mathrm{cl}(\mathcal{A})$, which ends the proof. □

Lemma 31.3 *Let $\mathcal{A}$ be a subalgebra of $C(X)$.*
Then the algebra $\mathrm{cl}(\mathcal{A})$ is a sublattice of $C(X)$, i.e., for any two functions f and g from $\mathrm{cl}(\mathcal{A})$, the relations

$$\sup(f, g) \in \mathrm{cl}(\mathcal{A}), \qquad \inf(f, g) \in \mathrm{cl}(\mathcal{A})$$

hold true.

Proof Recall the following standard formulas:

$$\sup(f, g) = (1/2)(f + g + |f - g|), \qquad \inf(f, g) = (1/2)(f + g - |f - g|).$$

Keeping in mind Lemma 31.2, these two formulas yield at once the required result.
Lemma 31.3 has thus been proved. □

We shall say that an algebra $\mathcal{A} \subset C(X)$ is *admissible* if $\mathcal{A}$ separates points in X and does not vanish at any point of X.

Lemma 31.4 *Let $\mathcal{A}$ be an arbitrary admissible subalgebra of $C(X)$ and let x and y be two distinct points of X.*
Then, for any two real numbers α and β, there exist functions f and g from $\mathcal{A}$ such that

$$\alpha = rf(x) + tg(x), \qquad \beta = rf(y) + tg(y),$$

where r and t are some real numbers.

Proof First observe that, since $\mathcal{A}$ separates points of X, there exists an f from $\mathcal{A}$ satisfying $f(x) \neq f(y)$. Obviously, for this f we have only two possible cases.

1. $f(x)f(y) = 0$. In this case, assume without loss of generality that $f(x) = 0$. Then $f(y) \neq 0$. Since $\mathcal{A}$ does not vanish on x, there exists a g from $\mathcal{A}$ such that $g(x) \neq 0$. So the functions f and g are as required, because

$$f(x)g(y) - f(y)g(x) \neq 0.$$

2. $f(x)f(y) \neq 0$. In this case, take a function g from $\mathcal{A}$ satisfying the relation $g(x) \neq g(y)$ and consider the following two equalities:

$$f(x)g(y) - f(y)g(x) = 0, \qquad f(x)g^2(y) - f(y)g^2(x) = 0.$$

An easy verification shows that at least one of these equalities is false. So either the functions f and g are as required or the functions f and g^2 are appropriate, which completes the proof.

□

Remark 31.2 Actually, Lemma 31.4 states that if $\mathcal{A}$ is an admissible subalgebra of $C(X)$ and x and y are distinct points of X, then for any two real numbers α and β, there exists a function h from $\mathcal{A}$ such that $h(x) = \alpha$ and $h(y) = \beta$. In addition, since $\mathcal{A}$ does not vanish at x, for any real number γ there exists a function g from $\mathcal{A}$ satisfying $g(x) = \gamma$.

Lemma 31.5 *Suppose that $\mathcal{A}$ is an admissible subalgebra of $C(X)$. Let x be a point of X, let ε be a strictly positive real number, and let $f \in C(X)$.*
Then there exists a function $h_x \in \text{cl}(\mathcal{A})$ such that $h_x(x) = f(x)$ and $h_x(z) \geq f(z) - \varepsilon$ for all points $z \in X$.

Proof According to Lemma 31.4 and Remark 31.2, for any point $y \in X$, there exists a function $\psi_y \in \mathcal{A}$ satisfying the relations

$$\psi_y(x) = f(x), \qquad \psi_y(y) \geq f(y) - \varepsilon/2.$$

Consider the family of functions $\{\psi_y : y \in X\}$. Evidently, for each $y \in X$, there exists an open neighborhood $U(y)$ such that

$$(\forall z \in U(y))(\psi_y(z) \geq f(z) - \varepsilon).$$

The above-mentioned neighborhoods cover the compact space X, so there are finitely many neighborhoods $U(y_1), U(y_2), \ldots, U(y_n)$ of this type, which also constitute a covering of X. Defining

$$h_x = \sup(\psi_{y_1}, \psi_{y_2}, \ldots, \psi_{y_n}),$$

we see that $h_x(x) = f(x)$ and $h_x(z) \geq f(z) - \varepsilon$ whenever $z \in X$.

Finally, in view of Lemma 31.3, the same function h_x belongs to $\text{cl}(\mathcal{A})$.
This ends the proof of Lemma 31.5.

□

31 The Stone–Weierstrass Theorem

Taking into account the above lemmas, we now are able to formulate and prove the celebrated Stone–Weierstrass theorem (see [172]).

Theorem 31.1 *Let X be a nonempty compact space and let $\mathcal{A}$ be a subalgebra of the algebra $C(X)$.*

Then the following two assertions are equivalent:

(1) $\mathrm{cl}(\mathcal{A}) = C(X)$;
(2) $\mathcal{A}$ *is an admissible algebra.*

Proof As has already been mentioned earlier (see Remark 31.1), the implication (1) $\Rightarrow$ (2) holds true.

Let us establish the reverse implication (2) $\Rightarrow$ (1). For this purpose, suppose (2) and take an arbitrary function $f \in C(X)$ and an arbitrary real number $\varepsilon > 0$.

In view of Lemma 31.5, for each point $x \in X$, there exists a function h_x from $\mathrm{cl}(\mathcal{A})$ such that

$$h_x(x) = f(x), \qquad (\forall y \in X)(h_x(y) \geq f(y) - \varepsilon).$$

Obviously, there exists an open neighborhood $V(x)$ such that

$$(\forall y \in V(x))(h_x(y) \leq f(y) + \varepsilon).$$

The above-mentioned neighborhoods cover the compact space X, so there are finitely many neighborhoods $V(x_1), V(x_2), \ldots, V(x_m)$ of this type, which also produce a covering of X. Defining

$$\phi = \inf(h_{x_1}, h_{x_2}, \ldots, h_{x_m}),$$

we easily infer that $\phi \in \mathrm{cl}(\mathcal{A})$ and

$$f(y) - \varepsilon \leq \phi(y) \leq f(y) + \varepsilon$$

whenever $y \in X$. This shows us that the algebra $\mathcal{A}$ is everywhere dense in the Banach space $C(X)$.

The Stone–Weierstrass theorem has thus been proved. □

From Theorem 31.1, one immediately gets the following statement (cf. [35]).

Theorem 31.2 *Let X be a nonempty compact space and let $\mathcal{A}$ be a subalgebra of $C(X)$ which separates points of X and contains at least one nonzero constant function on X.*

Then the equality $\mathrm{cl}(\mathcal{A}) = C(X)$ holds true.

Proof From the assumption that $\mathcal{A}$ contains a nonzero constant function on X it follows that $\mathcal{A}$ does not vanish on the points of X. So Theorem 31.1 is applicable in this case, which yields the desired result. □

Remark 31.3 As a rule, subalgebras of $C(X)$ considered in mathematical analysis contain all real-valued constant functions on X.

Exercises

1. Verify that if a family $\mathcal{F} \subset C(X)$ vanishes at some point of X, then $\mathcal{F}$ cannot be everywhere dense in $C(X)$.
   For this purpose, observe that no nonzero constant real-valued function on X can be uniformly approximated by functions from $\mathcal{F}$.
2. Let X be a nonempty compact topological space.
   Verify that if a family $\mathcal{F} \subset C(X)$ does not separate points of X, then $\mathcal{F}$ cannot be everywhere dense in $C(X)$.
   For this purpose, consider two distinct points x and y in X such that $f(x) = f(y)$ for all $f \in \mathcal{F}$. There exists a function $g \in C(X)$ satisfying $g(x) \neq g(y)$. Demonstrate that this g cannot be uniformly approximated by functions from $\mathcal{F}$.
3. Let X be a nonempty compact space and let $\mathcal{A}$ be an arbitrary subalgebra of $C(X)$.
   Check that $\text{cl}(\mathcal{A})$ is also a subalgebra of $C(X)$.
4. Let X and Y be two nonempty compact spaces and let $Z = X \times Y$ denote their topological product.
   Show that, for any function $f \in C(Z)$ and for every real $\varepsilon > 0$, there exist a natural number n and some functions
   $$\{g_1, g_2, \ldots, g_n\} \subset C(X), \qquad \{h_1, h_2, \ldots, h_n\} \subset C(Y)$$
   such that
   $$|f(x, y) - \sum \{g_i(x)h_i(y) : 1 \leq i \leq n\}| < \varepsilon$$
   whenever $(x, y) \in Z$.
   For this purpose, apply the Stone–Weierstrass theorem to $C(Z)$.
5. Let K be a nonempty compact subset of the Euclidean space $\mathbb{R}^m$ and let f be an arbitrary function from $C(K)$.
   Prove that, for any real $\varepsilon > 0$, there exists a polynomial p of m real variables, all coefficients of which are real and which satisfies the relation
   $$||f - p|K|| < \varepsilon.$$
   For this purpose, apply the Stone–Weierstrass theorem to $C(K)$.
6. Let X be a nonempty non-compact locally compact topological space and let $C_0(X)$ denote the family of all those real-valued continuous functions on X which vanish at infinity.

Verify that:

(a) $C_0(X)$ is an algebra of functions, i.e., $C_0(X)$ is a vector space over $\mathbb{R}$ and
$$(\forall f \in C_0(X))(\forall g \in C_0(X))(f \cdot g \in C_0(X));$$

(b) the family $\mathcal{F}$ of all those functions from $C_0(X)$ which possess a compact support is a subalgebra of $C_0(X)$;

(c) $\mathcal{F}$ is everywhere dense in $C_0(X)$ (where $C_0(X)$ is equipped with the standard sup-norm).

Argue as follows. The assertions (a) and (b) are almost trivial.

To show the validity of (c), choose arbitrarily $f \in C_0(X)$ and take any real $\varepsilon > 0$. There exists a compact set $K \subset X$ such that
$$(\forall x \in X \setminus K)(|f(x)| < \varepsilon/2).$$

In view of the local compactness of X, for any point $x \in K$, there is a relatively compact open neighborhood $U(x)$ of x (i.e., the closure of $U(x)$ is compact). Consequently, there are finitely many such open neighborhoods collectively covering the compact set K. Denote them by
$$U(x_1), \ U(x_2), \ \ldots, \ U(x_n)$$
and put
$$V = U(x_1) \cup U(x_2) \cup \cdots \cup U(x_n).$$

Observe that V itself is a relatively compact open set in X. Further, let $\phi : X \to [0, 1]$ be a continuous function such that $\phi|K = 1$ and $\phi|(X \setminus V) = 0$. Define $f_0 = \phi \cdot f$ and show that:

(i) f_0 is continuous and has a compact support;
(ii) $\|f - f_0\| < \varepsilon$.

Conclude that (i) and (ii) give the required result.

7. Preserving the notation and assumptions of Exercise 6, try to prove the direct analog of Theorem 31.1 for $C_0(X)$ and its admissible subalgebras.

For this purpose, consider any admissible subalgebra $\mathcal{A}$ of $C_0(X)$. Let $X^* = X \cup \{z\}$ denote Alexandrov's compactification of X. Then $C_0(X)$ can be treated as the family of all those functions $f \in C(X^*)$ which satisfy the equality $f(z) = 0$. Further, define
$$\mathcal{A}^* = \{f + t : f \in \mathcal{A},\ t \in \mathbb{R}\}$$
and verify that $\mathcal{A}^*$ is an admissible algebra in $C(X^*)$. Therefore, according to Theorem 31.1, $\mathcal{A}^*$ is everywhere dense in $C(X^*)$. This implies that, for any

function $g \in C_0(X)$ and for any real $\varepsilon > 0$, there exist $h \in \mathcal{A}$ and $t \in \mathbb{R}$ satisfying the inequality

$$\|g - (h+t)\| < \varepsilon/2.$$

In particular, one has

$$|g(z) - (h(z) + t)| = |t| < \varepsilon/2.$$

Conclude from the above relation that $\|g - h\| < \varepsilon$, so $\mathcal{A}$ is everywhere dense in the function space $C_0(X)$.

Remark 31.4 Consider the vector space $C_b(\mathbb{R})$ $(= C_b(\mathbb{R}, \mathbb{R}))$ of all real-valued bounded continuous functions on $\mathbb{R}$ and equip $C_b(\mathbb{R})$ with the standard sup-norm. So, we obtain a certain Banach algebra which properly contains the algebra $C_0(\mathbb{R})$. The family $\mathcal{F}$ of all those functions from $C_0(\mathbb{R})$ which possess a compact support is an admissible subalgebra of $C_0(\mathbb{R})$ and hence of $C_b(\mathbb{R})$. In other words, $\mathcal{F}$ separates the points of $\mathbb{R}$ and does not vanish at any point of $\mathbb{R}$. However, the closure of $\mathcal{F}$ is $C_0(\mathbb{R})$, which differs from $C_b(\mathbb{R})$. We thus see that, in general, the direct analog of Theorem 31.1 does not hold for the Banach algebra $C_b(X)$, where X is a non-compact locally compact topological space (naturally, $C_b(X)$ denotes here the family of all real-valued bounded continuous functions on X).

8. Let $\{X_i : i \in I\}$ be an uncountable family of nonempty compact spaces and let X be their topological product.
   Demonstrate that, for every function $f \in C(X)$, there exists a countable subset $J = J_f$ of I such that one has $f(x) = f(y)$ whenever $x \in X$ and $y \in X$ satisfy the equality $\text{pr}_J(x) = \text{pr}_J(y)$.
   For this purpose, denote by $\mathcal{A}$ the family of all those functions from $C(X)$ which depend on finitely many arguments and check that $\mathcal{A}$ is an admissible subalgebra of $C(X)$.

9. Work in **ZF** & **DC** theory and prove the following statement:
   If $(E, \|\cdot\|)$ is a real infinite-dimensional separable Banach space, then the existence of an algebraic basis of the vector space E implies the existence of a subset of $\mathbb{R}$ not having the Baire property.
   Infer from this statement that, for the existence of an algebraic basis of E, an uncountable version of the Axiom of Choice is necessary.
   Argue step by step.

   (a) Using the completeness of E, first of all verify that every algebraic basis of E is uncountable (for this purpose, keep in mind that E is of second Baire category on itself).
   (b) Let $\{e_i : i \in I\}$ be an algebraic basis of E. Taking into account the separability of E, prove that there exists a countable subset J of I such

31 The Stone–Weierstrass Theorem

that any linear continuous functional $f : E \to \mathbb{R}$ is completely determined by its restriction to the set $\{e_i : i \in J\}$.

(c) Choose some index $i \in I \setminus J$ and consider two linear functionals

$$g : E \to \mathbb{R}, \qquad h : E \to \mathbb{R}$$

satisfying the relations

$$(\forall j \in J)(g(e_j) = h(e_j) = 1), \qquad g(e_i) = 1, \qquad h(e_i) = -1.$$

Check that either g is everywhere discontinuous on E or h is everywhere discontinuous on E.

(d) Supposing that the functional g is everywhere discontinuous on E, deduce that g does not have the Baire property, so there is a subset of E lacking this property. Finally, use the existence of a Borel isomorphism ϕ between the topological spaces $\mathbb{R}$ and E, such that ϕ preserves the Baire property, and obtain the required result.

Remark 31.5 It is not difficult to find (in **ZF** theory) a real normed vector space $(E, \|\cdot\|)$ such that the algebraic dimension of E is equal to ω and there exists a Borel measurable everywhere discontinuous linear functional on E. Also, it follows from Exercise 9 that, without the aid of an uncountable form of **AC**, it is impossible to point out at least one algebraic basis of the classical function space $C([0, 1])$ (and the same can be said of the standard real Hilbert space ℓ_2).

10. Show that:

    (a) the algebraic dimension of the vector space $C([0, 1])$ is **c**;
    (b) the algebraic dimension of the vector space ℓ_2 is **c**.

    For (a), consider on $[0, 1]$ the family of real-valued functions

    $$f_t(x) = |t - x| \qquad (x \in [0, 1]),$$

    where $0 < t < 1$, and verify that these functions are linearly independent.
    For (b), use Sierpiński's classical result in combinatorics which states that there exists (in **ZF** theory) a family $\{N_i : i \in I\}$ of infinite subsets of $\mathbb{N}$ such that $\text{card}(I) = \mathbf{c}$ and the set $N_i \cap N_j$ is finite for any two distinct indices i and j from I.

11. Equip the Cantor space $C = \{0, 1\}^{\mathbb{N}}$ with a metric ρ defined by the formula

    $$\rho(x, y) = \sum \{|x_n - y_n|/3^n : n \in \mathbb{N}\} \qquad (x \in C, \; y \in C).$$

    Verify the following two assertions for this metric:

    (a) ρ produces the topology of C;

(b) if x, y, z are any points of C such that $\rho(x, y) = \rho(x, z)$, then $y = z$.

Infer from (b) that if Z is a nonempty closed subset of C and x is an arbitrary point of C, then there exists a unique point $z \in Z$ for which $\rho(x, z) = \rho(x, Z)$. Starting with this fact, deduce that Z is a retract of C, i.e., there exists a continuous mapping ϕ from C to Z such that $\phi(z) = z$ whenever $z \in Z$.

Further, construct effectively a surjective continuous mapping from C onto $[0, 1]$ and, using a canonical homeomorphism between C and $C^{\mathbb{N}}$, conclude that there exists a surjective continuous mapping from C onto Hilbert's cube $[0, 1]^{\mathbb{N}}$.

12. Prove P.S. Alexandrov's theorem stating that any nonempty compact metric space X is a continuous image of Cantor's space C.

    Argue as follows. Take a countable family $\{f_n : n \in \mathbb{N}\}$ of continuous functions from X to $[0, 1]$ which separates the points of X. Define a mapping f from X to Hilbert's cube $[0, 1]^{\mathbb{N}}$ by putting

    $$f(x) = (f_0(x), f_1(x), \ldots, f_n(x), \ldots) \qquad (x \in X).$$

    Obviously, f is injective and continuous. Therefore, $f(X)$ is homeomorphic to X. According to Exercise 11, there exists a surjective continuous mapping ψ from C onto $[0, 1]^{\mathbb{N}}$. Observe now that $Z = \psi^{-1}(f(X))$ is a nonempty closed subset of C. By virtue of the same Exercise 11, one has a retraction $\phi : C \to Z$. Now, it becomes clear that the composition $\psi \circ \phi$ is a surjective continuous mapping from C onto $f(X)$. Since $f(X)$ and X are homeomorphic, one gets the required result.

13. Let E be a real topological vector space and let Y be a nonempty compact metrizable convex subspace of E.

    Demonstrate that there exists a surjective continuous mapping from the unit segment $[0, 1]$ onto Y.

    For this purpose, use Alexandrov's theorem formulated in Exercise 12.

14. Show that, for any metric space (X, d), there exists a real Banach space $(E, \|\cdot\|)$ such that

    (a) X can be isometrically embedded in E;
    (b) the topological weight of E (i.e., $w(E)$) does not exceed $w(X) + \omega$ (in particular, if X is separable, then E is also separable).

    Argue as follows. Assume, without loss of generality, that X is nonempty and choose a point $t \in X$. Further, for any point $x \in X$, introduce a function $f_x : X \to \mathbb{R}$ by the formula

    $$f_x(y) = d(x, y) - d(t, y) \qquad (y \in X).$$

    Check that f_x is continuous and that the relation

    $$|f_x(y)| \leq |d(x, y) - d(t, y)| \leq d(x, t)$$

holds. Consequently, f_x is a real-valued bounded function on X.

Let $C_b(X)$ ($= C_b(X, \mathbb{R})$) denote again the Banach space, with respect to the standard sup-norm, of all real-valued bounded continuous functions on X. Consider a mapping $\phi : X \to C_b(X)$ defined by the formula

$$\phi(x) = f_x \qquad (x \in X).$$

Verify that ϕ is an isometric embedding of X into $C_b(X)$.

Finally, let E be the closed vector subspace of $C_b(X)$ generated by the set $\phi(X)$. Then E is a Banach space and

$$w(E) \leq w(\phi(X)) + \omega = w(X) + \omega.$$

This yields the required result.

15. Prove the Banach–Mazur theorem stating that the space $C([0, 1])$ equipped with the standard sup-norm is universal for the class of all separable metric spaces (i.e., $C([0, 1])$ contains an isometric copy of any separable metric space).

For this purpose, argue as follows. Let X be an arbitrary separable metric space. According to Exercise 14, there exists a separable Banach space E containing an isometric copy of X. Hence it suffices to show that E can be isometrically embedded in $C([0, 1])$. For this E, consider its dual space E^* consisting of all real-valued continuous linear functionals on E, and let

$$B = \{f \in E^* : ||f|| \leq 1\}.$$

Further, equip E^* with the weakest topology $\mathcal{T}$ with respect to which all elements from E (treated as linear functionals on E^*) become continuous. Then, keeping in mind that E is separable, infer that B endowed with the topology induced by $\mathcal{T}$ becomes a compact metrizable space (cf. Chap. 21, the text preceding Theorem 21.3). By virtue of Exercise 13, there exists a continuous surjective mapping g from $[0, 1]$ onto B. Now, take an element x from E and define a function $f_x : [0, 1] \to \mathbb{R}$ by the formula

$$f_x(t) = g(t)(x) \qquad (t \in [0, 1]).$$

In view of the continuity of g, the function f_x is continuous on $[0, 1]$. If y is any other element from E, then, for each $t \in [0, 1]$, one may write

$$|f_x(t) - f_y(t)| = |g(t)(x - y)| \leq ||x - y||.$$

On the other hand, a simple consequence of the Hahn–Banach theorem states that if $x \neq y$, then there is a continuous linear functional $u : E \to \mathbb{R}$ satisfying the following two relations:

$$||u|| = 1, \qquad u(x - y) = ||x - y||.$$

In particular, $u \in B$ and, since g is a surjection from $[0, 1]$ onto B, there exists a point $t_0 \in [0, 1]$ such that $u = g(t_0)$. Then

$$|f_x(t_0) - f_y(t_0)| = |u(x - y)| = ||x - y||,$$

which shows that $||f_x - f_y|| = ||x - y||$ and, consequently, the mapping Φ given by the formula

$$\Phi(x) = f_x \qquad (x \in E)$$

is an isometric embedding of E into $C([0, 1])$.

Chapter 32
More on the Function Space $C(X)$

As in the previous chapter, the family $C(X)$ of all real-valued continuous functions on a nonempty compact topological space X is considered with its standard sup-norm, i.e.,

$$||f|| = \sup\{|f(x)| : x \in X\} \qquad (f \in C(X)).$$

Also, $C(X)$ is endowed with the natural algebraic structure determined by the pointwise addition and multiplication of functions. Recall that these two operations enable one to treat $C(X)$ as a commutative Banach algebra with its unit χ_X (the characteristic function of X), which is identically equal to 1.

Theorem 32.1 *The following three assertions are equivalent:*

(1) *the Banach space $C(X)$ is separable;*
(2) *the compact space X is metrizable;*
(3) *the compact space X has a countable base.*

Proof Suppose (1). Let $\{f_n : n \in \mathbb{N}\}$ be a countable family of functions from $C(X)$ everywhere dense with respect to the norm $||\cdot||$. Observe that if x and y are any two distinct points of X, then there exists a function $f \in C(X)$ such that $f(x) = 1$ and $f(y) = 0$. In view of the everywhere density of the above family in $C(X)$, there exists a natural number n such that $f_n(x) > 1/2$ and $f_n(y) < 1/2$. So the family $\{f_n : n \in \mathbb{N}\}$ separates points in $C(X)$. Now, consider a mapping F from X into the topological product space $\mathbb{R}^\mathbb{N}$, defined by the formula

$$F(x) = \{f_n(x) : n \in \mathbb{N}\} \qquad (x \in X).$$

This mapping is injective and continuous. Since X is compact, the image $F(X)$ is homeomorphic to X. But $F(X)$ is a subset of the metrizable topological space $\mathbb{R}^\mathbb{N}$. Consequently, X is metrizable too. This shows that (2) holds true.

Suppose (2), which means that X is compact and metrizable simultaneously. It can easily be seen that any compact metric space is separable (moreover, totally bounded), so has a countable base. Therefore, we infer that (3) is fulfilled.

Finally, suppose (3). Let $\{U_n : n \in \mathbb{N}\}$ denote a countable base of X, consisting of nonempty open sets. For any pair (U_n, U_m) of members of this base, which satisfy the relation $\mathrm{cl}(U_n) \cap \mathrm{cl}(U_m) = \emptyset$, introduce a continuous function $f_{(n,m)} : X \to \mathbb{R}$ having the following property:

$$f_{(n,m)}|U_n = 1, \qquad f_{(n,m)}|U_m = 0.$$

Let $\{f_i : i \in I\}$ be the family of all such functions. Clearly, this family is at most countable. A straightforward verification shows that $\{f_i : i \in I\}$ separates points of X and, of course, the same is true for the countable algebra of functions (over $\mathbb{Q}$) generated by $\{f_i : i \in I\}$. Denote this algebra by $\mathcal{A}$. Also, for each point $x \in X$, there exists an $i \in I$ such that

$$f_i(x) = 1 \neq 0.$$

So $\mathcal{A}$ does not vanish at x. Now, using the Stone–Weierstrass theorem (see Chap. 31), we can conclude that $\mathcal{A}$ is everywhere dense in $C(X)$. Therefore, $C(X)$ is separable, i.e., (1) holds true.

Theorem 32.1 has thus been proved. □

Let $(R, +, \cdot)$ be a ring (in the usual algebraic sense).

Recall that $r \in R$ is an *idempotent element* of R if the equality $r^2 = r$ holds.

Evidently, any ring R has the idempotent 0_R. If in R there exists a unit 1_R, then 1_R is another idempotent element in R.

Naturally, both 0_R and 1_R are called the *trivial idempotents* of R.

Any other idempotent element of R is *nontrivial*.

Theorem 32.2 *The following two assertions are equivalent:*

(1) *a nonempty compact space X is connected;*
(2) *the ring $C(X)$ does not have nontrivial idempotent elements.*

Proof Assume (1) and argue to the contrary, i.e., suppose that there exists a function $f \in C(X)$ such that $f^2 = f$ and f differs from $0_{C(X)}$ and $1_{C(X)}$. The equality $f^2 = f$ immediately implies that $0 \leq f \leq 1$. Also, the same equality shows that there is no $x \in X$ such that $0 < f(x) < 1$. Consequently, the range of f must be identical to $\{0, 1\}$. Then the two nonempty closed sets $f^{-1}(0)$ and $f^{-1}(1)$ form a partition of X contradicting (1). The obtained contradiction establishes the implication (1) $\Rightarrow$ (2).

Assume (2). Again, supposing to the contrary that X is not connected, take some partition $\{A, B\}$ of X into two open sets and introduce a function $f : X \to \mathbb{R}$ by the formula:

$$f(x) = \begin{cases} 0 & \text{if } x \in A, \\ 1 & \text{if } x \in B. \end{cases}$$

Clearly, $f \in C(X)$ and $f^2 = f$. At the same time, f is a nontrivial idempotent in $C(X)$ contradicting (2). The obtained contradiction establishes the implication (2) $\Rightarrow$ (1).

Theorem 32.2 has thus been proved. □

Remark 32.1 The two results presented above show that some properties of $C(X)$ are closely related to corresponding properties of X. Moreover, by Theorem 32.2 a certain purely algebraic property of the ring $C(X)$ is equivalent to a certain purely topological property of X. Such a situation is not occasional and has very deep roots (see, e.g., the Gelfand–Kolmogorov theorem below).

Consider any algebraic homomorphism ϕ of the ring $C(X)$ into the standard ring $\mathbb{R}$ of all real numbers. Keeping in mind that both $C(X)$ and $\mathbb{R}$ are rings with units, we may assert that ϕ is not identically zero (because, by definition, we must have $\phi(\chi_X) = 1$). Also, it is not hard to show that ϕ turns out to be a positive linear functional from $C(X)$ into $\mathbb{R}$ (see Exercise 5 of the present chapter).

So, the set $\mathcal{I} = \phi^{-1}(0)$ is an *ideal* in the ring $C(X)$. The latter phrase means that $\mathcal{I}$ is a subgroup of the additive group of $C(X)$ and, for every $f \in C(X)$, one has $f \cdot \mathcal{I} \subset \mathcal{I}$. Let us verify that $\mathcal{I}$ is a maximal proper ideal in $C(X)$. For this purpose, take any function $g \in C(X) \setminus \mathcal{I}$. Then $\phi(g) \neq 0$. Consequently, for the constant function $g' \in C(X)$ whose value is $\phi(g)$, one can write

$$\phi(g' - g) = \phi(g) - \phi(g) = 0,$$

whence it follows that $g' - g \in \mathcal{I}$ or, equivalently, $g' \in g + \mathcal{I}$.

Therefore, $\chi_X \in g/\phi(g) + \mathcal{I}$ and we see that the ideal generated by $\{g\} \cup \mathcal{I}$ coincides with $C(X)$, which implies the maximality of $\mathcal{I}$.

Theorem 32.3 *Let X be a nonempty compact space and let $\mathcal{I}$ be a maximal proper ideal in the ring $C(X)$.*

Then there exists a uniquely determined point $x_0 = x_0(\mathcal{I})$ such that the equality $\mathcal{I} = \{f \in C(X) : f(x_0) = 0\}$ holds true.

Proof Suppose otherwise, i.e., for any point $x \in X$, there is a function $f_x \in \mathcal{I}$ satisfying $f_x(x) \neq 0$. Since $\mathcal{I}$ is a subgroup of the additive group of $C(X)$, we may assume without loss of generality that $f_x(x) > 0$. Obviously, there exists an open neighborhood $U(x)$ of x such that

$$(\forall y \in U(x))(f_x(y) > 0).$$

From the compactness of X we infer that there exists some finite family $(x_1, x_2, \ldots, x_n)$ of points of X for which

$$U(x_1) \cup U(x_2) \cup \cdots \cup U(x_n) = X.$$

Let $(g_1, g_2, \ldots, g_n)$ be a family of real-valued non-negative continuous functions on X such that:

(a) $\operatorname{supp}(g_i) \subset U(x_i)$ for each $i \in \{1, 2, \ldots, n\}$;
(b) $\sum \{g_i(x) : i \in \{1, 2, \ldots, n\}\} = 1$ for any point $x \in X$.

The existence of $(g_1, g_2, \ldots, g_n)$ has already been established in this lecture course (see Exercise 2 from Chap. 28). Now, consider the function

$$f = \sum \{f_{x_i} g_i : i \in \{1, 2, \ldots, n\}\}.$$

A straightforward verification shows that:

(c) f is continuous, i.e., $f \in C(X)$;
(d) f is strictly positive on X;
(e) f belongs to the ideal $\mathcal{I}$.

Therefore, we get $\chi_X = (1/f) \cdot f \in \mathcal{I}$, which is impossible. The obtained contradiction gives us that, for some point $x_0 \in X$, the equality

$$\mathcal{I} = \{f \in C(X) : f(x_0) = 0\}$$

must be fulfilled. The uniqueness of x_0 follows from the maximality of $\mathcal{I}$, taking into account that $C(X)$ separates points in X.

Theorem 32.3 has thus been proved. $\square$

Theorem 32.4 *Let X be a nonempty compact space and let ϕ be a homomorphism of the ring $C(X)$ into the standard ring $\mathbb{R}$.*

Then there exists a unique point $x_0 = x_0(\phi)$ in X such that $\phi(f) = f(x_0)$ for all $f \in C(X)$.

Proof Consider the set $\mathcal{I} = \phi^{-1}(0)$. We know that this set is a maximal proper ideal in the ring $C(X)$. According to Theorem 32.3, there is a unique point $x_0 \in X$ such that

$$(\forall g \in C(X))(g \in \mathcal{I} \Leftrightarrow g(x_0) = 0).$$

Now, take an arbitrary function $f \in C(X)$ and consider the function

$$g = f - \phi(f) \chi_X.$$

Obviously, $\phi(g) = 0$, so $g \in \mathcal{I}$ and $g(x_0) = 0$. Consequently, we may write

$$0 = g(x_0) = f(x_0) - \phi(f), \qquad \phi(f) = f(x_0),$$

which yields the required result. The uniqueness of x_0 follows from the fact that $C(X)$ separates points in X.

Theorem 32.4 has thus been proved. □

Example 32.1 Let X and Y be any two nonempty compact spaces and let $H = H_{Y,X}$ be a continuous mapping from Y to X. This mapping canonically produces the mapping

$$u_H : C(X) \to C(Y)$$

defined by the formula

$$u_H(f) = f \circ H \qquad (f \in C(X)).$$

An easy verification shows that u_H turns out to be a ring homomorphism (in the algebraic sense) from $C(X)$ to $C(Y)$.

Moreover, putting $Y = X$, one can deduce that

$$u_H : C(X) \to C(X)$$

is the identity automorphism of the ring $C(X)$ if and only if H is the identity homeomorphism of X onto itself. To see this fact, suppose first that H is the identity homeomorphism of X onto itself. Then we obviously have $f = f \circ H$ for every function $f \in C(X)$, so u_H is the identity automorphism of the ring $C(X)$. On the other hand, suppose that H is not the identity homeomorphism of X onto itself. Then there exists at least one point x in X such that $H(x) = x'$, where x' differs from x. Also, there exists a function $f \in C(X)$ for which $f(x) \neq f(x')$. So we may write

$$f(x) \neq f(x') = (f \circ H)(x) = (u_H(f))(x), \qquad f \neq u_H(f).$$

This gives us that u_H is not the identity automorphism of $C(X)$.

Example 32.2 Let X, Y, and Z be any three nonempty compact spaces, let $H = H_{Y,X}$ be a continuous mapping from Y to X, and let $G = G_{Z,Y}$ be a continuous mapping from Z to Y. According to Example 32.1, we get two ring homomorphisms

$$u_H : C(X) \to C(Y), \qquad v_G : C(Y) \to C(Z)$$

and their composition $v_G \circ u_H : C(X) \to C(Z)$, which trivially is a ring homomorphism. It is not difficult to check that this composition is defined by the formula

$$(v_G \circ u_H)(f) = f \circ H \circ G \qquad (f \in C(X)).$$

In the language of category theory the latter formula means that we have a contravariant functor from the class of all nonempty compact spaces (with continuous mappings as morphisms) to the class of all commutative rings with units (where morphisms are homomorphisms of rings).

Example 32.3 Let X and Y be two nonempty compact spaces and let

$$u : C(X) \to C(Y)$$

be a surjective ring homomorphism. Let us define some continuous mapping $H_u : Y \to X$ associated with the given u.

For this purpose, take any point $y \in Y$. As we know, this y determines a maximal proper ideal $\mathcal{I}_y$ in the ring $C(Y)$. The u-pre-image of $\mathcal{I}_y$ is a maximal proper ideal $\mathcal{I}$ in the ring $C(X)$ (see Exercise 2). In its turn, $\mathcal{I}$ is determined by a certain (unique) point $x \in X$. So $\mathcal{I} = \mathcal{I}_x$ and we may put $H_u(y) = x$.

It immediately follows from the definition of H_u that, for any $f \in C(X)$, the relation

$$f(x) = 0 \Leftrightarrow u(f)(y) = 0$$

holds true. Since u is a surjection, we also have

$$g(y) = 0 \Leftrightarrow f(x) = 0,$$

where g is an arbitrary function from $C(Y)$ and f satisfies $u(f) = g$.

The above two equivalences readily imply the equality

$$u(f)(y) = f(x)$$

for each $f \in C(X)$. Indeed, consider the function $f_1 = f - f(x)\chi_X$ from $C(X)$. Then $f_1(x) = 0$, so we must have $u(f_1)(y) = 0$, i.e.,

$$u(f)(y) - u(f(x)\chi_X)(y) = u(f)(y) - f(x) = 0, \qquad u(f)(y) = f(x),$$

as required (cf. Exercise 5 of this chapter).

It remains to show that the mapping H_u is continuous at each point y of the space Y. The simplest way to establish this fact is to use so-called *generalized sequences* (*nets*) of points in a topological space (see, e.g., [38, 40, 84]).

Let $\{y_\xi : \xi \in \Xi\}$ be any generalized sequence of points of Y, converging to y, and let $\{x_\xi : \xi \in \Xi\}$ be the corresponding generalized sequence of points of X, where $x_\xi = H_u(y_\xi)$ for all $\xi \in \Xi$. Suppose to the contrary that $\{x_\xi : \xi \in \Xi\}$ does not converge to $x = H_u(y)$. On the other hand, since X is compact, $\{x_\xi : \xi \in \Xi\}$ must contain a generalized subsequence (subnet) which converges to some point $x' \in X$ distinct from x. We may assume, without loss of generality, that $\{x_\xi : \xi \in \Xi\}$ itself converges to x'. Keeping in mind the formula

$$u(f)(y_\xi) = f(x_\xi) \qquad (f \in C(X),\ \xi \in \Xi)$$

and taking the limits of both sides of this formula, we get

$$u(f)(y) = f(x') \qquad (f \in C(X)),$$

which implies

$$f(x) = u(f)(y) = f(x')$$

for every $f \in C(X)$. But the latter is impossible in view of $x \neq x'$.

The obtained contradiction yields the continuity of H_u.

Clearly, if X and Y are nonempty homeomorphic compact spaces, then the rings $C(X)$ and $C(Y)$ are isomorphic as purely algebraic objects. The fundamental converse statement is due to Gelfand and Kolmogorov (see [54]).

Theorem 32.5 *Let X and Y be two nonempty compact spaces and let the rings $C(X)$ and $C(Y)$ be isomorphic to each other (as algebraic objects).*

Then the spaces X and Y are homeomorphic.

Proof Let $u : C(X) \to C(Y)$ be an algebraic isomorphism between these two rings. Obviously, both ring homomorphisms

$$u : C(X) \to C(Y), \qquad u^{-1} : C(Y) \to C(X)$$

are surjective (i.e., are epimorphisms). By virtue of Example 32.3, we get two continuous mappings

$$H_u : Y \to X, \qquad H_{u^{-1}} : X \to Y,$$

which satisfy the relations

$$(\forall f \in C(X))(u(f) = f \circ H_u), \qquad (\forall g \in C(Y))(u^{-1}(g) = g \circ H_{u^{-1}}).$$

Observe now that, by Example 32.1, the composition $H_u \circ H_{u^{-1}}$ corresponds to the identity mapping of $C(X)$, and the composition $H_{u^{-1}} \circ H_u$ corresponds to the identity mapping of $C(Y)$. In view of this circumstance, we must have the following two equalities:

$$H_u \circ H_{u^{-1}} = \mathrm{Id}_X, \qquad H_{u^{-1}} \circ H_u = \mathrm{Id}_Y.$$

Therefore, H_u is a homeomorphism of Y onto X, and $H_{u^{-1}}$ is the inverse homeomorphism of X onto Y.

This completes the proof of Theorem 32.5. □

Remark 32.2 For any two nonempty compact topological spaces X and Y, one can consider $C(X)$ and $C(Y)$ as commutative groups with respect to the pointwise addition of real-valued functions. In contrast to Theorem 32.5, very simple examples show that the existence of an isomorphism between these two groups does not imply the existence of a homeomorphism between X and Y. Moreover, it may happen that the above-mentioned groups are isomorphic, but X and Y are not equinumerous (cf. Exercise 10 of the present chapter).

Exercises

1. Work in **ZF** & **DC** theory and prove that the topological product of a countable family of metrizable topological spaces is metrizable.
2. Check that a proper ideal I in a commutative ring R with unit 1_R is maximal if and only if, for any element $r \in R \setminus I$, one has a representation

$$1_R = ar + b,$$

where $a \in R$ and $b \in I$ depend on r.
Let S be another commutative ring with unit 1_S and let $u : R \to S$ be a homomorphism.
Show that if $\mathcal{J}$ is an ideal in S, then its u-pre-image $\{u^{-1}(s) : s \in \mathcal{J}\}$ is an ideal in R. In particular, the kernel $\ker(u) = u^{-1}(0_S)$ of u is an ideal in R.
Moreover, demonstrate that if $\mathcal{J}$ is a maximal proper ideal in S and u is an epimorphism of R onto S, then the u-pre-image of $\mathcal{J}$ is a maximal proper ideal in R.
For this purpose, use the above characterization of maximal proper ideals in R.
3. Let R be again a commutative ring with unit 1_R and let I be a maximal proper ideal in R.
Verify that the following assertion holds:

(*) For any two elements a and b of R, the relation $ab \in I$ implies either $a \in I$ or $b \in I$.

For this purpose, use Exercise 2.

Remark 32.3 Any proper ideal I satisfying (*) is called a *prime ideal*. In view of Exercise 3, every maximal proper ideal is prime. The converse assertion is not true, in general.

4. Give a detailed explanation of the uniqueness of a point x_0 in Theorem 32.3.
5. Let X be a nonempty topological space and let $C_b(X)$ denote the family of all real-valued bounded continuous functions on X, equipped with the sup-norm

$$||f|| = \sup\{|f(x)| : x \in X\} \quad (f \in C_b(X)).$$

Consider $C_b(X)$ as a commutative ring with respect to the standard operations of pointwise addition and multiplication of functions.

Show that every ring homomorphism $\phi : C_b(X) \to \mathbb{R}$ is continuous.

Argue as follows. First, take any non-negative function $f \in C_b(X)$ and associate to f the non-negative function $g \in C_b(X)$ such that $f = g^2$. Infer that

$$\phi(f) = \phi(g^2) = \phi(g) \cdot \phi(g) \geq 0$$

and then deduce that, for any two functions h_1 and h_2 from $C_b(X)$, the inequality $h_1 \leq h_2$ implies $\phi(h_1) \leq \phi(h_2)$.

Further, take arbitrarily a real $t \in \mathbb{R}$ and a function $h \geq 0$ from $C_b(X)$. Consider any two rational numbers p and q such that $p \leq t \leq q$. Then

$$p\phi(h) = \phi(ph) \leq \phi(th) \leq \phi(qh) = q\phi(h),$$

whence it follows that $\phi(th) = t\phi(h)$. Conclude that ϕ is a positive linear functional and, therefore, ϕ is continuous on $C(X)$ (cf. Theorem 27.3 from Chap. 27).

6. Let X and Y be nonempty compact topological spaces.

Prove that the following two assertions are equivalent:

(a) there exists a surjective homomorphism of the ring $C(X)$ onto the ring $C(Y)$;
(b) there exists a subspace of X homeomorphic to Y.

For this purpose argue similarly to the proof of Theorem 32.5, keeping in mind the fact that any compact space is normal.

7. Let $(R, +, \cdot)$ be a commutative ring with unit 1_R.

This algebraic structure is called a *Noether ring* if every increasing (by inclusion) sequence $\{I_n : n \in \mathbb{N}\}$ of proper ideals in R is stationary, i.e. there exists an $m \in \mathbb{N}$ such that $I_m = I_n$ for all $n > m$.

Demonstrate that the following three assertions are equivalent for R:

(a) R is a Noether ring;
(b) for each nonempty family of proper ideals in R, there is a member of the family not properly contained in any other member;
(c) every proper ideal in R is finitely generated.

For this purpose, verify the implications (a) $\Rightarrow$ (b) $\Rightarrow$ (c) $\Rightarrow$ (a). Also, check that:

(i) any field trivially is a Noether ring;
(ii) the ring $(\mathbb{Z}, +, \cdot)$ of all integers is a Noether ring;
(iii) if E is an infinite set, then the ring $(\mathcal{P}(E), \cap, \triangle)$ is not Noether (here $\mathcal{P}(E)$ denotes, as usual, the family of all subsets of E, the operation $\cap$ is the intersection of two sets, and the operation $\triangle$ is the symmetric difference of two sets).

Remark 32.4 In connection with (b) of Exercise 7, it should be mentioned that, according to the Kuratowski–Zorn lemma, the family of all proper ideals in an arbitrary commutative ring S with 1_S is inductive (by the standard inclusion relation), but S does not need to be a Noether ring.

8. Prove Hilbert's theorem on finitely generated ideals, which is formulated as follows.

    If R is a Noether ring, x is a variable, and $R(x)$ denotes the ring of all polynomials of x (over R), then $R(x)$ is also a Noether ring.

    Argue to the contrary, i.e., suppose that there exists an ideal $\mathcal{I}$ in $R(x)$ which is not finitely generated. Define by recursion a certain infinite sequence of elements of $\mathcal{I}$.

    First, take a polynomial $f_0 \in \mathcal{I}$ whose degree $\deg(f_0)$ is the least.

    If the polynomials $f_0, f_1, \ldots, f_n$ have already been defined, then take as f_{n+1} a polynomial from $\mathcal{I}$ having the least degree and not belonging to the ideal generated by $(f_0, f_1, \ldots, f_n)$. Obviously, the inequalities

    $$\deg(f_0) \leq \deg(f_1) \leq \cdots \leq \deg(f_n) \leq \cdots$$

    hold true. Further, for each natural number n, let $a_n \in R$ be the major coefficient of the polynomial f_n. Consider the ideal in R generated by $\{a_n : n \in \mathbb{N}\}$. This ideal is either improper (i.e., coincides with the whole R) or is proper and finitely generated. In any of these two cases there exists an $m \in \mathbb{N}$ such that

    $$a_{m+1} = b_0 a_0 + b_1 a_1 + \cdots + b_m a_m$$

    for some elements $b_0, b_1, \ldots, b_m$ of R. Now, introduce the polynomial

    $$f(x) = b_0 x^{\deg(f_{m+1}) - \deg(f_0)} f_0(x) +$$
    $$b_1 x^{\deg(f_{m+1}) - \deg(f_1)} f_1(x) + \cdots + b_m x^{\deg(f_{m+1}) - \deg(f_m)} f_m(x).$$

    The degree of f is equal to $\deg(f_{m+1})$ and the major coefficient of f is exactly a_{m+1}. At the same time, the definition of f implies that f belongs to the ideal $\mathcal{I}_m$ generated by $(f_0, f_1, \ldots, f_m)$. Since f_{m+1} does not belong to $\mathcal{I}_m$, the difference $f - f_{m+1}$ also does not belong to $\mathcal{I}_m$. But the degree of $f - f_{m+1}$ is strictly less than the degree of f_{m+1}, which contradicts the definition of f_{m+1}. The obtained contradiction yields the required result.

    Remark 32.5 In fact, Hilbert's above-mentioned theorem remains valid in the case of any non-commutative Noether ring R, for which the notions of left and of right ideals are considered. In this more general situation, the proof is almost the same as in the commutative case.

9. Let R be a Noether ring and let $x_1, x_2, \ldots, x_n$ be finitely many (mutually independent) variables.

Show that the ring $R(x_1, x_2, \ldots, x_n)$ of all polynomials of these variables (over R) is also Noether.

For this purpose, use Exercise 8 and the method of induction on n.

10. Take the two sets $\{0\}$ and $\{0, 1\}$ and equip each of them with the discrete topology. Consider the corresponding function spaces $C(\{0\})$ and $C(\{0, 1\})$ not as commutative rings, but as commutative groups with respect to the standard pointwise addition of functions.

    Verify that:

    (a) $C(\{0\})$ and $C(\{0, 1\})$ are isomorphic as groups (and even as vector spaces over the field $\mathbb{Q}$ of all rational numbers);

    (b) in **ZF** & **DC** theory the existence of the above-mentioned isomorphism between $C(\{0\})$ and $C(\{0, 1\})$ implies the existence of a subset of $\mathbb{R}$ nonmeasurable in the Lebesgue sense.

    Conclude that, for two nonempty compact topological spaces X and Y, an isomorphism between the additive groups $C(X)$ and $C(Y)$ does not guarantee a homeomorphism between X and Y.

11. Consider the ordinal numbers ω_1 and $\omega_1 + 1$ equipped with their standard order topologies.

    Check that the associated rings $C(\omega_1)$ and $C(\omega_1 + 1)$ are canonically isomorphic (consequently, if at least one of topological spaces X and Y in the formulation of Theorem 32.5 is non-compact, then the assertion of the theorem fails to be true in general).

    For this purpose, use Exercise 9 from Chap. 28.

Chapter 33
Uniformization of Plane Sets by Relatively Measurable Functions

According to the standard definition (see, e.g., Exercise 4 of Chap. 19), a topological space E is called *universal measure zero* (or, briefly, *absolute null*) if there exists no nonzero σ-finite Borel continuous measure on E.

In this definition it is assumed that all singletons in E are Borel subsets of E, and a measure on E is called *continuous* (or *diffuse*) if it vanishes at each singleton in E.

For our further purposes, we need the following simple lemma concerning absolute null subsets in the product space of two topological spaces.

Lemma 33.1 *Let E_1 and E_2 be two topological spaces, let X be an absolute null subset of E_1, and let for every point $x \in X$ a nonempty subset Y_x of E_2 be given, which is absolute null in E_2.*

Then the set $Z = \cup\{\{x\} \times Y_x : x \in X\}$ is absolute null in the topological product space $E_1 \times E_2$.

Proof Suppose for a moment otherwise, i.e., the set Z is not absolute null in $E_1 \times E_2$. Then there exists a probability continuous measure μ on the Borel σ-algebra $\mathcal{B}(Z)$. Consider the canonical projection

$$\mathrm{pr}_1 : E_1 \times E_2 \to E_1$$

and observe that $\mathrm{pr}_1(Z) = X$ and $\mathrm{pr}_1^{-1}(x) = \{x\} \times E_2$ for every $x \in X$.

Now, define a function ν by the formula

$$\nu(B) = \mu(\mathrm{pr}_1^{-1}(B) \cap Z) \qquad (B \in \mathcal{B}(X)).$$

Then a straightforward verification gives us that ν is a Borel probability continuous measure on X (because the set Y_x is absolute null for any $x \in X$). But the existence of ν contradicts the assumption that X is an absolute null subset of E_1.

The obtained contradiction ends the proof. □

Remark 33.1 Actually, Lemma 33.1 slightly generalizes the well-known fact that the topological product of two absolute null spaces is again an absolute null space. The same remains true for the topological product of any finite family of absolute null spaces. But, in general, the above-mentioned fact fails to be valid for topological products of countably many absolute null spaces.

Remark 33.2 Recall that it is provable within **ZFC** theory that there are uncountable absolute null subspaces of $\mathbb{R}$ (see, for instance, [26, 76, 109, 121, 189]). Moreover, any Luzin subset of $\mathbb{R}$ is absolute null (see Exercise 4 of Chap. 19). But the existence of Luzin sets needs additional set-theoretical assumptions, because if one assumes **MA** & ¬**CH**, then the non-existence of Luzin sets on $\mathbb{R}$ readily follows.

Let now E be an arbitrary nonempty ground (base) set and let f be a function from E into $\mathbb{R}$.

In this chapter, f will be called *relatively measurable* (on E) if there exists a nonzero σ-finite continuous (i.e., vanishing at all singletons in E) measure μ defined on some σ-algebra of subsets of E and such that f is measurable with respect to μ.

Otherwise, f will be called *absolutely nonmeasurable* (cf. [92, 95, 99, 100], and Chap. 17 of the present book).

As a rule, in the sequel any given function will be identified with its graph.

Remark 33.3 As was shown by Sierpiński and Zygmund in [168], there exists a function $h : \mathbb{R} \to \mathbb{R}$ such that, for every set $X \subset \mathbb{R}$ of cardinality **c**, the restricted function $h|X$ is not continuous on X. It follows from this property of h that if μ is the completion of a nonzero σ-finite continuous Borel measure on $\mathbb{R}$, then h is nonmeasurable with respect to μ. In a certain sense, absolutely nonmeasurable functions acting from $\mathbb{R}$ into $\mathbb{R}$ are similar to Sierpiński–Zygmund functions. Namely, if a function $f : \mathbb{R} \to \mathbb{R}$ is absolutely nonmeasurable and Y is any non-absolute null subset of $\mathbb{R}$, then the restricted function $f|Y$ is not continuous (and even is not Borel) on Y. Notice also that:

(a) there exist Sierpiński–Zygmund functions which are not absolutely nonmeasurable;
(b) the assertion that there exist absolutely nonmeasurable functions from $\mathbb{R}$ into $\mathbb{R}$ which are not Sierpiński–Zygmund functions is consistent with **ZFC** theory (e.g., holds true under **MA**);
(c) the assertion that there exist absolutely nonmeasurable functions from $\mathbb{R}$ into $\mathbb{R}$ which simultaneously are Sierpiński–Zygmund functions is consistent with **ZFC** theory (e.g., holds true under the same **MA**).

It should be underlined that (a) is provable within **ZFC** theory, but the assertion of (b) as well as the assertion of (c) cannot be established without using additional set-theoretical hypotheses.

Our further argument is essentially based on a certain characterization of absolutely nonmeasurable functions.

33 Uniformization of Plane Sets by Relatively Measurable Functions

Theorem 33.1 *For a real-valued function f defined on a ground set E, the following two conditions are equivalent:*

(1) *f is absolutely nonmeasurable;*
(2) *the range of f is an absolute null subset of $\mathbb{R}$ and the set $f^{-1}(r)$ is at most countable for each $r \in \mathbb{R}$.*

Proof Suppose first that (1) holds true, i.e., f is absolutely nonmeasurable. We will demonstrate that (2) is also fulfilled.

Assume on the contrary that, for some point $t_0 \in \mathbb{R}$, the inequality

$$\mathrm{card}(f^{-1}(t_0)) > \omega$$

holds. In this case, it is not difficult to define a complete continuous probability measure μ_0 on E such that

$$f^{-1}(t_0) \in \mathrm{dom}(\mu_0), \qquad \mu_0(f^{-1}(t_0)) = 1.$$

Consequently, for μ_0 and for any set $T \subset \mathbb{R}$, we have:

$$\mu_0(f^{-1}(T)) = \begin{cases} 1 & \text{if } t_0 \in T; \\ 0 & \text{if } t_0 \notin T. \end{cases}$$

This fact immediately implies that f is μ_0-measurable and hence f is relatively measurable, which contradicts (1). The obtained contradiction shows that the relation

$$\mathrm{card}(f^{-1}(r)) \leq \omega$$

must be satisfied for any $r \in \mathbb{R}$. Now, let us check that the set $\mathrm{ran}(f)$ is absolute null. Indeed, assuming again on the contrary that $\mathrm{ran}(f)$ is not absolute null, consider some Borel continuous probability measure ν on $\mathrm{ran}(f)$ and denote

$$S = \{f^{-1}(Z) : Z \in \mathrm{dom}(\nu)\}.$$

Evidently, S is a σ-algebra of subsets of E and the family of countable sets $\{f^{-1}(r) : r \in \mathrm{ran}(f)\}$ forms a disjoint covering of E. We put

$$\mu(f^{-1}(Z)) = \nu(Z) \qquad (Z \in \mathrm{dom}(\nu)).$$

In this manner, the probability measure μ on S is well defined and, by virtue of the definition of μ, the function f becomes μ-measurable. Clearly, the completion μ' of μ is a continuous measure on E and f remains measurable with respect to μ'. However, this contradicts our assumption that f is absolutely nonmeasurable. The obtained contradiction yields the validity of (2).

Now, assume that condition (2) is fulfilled for the given function f. Let us establish that condition (1) also holds, i.e., f is absolutely nonmeasurable. Suppose to the contrary that there exists a nonzero σ-finite continuous measure μ on E such that f is measurable with respect to μ. We may assume, without loss of generality, that μ is a probability measure. Denoting by $\mathcal{B}(\operatorname{ran}(f))$ the Borel σ-algebra of the subspace $\operatorname{ran}(f)$ of $\mathbb{R}$, we may put

$$\nu(Z) = \mu(f^{-1}(Z)) \qquad (Z \in \mathcal{B}(\operatorname{ran}(f))).$$

So we get a Borel probability measure ν on the space $\operatorname{ran}(f) \subset \mathbb{R}$ and we see that ν is continuous in view of the inequality $\operatorname{card}(f^{-1}(r)) \leq \omega$ for any $r \in \mathbb{R}$. But this contradicts the fact that $\operatorname{ran}(f)$ is absolute null. The obtained contradiction gives us the validity of (1).

Theorem 33.1 has thus been proved. $\square$

Remark 33.4 Let E be a ground set with $\operatorname{card}(E) > \mathbf{c}$ and let f be a function acting from E into $\mathbb{R}$. As a direct consequence of Theorem 33.1, we obtain that f is relatively measurable on E.

Remark 33.5 A model of **ZFC** theory is known in which $\mathbf{c} > \omega_1$ and the cardinality of any absolute null subset of $\mathbb{R}$ does not exceed ω_1 (see, e.g., [76]). Keeping in mind this result and Theorem 33.1, one can conclude that the existence of an absolutely nonmeasurable function $f : \mathbb{R} \to \mathbb{R}$ is not provable within **ZFC** theory. On the other hand, the existence of such f can be established assuming **MA** (see Exercise 2 of this chapter).

If G is a subset of the plane $\mathbb{R}^2$ having some nice measurability properties, then the problem of the existence of a function $g : \operatorname{pr}_1(G) \to \mathbb{R}$ entirely contained in G and having analogous nice measurability properties naturally arises. This g may be treated as an appropriate uniformization (or an appropriate selector) of G.

The uniformization problem was first formulated in the framework of classical descriptive set theory (cf. [119, 121]). This important problem was extensively studied from various points of view (see, for instance, [33, 47, 76, 82, 112, 121]).

Example 33.1 In the Euclidean plane $\mathbb{R}^2$ consider a set Z such that:

(i) all vertical and horizontal sections of Z are at most countable;
(ii) $\operatorname{pr}_1(Z)$ is an uncountable absolute null set in $\mathbb{R}$;
(iii) $\operatorname{pr}_2(Z)$ is an uncountable absolute null set in $\mathbb{R}$.

It is not difficult to see that there are many sets $Z \subset \mathbb{R}^2$ having the above properties.

By virtue of Lemma 33.1, Z is an absolute null set in $\mathbb{R}^2$. Keeping in mind Theorem 33.1, one can assert that:

(a) any function $f : \operatorname{pr}_1(Z) \to \mathbb{R}$ whose graph is contained in Z is absolutely nonmeasurable;

(b) any function $g : \mathrm{pr}_2(Z) \to \mathbb{R}$ whose graph is contained in Z^{-1} is absolutely nonmeasurable.

In view of Example 33.1, we see that if a set $G \subset \mathbb{R}^2$ is absolute null, then it may happen that, for this G, there is no relatively measurable uniformization (likewise for its reverse G^{-1}). So, in the sequel we will concentrate our attention on those sets $G \subset \mathbb{R}^2$ which are not absolute null. We will see below that, under **MA**, in this case either G or its reverse G^{-1} admits a uniformization by a relatively measurable function.

Lemma 33.2 *Assume* **MA**. *Let G be a subset of $\mathbb{R}^2$ equipped with a nonzero σ-finite Borel measure μ. Suppose also that all vertical sections of G are of μ-measure zero.*

Then there exists a function $g : \mathrm{pr}_1(G) \to \mathbb{R}$ satisfying the following two conditions:

(1) *g is entirely contained in G;*
(2) *g is μ-thick in G (i.e., g has common points with every μ-measurable subset Z of G such that $\mu(Z) > 0$).*

Proof It follows from **MA** that, for the given set G and for the measure μ, the following two relations hold:

(a) $\mathrm{card}(\mathrm{pr}_1(G)) = \mathbf{c}$;
(b) if $\{Z_i : i \in I\}$ is a family of μ-measure zero sets such that $\mathrm{card}(I) < \mathbf{c}$, then $\mu^*(\cup\{Z_i : i \in I\}) = 0$ (where μ^* is the outer measure produced by μ).

Also, since G is a separable metric space, one has the following relation:

(c) the family of all μ-measurable sets of strictly positive μ-measure has cardinality not exceeding $\mathbf{c}$.

Starting with (a), (b), (c) and using the method of transfinite recursion, one can define a function $g : \mathrm{pr}_1(G) \to \mathbb{R}$ satisfying (1) and (2). Details of the transfinite construction of g are left to the reader. □

Lemma 33.3 *Preserve the assumptions of Lemma 33.2 and suppose in addition that all horizontal sections of G are of μ-measure zero.*

Then any function g described in Lemma 33.2 is relatively measurable.

Proof Without loss of generality, we may suppose that μ is a probability measure. Consider the set $\mathrm{pr}_2(G)$ equipped with its Borel σ-algebra $\mathcal{B}(\mathrm{pr}_2(G))$. For each set $Y \in \mathcal{B}(\mathrm{pr}_2(G))$, put

$$\nu(Y) = \mu(G \cap \mathrm{pr}_2^{-1}(Y)).$$

Obviously, ν is a Borel probability measure on $\mathrm{pr}_2(G)$ and, since all horizontal sections of G are of μ-measure zero, this ν is continuous. Notice now that, in view of (2) of Lemma 33.2, the set $\mathrm{ran}(g)$ is ν-thick in $\mathrm{pr}_2(G)$, whence it follows that ran(g) cannot be absolute null. Therefore, according to Theorem 33.1, the function g is relatively measurable.

Lemma 33.3 has thus been proved. □

Lemma 33.4 *Let G be a subset of $\mathbb{R}^2$ such that some horizontal section of G is not absolute null.*

Then there exists a relatively measurable function $f : \mathrm{pr}_1(G) \to \mathbb{R}$ entirely contained in G.

Proof Actually, it is not difficult to see that a function $f \subset G$, the existence of which is stated by Lemma 33.4, can have a property much stronger than relative measurability. Namely, there is a probability continuous measure θ on some σ-algebra of subsets of $\mathrm{pr}_1(G)$ such that f turns out to be constant θ-almost everywhere on $\mathrm{pr}_1(G)$. The details of this stronger fact are omitted and are left to the reader. □

Now, we can formulate and prove the following theorem.

Theorem 33.2 *Assume MA and let G be a subset of $\mathbb{R}^2$ which is not absolute null. Then the disjunction of the following two assertions holds:*

(1) *there exists a relatively measurable function $f : \mathrm{pr}_1(G) \to \mathbb{R}$ entirely contained in G;*
(2) *there exists a relatively measurable function $g : \mathrm{pr}_2(G) \to \mathbb{R}$ entirely contained in G^{-1}.*

Proof Only three cases are possible for the given set G.

(a) There exists a horizontal section of G which is not absolute null.
    In this case, we apply Lemma 33.4 to G and obtain the required result.
(b) There exists a vertical section of G which is not absolute null.
    In this case, we apply Lemma 33.4 to G^{-1} and also obtain the required result.
(c) All horizontal and vertical sections of G are absolute null.
    In this case, let μ be a Borel probability continuous measure on G. Clearly, all horizontal and vertical sections of G are of μ-measure zero. Then, applying Lemma 33.3, we conclude that there exists a relatively measurable uniformization g of G.

This completes the proof of Theorem 33.2. □

Example 33.2 In the plane $\mathbb{R}^2$ consider the product set $G = \mathbb{R} \times Y$, where Y is any countably infinite set in $\mathbb{R}$. Clearly, this G is a Borel non-absolute null subset of $\mathbb{R}^2$. Also, it is easy to see that all functions $g : Y \to \mathbb{R}$ are absolutely nonmeasurable. Consequently, in the formulation of Theorem 33.2 the disjunction of assertions (1) and (2) is necessary.

Example 33.3 According to a well-known result of Sierpiński [159], there exists a bijective function $f : \mathbb{R} \to \mathbb{R}$ such that its graph is thick with respect to the Lebesgue standard measure λ_2 on $\mathbb{R}^2$. So the graph of f is not absolute null, although all vertical and horizontal sections of f are singletons. Both functions

f and f^{-1} are relatively measurable. Moreover, there exist measures μ and ν on $\mathbb{R}$ which satisfy the following three relations:

(a) both μ and ν are extensions of the Lebesgue standard measure λ on $\mathbb{R}$;
(b) f is measurable with respect to μ;
(c) f^{-1} is measurable with respect to ν.

Example 33.4 Consider any subset G of $\mathbb{R}^2$ having the following two properties:

(a) all vertical and horizontal sections of G are at most countable;
(b) G is not absolute null.

Note that, by virtue of Lemma 33.1, the sets $\text{pr}_1(G)$ and $\text{pr}_2(G)$ are not absolute null, so both of them are uncountable. From the proof of Theorem 33.2 one can see that, under **MA**, there exists a relatively measurable uniformization of G and, simultaneously, there exists a relatively measurable uniformization of G^{-1}.

Remark 33.6 We do not know whether an additional set-theoretical assumption is necessary in the statement of Theorem 33.2.

Exercises

1. Demonstrate that if a function $f : \mathbb{R} \to \mathbb{R}$ is absolutely nonmeasurable and Y is any non-absolute null subset of $\mathbb{R}$, then the restricted function $f|Y$ is not Borel on Y.
2. A subset L of $\mathbb{R}$ is called a *generalized Luzin set* if $\text{card}(L) = \mathbf{c}$ and, for each first category subset X of $\mathbb{R}$, the inequality $\text{card}(X \cap L) < \mathbf{c}$ holds true.
   Supposing **MA**, prove the existence of generalized Luzin sets in $\mathbb{R}$.
   For this purpose, use a transfinite construction similar to the classical construction of Luzin sets described in Exercise 2 of Chap. 19.
   Moreover, under the same **MA** verify that:

   (a) any generalized Luzin set is absolute null;
   (b) if L is a generalized Luzin set and f is a mapping from $\mathbb{R}$ into L such that

   $$(\forall r \in L)(\text{card}(f^{-1}(r)) \leq \omega),$$

   then f is an absolutely nonmeasurable function.
3. Assuming **MA**, give an example of an injective absolutely nonmeasurable function $g : \mathbb{R} \to \mathbb{R}$ which is continuous on some subset of $\mathbb{R}$ having cardinality $\mathbf{c}$.
   Conclude that this g is not a Sierpiński–Zygmund function.
4. Show that there exists a Sierpiński–Zygmund function acting from $\mathbb{R}$ into $\mathbb{R}$ which is not absolutely nonmeasurable.

For this purpose, using the method of transfinite recursion, construct a Sierpiński–Zygmund function whose graph is λ_2-thick in $\mathbb{R}^2$.

5. Denote by $\omega^{<\omega}$ the family of all finite sequences of natural numbers.

Let $\mathcal{L}$ be a class of subsets of a ground set E. Put $X_\emptyset = E$ and, for each nonempty finite sequence $(n_1, n_2, \ldots, n_k)$ of natural numbers, take arbitrarily a set $X_{n_1,n_2,\ldots,n_k}$ from $\mathcal{L}$. Then the set

$$X = \bigcup_{t \in \omega^\omega} X_{t_0} \cap X_{t_0,t_1} \cap \cdots \cap X_{t_0,t_1,\ldots,t_k} \cap \cdots$$

is called the result of *A-operation* over the countable family of sets

$$\{X_{n_1,n_2,\ldots,n_k} : (n_1, n_2, \ldots, n_k) \in \omega^{<\omega}\},$$

and the corresponding notation

$$X = A(\{X_{n_1,n_2,\ldots,n_k} : (n_1, n_2, \ldots, n_k) \in \omega^{<\omega}\})$$

is commonly used. The same X is also called the *analytic set* over this countable family of sets from $\mathcal{L}$.

Observe that if $\mathcal{L}$ is closed under finite intersections, then in the above definition of X one may assume, without loss of generality, that the family

$$\{X_{n_1,n_2,\ldots,n_k} : (n_1, n_2, \ldots, n_k) \in \omega^{<\omega}\} \subset \mathcal{L}$$

is regular, i.e.,

$$X_{n_1,n_2,\ldots,n_k,n_{k+1}} \subset X_{n_1,n_2,\ldots,n_k} \quad (k < \omega).$$

In the sequel, $\mathcal{A}(\mathcal{L})$ will denote the analytic class generated by $\mathcal{L}$, i.e., the class of analytic sets over all (appropriately enumerated) countable families of members of $\mathcal{L}$.

Evidently, the inclusion $\mathcal{L} \subset \mathcal{A}(\mathcal{L})$ always holds.

Check that:

(i) if $\mathcal{L}$ is closed under finite intersections and finite unions, then the analytic class $\mathcal{A}(\mathcal{L})$ is closed under countable intersections and countable unions;

(ii) any Borel subset of a Polish space E belongs to the analytic class $\mathcal{A}(\mathcal{L}_E)$, where $\mathcal{L}_E$ is the family of all closed subsets of E;

(iii) for the same class $\mathcal{L}_E$, a nonempty set $X \subset E$ belongs to $\mathcal{A}(\mathcal{L}_E)$ if and only if X is a continuous image of the canonical Baire space $\mathbb{N}^\mathbb{N}$;

(iv) if X belongs to $\mathcal{A}(\mathcal{L}_E)$ and Y is a continuous image of X in some Polish space F, then Y belongs to $\mathcal{A}(\mathcal{L}_F)$, where $\mathcal{L}_F$ is the family of all closed subsets of F;

(v) if T is an uncountable analytic subset of a Polish space, then T contains a homeomorphic copy of the Cantor space $\{0, 1\}^\omega$ and hence the equality

$$\operatorname{card}(T) = \mathbf{c}$$

is fulfilled (the theorem of Alexandrov and Hausdorff).

Argue as follows. First, verify that (i) implies (ii). Also, it is not difficult to show the validity of (iii) and (iv).

For (i), suppose that $\mathcal{L}$ is closed with respect to finite unions and finite intersections (in particular, $\emptyset$ and E belong to $\mathcal{L}$). Consider an arbitrary countable family $\{W_i : i < \omega\}$, all members of which belong to $\mathcal{A}(\mathcal{L})$. Then, for every $i < \omega$, one has

$$W_i = A(\{X_s^i : s \in \omega^{<\omega}\}),$$

where $\{X_s^i : s \in \omega^{<\omega}\} \subset \mathcal{L}$ is a regular family of sets. For any natural number i and for each $(i, n_1, \ldots, n_k) \in \omega^{k+1}$, define

$$Y_{i,n_1,n_2,\ldots,n_k} = X_{n_1,n_2,\ldots,n_k}^i.$$

Applying the A-operation to the obtained family $\{Y_s : s \in \omega^{<\omega}\} \subset \mathcal{L}$, infer that

$$\cup \{W_i : i < \omega\} \in \mathcal{A}(\mathcal{L}).$$

Further, for any $(n_1, n_2, \ldots, n_k) \in \omega^k$, put

$$Z_{n_1,n_2,\ldots,n_k} = \bigcap_{i<k} (\cup \{X_{m_1,m_2,\ldots,m_k}^i : m_1 \leq n_1, m_2 \leq n_2, \ldots, m_k \leq n_k\}).$$

Applying the A-operation to the obtained family $\{Z_s : s \in \omega^{<\omega}\} \subset \mathcal{L}$, derive that

$$\cap \{W_i : i < \omega\} \in \mathcal{A}(\mathcal{L}),$$

which yields (i).

For (v), use (iii) and take a surjective continuous mapping $f : E \to T$, where E is a complete separable metric space and T is an uncountable analytic subset of a Polish space. Considering condensation points of T, define by recursion some closed balls $B_{i_1,i_2,\ldots,i_n}$ in E, where

$$i_1 \in \{0, 1\}, \quad i_2 \in \{0, 1\}, \quad \ldots, \quad i_n \in \{0, 1\} \qquad (n \in \mathbb{N})$$

and the following conditions are fulfilled for these balls:

(a) $\text{diam}(B_{i_1,i_2,\ldots,i_n}) \leq 1/2^n$;
(b) $B_{i_1,i_2,\ldots,i_n,i_{n+1}} \subset B_{i_1,i_2,\ldots,i_n}$;
(c) $f(B_{i_1,i_2,\ldots,i_n,0}) \cap f(B_{i_1,i_2,\ldots,i_n,1}) = \emptyset$;
(d) the set $f(\text{int}(B_{i_1,i_2,\ldots,i_n}))$ is uncountable.

Then put

$$C = \cap\{\cup\{B_{i_1,i_2,\ldots,i_n} : i_1 \in \{0,1\}, i_2 \in \{0,1\}, \ldots, i_n \in \{0,1\}\} : n \in \mathbb{N}\}$$

and verify that C is a homeomorphic copy of the Cantor space and f is injective on C.

6. Let E be a base set and let $\mathcal{L}$ be some class of subsets of E. Suppose also that a function $\nu : \mathcal{P}(E) \to \mathbb{R}$ is given.
This ν is called a *capacity* on E for the class $\mathcal{L}$ (or with respect to $\mathcal{L}$) if the following three conditions hold:

(a) the relation $X \subset Y \subset E$ implies $\nu(X) \leq \nu(Y)$;
(b) if a sequence $\{X_n : n < \omega\}$ of subsets of E is increasing by inclusion, then $\nu(\cup\{X_n : n < \omega\}) = \lim_{n\to\infty} \nu(X_n)$;
(c) if a sequence $\{Y_n : n < \omega\} \subset \mathcal{L}$ is decreasing by inclusion, then $\nu(\cap\{Y_n : n < \omega\}) = \lim_{n\to\infty} \nu(Y_n)$.

Let now $(E, \mathcal{S}, \mu)$ be a measure space equipped with a finite countably additive measure μ. Take as $\mathcal{L}$ any class contained in the σ-algebra $\mathcal{S}$ and define a function ν by the formula

$$\nu(X) = \mu^*(X) \qquad (X \in \mathcal{P}(E)),$$

where μ^* denotes, as usual, the outer measure produced by μ.
Check that ν is a capacity with respect to the class $\mathcal{L}$.

Remark 33.7 As a rule, the class $\mathcal{L}$ is taken so that $\mathcal{S} = \sigma(\mathcal{L})$. In other words, in various considerations it is assumed that $\mathcal{L}$ generates the σ-algebra $\mathcal{S}$.

7. For a class $\mathcal{L}$ of subsets of a ground set E, denote by $\mathcal{L}_\delta$ the class of all intersections of countable families of sets from $\mathcal{L}$.
With this notation, prove Choquet's theorem on capacities formulated as follows:
Let $\mathcal{L}$ be some class of subsets of E which is closed under finite unions and finite intersections, and let ν be any capacity on E with respect to $\mathcal{L}$. Then, for every set X from the analytic class $\mathcal{A}(\mathcal{L})$, the equality

$$\nu(X) = \sup\{\nu(Y) : Y \in \mathcal{L}_\delta \ \& \ Y \subset X\}$$

is fulfilled.
To demonstrate the validity of this theorem, argue step by step. Pick a set $X \in \mathcal{A}(\mathcal{L})$. Obviously,

$$X = \bigcup_{t \in \omega^\omega} \bigcap_{k < \omega} F_{t_0,\ldots,t_k},$$

where $\{F_s : s \in \omega^{<\omega}\}$ is some countable system of sets from the given class $\mathcal{L}$. Since $\mathcal{L}$ is closed under finite intersections, one can assume without loss of generality that the above-mentioned system of sets is regular (see Exercise 5). For every infinite sequence $t = (t_0, t_1, \ldots, t_k, \ldots) \in \omega^\omega$, put

$$F_t = F_{t_0} \cap F_{t_0,t_1} \cap \cdots \cap F_{t_0,t_1,\ldots,t_k} \cap \cdots$$

and verify that

$$X = \bigcup_{n < \omega} (\bigcup_{\{t : t_0 \leq n\}} F_t).$$

By definition of ν, for any real number $\varepsilon > 0$, there exists a natural index n_0 such that $\nu(X) - \varepsilon < \nu(G_0)$, where G_0 is defined by the formula

$$G_0 = \bigcup_{\{t : t_0 \leq n_0\}} F_t.$$

Suppose that a finite sequence $(n_0, n_1, \ldots, n_k)$ of natural numbers and a finite sequence $(G_0, G_1, \ldots, G_k)$ of sets are constructed so that

$$G_r = \bigcup_{\{t : t_0 \leq n_0, \ldots, t_r \leq n_r\}} F_t \quad (r = 0, \ldots, k)$$

and the relations

$$\nu(X) - \varepsilon < \nu(G_r) \quad (r = 0, \ldots, k)$$

are satisfied. Consider the set G_k. Obviously, one may write

$$G_k = \bigcup_{n < \omega} (\bigcup_{\{t : t_0 \leq n_0, \ldots, t_k \leq n_k, t_{k+1} \leq n\}} F_t).$$

From the inequality $\nu(X) - \varepsilon < \nu(G_k)$ and from the basic property of ν it follows that there exists a natural index n_{k+1} such that, for the set

$$G_{k+1} = \bigcup_{\{t : t_0 \leq n_0, \ldots, t_k \leq n_k, t_{k+1} \leq n_{k+1}\}} F_t,$$

the inequality $\nu(X) - \varepsilon < \nu(G_{k+1})$ also holds true.

Proceeding in this manner, it becomes possible to construct two infinite sequences

$$(n_0, n_1, \ldots, n_k, \ldots), \qquad (G_0, G_1, \ldots, G_k, \ldots).$$

Further, for every natural number k, put

$$H_k = \bigcup_{t_0 \leq n_0, \ldots, t_k \leq n_k} F_{t_0, \ldots, t_k}.$$

Evidently, $H_k \in \mathcal{L}$. Moreover, check that $\{H_k : k < \omega\}$ is a decreasing sequence (with respect to the inclusion relation) and

$$G_k \subset H_k \qquad (k = 0, 1, 2, \ldots).$$

The latter fact implies that

$$\nu(X) - \varepsilon < \inf\{\nu(G_k) : k < \omega\} \leq \lim_{k \to \infty} \nu(H_k) = \nu(\cap\{H_k : k < \omega\}).$$

Now, using the regularity of the system $\{F_t : t \in \omega^{<\omega}\}$, infer the inclusion

$$\cap \{H_k : k < \omega\} \subset X.$$

Finally, denoting $\cap \{H_k : k < \omega\}$ by H, conclude that the set H belongs to the class $\mathcal{L}_\delta$ and

$$H \subset X, \qquad \nu(X) - \varepsilon \leq \nu(H).$$

Since $\varepsilon > 0$ was chosen arbitrarily small, the required result follows.

Remark 33.8 Let $(E, \mathcal{S}, \mu)$ be a space with a finite (or, more generally, σ-finite) countably additive measure μ, let $\mathcal{L} \subset \mathcal{S}$, and let μ' denote the completion of μ.

According to Exercise 6, the outer measure μ^* is a capacity for the class $\mathcal{L}$. Therefore, by Choquet's theorem, every analytic set over $\mathcal{L}$ is measurable with respect to μ'. It is also clear that if the original measure μ is complete, then every analytic set over $\mathcal{L}$ turns out to be μ-measurable.

Recall that the analytic sets in a Polish topological space E are precisely those sets which belong to the class $\mathcal{A}(\mathcal{L}_E)$, where $\mathcal{L}_E$ denotes the family of all closed subsets of E.

Another consequence of Choquet's theorem can be formulated as follows:

If E is a Polish space and μ is a σ-finite countably additive Borel measure on E, then any analytic set in E turns out to be μ'-measurable (where μ' denotes again the completion of μ).

In other words, any analytic set X in E is absolutely (universally) measurable with respect to the class of the completions of all σ-finite countably additive Borel measures on E. Equivalently, X is a Radon space (and its complement $E \setminus X$ is a Radon space). Recall also that X possesses the Baire property in E (see, e.g., [26, 76, 82, 109, 119]).

8. Let $(E, \mathcal{S}, \mu)$ be a space equipped with a σ-finite countably additive complete measure μ, let K be a nonempty locally compact topological space with a countable base, and let $(K, \mathcal{B}(K))$ be the measurable space canonically associated with K.
Demonstrate that if a set Z belongs to the product σ-algebra $\mathcal{S} \otimes \mathcal{B}(K)$, then the set $\mathrm{pr}_1(Z)$ belongs to the σ-algebra $\mathcal{S}$.
For this purpose, first reduce the consideration to the case when μ is a probability measure. Denote by $Comp(K)$ the family of all compact subsets of the space K. Evidently, the class $Comp(K)$ is closed under finite unions and nonempty finite intersections.
Check that the Borel σ-algebra $\mathcal{B}(K)$ is generated by $Comp(K)$.
In $E \times K$ consider the class $\mathcal{L}$ of all finite unions of the form

$$\cup \{X_i \times Y_i : 1 \leq i \leq m\},$$

where m is an arbitrary natural number, X_i are elements of $\mathcal{S}$ and Y_i are members of $Comp(K)$. Notice that $\mathcal{L}$ is closed under the operations of finite unions and finite intersections.
Verify that the σ-algebra generated by $\mathcal{L}$ coincides with the product σ-algebra $\mathcal{S} \otimes \mathcal{B}(K)$.
Further, define a real-valued function ν on the family of all subsets of $E \times K$ by the formula

$$\nu(Z) = \mu^*(\mathrm{pr}_1(Z)) \qquad (Z \subset E \times K)$$

and establish that this ν is a capacity on $E \times K$ with respect to $\mathcal{L}$. Indeed, if one has the relations

$$Z_1 \subset E \times K, \qquad Z_2 \subset E \times K, \qquad Z_1 \subset Z_2,$$

then $\nu(Z_1) \leq \nu(Z_2)$. Moreover, if $\{Z_n : n < \omega\}$ is an increasing (by inclusion) sequence of subsets of $E \times K$ and $Z = \cup\{Z_n : n < \omega\}$, then

$$\mathrm{pr}_1(Z) = \cup\{\mathrm{pr}_1(Z_n) : n < \omega\},$$
$$\nu(Z) = \mu^*(\mathrm{pr}_1(Z)) = \lim_{n \to \infty} \mu^*(\mathrm{pr}_1(Z_n)) = \lim_{n \to \infty} \nu(Z_n).$$

Also, it should be shown that if $\{Z_n : n < \omega\}$ is a decreasing (by inclusion) sequence of sets from the class $\mathcal{L}$, then

$$\nu(\cap\{Z_n : n < \omega\}) = \lim_{n \to \infty} \nu(Z_n).$$

To show the latter equality, first prove that

$$\mathrm{pr}_1(\cap\{Z_n : n < \omega\}) = \cap\{\mathrm{pr}_1(Z_n) : n < \omega\}.$$

In fact, it suffices to verify the inclusion

$$\cap\{\mathrm{pr}_1(Z_n) : n < \omega\} \subset \mathrm{pr}_1(\cap\{Z_n : n < \omega\}),$$

because the converse inclusion does always hold. Let x be an arbitrary element of $\cap\{\mathrm{pr}_1(Z_n) : n < \omega\}$. This means that, for any natural number n, there exists a point y_n such that

$$(x, y_n) \in X_{n,i(n)} \times Y_{n,i(n)},$$

where $X_{n,i(n)} \times Y_{n,i(n)}$ is some component of the representation of Z_n in the form

$$Z_n = \cup\{X_{n,i} \times Y_{n,i} : 1 \leq i \leq m(n)\}.$$

In particular, one has $y_n \in Y_{n,i(n)}$ and hence $Y_{n,i(n)} \neq \emptyset$. Moreover, since the sequence $\{Z_n : n < \omega\}$ is decreasing by inclusion, one may assume without loss of generality that the families of sets

$$\{X_{n+1,i} : 1 \leq i \leq m(n+1)\}, \qquad \{Y_{n+1,i} : 1 \leq i \leq m(n+1)\}$$

are respectively inscribed in the preceding families

$$\{X_{n,i} : 1 \leq i \leq m(n)\}, \qquad \{Y_{n,i} : 1 \leq i \leq m(n)\}.$$

Deduce from the latter that there exists a decreasing sequence of sets $\{X_{n,i(n)} \times Y_{n,i(n)} : n < \omega\}$ such that all sets $Y_{n,i(n)}$ are nonempty and the relation $x \in X_{n,i(n)}$ holds for every $n < \omega$. Since all $Y_{n,i(n)}$ are compact and decrease by inclusion, one can assert that

$$\cap\{Y_{n,i(n)} : n < \omega\} \neq \emptyset.$$

Let y be an element of $\cap\{Y_{n,i(n)} : n < \omega\}$. Then it is clear that

$$(x, y) \in \cap\{Z_n : n < \omega\}$$

and, therefore, $x \in \mathrm{pr}_1(\cap\{Z_n : n < \omega\})$.

Now, since all sets from the family $\{\mathrm{pr}_1(Z_n) : n < \omega\}$ are μ-measurable, one can write

$$\nu(\cap\{Z_n : n < \omega\}) = \mu^*(\mathrm{pr}_1(\cap\{Z_n : n < \omega\}))$$
$$= \mu(\cap\{\mathrm{pr}_1(Z_n) : n < \omega\})$$
$$= \lim_{n \to \infty} \mu(\mathrm{pr}_1(Z_n)) = \lim_{n \to \infty} \nu(Z_n).$$

Thus, the function ν is a capacity on $E \times K$ with respect to the class $\mathcal{L}$. According to Choquet's theorem, any set from the analytic class $\mathcal{A}(\mathcal{L})$ is ν-approximable from below by sets belonging to the class $\mathcal{L}_\delta$.
Keeping in mind that K is a locally compact topological space with a countable base, check the inclusion $\mathcal{S} \otimes \mathcal{B}(K) \subset \mathcal{A}(\mathcal{L})$.
Consequently, for any set $Z \in \mathcal{S} \otimes \mathcal{B}(K)$, the equality

$$\nu(Z) = \sup\{\nu(D) : D \in \mathcal{L}_\delta \,\&\, D \subset Z\}$$

holds, i.e.,

$$\mu^*(\mathrm{pr}_1(Z)) = \sup\{\mu(\mathrm{pr}_1(D)) : D \in \mathcal{L}_\delta \,\&\, D \subset Z\}.$$

This yields the μ-measurability of the set $\mathrm{pr}_1(Z)$.

9. Denote $\{t \in \mathbb{R} : t \geq 0\}$ by $\mathbb{R}_+$ and equip $\mathbb{R}_+$ with its standard order topology. Clearly, $\mathbb{R}_+$ is a locally compact space with a countable base.
Let $(E, \mathcal{S}, \mu)$ be as in Exercise 8 and let $Z \in \mathcal{S} \otimes \mathcal{B}(\mathbb{R}_+)$.
Verify that a function $d_Z : \mathrm{pr}_1(Z) \to \mathbb{R}_+$ defined by the formula

$$d_Z(e) = \inf\{t \in \mathbb{R}_+ : (e, t) \in Z\} \quad (e \in \mathrm{pr}_1(Z))$$

is measurable with respect to the σ-algebras $\mathcal{B}(\mathbb{R}_+)$ and $\mathcal{S}$.
Argue as follows. Observe that, for any real number $t \geq 0$, the set

$$d_Z^{-1}([0, t[) = \{e \in E : d_Z(e) < t\}$$

coincides with the set $\mathrm{pr}_1(Z \cap (E \times [0, t[))$. The set $Z \cap (E \times [0, t[)$ is measurable with respect to the σ-algebra $\mathcal{S} \otimes \mathcal{B}(\mathbb{R}_+)$. So, according to Exercise 8, the set $d_Z^{-1}([0, t[)$ is measurable with respect to $\mathcal{S}$. This implies that the function d_Z is measurable with respect to $\mathcal{B}(\mathbb{R}_+)$ and $\mathcal{S}$.
Deduce from the above result that there exists a μ-measurable function g acting from $\mathrm{pr}_1(Z)$ into $\mathbb{R}_+$ whose graph is entirely contained in Z.
For this purpose, consider the class $\mathcal{L}$ of all finite unions of the form

$$\cup\{X_i \times Y_i : 1 \leq i \leq m\},$$

where m is an arbitrary natural number, X_i are elements of S and Y_i are members of $Comp(\mathbb{R}_+)$. In view of Exercise 8, for any real $\varepsilon > 0$, there exists a set $D \in \mathcal{L}_\delta$ such that

$$D \subset Z, \qquad \mu(\mathrm{pr}_1(Z) \setminus \mathrm{pr}_1(D)) < \varepsilon.$$

Observe that all vertical sections of D are compact subsets of $\mathbb{R}_+$. Define a function d_D on the set $\mathrm{pr}_1(D)$ by the similar formula

$$d_D(e) = \inf\{t \in \mathbb{R}_+ : (e, t) \in D\} \qquad (e \in \mathrm{pr}_1(D)).$$

Then the graph of d_D is contained in D (hence in Z).
Finally, use an analogous argument for the set $Z \setminus (\mathrm{pr}_1(D) \times \mathbb{R}_+)$ and after countably many steps come to the required result.

10. Let K be a locally compact space with a countable base.
    Show that K is a Polish topological space.
    For this purpose, first check that K has a countable base consisting of relatively compact open sets. Then consider Alexandrov's compactification

$$K^* = K \cup \{p\}$$

of K, where a point p does not belong to K, and verify that p possesses in K^* a countable fundamental system of neighborhoods. Deduce from these facts that K^* is a compact space with a countable base, so is metrizable by a complete metric. Taking into account that K is an open subset of K^*, conclude that K is a Polish space.

11. Prove that any uncountable Polish topological space is Borel isomorphic to $\mathbb{R}_+$.
    For this purpose, argue step by step. Let P be an arbitrary uncountable Polish space.

    (a) Check that the product topological space $\mathbb{R}^{\mathbb{N}}$ contains a Borel isomorphic copy of P (and this copy is a Borel subset of $\mathbb{R}^{\mathbb{N}}$).

    For (a), assume that P is equipped with a complete metric and take into account the Banach–Mazur theorem, according to which the space $C([0, 1])$ contains an isometric copy of P (see Exercise 15 of Chap. 31). So one may suppose that P is contained in $C([0, 1])$. Now, in view of the completeness of P, it is a closed subset of $C([0, 1])$. In its turn $C([0, 1])$ can be considered as a Borel subspace of $\mathbb{R}^{\mathbb{N}}$ (cf. Exercise 24 from Chap. 30). Therefore, P is also a Borel set in $\mathbb{R}^{\mathbb{N}}$.

    (b) Verify that the canonical Baire space $\mathbb{N}^{\mathbb{N}}$ contains a Borel isomorphic copy of $\mathbb{R}$ (and this copy is a Borel subset of $\mathbb{N}^{\mathbb{N}}$).

    For (b), keep in mind that $\mathbb{R}$ consists of the rational and irrational numbers and the set $\mathbb{R} \setminus \mathbb{Q}$ is homeomorphic to $\mathbb{N}^{\mathbb{N}}$.

(c) Deduce from (b) that $\mathbb{N}^{\mathbb{N}}$ contains a Borel isomorphic copy of $\mathbb{R}^{\mathbb{N}}$, so contains a Borel isomorphic copy of P (and this copy is a Borel subset of $\mathbb{N}^{\mathbb{N}}$).

(d) Show that P contains a homeomorphic copy of Cantor's space

$$C = \{0, 1\}^{\omega},$$

so P contains a homeomorphic copy of $\mathbb{N}^{\mathbb{N}}$ (and this copy is a Borel subset of P).

Keeping in mind (c) and (d) and using the Cantor–Bernstein type argument (cf. Chap. 3), infer that P is Borel isomorphic to $\mathbb{N}^{\mathbb{N}}$.

Finally, conclude that P and $\mathbb{R}_+$ are also Borel isomorphic.

Remark 33.9 Arguing in a similar manner, a more general result of classical descriptive set theory can be established. Namely, if B_1 and B_2 are Borel subsets of Polish spaces P_1 and P_2 respectively, then the following two assertions are equivalent:

(1) $\mathrm{card}(B_1) = \mathrm{card}(B_2)$;
(2) there exists a Borel isomorphism between the spaces B_1 and B_2 (equipped with the induced topologies).

Another, closely related classical result in descriptive set theory is formulated as follows:

Let P be a Polish space, let B be a Borel subset of a Polish space, and let f be an injective Borel mapping from B to P. Then $f(B)$ is a Borel subset of P.

A proof of this second statement can be found, e.g., in the works [82, 109, 121].

12. Let E_1 and E_2 be two Polish spaces, X be an analytic subset of E_1 and let $g : E_1 \to E_2$ be a Borel mapping.

 Show that $g(X)$ is an analytic subset of E_2.

 Argue in the following manner. In the product space $E_1 \times E_2$ consider the graph of g, i.e., consider the set

$$G = \{(x, y) \in E_1 \times E_2 : g(x) = y\}.$$

First, show that G is a Borel subset of $E_1 \times E_2$. For this purpose, denote by ρ any metric in E_2 generating the topology of E_2 and introduce a function

$$\phi : E_1 \times E_2 \to [0, +\infty[$$

defined by the formula

$$\phi(x, y) = \rho(g(x), y) \qquad ((x, y) \in E_1 \times E_2).$$

Check that ϕ is a Borel mapping and $G = \phi^{-1}(0)$, which yields that G is a Borel set in $E_1 \times E_2$. Finally, consider the canonical projection

$$\mathrm{pr}_2 : E_1 \times E_2 \to E_2$$

and apply Exercise 5 to the continuous mapping pr_2 and to the analytic set $(X \times E_2) \cap G$.

13. Let E_1 be a Polish space equipped with a σ-finite countably additive Borel measure μ, let E_2 be another Polish space, and let Z be an analytic subset of the product topological space $E = E_1 \times E_2$.
    Denoting by μ' the completion of μ, demonstrate that there exists a function $g : \mathrm{pr}_1(Z) \to E_2$ satisfying the following two conditions:

    (a) the graph of g is entirely contained in Z;
    (b) g is μ'-measurable (as a partial function from E_1 into E_2).

    Argue step by step. First, assume without loss of generality that the space E_2 is uncountable. Use Exercise 11 and consider a Borel isomorphism ψ of E_2 onto $\mathbb{R}_+$. Then define a mapping Ψ from $E_1 \times E_2$ to $E_1 \times \mathbb{R}_+$ by the formula

    $$\Psi(x, y) = (x, \psi(y)) \qquad ((x, y) \in E_1 \times E_2).$$

    Check that Ψ is a Borel isomorphism of $E_1 \times E_2$ onto $E_1 \times \mathbb{R}_+$.
    According to Exercise 12, the set $\Psi(Z)$ is analytic in $E_1 \times \mathbb{R}_+$. Also, by virtue of Exercise 9, there exists a function

    $$f : \mathrm{pr}_1(\Psi(Z)) \to \mathbb{R}_+$$

    such that the graph of f is entirely contained in $\Psi(Z)$ and f is μ'-measurable (as a partial function from E_1 into $\mathbb{R}_+$). Put $g = \Psi^{-1}(f)$ and verify that g is as required.

Remark 33.10 Note that there are many other theorems on measurable uniformizations (measurable selectors) which essentially differ from the result presented in the last exercise. Some of them may be found, e.g., in [47, 76, 82, 112], and [121]. Nevertheless, the statement formulated in Exercise 13 suffices for most needs of real analysis and measure theory.

Appendix A
Lebesgue Integration of Real-Valued Functions

There are many approaches to introducing the Lebesgue integral of those real-valued functions which are defined on a space equipped with a finite (or σ-finite) countably additive measure. Here we would like to present only one (in fact, fairly standard) construction of the Lebesgue integral (cf. [17, 20, 38, 63, 69, 104, 134, 152, 177]).

Let E be a ground set and let S be a σ-algebra of subsets of E.

Recall that a function $f : E \to \mathbb{R}$ is *measurable with respect to* S (briefly, S-*measurable*) if, for any open subset U of $\mathbb{R}$, the set $f^{-1}(U)$ belongs to S.

Clearly, f is S-measurable if and only if, for every interval $\Delta \subset \mathbb{R}$ with rational endpoints, the set $f^{-1}(\Delta)$ belongs to S.

Also, it can easily be seen that the family of all S-measurable functions forms a vector space over $\mathbb{R}$ and, moreover, is an algebra over $\mathbb{R}$ (i.e., is closed under the operation of pointwise multiplication of functions). The same family is a lattice with respect to the pointwise partial ordering $\leq$ and the canonical binary operations

$$(f, g) \to \sup(f, g), \quad (f, g) \to \inf(f, g).$$

In addition, it is not hard to show that if $g : E \to \mathbb{R}$ is the pointwise limit of a sequence of S-measurable functions, then g is S-measurable, too.

Let E be endowed with a σ-finite countably additive measure μ and let $g : E \to \mathbb{R}$. This g is called *measurable with respect to* μ (briefly, μ-*measurable*) if g is $\text{dom}(\mu)$-measurable.

Let μ be a complete measure on E and let h be a partial function from E into $\mathbb{R}$ defined μ-almost everywhere on E, i.e., $\mu(E \setminus \text{dom}(h)) = 0$.

We again say that h is *measurable with respect to* μ if this h is S-measurable, where $S = \{\text{dom}(h) \cap X : X \in \text{dom}(\mu)\}$.

Let $h_1 : E \to \mathbb{R}$ and $h_2 : E \to \mathbb{R}$ be two partial functions both measurable with respect to μ.

We say that h_1 and h_2 are μ-equivalent if

$$\mu(\{x \in E : h_1(x) \neq h_2(x)\}) = 0.$$

It can readily be checked that the introduced notion of μ-equivalence of partial functions is a certain equivalence relation. As a rule, in measure theory all μ-equivalent partial functions are identified and, instead of them, the corresponding equivalence classes are often considered.

Below, we will be dealing only with countably additive measures defined on various σ-algebras of subsets of E.

The following classical statement is due to D. Egorov (see [39]).

Theorem A.1 *Let E be a ground set equipped with a finite measure μ and let $\{f_n : n \in \mathbb{N}\}$ be a sequence of real-valued μ-measurable functions on E which converges pointwise to a function f.*

Then, for every real $\varepsilon > 0$, there exists a μ-measurable set X such that:

(1) $\mu(E \setminus X) < \varepsilon$;
(2) *the sequence of functions $\{f_n | X : n \in \mathbb{N}\}$ converges uniformly to $f | X$.*

Proof For any two natural numbers n and m, denote

$$E_{n,m} = \{x \in E : (\forall k \geq n)(|f_k(x) - f(x)| \leq 1/2^m)\}.$$

A straightforward verification gives us that:

(a) all sets $E_{n,m}$ are μ-measurable;
(b) $E_{n,m} \subset E_{n+1,m}$ for all n and m from $\mathbb{N}$;
(c) for each $m \in \mathbb{N}$, the equality $\cup \{E_{n,m} : n \in \mathbb{N}\} = E$ holds true.

Consequently, for every $m \in \mathbb{N}$, there exists an $n(m) \in \mathbb{N}$ such that

$$\mu(E \setminus E_{n(m),m}) < \varepsilon/2^{m+1}.$$

Now, let us define $X = \cap \{E_{n(m),m} : m \in \mathbb{N}\}$ and let us check that the set X is as required.

Indeed, the relation $\mu(E \setminus X) < \varepsilon$ is almost trivial in view of the definition of X.

Further, take any real $\delta > 0$. There exists an $m \in \mathbb{N}$ for which $1/2^m < \delta$. Then, for all $x \in E_{n(m),m}$, we have

$$(\forall k \geq n(m))(|f_k(x) - f(x)| \leq 1/2^m < \delta).$$

Since $X \subset E_{n(m),m}$, the previous relation holds for every $x \in X$, which establishes the uniform convergence of the sequence $\{f_n | X : n \in \mathbb{N}\}$ to $f | X$.

Egorov's theorem has thus been proved. □

Remark A.1 There are substantially more general versions of Egorov's theorem. One of them is formulated and proved in [100].

A Lebesgue Integration of Real-Valued Functions

Let E be again a set equipped with a finite measure μ and let ϕ be a real-valued function on E.

We say that ϕ is a *step-function* (with respect to μ) if there exists a finite partition $\{X_k : k \in K\}$ of E into μ-measurable sets such that the restriction of ϕ to each X_k is a constant function whose range is some $\{t_k\}$, where $t_k \in \mathbb{R}$.

Remark A.2 If ϕ is a step-function on (E, μ), then, in general, there are various representations of ϕ in the form

$$\phi = \sum \{t_k \chi_{X_k} : k \in K\},$$

where $\{X_k : k \in K\}$ is a finite partition of E associated with ϕ. Moreover, it can easily be seen that if ϕ and ψ are any two step-functions on E, then there are representations of ϕ and ψ, respectively, which have the same associated partition of E.

We shall denote by $\mathcal{F}_s(E, \mu)$ the family of all step-functions on (E, μ).

Obviously, $\mathcal{F}_s(E, \mu)$ is a certain algebra of μ-measurable functions on E and is partially ordered by the relation $\phi \leq \psi$, where, as usual,

$$\phi \leq \psi \Leftrightarrow (\forall x \in E)(\phi(x) \leq \psi(x)).$$

Moreover, $\mathcal{F}_s(E, \mu)$ is a lattice with respect to the two binary operations $\sup(\cdot, \cdot)$ and $\inf(\cdot, \cdot)$.

Now, we define $I(\phi) = \sum \{t_k \mu(X_k) : k \in K\}$ for a step-function ϕ of the above form, i.e., for $\phi = \sum \{t_k \chi_{X_k} : k \in K\}$.

The function I is well-defined, which means that it does not depend on different representations of ϕ, and the mapping

$$\phi \to I(\phi) \qquad (\phi \in \mathcal{F}_s(E, \mu))$$

is a positive linear functional on the partially ordered vector space $\mathcal{F}_s(E, \mu)$.

Our goal is to extend the introduced functional to the much bigger family of all real-valued bounded μ-measurable functions on E.

Let $f : E \to \mathbb{R}$ be an arbitrary bounded function.

We shall say that f is μ-*integrable* (on E) if, for any real $\varepsilon > 0$, two step-functions ϕ_1 and ϕ_2 on E can be found such that

$$\phi_1 \leq f \leq \phi_2, \qquad I(\phi_2) - I(\phi_1) < \varepsilon.$$

In this case, we obviously have the equality

$$\sup\{I(\phi) : \phi \in \mathcal{F}_s(E, \mu) \,\&\, \phi \leq f\} = \inf\{I(\psi) : \psi \in \mathcal{F}_s(E, \mu) \,\&\, f \leq \psi\}.$$

So, for a μ-integrable function f, we can put

$$I(f) = \sup\{I(\phi) : \phi \in \mathcal{F}_s(E, \mu) \ \& \ \phi \leq f\}$$
$$= \inf\{I(\psi) : \psi \in \mathcal{F}_s(E, \mu) \ \& \ f \leq \psi\}.$$

The extended functional I on the family of all μ-integrable functions is also linear and positive. For instance, let us check the additivity of I. Consider any two μ-integrable functions f and g and take a real $\varepsilon > 0$. There exist four step-functions $\phi_1, \psi_1, \phi_2, \psi_2$ on E such that

$$\phi_1 \leq f \leq \psi_1, \quad I(\psi_1) - I(\phi_1) < \varepsilon/2,$$
$$\phi_2 \leq g \leq \psi_2, \quad I(\psi_2) - I(\phi_2) < \varepsilon/2.$$

So we may write $\phi_1 + \phi_2 \leq f + g \leq \psi_1 + \psi_2$ and

$$I(\psi_1 + \psi_2) - I(\phi_1 + \phi_2) < \varepsilon,$$

which implies the μ-integrability of $f + g$. The same argument gives us the equality $I(f + g) = I(f) + I(g)$.

The next theorem states that every real-valued bounded μ-measurable function on E turns out to be μ-integrable.

Theorem A.2 *If a real-valued bounded function f on a finite measure space (E, μ) is μ-measurable, then f is μ-integrable.*

Proof Let $[a, b[$ be an interval in $\mathbb{R}$ entirely containing the range of f. For any nonzero natural number n, take the intervals $\Delta_1, \Delta_2, \ldots, \Delta_n$, where

$$\Delta_k = [a + (k-1)(b-a)/n, a + k(b-a)/n[\quad (k = 1, 2, \ldots, n),$$

and define:

$$X_k = f^{-1}(\Delta_k) \quad (k = 1, 2, \ldots, n),$$
$$\phi = a\chi_{X_1} + (a + (b-a)/n)\chi_{X_2} + \cdots + (a + (n-1)(b-a)/n)\chi_{X_n},$$
$$\psi = (a + (b-a)/n)\chi_{X_1} + (a + 2(b-a)/n)\chi_{X_2} + \cdots + b\chi_{X_n}.$$

The introduced step-functions ϕ and ψ on E are such that

$$\phi \leq f \leq \psi, \quad I(\psi) - I(\phi) = \mu(E)(b-a)/n.$$

If n is sufficiently large, the difference $I(\psi) - I(\phi)$ becomes arbitrarily small, which shows us that the function f is μ-integrable. □

Lemma A.1 *Let ϕ and ψ be two step-functions on a finite measure space (E, μ) such that $\phi \leq \psi$, and let d be a positive real number.*
Then the inequality

A Lebesgue Integration of Real-Valued Functions

$$I(\psi) - I(\phi) \geq d \cdot \mu(\{x \in E : \psi(x) - \phi(x) \geq d\})$$

is fulfilled.

We omit the easy proof of the above lemma (keep in mind Remark A.2).

Under some natural condition on μ, the converse to Theorem A.2 also holds. Namely, we have the following theorem:

Theorem A.3 *Let a finite measure μ on E be complete. Then every real-valued μ-integrable function f on E is μ-measurable.*

Proof Consider an arbitrary real-valued μ-integrable function f on E.

There exist an increasing sequence $\{\phi_n : n \in \mathbb{N}\}$ of step-functions on E and a decreasing sequence $\{\psi_n : n \in \mathbb{N}\}$ of step-functions on E such that

$$(\forall n \in \mathbb{N})(\phi_n \leq f \leq \psi_n),$$

$$\lim_{n \to \infty} I(\phi_n) = \lim_{n \to \infty} I(\psi_n) = I(f).$$

Let $\phi = \lim_{n \to \infty} \phi_n$ and $\psi = \lim_{n \to \infty} \psi_n$.

The functions ϕ and ψ are μ-measurable (as pointwise limits of two sequences of μ-measurable functions) and $\phi \leq f \leq \psi$.

We assert that $\mu(\{x \in E : \psi(x) > \phi(x)\}) = 0$.

Suppose otherwise, i.e., $\mu(\{x \in E : \psi(x) > \phi(x)\}) > 0$. Then there are two real numbers $\varepsilon > 0$ and $\delta > 0$ such that

$$\mu(\{x \in E : \psi(x) - \phi(x) \geq \varepsilon\}) \geq \delta.$$

This inequality obviously implies that

$$\mu(\{x \in E : \psi_n(x) - \phi_n(x) \geq \varepsilon\}) \geq \delta$$

for all natural numbers n. Using Lemma A.1, we infer that

$$I(\psi_n) - I(\phi_n) \geq \varepsilon \cdot \delta \qquad (n \in \mathbb{N}),$$

which contradicts the equality $\lim_{n \to \infty} I(\phi_n) = \lim_{n \to \infty} I(\psi_n)$.

The obtained contradiction enables us to infer that the μ-measurable functions ϕ and ψ are μ-almost identical. Since $\phi \leq f \leq \psi$ and μ is complete, we conclude that f is μ-measurable.

Theorem A.3 has thus been proved. □

Lemma A.2 *Let (E, μ) be a finite measure space, d be a positive real number, and let $\{f_n : n \in \mathbb{N}\}$ be a sequence of real-valued μ-measurable functions on E such that*

$$(\forall n \in \mathbb{N})(\forall x \in E)(|f_n(x)| \leq d).$$

Suppose also that the pointwise limit $f = \lim_{n\to\infty} f_n$ exists.
Then the equality $\lim_{n\to\infty} I(f_n) = I(f)$ holds true.

Proof Clearly, $|f(x)| \leq d$ for all $n \in \mathbb{N}$ and for all $x \in E$. Take an arbitrary real $\varepsilon > 0$. According to Egorov's theorem, there exists a decomposition $\{X, Y\}$ of E into two μ-measurable sets such that

(a) $\mu(Y) < \varepsilon$;
(b) the sequence $\{f_n | X : n \in \mathbb{N}\}$ converges uniformly on X.

Let $n(\varepsilon)$ be a natural number which satisfies the relation

$$(\forall n \geq n(\varepsilon))(\forall x \in X)(|f_n(x) - f(x)| \leq \varepsilon).$$

Then, for any $n \geq n(\varepsilon)$, we may write

$$|I(f) - I(f_n)| = |I(f - f_n)|$$
$$\leq |I(\chi_X(f - f_n))| + |I(\chi_Y(f - f_n))|$$
$$\leq \varepsilon \mu(E) + 2d\varepsilon,$$

whence it follows that $\lim_{n\to\infty} I(f_n) = I(f)$.
This completes the proof of Lemma A.2. □

We have already obtained the linear positive functional I on the vector space $\mathcal{F}_b(E, \mu)$ of all real-valued bounded μ-measurable functions on (E, μ), where $\mu(E) < +\infty$.

We are now going to extend this I preserving its linearity and positivity. Actually, we intend to include in the domain of the extended functional some class of real-valued μ-measurable functions on (E, μ) which are not necessarily bounded.

Let $h \geq 0$ be a real-valued μ-measurable function on E.

Suppose that there exists an increasing sequence $\{f_n : n \in \mathbb{N}\}$ of real-valued non-negative bounded μ-measurable functions on E such that

$$\lim_{n\to\infty} f_n = h, \qquad \lim_{n\to\infty} I(f_n) < +\infty.$$

Then we say that h is μ-integrable and define $I(h) = \lim_{n\to\infty} I(f_n)$.
To prove that this extension of I is well-defined, we need the following lemma.

Lemma A.3 *Under the above assumptions, using the same notation, consider any other increasing sequence $\{g_n : n \in \mathbb{N}\}$ of real-valued non-negative bounded μ-measurable functions on E satisfying $\lim_{n\to\infty} g_n = h$.*
Then the equality $\lim_{n\to\infty} I(f_n) = \lim_{n\to\infty} I(g_n)$ holds.

Proof It suffices to demonstrate that

A Lebesgue Integration of Real-Valued Functions

$$I(f_m) \leq \lim_{n \to \infty} I(g_n),$$

$$I(g_m) \leq \lim_{n \to \infty} I(f_n)$$

for any natural number m. We only will show the validity of

$$I(f_m) \leq \lim_{n \to \infty} I(g_n),$$

because the second inequality can be established in the same manner.

Let us define

$$\phi_n = \inf(g_n, f_m) \qquad (n \in \mathbb{N}).$$

It can easily be verified that $\phi_n \leq f_m$ and $\lim_{n \to \infty} \phi_n = f_m$.

Using Lemma A.2, we get $I(f_m) = \lim_{n \to \infty} I(\phi_n)$. At the same time, we have $\phi_n \leq g_n$ for all $n \in \mathbb{N}$. Consequently,

$$I(f_m) = \lim_{n \to \infty} I(\phi_n) \leq \lim_{n \to \infty} I(g_n),$$

which ends the proof. □

Let us remark that the extended functional I preserves the additivity property, i.e., if $f \geq 0$ and $g \geq 0$ are μ-integrable, then $f + g$ is also μ-integrable and

$$I(f + g) = I(f) + I(g).$$

Indeed, it suffices to refer to the additivity of I on the family $\mathcal{F}_b(E, \mu)$ and to apply the corresponding limit process.

This enables us to extend I to a well-defined linear positive functional on a certain class of real-valued μ-measurable functions which, in general, change their sign.

Let h be a real-valued μ-measurable function on E. Evidently, there are various representations of h in the form $h = h_1 - h_2$, where h_1 and h_2 are two real-valued non-negative μ-measurable functions on E.

This h is called μ-*integrable* if both values $I(h_1)$ and $I(h_2)$ exist by the previous definition (for at least one representation of h).

In such a case, the difference $I(h_1) - I(h_2)$ is called the μ-*integral* of h (for the given finite measure space (E, μ)).

The introduced concept is well-defined, as immediately follows from the remark made above.

Observe that h is μ-integrable if and only if $|h|$ is μ-integrable.

The commonly adopted notation for the μ-integral of h is $\int_E h(x)\,d\mu(x)$ and, if E is fixed in the considerations, the symbol $\int h(x)\,d\mu(x)$ is also used.

The family of all real-valued μ-integrable functions on E is denoted by $L_1(E, \mu)$. This family is a vector space over $\mathbb{R}$ and, in general, is not an algebra over $\mathbb{R}$.

Lemma A.4 *Let (E, μ) be a finite measure space and let f be a real-valued μ-integrable function on E.*

Then, for each real $\varepsilon > 0$, there exists a real $\delta > 0$ such that

$$\int_Z |f(x)|\,d\mu(x) = \int_E |\chi_Z(x) f(x)|\,d\mu(x) < \varepsilon$$

whenever Z is a μ-measurable subset of E with $\mu(Z) < \delta$.

Proof Without loss of generality, we may assume that $f \geq 0$. Let ϕ be a real-valued bounded μ-measurable function on E satisfying the relations

$$0 \leq \phi \leq f, \qquad \int_E (f(x) - \phi(x))\,d\mu(x) < \varepsilon/2.$$

Since ϕ is bounded, we have $(\forall x \in E)(\phi(x) \leq d)$ for some real constant $d > 0$. Take $\delta = \varepsilon/2d$ and let Z be an arbitrary μ-measurable subset of E with $\mu(Z) < \delta$. We may write

$$\int_Z f(x)\,d\mu(x) = \int_E (f(x) - \phi(x))\chi_Z(x)\,d\mu(x) + \int_E \phi(x)\chi_Z(x)\,d\mu(x)$$
$$< \varepsilon/2 + d \cdot \varepsilon/2d$$
$$= \varepsilon,$$

which completes the proof of Lemma A.4. □

Remark A.3 Actually, in Lemma A.4 we have a signed measure ν on E defined by the formula

$$\nu(X) = \int_X f(x)\,d\mu(x) \qquad (X \in \mathrm{dom}(\mu)),$$

and this lemma asserts that ν is absolutely continuous with respect to μ (cf. Chap. 26).

The next important result is due to Lebesgue and is called the dominated convergence theorem. It substantially generalizes Lemma A.2 and is useful in many topics of mathematical analysis (especially, in integration theory).

Theorem A.4 *Suppose that (E, μ) is a finite measure space and a sequence $\{f_n : n \in \mathbb{N}\}$ of real-valued μ-measurable functions on E is given. Suppose also that the following two conditions are fulfilled:*

A Lebesgue Integration of Real-Valued Functions

(1) *there exists a real-valued μ-integrable function ϕ on E such that*

$$(\forall n \in \mathbb{N})(\forall x \in E)(|f_n(x)| \leq |\phi(x)|);$$

(2) *the sequence $\{f_n : n \in \mathbb{N}\}$ converges pointwise to some function f on E.*

Then the function f is μ-integrable and the equality

$$\lim_{n \to \infty} \int_E f_n(x) \, d\mu(x) = \int_E f(x) \, d\mu(x)$$

holds true.

Proof Conditions (1) and (2) immediately imply that $|f(x)| \leq |\phi(x)|$ for all x from E, so f is μ-integrable. Take a real $\varepsilon > 0$. According to Lemma A.4, there exists a real $\delta > 0$ such that

$$\int_Z |\phi(x)| \, d\mu(x) < \varepsilon$$

whenever Z is a μ-measurable subset of E with $\mu(Z) < \delta$. Obviously, we also have

$$\int_Z |f(x)| \, d\mu(x) < \varepsilon, \qquad (\forall n \in \mathbb{N})(\int_Z |f_n(x)| \, d\mu(x) < \varepsilon)$$

for any μ-measurable set Z with $\mu(Z) < \delta$.

Further, according to Egorov's theorem, there exists a μ-measurable set Y with $\mu(Y) < \delta$ such that the sequence $\{f_n | (E \setminus Y) : n \in \mathbb{N}\}$ converges uniformly on $E \setminus Y$. Let a natural number $n(\varepsilon)$ be such that

$$(\forall n > n(\varepsilon))(\forall x \in E \setminus Y)(|f_n(x) - f(x)| < \varepsilon).$$

Then, for each natural number $n > n(\varepsilon)$, we may write

$$\int_E |f_n(x) - f(x)| \, d\mu(x) \leq \int_Y |f_n(x) - f(x)| \, d\mu(x) + \int_{E \setminus Y} |f_n(x) - f(x)| \, d\mu(x)$$

$$\leq \int_Y |f_n(x)| \, d\mu(x) + \int_Y |f(x)| \, d\mu(x)$$

$$+ \int_{E \setminus Y} |f_n(x) - f(x)| \, d\mu(x)$$

$$\leq 2\varepsilon + \varepsilon \mu(E \setminus Y) \leq \varepsilon(2 + \mu(E)),$$

which shows us that

$$\lim_{n \to \infty} \int_E |f_n(x) - f(x)| \, d\mu(x) = 0$$

and, consequently,

$$\lim_{n\to\infty} \int_E f_n(x)\,\mathrm{d}\mu(x) = \int_E f(x)\,\mathrm{d}\mu(x).$$

The Lebesgue theorem has thus been proved. □

Some other standard facts concerning real-valued μ-integrable functions on a measure space (E, μ) are presented in the exercises below.

Remark A.4 Let E be a ground set equipped with a σ-finite measure μ and let $\{X_i : i \in I\}$ be a countable partition of E into μ-measurable sets such that $\mu(X_i) < +\infty$ for all $i \in I$.

Consider a μ-measurable function $g : E \to \mathbb{R}$. For each index $i \in I$, denote by g_i the restriction of g to X_i and denote by μ_i the restriction of μ to the σ-algebra of all μ-measurable subsets of X_i.

Suppose that the following two conditions are satisfied:

(1) for any $i \in I$, the function g_i is μ_i-integrable on X_i;
(2) $\sum\{\int_{X_i} |g_i(x)|\,\mathrm{d}\mu_i(x) : i \in I\} < +\infty$.

In this case, one says that g is μ-*integrable* on E and the value

$$\sum\{\int_{X_i} g_i(x)\,\mathrm{d}\mu_i(x) : i \in I\}$$

is called the μ-*integral* of g.

One can easily check that the introduced notion of μ-integral of g does not depend on the choice of a countable partition $\{X_i : i \in I\}$ of E. In other words, the μ-integral is well-defined for the given σ-finite measure μ.

Also, for a finite measure space, the notion introduced above coincides with the previous notion of μ-integral.

Exercises

1. Give an example of a real-valued Lebesgue nonmeasurable function f on the segment $[0, 1]$ such that $|f|$ is a constant (hence Lebesgue measurable) function on $[0, 1]$.
2. Let E be a ground set endowed with a non-finite σ-finite measure μ and let $\{f_n : n \in \mathbb{N}\}$ be a sequence of real-valued μ-measurable functions on E pointwise convergent to some function f.

   Prove that, for any real $r > 0$, there exists a μ-measurable set $X \subset E$ such that $\mu(X) > r$ and the sequence of functions $\{f_n|X : n \in \mathbb{N}\}$ uniformly converges to $f|X$.
3. Let λ be the standard Lebesgue measure on the real line $\mathbb{R}$.

Give an example of a sequence $\{g_n : n \in \mathbb{N}\}$ of real-valued uniformly bounded λ-measurable functions on $\mathbb{R}$ having the following two properties:

(a) $\{g_n : n \in \mathbb{N}\}$ pointwise converges on $\mathbb{R}$;
(b) there exists no λ-measurable set $X \subset \mathbb{R}$ with $\lambda(X) < +\infty$ such that the sequence $\{g_n | (\mathbb{R} \setminus X) : n \in \mathbb{N}\}$ uniformly converges on $\mathbb{R} \setminus X$.

4. For a set E equipped with a finite measure μ and for a real-valued μ-measurable function f on E, verify that:

(a) if f^2 is μ-integrable on E, then f is also μ-integrable on E;
(b) it may happen that f is μ-integrable on E, but f^2 is not μ-integrable on E.

5. Let (E, μ) be a σ-finite measure space and let $\{f_n : n \in \mathbb{N}\}$ be an increasing sequence of real-valued non-negative μ-integrable functions on E such that

$$\lim_{n \to \infty} \int_E f_n(x) \, d\mu(x) < +\infty.$$

Prove that there is a μ-measurable set $Z \subset E$ such that:

(a) $\mu(E \setminus Z) = 0$;
(b) for each point $x \in Z$, the limit $f(x) = \lim_{n \to \infty} f_n(x)$ exists;
(c) $\lim_{n \to \infty} \int_E f_n(x) \, d\mu(x) = \int_Z f(x) \, d\mu(x)$.

In particular, if $\{g_n : n \in \mathbb{N}\}$ is a sequence of real-valued non-negative μ-integrable functions on E, then the following two assertions are equivalent:

(a) the series $\sum \{g_n : n \in \mathbb{N}\}$ converges μ-almost everywhere to some μ-integrable function $g : E \to \mathbb{R}$;
(b) $\sum \{\int_E g_n(x) \, d\mu(x) : n \in \mathbb{N}\} < +\infty$.

In addition, if (a) holds, then

$$\int_E g(x) \, d\mu(x) = \sum \{\int_E g_n(x) \, d\mu(x) : n \in \mathbb{N}\}.$$

6. Let (E, μ) be a finite measure space and let $\mathcal{M}$ denote the family of all real-valued μ-measurable functions on E. As usual, identify any two functions from $\mathcal{M}$ which coincide μ-almost everywhere on E, and put

$$d(f, g) = \int_E (|f(x) - g(x)| / (1 + |f(x) - g(x)|)) \, d\mu(x)$$

for all f and g from $\mathcal{M}$.

Check that d is a metric on $\mathcal{M}$ and that the metric space $(\mathcal{M}, d)$ is complete.

Remark A.5 Put $E = [0, 1]$ and $\mu = \lambda$, where λ is the Lebesgue probability measure on $[0, 1]$. In this case, $\mathcal{M}$ of Exercise 6 turns out to be a Polish topological space. Moreover, the following three assertions are true for the space $\mathcal{M}$ associated with $([0, 1], \lambda)$:

(a) there are $\mathbf{c}$ many linear positive continuous mappings from $\mathcal{M}$ into itself;
(b) there are $2^{\mathbf{c}}$ many linear bijective mappings from $\mathcal{M}$ onto itself;
(c) the vector space $\mathcal{M}^*$ of all real-valued linear continuous functionals on $\mathcal{M}$ is trivial, i.e., $\mathcal{M}^* = \{0\}$.

7. Let E be a ground set equipped with a finite (or σ-finite) measure μ. For a real number $p \geq 1$, let $L_p(E, \mu)$ be the family of all those real-valued μ-measurable functions f on E which satisfy the inequality

$$\int_E |f(x)|^p \, d\mu(x) < +\infty.$$

This family is called the *Lebesgue p-space associated with* μ (of course, μ-equivalent functions are identified). Use the standard notation

$$||f|| = \left(\int_E |f(x)|^p \, d\mu(x) \right)^{1/p} \qquad (f \in L_p(E, \mu))$$

and demonstrate that:

(a) $L_p(E, \mu)$ is a vector space over $\mathbb{R}$;
(b) the functional $|| \cdot ||$ is a norm on $L_p(E, \mu)$;
(c) $(L_p(E, \mu), || \cdot ||)$ is a Banach space.

In particular, for $p = 2$, one has the Hilbert space $L_2(E, \mu)$ of all real-valued square μ-integrable functions on E.

8. Verify that the notion of μ-integral introduced in Remark A.4 is well-defined. Also, examine which properties of the μ-integral for a finite measure space (e.g., properties discussed in this appendix) remain valid in the case of the μ-integral for a σ-finite measure space.

9. Let (E, μ) be a measure space, $\{f_n : n \in \mathbb{N}\}$ be a sequence of real-valued μ-measurable functions on E, and let f be a real-valued μ-measurable function on E.
By definition, $\{f_n : n \in \mathbb{N}\}$ *converges to f in measure* μ if, for any real $\varepsilon > 0$, the relation

$$\lim_{n \to \infty} \mu(\{x \in E : |f_n(x) - f(x)| \geq \varepsilon\}) = 0$$

holds true.
Prove that:

(a) if $\mu(E) < +\infty$ and $\{f_n : n \in \mathbb{N}\}$ converges μ-almost everywhere to f, then $\{f_n : n \in \mathbb{N}\}$ converges to f in measure μ;

A Lebesgue Integration of Real-Valued Functions   387

(b) if $\mu(E) = +\infty$, then the pointwise convergence of $\{f_n : n \in \mathbb{N}\}$ to f does not imply, in general, the convergence of $\{f_n : n \in \mathbb{N}\}$ to f in measure μ.

For (a), use Egorov's theorem.

Also, give an example of a sequence $\{g_n : n \in \mathbb{N}\}$ of real-valued uniformly bounded μ-measurable functions on E which converges in measure μ (where $0 < \mu(E) < +\infty$), but diverges at all points of E.

10. Let (E, μ) be a finite measure space, $\{f_n : n \in \mathbb{N}\}$ be a sequence of real-valued μ-measurable functions on E, and let f be a real-valued μ-measurable function on the same E.

    Show that $\{f_n : n \in \mathbb{N}\}$ converges to f in measure μ if and only if

    $$\lim_{n \to \infty} d(f_n, f) = 0,$$

    where d is the metric described in Exercise 6.

11. Let (E, μ) be a measure space, $\{f_n : n \in \mathbb{N}\}$ be a sequence of real-valued μ-measurable functions on E, and let f be a real-valued μ-measurable function on E.

    Demonstrate that if $\{f_n : n \in \mathbb{N}\}$ converges to f in measure μ, then this sequence contains a subsequence converging pointwise μ-almost everywhere on E to the same f.

    For this purpose, use an argument similar to the proof of Egorov's theorem. Namely, for any two natural numbers n and m, put

    $$E_{n,m} = \{x \in E : |f_n(x) - f(x)| \geq 1/(m+1)\}.$$

    Since $\{f_n : n \in \mathbb{N}\}$ converges to f in measure μ, for each $m \in \mathbb{N}$ there exists an $n(m) \in \mathbb{N}$ such that

    $$(\forall n \geq n(m))(\mu(E_{n,m}) < 1/2^m).$$

    Without loss of generality, one may assume that

    $$n(0) < n(1) < \cdots < n(m) < \cdots .$$

    Further, for any $m \in \mathbb{N}$, denote $\cup \{E_{n(k),k} : k \geq m\}$ by X_m and define

    $$X = \cap \{X_m : m \in \mathbb{N}\}.$$

    Check that $\mu(X) = 0$. Finally, verify that the subsequence $\{f_{n(m)} : m \in \mathbb{N}\}$ of $\{f_n : n \in \mathbb{N}\}$ converges pointwise to f on the set $E \setminus X$.

12. Try to describe all those σ-finite measure spaces (E, μ) which have the following property:

Any sequence $\{g_n : n \in \mathbb{N}\}$ of real-valued μ-measurable functions on E converges in measure μ to a function $g : E \to \mathbb{R}$ if and only if the same $\{g_n : n \in \mathbb{N}\}$ converges μ-almost everywhere to g.

For this purpose, keep in mind the notion of an atom of a measure (cf. Exercise 8 from Chap. 24).

13. Let μ and ν be two finite measures on a ground set E such that ν is absolutely continuous with respect to μ, let $\phi = d\nu/d\mu$ (see Chap. 26), and let f be an arbitrary ν-integrable function on E.

    Prove that $\int_E f(x)\, d\nu(x) = \int_E f(x)\phi(x)\, d\mu(x)$.

    For this purpose, first consider the case when the given f is a step-function on E.

Appendix B
Product Measures

This appendix is devoted to the standard construction of the product of a family of countably additive measures (of course, under certain natural assumptions on members of the family).

Naturally, we begin with the special case when a given family consists of exactly two measures (cf. [17, 38, 63, 69, 104, 134, 141, 154]).

Let (E_1, μ_1) and (E_2, μ_2) be two finite (or σ-finite) measure spaces and let $E = E_1 \times E_2$ be the Cartesian product of E_1 and E_2. Consider the family $\mathcal{S}_0(E)$ of all those sets Z in E which are representable in the form

$$Z = \cup\{A_i \times B_i : i \in I\},$$

where I is a finite set, $A_i \in \mathrm{dom}(\mu_1)$ and $B_i \in \mathrm{dom}(\mu_2)$ for all indices $i \in I$.

It is convenient to say in the sequel that a set of the form $A \times B$, where $A \in \mathrm{dom}(\mu_1)$ and $B \in \mathrm{dom}(\mu_2)$, is a (measurable) rectangle in E.

Obviously, the empty set is a rectangle in E and the intersection of any countable family of rectangles in E is also a rectangle in E.

Lemma B.1 *$\mathcal{S}_0(E)$ is an algebra of subsets of E.*

Proof A straightforward verification shows that

$$(E_1 \times E_2) \setminus (A \times B) = ((E_1 \setminus A) \times E_2) \cup (E_1 \times (E_2 \setminus B)),$$

$$(\cup\{A_i \times B_i : i \in I\}) \cap (\cup\{C_j \times D_j : j \in J\})$$

$$= \cup\{(A_i \cap C_j) \times (B_i \cap D_j) : (i, j) \in I \times J\}.$$

The assertion of Lemma B.1 directly follows from the above formulas. □

Lemma B.2 *Every set $Z \in \mathcal{S}_0(E)$ can be represented in the form*

$$Z = \cup\{A_i \times B_i : i \in I\},$$

where I is a finite set, $A_i \in \mathrm{dom}(\mu_1)$ and $B_i \in \mathrm{dom}(\mu_2)$ for all indices $i \in I$, and $\{A_i \times B_i : i \in I\}$ is a disjoint family of sets.

We omit the easy proof of Lemma B.2 and leave it to the reader (here it makes sense to use induction on $\mathrm{card}(I)$).

Now, take an arbitrary set $Z \in S_0(E)$. According to Lemma B.2, this Z is representable in the form

$$Z = \cup\{A_i \times B_i : i \in I\},$$

where I is a finite set, $A_i \in \mathrm{dom}(\mu_1)$ and $B_i \in \mathrm{dom}(\mu_2)$ for all indices $i \in I$, and $\{A_i \times B_i : i \in I\}$ is a disjoint family of sets. Let us put

$$\mu_0(Z) = \sum\{\mu_1(A_i)\mu_2(B_i) : i \in I\}.$$

Our purpose is to prove that μ_0 is a well-defined non-negative function on the algebra $S_0(E)$ and, moreover, that μ_0 is countably additive on the same $S_0(E)$.

Lemma B.3 *Suppose that the following condition is fulfilled:*

(*) *If, for any rectangle $X \times Y$, one has a representation*

$$X \times Y = \cup\{X_k \times Y_k : k \in K\},$$

where $\{X_k \times Y_k : k \in K\}$ is a disjoint finite family of rectangles, then the equality

$$\mu_1(X)\mu_2(Y) = \sum\{\mu_1(X_k)\mu_2(Y_k) : k \in K\}$$

holds true.

In this case, the introduced value $\mu_0(Z)$ does not depend on the above representation of Z and the obtained function μ_0 is finitely additive on $S_0(E)$.

Proof Consider any two representations of Z in the form

$$Z = \cup\{A_i \times B_i : i \in I\} = \cup\{C_j \times D_j : j \in J\},$$

where $\{A_i \times B_i : i \in I\}$ and $\{C_j \times D_j : j \in J\}$ are disjoint finite families of rectangles. Obviously, we may write

$$A_i \times B_i = \cup\{(A_i \cap C_j) \times (B_i \cap D_j) : j \in J\} \quad (i \in I),$$
$$C_j \times D_j = \cup\{(A_i \cap C_j) \times (B_i \cap D_j) : i \in I\} \quad (j \in J),$$

where the families of rectangles

$$\{(A_i \cap C_j) \times (B_i \cap D_j) : j \in J\} \quad (i \in I),$$
$$\{(A_i \cap C_j) \times (B_i \cap D_j) : i \in I\} \quad (j \in J)$$

are also disjoint. Using (*), we readily get

$$\sum \{\mu_1(A_i \cap C_j)\mu_2(B_i \cap D_j) : (i,j) \in I \times J\} =$$
$$\sum \{\mu_1(C_j)\mu_2(D_j) : j \in J\} = \sum \{\mu_1(A_i)\mu_2(B_i) : i \in I\},$$

which shows that μ_0 is well-defined (under the condition (*)). □

An analogous argument gives us that, under the same condition (*), μ_0 is a finitely additive function on the algebra $S_0(E)$.

Clearly, μ_0 is non-negative and $\mu_0(\emptyset) = 0$.

The next auxiliary lemma justifies the condition (*) and, in fact, yields the countable additivity of μ_0 on $S_0(E)$.

Lemma B.4 *If $A \times B$ is any rectangle in E and $\{A_n \times B_n : n \in \mathbb{N}\}$ is a disjoint countable family of rectangles in E satisfying the relation*

$$A \times B = \cup \{A_n \times B_n : n \in \mathbb{N}\},$$

then the equality $\mu_1(A)\mu_2(B) = \sum\{\mu_1(A_n)\mu_2(B_n) : n \in \mathbb{N}\}$ is fulfilled.

Proof First, observe that

$$\chi_A(x)\chi_B(y) = \sum \{\chi_{A_n}(x)\chi_{B_n}(y) : n \in \mathbb{N}\}$$

for each $(x,y) \in E = E_1 \times E_2$, where $\chi_A, \chi_B, \chi_{A_n}, \chi_{B_n}$ denote, as usual, the characteristic functions of A, B, A_n, B_n respectively.

For a fixed $x \in E_1$, we may compute the Lebesgue integral (with respect to y) of both sides of the above equality. So, we obtain

$$\chi_A(x) \int_{E_2} \chi_B(y) \, d\mu_2(y) = \sum \{\chi_{A_n}(x) \int_{E_2} \chi_{B_n}(y) \, d\mu_2(y) : n \in \mathbb{N}\}$$

or, equivalently,

$$\mu_2(B)\chi_A(x) = \sum \{\mu_2(B_n)\chi_{A_n}(x) : n \in \mathbb{N}\}.$$

Applying once more the Lebesgue integration (with respect to x) for both sides of the last formula, we get

$$\mu_2(B) \int_{E_1} \chi_A(x) \, d\mu_1(x) = \sum \{\mu_2(B_n) \int_{E_1} \chi_{A_n}(x) \, d\mu_1(x) : n \in \mathbb{N}\}$$

or, equivalently,

$$\mu_1(A)\mu_2(B) = \sum \{\mu_1(A_n)\mu_2(B_n) : n \in \mathbb{N}\}.$$

This completes the proof of Lemma B.4. □

Let us denote by $\mathcal{S}(E)$ the σ-algebra generated by the algebra $\mathcal{S}_0(E)$. According to Carathéodory's classical theorem (see Chap. 17), there exists a unique countably additive measure μ on $\mathcal{S}(E)$ extending μ_0.

This measure is called the *product measure* of μ_1 and μ_2 and is denoted by the symbol $\mu_1 \otimes \mu_2$. So, by definition of $\mu_1 \otimes \mu_2$, we have

$$\mathrm{dom}(\mu_1 \otimes \mu_2) = \sigma(\{X \times Y : X \in \mathrm{dom}(\mu_1) \ \& \ Y \in \mathrm{dom}(\mu_2)\}),$$

where $\sigma(\{X \times Y : X \in \mathrm{dom}(\mu_1) \ \& \ Y \in \mathrm{dom}(\mu_2)\})$ stands for the σ-algebra generated by the family of sets $\{X \times Y : X \in \mathrm{dom}(\mu_1) \ \& \ Y \in \mathrm{dom}(\mu_2)\}$.

Remark B.1 Quite often, the completion of $\mu_1 \otimes \mu_2$ is also called the product measure of μ_1 and μ_2.

More generally, if $(E_1, \mathcal{S}_1)$ and $(E_2, \mathcal{S}_2)$ are any two measurable spaces, then the symbol $\mathcal{S}_1 \otimes \mathcal{S}_2$ denotes the σ-algebra in $E = E_1 \times E_2$ generated by the family of sets $\{X \times Y : X \in \mathcal{S}_1 \ \& \ Y \in \mathcal{S}_2\}$.

It is natural to say that $\mathcal{S}_1 \otimes \mathcal{S}_2$ is the *product σ-algebra* of $\mathcal{S}_1$ and $\mathcal{S}_2$.

Suppose now that a function $f : E \to \mathbb{R}$ is measurable with respect to $\mathcal{S}_1 \otimes \mathcal{S}_2$. Then it can easily be shown that:

(i) for each $x \in E_1$, the function $f_x : E_2 \to \mathbb{R}$ defined by

$$f_x(y) = f(x, y) \qquad (y \in E_2)$$

is measurable with respect to $\mathcal{S}_2$;

(ii) for each $y \in E_2$, the function $f_y : E_1 \to \mathbb{R}$ defined by

$$f_y(x) = f(x, y) \qquad (x \in E_1)$$

is measurable with respect to $\mathcal{S}_1$.

In particular, if a set Z belongs to $\mathcal{S}_1 \otimes \mathcal{S}_2$, then for any $x \in E_1$ the x-section of Z belongs to $\mathcal{S}_2$ and for any $y \in E_2$ the y-section of Z belongs to $\mathcal{S}_1$.

In connection with (i) and (ii), it is useful to compare Exercise 5 of this chapter.

Lemma B.5 *Let (E_1, μ_1) and (E_2, μ_2) be two finite measure spaces, let $E = E_1 \times E_2$, and let $\mu = \mu_1 \otimes \mu_2$.*

For every μ-measurable set $Z \subset E$, the following formula holds:

B Product Measures

$$\int_E \chi_Z(x,y)\,\mathrm{d}\mu(x,y) = \int_{E_1}(\int_{E_2} \chi_{Z_x}(y)\,\mathrm{d}\mu_2(y))\,\mathrm{d}\mu_1(x)$$
$$= \int_{E_2}(\int_{E_1} \chi_{Z_y}(x)\,\mathrm{d}\mu_1(x))\,\mathrm{d}\mu_2(y),$$

where $Z_x = \{y : (x,y) \in Z\}$ and $Z_y = \{x : (x,y) \in Z\}$.

Proof A simple verification shows that the above formula is fulfilled for all sets Z from the algebra $S_0(E)$. Now, applying the properties of Lebesgue integral, we readily deduce that:

(a) if the formula holds true for each member of an increasing (by inclusion) sequence $\{Z_n : n \in \mathbb{N}\}$ of μ-measurable sets, then the same formula is true for the set $\cup \{Z_n : n \in \mathbb{N}\}$;
(b) if the formula holds true for each member of a decreasing (by inclusion) sequence $\{Z_n : n \in \mathbb{N}\}$ of μ-measurable sets, then the same formula is true for the set $\cap \{Z_n : n \in \mathbb{N}\}$.

Thus, the formula is fulfilled for every member of the monotone class generated by $S_0(E)$. As is well known, this class coincides with the σ-algebra $S(E)$, which completes the proof. □

Now, we are able to formulate and prove a special case of Fubini's theorem (recall that this theorem is one of the central results in the theory of Lebesgue integration).

Theorem B.1 *Let (E_1, μ_1) and (E_2, μ_2) be two finite measure spaces, let $E = E_1 \times E_2$, and let $\mu = \mu_1 \otimes \mu_2$.*
Suppose that $f : E \to \mathbb{R}$ is a bounded μ-measurable function on E.
Then the following relations hold:

(1) *for each $x \in E_1$, the function $f_x : E_2 \to \mathbb{R}$ defined by*

$$f_x(y) = f(x,y) \qquad (y \in E_2)$$

is μ_2-integrable on E_2, and the function

$$x \to \int_{E_2} f_x(y)\,\mathrm{d}\mu_2(y) \qquad (x \in E_1)$$

is μ_1-integrable on E_1;
(2) *for each $y \in E_2$, the function $f_y : E_1 \to \mathbb{R}$ defined by*

$$f_y(x) = f(x,y) \qquad (x \in E_1)$$

is μ_1-integrable on E_1, and the function

$$y \to \int_{E_1} f_y(x) \, d\mu_1(x) \qquad (y \in E_2)$$

is μ_2-integrable on E_2;
(3) *the equalities*

$$\int_E f(x, y) \, d\mu(x, y) = \int_{E_1} (\int_{E_2} f_x(y) \, d\mu_2(y)) \, d\mu_1(x)$$
$$= \int_{E_2} (\int_{E_1} f_y(x) \, d\mu_1(x)) \, d\mu_2(y)$$

are true.

Proof Since the given μ-measurable function f is bounded, there exists a sequence $\{f_n : n \in \mathbb{N}\}$ of uniformly bounded μ-measurable step-functions on E such that

$$f(x, y) = \lim_{n \to \infty} f_n(x, y) \qquad ((x, y) \in E).$$

Observe that, in view of Lemma B.5, for any step-function f_n all relations (1), (2), (3) are trivially fulfilled. It follows from this fact that:

(a) for each $x \in E_1$, the function $f_x : E_2 \to \mathbb{R}$ equal to $\lim_{n \to \infty} (f_n)_x$ is μ_2-integrable on E_2 and the function

$$x \to \int_{E_2} f_x(y) \, d\mu_2(y) \qquad (x \in E_1)$$

is μ_1-integrable on E_1;
(b) for each $y \in E_2$, the function $f_y : E_1 \to \mathbb{R}$ equal to $\lim_{n \to \infty} (f_n)_y$ is μ_1-integrable on E_1, and the function

$$y \to \int_{E_1} f_y(x) \, d\mu_1(x) \qquad (y \in E_2)$$

is μ_2-integrable on E_2.

Finally, using the well-known properties of the Lebesgue integral (see, e.g., Appendix A), we may write

$$\int_E f(x, y) \, d\mu(x, y) = \lim_{n \to \infty} \int_E f_n(x, y) \, d\mu(x, y),$$

$$\int_E f_n(x, y) \, d\mu(x, y) = \int_{E_1} (\int_{E_2} (f_n)_x(y) \, d\mu_2(y)) \, d\mu_1(x)$$

$$= \int_{E_2} \left(\int_{E_1} (f_n)_y(x) \, d\mu_1(x) \right) d\mu_2(y).$$

Letting n tend to infinity in the last formula, we obtain relation (3) for the given function f, which ends the proof. □

Remark B.2 A more general version of Fubini's theorem is concerned with any two σ-finite measure spaces (E_1, μ_1) and (E_2, μ_2), where an arbitrary real-valued $(\mu_1 \otimes \mu_2)$-integrable function

$$f : E_1 \times E_2 \to \mathbb{R}$$

is given. The formulation and proof of this version are slightly different from the above ones, because a few additional technical nuances need to be taken into account (cf. Exercise 9 of the present chapter).

It is easy to extend the construction of a product measure to the case when a nonempty finite family of σ-finite measure spaces is given. The argument remains almost the same. So we omit the details and only formulate the corresponding result.

Theorem B.2 *Let (E_1, μ_1), (E_2, μ_2), ..., (E_n, μ_n) be a nonempty finite family of σ-finite measure spaces and let $E = E_1 \times E_2 \times \ldots \times E_n$.*
Then there exists a unique σ-finite measure μ on E such that:

(1) *the domain of μ coincides with the product σ-algebra*

$$\mathrm{dom}(\mu_1) \otimes \mathrm{dom}(\mu_2) \otimes \ldots \otimes \mathrm{dom}(\mu_n);$$

(2) *for any sets $X_i \in \mathrm{dom}(\mu_i)$, where $i = 1, 2, \ldots, n$, the equality*

$$\mu(\prod \{X_i : 1 \le i \le n\}) = \prod \{\mu_i(X_i) : 1 \le i \le n\}$$

holds true.

The notation $\mu_1 \otimes \mu_2 \otimes \ldots \otimes \mu_n$ is commonly used for this measure μ.

Also, it can readily be seen that the corresponding analog of Fubini's theorem is valid for $\mu_1 \otimes \mu_2 \otimes \cdots \otimes \mu_n$.

Remark B.3 Intending to define the product measure for an arbitrary infinite family of measures, one encounters certain technical difficulties. This general case is more delicate and is touched upon in some exercises below.

Exercises

1. Give a detailed proof of Lemma B.1.

2. Let (E_1, S_1) and (E_2, S_2) be any two measurable spaces and let $E = E_1 \times E_2$. Consider a subset Z of E having the form

$$Z = \cup \{A_i \times B_i : i \in I\},$$

where I is a finite set, $A_i \in S_1$ and $B_i \in S_2$ for all $i \in I$.
Verify that the same Z can be represented in the form

$$Z = \cup \{C_j \times D_j : j \in J\},$$

where J is a finite set, $C_j \in S_1$ and $D_j \in S_2$ for all indices $j \in J$, and $\{C_j \times D_j : j \in J\}$ is a disjoint family of sets.
For this purpose, apply the method of induction on card(I).

3. Let (E_1, S_1) and (E_2, S_2) be again two measurable spaces and let $E = E_1 \times E_2$.
Demonstrate that the σ-algebra $S = S_1 \otimes S_2$ generated in E by the family of sets $\{X \times Y : X \in S_1 \,\&\, Y \in S_2\}$ coincides with the σ-algebra generated by

$$\{E_1 \times Y : Y \in S_2\} \cup \{X \times E_2 : X \in S_1\}.$$

In addition to this, check that S is the smallest (by inclusion) σ-algebra of subsets of E such that both mappings

$$\mathrm{pr}_1 : (E, S) \to (E_1, S_1), \qquad \mathrm{pr}_2 : (E, S) \to (E_2, S_2)$$

are measurable (i.e., for each set X from S_1, the set $\mathrm{pr}_1^{-1}(X)$ belongs to S and, for each set Y from S_2, the set $\mathrm{pr}_2^{-1}(Y)$ belongs to S).

4. Let λ denote, as usual, the standard Lebesgue measure on $\mathbb{R}$ and let λ_2 denote the standard two-dimensional Lebesgue measure on the plane $\mathbb{R}^2$ (actually, λ_2 is the completion of the product measure $\lambda \otimes \lambda$).
Prove that there exists a bijective function $f : \mathbb{R} \to \mathbb{R}$ whose graph $\Gamma(f)$ is λ_2-thick in $\mathbb{R}^2$, i.e., the equality

$$(\lambda_2)_*(\mathbb{R}^2 \setminus \Gamma(f)) = 0$$

holds true, where $(\lambda_2)_*$ is the inner measure produced by λ_2.
For this purpose, construct the required function f by the method of transfinite recursion.

5. Using the notation of Exercise 4, show that there exists a function $g : \mathbb{R}^2 \to \mathbb{R}$ having the following three properties:

(a) g is not λ_2-measurable;
(b) for each $x \in \mathbb{R}$, the function $g_x : \mathbb{R} \to \mathbb{R}$ defined by

$$g_x(y) = g(x, y) \qquad (y \in \mathbb{R})$$

is the characteristic function of a singleton in $\mathbb{R}$ (hence is λ-measurable);
(c) for each $y \in \mathbb{R}$, the function $g_y : \mathbb{R} \to \mathbb{R}$ defined by

$$g_y(x) = g(x, y) \qquad (x \in \mathbb{R})$$

is also the characteristic function of a singleton in $\mathbb{R}$ (hence is λ-measurable as well).

6. Give a proof of Theorem B.2.
   For this purpose, use induction on n.
7. Let (E, μ) be a σ-finite measure space, let $(\mathbb{R}, \lambda)$ be the standard Lebesgue measure space, and let $f : E \to \mathbb{R}$ be a non-negative μ-integrable function. Demonstrate that the set $\{(x, t) \in E \times \mathbb{R} : 0 \le t \le f(x)\}$ is measurable with respect to the completion of $\mu \otimes \lambda$ and

$$\int_E f(x)\, d\mu(x) = (\mu \otimes \lambda)(\{(x, t) \in E \times \mathbb{R} : 0 \le t \le f(x)\}).$$

This formula gives a certain geometric interpretation of the Lebesgue μ-integral of f.

8. Let $(E_1, \mathcal{S}_1)$ and $(E_2, \mathcal{S}_2)$ be two measurable spaces, let $E = E_1 \times E_2$, and let $\mathcal{S} = \mathcal{S}_1 \otimes \mathcal{S}_2$.
   For any functions $\phi : E_1 \to \mathbb{R}$ and $\psi : E_2 \to \mathbb{R}$, consider the functions $f : E \to \mathbb{R}$ and $g : E \to \mathbb{R}$ such that

$$f(x, y) = \phi(x) + \psi(y) \qquad ((x, y) \in E),$$
$$g(x, y) = \phi(x) \cdot \psi(y) \qquad ((x, y) \in E).$$

Verify the following two assertions:

(a) f is $\mathcal{S}$-measurable $\Leftrightarrow$ (ϕ is $\mathcal{S}_1$-measurable and ψ is $\mathcal{S}_2$-measurable);
(b) if ϕ is $\mathcal{S}_1$-measurable and ψ is $\mathcal{S}_2$-measurable, then g is $\mathcal{S}$-measurable (but not conversely).

Further, let μ_1 be a σ-finite measure with $\text{dom}(\mu_1) = \mathcal{S}_1$, let μ_2 be a σ-finite measure with $\text{dom}(\mu_2) = \mathcal{S}_2$, and let $\mu = \mu_1 \otimes \mu_2$.
Check that if ϕ is μ_1-integrable and ψ is μ_2-integrable, then g is μ-integrable and

$$\int_E g(x, y)\, d\mu(x, y) = \int_{E_1} \phi(x)\, d\mu_1(x) \cdot \int_{E_2} \psi(y)\, d\mu_2(y).$$

In the case $\mu_1(E_1) < +\infty$ and $\mu_2(E_2) < +\infty$, formulate and prove the analogous result for f.

9. Let (E_1, μ_1) and (E_2, μ_2) be any two σ-finite measure spaces, let $E = E_1 \times E_2$, and let $\mu = \mu_1 \otimes \mu_2$.

Suppose that $f : E \to \mathbb{R}$ is a partial μ-integrable function such that

$$\mu^*(E \setminus \text{dom}(f)) = 0.$$

Prove the following assertions:

(i) for μ_1-almost all $x \in E_1$, the partial function $f_x : E_2 \to \mathbb{R}$ defined by

$$f_x(y) = f(x, y) \qquad (y \in \text{dom}(f_x))$$

is μ_2-integrable on E_2, and the partial function

$$x \to \int_{E_2} f_x(y) \, d\mu_2(y)$$

is μ_1-integrable on E_1;

(ii) for μ_2-almost all $y \in E_2$, the partial function $f_y : E_1 \to \mathbb{R}$ defined by

$$f_y(x) = f(x, y) \qquad (x \in \text{dom}(f_y))$$

is μ_1-integrable on E_1, and the partial function

$$y \to \int_{E_1} f_y(x) \, d\mu_1(x)$$

is μ_2-integrable on E_2;

(iii) the equalities

$$\int_E f(x, y) \, d\mu(x, y) = \int_{E_1} \left(\int_{E_2} f_x(y) \, d\mu_2(y) \right) d\mu_1(x)$$

$$= \int_{E_2} \left(\int_{E_1} f_y(x) \, d\mu_1(x) \right) d\mu_2(y)$$

are fulfilled.

This is the formulation of Fubini's theorem for the product of two given σ-finite measures μ_1 and μ_2.

Reduce the proof of the formulated version of Fubini's theorem to the case where both μ_1 and μ_2 are finite measures and f is a real-valued bounded μ-measurable function.

10. Let (E_1, μ_1) and (E_2, μ_2) be any two σ-finite measure spaces, let $E = E_1 \times E_2$, and let $\mu = \mu_1 \otimes \mu_2$.
Suppose that $f : E \to \mathbb{R}$ is a μ-measurable function on E satisfying at least one of the following two inequalities:

(a) $\int_{E_1}(\int_{E_2}|f_x(y)|\,\mathrm{d}\mu_2(y))\,\mathrm{d}\mu_1(x) < +\infty$,
(b) $\int_{E_2}(\int_{E_1}|f_y(x)|\,\mathrm{d}\mu_1(x))\,\mathrm{d}\mu_2(y) < +\infty$.

Demonstrate that f is a μ-integrable function on E.
For this purpose, assume on the contrary that

$$\int_E |f(x,y)|\,\mathrm{d}\mu(x,y) = +\infty$$

and, using an appropriate increasing sequence $\{f_n : n \in \mathbb{N}\}$ of non-negative μ-integrable functions on E converging pointwise to $|f|$, obtain a contradiction with the disjunction of (a) and (b).

11. Let $\{(E_n, \mu_n) : n = 1, 2, \ldots\}$ be a countably infinite family of probability measure spaces and let S be the smallest σ-algebra of subsets of the product set

$$E = \prod\{E_n : n = 1, 2, \ldots\},$$

such that every canonical projection

$$\mathrm{pr}_n : E \to E_n \qquad (n = 1, 2, \ldots)$$

is measurable with respect to S and $\mathrm{dom}(\mu_n)$.

Work in **ZF** & **DC** theory and prove the Lomnicki–Ulam theorem, which states that there exists a unique probability measure μ on E such that $\mathrm{dom}(\mu) = S$ and

$$\mu(\prod\{A_n : n = 1, 2, \ldots\}) = \prod\{\mu_n(A_n) : n = 1, 2, \ldots\}$$

for any set of the form $\prod\{A_n : n = 1, 2, \ldots\}$, where each A_n belongs to $\mathrm{dom}(\mu_n)$ and the relation $A_n \neq E_n$ holds true only for finitely many natural indices n.

Argue as follows. First, denote by S_0 the algebra generated by all sets in E having the indicated form, and check that the above equality enables one to define on S_0 a finitely additive measure μ. Then verify that μ is countably additive as well.

Indeed, it suffices to show that μ is upper semicontinuous at $\emptyset$, i.e., if a real ε is strictly positive and $\{Z_m : m \in \mathbb{N}\}$ is any decreasing (by inclusion) sequence of members of S_0 satisfying the relation

$$(\forall m \in \mathbb{N})(\mu(Z_m) \geq \varepsilon),$$

then $\cap\{Z_m : m \in \mathbb{N}\} \neq \emptyset$. For this purpose, define by recursion a sequence $\{y_n : n = 1, 2, \ldots\} \subset E$ so that, for any natural number m, the relations

$$\nu_n(Z_m(y_1, y_2, \ldots, y_n)) \geq \varepsilon/2^n \qquad (n = 1, 2, \ldots)$$

hold, where ν_n is the direct analog of μ for the partial product set

$$\prod \{E_k : k = n+1, n+2, \ldots\}$$

and $Z_m(y_1, y_2, \ldots, y_n)$ is the set of all those sequences

$$\{x_1, x_2, \ldots, x_k, \ldots\} \in E_{n+1} \times E_{n+2} \times \cdots \times E_{n+k} \times \cdots$$

which satisfy the relation

$$\{y_1, y_2, \ldots, y_n, x_1, x_2, \ldots, x_k, \ldots\} \in Z_m.$$

Let $\nu_0 = \mu$ and observe that

$$\nu_n = \mu_{n+1} \otimes \nu_{n+1} \qquad (n \in \mathbb{N}),$$
$$\nu_n(Z_m(y_1, y_2, \ldots, y_n)) = (\mu_{n+1} \otimes \nu_{n+1})(Z_m(y_1, y_2, \ldots, y_n)).$$

Suppose that, for some $n \in \mathbb{N}$ and for all $m \in \mathbb{N}$,

$$\nu_n(Z_m(y_1, y_2, \ldots, y_n)) \geq \varepsilon/2^n.$$

Applying Fubini's theorem to $\mu_{n+1} \otimes \nu_{n+1}$ and to the set $Z_m(y_1, y_2, \ldots, y_n)$, check that

$$\mu_{n+1}(\{y \in E_{n+1} : \nu_{n+1}(Z_m(y_1, y_2, \ldots, y_n, y)) \geq \varepsilon/2^{n+1}\}) \geq \varepsilon/2^{n+1}.$$

Notice that Fubini's theorem is applicable here, because of the special form of $Z_m(y_1, y_2, \ldots, y_n)$.

Since μ_{n+1} is a countably additive measure and the sequence of sets

$$\{y \in E_{n+1} : \nu_{n+1}(Z_m(y_1, y_2, \ldots, y_n, y)) \geq \varepsilon/2^{n+1}\} \qquad (m \in \mathbb{N})$$

is decreasing by inclusion (in view of $Z_{m+1} \subset Z_m$), there exists a $y \in E_{n+1}$ belonging to all of these sets. Therefore, putting $y_{n+1} = y$, one has

$$\nu_{n+1}(Z_m(y_1, y_2, \ldots, y_n, y_{n+1})) \geq \varepsilon/2^{n+1}$$

for each $m \in \mathbb{N}$.

As soon as the sequence $\{y_n : n = 1, 2, \ldots\}$ is recursively constructed, it is not difficult to infer that

$$\{y_n : n = 1, 2, \ldots\} \in \cap\{Z_m : m \in \mathbb{N}\},$$

which establishes the countable additivity of μ on $\mathcal{S}_0$ (and hence on $\mathcal{S}$).

The uniqueness of μ easily follows from its definition.

Remark B.4 The measure μ described in Exercise 11 is called the *product measure* of the given family of probability measures $\{\mu_n : n = 1, 2, \ldots\}$ and is denoted by $\otimes \{\mu_n : n = 1, 2, \ldots\}$.

12. Let I be an arbitrary nonempty set of indices and let

$$\{(E_i, \mu_i) : i \in I\}$$

be a family of σ-finite measure spaces such that, for some finite subset J of I, all μ_i ($i \in I \setminus J$) are probability measures.
Denote by $\mathcal{S}$ the smallest σ-algebra of subsets of $E = \prod\{E_i : i \in I\}$, for which every canonical projection

$$\mathrm{pr}_i : E \to E_i \quad (i \in I)$$

is measurable (with respect to $\mathcal{S}$ and $\mathrm{dom}(\mu_i)$).
Prove that there exists a unique σ-finite measure μ on E such that $\mathrm{dom}(\mu) = \mathcal{S}$ and

$$\mu(\prod\{A_i : i \in I\}) = \prod\{\mu_i(A_i) : i \in I\}$$

for any set of the form $\prod\{A_i : i \in I\}$, where each A_i belongs to $\mathrm{dom}(\mu_i)$ and the relation $A_i \neq E_i$ holds true only for finitely many indices $i \in I$.
Reduce the proof of the existence of μ to the case of a countable set I and apply the result of Exercise 11.

Remark B.5 Naturally, the above-mentioned measure μ is called the *product measure* of the given family of measures $\{\mu_i : i \in I\}$ and is denoted by $\otimes \{\mu_i : i \in I\}$.

13. With the notation of the previous exercise, demonstrate that:
    (a) if I is at most countable, then any set of the form $\prod\{A_i : i \in I\}$, where

    $$(\forall i \in I)(A_i \in \mathrm{dom}(\mu_i)),$$

    is μ-measurable and $\mu(\prod\{A_i : i \in I\}) = \prod\{\mu_i(A_i) : i \in I\}$;
    (b) if I is uncountable, then it may happen that there exists a set of the same form which is not measurable even with respect to the completion of μ.

14. Try to formulate and prove an analog of Fubini's theorem for the product measure $\mu = \otimes\{\mu_i : i \in I\}$, where I is an arbitrary nonempty set of indices, all measures μ_i are σ-finite, and there exists a finite subset J of I such that every μ_i ($i \in I \setminus J$) is a probability measure.
For this purpose, reduce the general case to Fubini's classical theorem.

15. Give an explanation of the following phenomenon:
   If one has a countable family $\{\mu_i : i \in I\}$ of nonzero σ-finite measures, then, in general, it is impossible to define the product measure μ of this family so that μ satisfies Suslin's condition.
16. Let $\mu_1, \mu_2, \nu_1, \nu_2$ be four σ-finite measures such that μ_1 is equivalent to μ_2 and ν_1 is equivalent to ν_2.
   Using Fubini's theorem, verify that the product measure $\mu_1 \otimes \nu_1$ is equivalent to the product measure $\mu_2 \otimes \nu_2$.
17. Let (E_1, S_1, μ_1) and (E_2, S_2, μ_2) be two measure spaces and let $E_1 \cap E_2 = \emptyset$. Consider the family S defined by

$$S = \{X \cup Y : X \in S_1, \ Y \in S_2\}$$

and put

$$\mu(X \cup Y) = \mu_1(X) + \mu_2(Y) \qquad (X \in S_1, \ Y \in S_2).$$

Check that the triple $(E_1 \cup E_2, S, \mu)$ is a measure space for which the following assertions hold:

(a) if μ_1 and μ_2 are finite (σ-finite), then μ is finite (σ-finite);
(b) if μ_1 and μ_2 are Radon measures, then μ is also a Radon measure;
(c) if μ_1 is G_1-invariant (G_1-quasi-invariant) with respect to some group G_1 of transformations of E_1 and μ_2 is G_2-invariant (G_2-quasi-invariant) with respect to some group G_2 of transformations of E_2, then μ is G-invariant (G-quasi-invariant), where G denotes the group of all those transformations g of $E_1 \cup E_2$ which have the property that $g|E_1$ belongs to G_1 and $g|E_2$ belongs to G_2.

Remark B.6 The measure μ defined in Exercise 17 is called the *direct sum* of two given measures μ_1 and μ_2.

18. Let $\{(E_i, S_i, \mu_i) : i \in I\}$ be a nonempty family of measure spaces such that the corresponding ground sets E_i ($i \in I$) are pairwise disjoint.
   Generalizing Exercise 17, introduce the notion of the direct sum (E, S, μ) of the family $\{(E_i, S_i, \mu_i) : i \in I\}$, where $E = \cup\{E_i : i \in I\}$.
   Show that a measure ν is σ-finite if and only if ν is representable as the direct sum of a countable family of finite measures.
19. Let (E, μ) be a probability measure space and let, for each $n \in \mathbb{N}$, the measure space (E_n, μ_n) be identical with (E, μ). Equip the product set $F = \prod\{E_n : n \in \mathbb{N}\}$ with the product measure

$$\nu = \otimes\{\mu_n : n \in \mathbb{N}\}.$$

Any permutation ϕ of $\mathbb{N}$ canonically produces the bijection g of F onto itself, defined by the formula

B Product Measures

$$g(\{e_n : n \in \mathbb{N}\}) = \{e_{\phi(n)} : n \in \mathbb{N}\} \qquad (\{e_n : n \in \mathbb{N}\} \in F).$$

Denote by G the group of those bijections g which correspond to permutations ϕ of $\mathbb{N}$ having the property

$$\operatorname{card}(\{n \in \mathbb{N} : \phi(n) \neq n\}) < \omega.$$

Verify that ν is a G-invariant measure on F.
Let Z be an arbitrary ν-measurable ν-almost G-invariant subset of F.
Prove the Hewitt–Savage theorem (zero-one law) which states that either $\nu(Z) = 0$ or $\nu(Z) = 1$.
For this purpose, argue as follows. Let $\mathcal{A}$ be the smallest algebra of sets in F for which all canonical projections

$$\operatorname{pr}_n : (F, \operatorname{dom}(\nu)) \to (E_n, \operatorname{dom}(\mu_n)) \qquad (n \in \mathbb{N})$$

are measurable. Observe that $\operatorname{dom}(\nu) = \sigma(\mathcal{A})$. Consequently, for any real $\varepsilon > 0$, there exists a set $A = A_\varepsilon$ belonging to $\mathcal{A}$ such that $\nu(Z \triangle A) < \varepsilon$. So

$$|\nu(Z) - \nu(A)| < \varepsilon, \qquad |\nu^2(Z) - \nu^2(A)| < 2\varepsilon.$$

Notice now that there is a $g \in G$ satisfying the relation

$$\nu(g(A) \cap A) = \nu(A) \cdot \nu(g(A)) = \nu^2(A).$$

Further, one may write

$$\nu((A \cap g(A)) \triangle (Z \cap g(Z))) \leq \nu(A \triangle Z) + \nu(g(A) \triangle g(Z)) < 2\varepsilon,$$

$$|\nu(A \cap g(A)) - \nu(Z \cap g(Z))| < 2\varepsilon.$$

Since the set Z is almost G-invariant with respect to ν, one has the equality

$$\nu(Z \cap g(Z)) = \nu(Z)$$

and so one obtains $|\nu^2(A) - \nu(Z)| < 2\varepsilon$. Finally,

$$|\nu^2(Z) - \nu(Z)| < 4\varepsilon.$$

Since ε can be arbitrarily small, $\nu^2(Z) = \nu(Z)$, which trivially implies the zero-one law.

Appendix C
Comparing the Riemann and Lebesgue Integrals

It is well known that, before Lebesgue introduced his integral, a standard tool in classical mathematical analysis was the Riemann integral and closely connected with it Jordan measure (more precisely, Jordan volume). In this appendix we recall the notion of Riemann integral and compare it with the Lebesgue integral (see, e.g., [44, 104, 134, 141, 152]). Actually, the Lebesgue integral is much more general than the Riemann integral, is useful in various delicate questions of modern real analysis, and its range of applications is enormously wide. In addition, one of the main weak features of the Riemann integral is the following: even rather simple real-valued bounded discontinuous functions on [0, 1] are not integrable in the Riemann sense. For instance, there are many real-valued bounded semicontinuous functions on [0,1] for which the process of Riemann integration does not work (see, e.g., Exercise 2 of this appendix).

Without essential loss of generality, we restrict our further considerations to the one-dimensional Riemann integral of real-valued bounded functions defined on a non-degenerate segment $[a, b]$ of $\mathbb{R}$. Notice that the argument presented below is applicable to the multi-dimensional Riemann integral.

Let $f : [a, b] \to \mathbb{R}$ be a bounded function, i.e.,

$$\sup\{|f(x)| : x \in [a, b]\} \leq \tau,$$

where τ is some strictly positive real constant.

Let $\mathcal{D} = \{\triangle_i : i = 1, 2, \ldots, n\}$ be an arbitrary finite decomposition of $[a, b]$ into subsegments. This phrase means that $\mathcal{D}$ is a covering of $[a, b]$ and the interiors of the subsegments $\triangle_i$ ($i = 1, 2, \ldots, n$) are pairwise disjoint.

One can associate with f and $\mathcal{D}$ the following two real numbers:

$$s_0(f, \mathcal{D}) = \sum \{\inf(f|\triangle_i) l(\triangle_i) : i = 1, 2, \ldots, n\},$$

$$s_1(f, \mathcal{D}) = \sum \{\sup(f|\triangle_i) l(\triangle_i) : i = 1, 2, \ldots, n\},$$

where $l(\Delta_i)$ denotes the length of Δ_i (clearly, $l(\Delta_i) = \lambda(\Delta_i)$, where λ stands for the ordinary Lebesgue measure on $\mathbb{R}$).

The value $s_0(f, \mathcal{D})$ is called the *lower Darboux number* (*lower Darboux sum*) associated with f and $\mathcal{D}$.

The value $s_1(f, \mathcal{D})$ is called the *upper Darboux number* (*upper Darboux sum*) associated with f and $\mathcal{D}$.

Obviously, one has the inequality $s_0(f, \mathcal{D}) \leq s_1(f, \mathcal{D})$.

According to the standard terminology, a decomposition $\mathcal{D}$ of $[a, b]$ is inscribed in a decomposition $\mathcal{D}'$ of $[a, b]$ if each member of $\mathcal{D}$ is contained in some member of $\mathcal{D}'$.

Observe that if $\mathcal{D}'$ and $\mathcal{D}''$ are any two decompositions of $[a, b]$, then they canonically produce the decomposition $\mathcal{D}$ of $[a, b]$ which is inscribed in both of them. For this $\mathcal{D}$, a straightforward verification gives

$$s_0(f, \mathcal{D}) \geq \max(s_0(f, \mathcal{D}'), s_0(f, \mathcal{D}'')),$$
$$s_1(f, \mathcal{D}) \leq \min(s_1(f, \mathcal{D}'), s_1(f, \mathcal{D}'')).$$

Therefore, we get $s_0(f, \mathcal{D}') \leq s_1(f, \mathcal{D}'')$.

Keeping this in mind, let us introduce the notation

$$s_0(f) = \sup\{s_0(f, \mathcal{D}) : \mathcal{D} \text{ is a decomposition of } [a, b]\},$$
$$s_1(f) = \inf\{s_1(f, \mathcal{D}) : \mathcal{D} \text{ is a decomposition of } [a, b]\}.$$

Then we have the inequality $s_0(f) \leq s_1(f)$.

The value $s_0(f)$ is called the *lower Riemann integral* of f and the value $s_1(f)$ is called the *upper Riemann integral* of f.

If $s_0(f) = s_1(f)$, then f is called *integrable in the Riemann sense* and, by definition, the value of the Riemann integral of f is equal to $s_0(f)$ ($= s_1(f)$).

Theorem C.1 *For a real-valued bounded function f on $[a, b]$, the following two assertions are equivalent:*

(1) *f is integrable in the Riemann sense;*
(2) *for any real $\varepsilon > 0$, there exists a decomposition $\mathcal{D}$ of $[a, b]$ such that the inequality $s_1(f, \mathcal{D}) - s_0(f, \mathcal{D}) < \varepsilon$ holds true.*

Proof Suppose (1) and take arbitrarily a real $\varepsilon > 0$. There exist two decompositions $\mathcal{D}'$ and $\mathcal{D}''$ of $[a, b]$ such that

$$s_1(f, \mathcal{D}') - s_0(f, \mathcal{D}'') < \varepsilon.$$

Let $\mathcal{D}$ be a decomposition of $[a, b]$ inscribed in both $\mathcal{D}'$ and $\mathcal{D}''$. Then, as stated earlier,

$$s_1(f, \mathcal{D}) \leq s_1(f, \mathcal{D}'), \qquad s_0(f, \mathcal{D}) \geq s_0(f, \mathcal{D}'').$$

The latter formulas immediately imply that $s_1(f, \mathcal{D}) - s_0(f, \mathcal{D}) < \varepsilon$, so (2) holds.

The converse implication (2) $\Rightarrow$ (1) is trivial, and Theorem C.1 has thus been proved. □

To reveal a connection between the Riemann integral and Lebesgue integral of a function f, we need some additional constructions and a few facts from Appendix A about real-valued Lebesgue measurable step-functions.

Let f be an arbitrary real-valued bounded function on $[a, b]$.

Consider again any decomposition $\mathcal{D} = \{\Delta_i : i = 1, 2, \ldots, n\}$ of $[a, b]$. For each index $i \in \{1, 2, \ldots, n\}$, let X_i be a Lebesgue measurable subset of the interior of Δ_i such that $\lambda(\Delta_i \setminus X_i) < \varepsilon/2n\tau$, where

$$\varepsilon > 0, \qquad \tau > 0, \qquad \sup\{|f(x)| : x \in [a, b]\} \leq \tau.$$

Further, put

$$Y_i = \Delta_i \setminus X_i, \qquad Y = \cup\{Y_i : i \in \{1, 2, \ldots, n\}\}.$$

Obviously, we have

$$Y = [a, b] \setminus \cup\{X_i : i \in \{1, 2, \ldots, n\}\},$$

$$\lambda(Y) = \sum\{\lambda(Y_i) : i \in \{1, 2, \ldots, n\}\} < \varepsilon/2\tau.$$

Define a λ-measurable step-function $\phi : [a, b] \to \mathbb{R}$ as follows:

$$\phi(x) = \begin{cases} \sup(f|X_i) & \text{if } x \in X_i; \\ \sup(f|Y) & \text{if } x \in Y. \end{cases}$$

This definition of ϕ implies at once that $f \leq \phi$. Moreover, for the Lebesgue integral $I(\phi)$, we may write

$$I(\phi) = \sum\{\sup(f|X_i)\lambda(X_i) : i = 1, 2, \ldots, n\} + \sup(f|Y)\lambda(Y)$$

$$= \sum\{\sup(f|X_i)\lambda(\Delta_i) : i = 1, 2, \ldots, n\}$$

$$\quad - \sum\{\sup(f|X_i)\lambda(Y_i) : i = 1, 2, \ldots, n\}$$

$$\quad + \sup(f|Y)\lambda(Y)$$

$$\leq \sum\{\sup(f|\Delta_i)\lambda(\Delta_i) : i = 1, 2, \ldots, n\} + \varepsilon.$$

Therefore, we get $I(\phi) \leq s_1(f, \mathcal{D}) + \varepsilon$.

Analogously, we define a λ-measurable step-function $\psi : [a, b] \to \mathbb{R}$ as follows:

$$\psi(x) = \begin{cases} \inf(f|X_i) & \text{if } x \in X_i; \\ \inf(f|Y) & \text{if } x \in Y. \end{cases}$$

The definition of ψ implies at once that $f \geq \psi$. Moreover, using an argument similar to the previous one, it is not hard to check that the Lebesgue integral $I(\psi)$ satisfies the inequality $I(\psi) \geq s_0(f, \mathcal{D}) - \varepsilon$.

Combining the two obtained inequalities, we infer that

$$I(\phi) - I(\psi) \leq s_1(f, \mathcal{D}) - s_0(f, \mathcal{D}) + 2\varepsilon.$$

Now, we can formulate and prove an important theorem concerning the close relationship between the Riemann and Lebesgue integrals.

Theorem C.2 *If a function $f : [a, b] \to \mathbb{R}$ is integrable in the Riemann sense, then f is also integrable in the Lebesgue sense, and the Riemann integral of f coincides with the Lebesgue integral of f.*

Consequently, every Riemann integrable function is Lebesgue measurable.

Proof Suppose that $f : [a, b] \to \mathbb{R}$ is an arbitrary Riemann integrable function. In view of Theorem C.1, this means that, for any real $\varepsilon > 0$, there exists a decomposition $\mathcal{D}$ of $[a, b]$ satisfying

$$s_1(f, \mathcal{D}) - s_0(f, \mathcal{D}) < \varepsilon.$$

As has been established above, for this $\mathcal{D}$ we can find two Lebesgue measurable step-functions ϕ and ψ such that

$$\psi \leq f \leq \phi,$$

$$s_0(f, \mathcal{D}) - \varepsilon \leq I(\psi) \leq I(\phi) \leq s_1(f, \mathcal{D}) + \varepsilon.$$

Therefore,

$$I(\phi) - I(\psi) \leq s_1(f, \mathcal{D}) - s_0(f, \mathcal{D}) + 2\varepsilon,$$

$$I(\phi) - I(\psi) < 3\varepsilon,$$

which shows us that f is integrable in the Lebesgue sense (see Appendix A).

It is also clear from this argument that the Riemann integral of f is identical to the Lebesgue integral $I(f)$ of f.

Finally, as we know (see again Appendix A), every real-valued Lebesgue integrable function is Lebesgue measurable. In particular, our function f is Lebesgue measurable.

This completes the proof of Theorem C.2. □

Let $f : [a, b] \to \mathbb{R}$ be a bounded function and let x be a point of $[a, b]$.

C Comparing the Riemann and Lebesgue Integrals

Recall that the symbol $O_f(x)$ denotes the oscillation of f at x and is defined as follows:

$$O_f(x) = \limsup_{y \to x} f(y) - \liminf_{y \to x} f(y).$$

Equivalently, $O_f(x)$ can be defined as

$$O_f(x) = \inf\{\operatorname{diam}(f(U)) : U \in \mathcal{U}(x)\},$$

where $\mathcal{U}(x)$ stands for the family of all neighborhoods of x and $\operatorname{diam}(f(U))$ stands for the diameter of $f(U)$ (see Chap. 5).

Lemma C.1 *Let $f : [a, b] \to \mathbb{R}$ be a bounded function and let X be a closed subset of $[a, b]$. Suppose that, for a given real $\varepsilon > 0$, the inequality $O_f(x) < \varepsilon$ holds true whenever $x \in X$.*

Then there exists a finite family $\{\Delta_i : i = 1, 2, \ldots, n\}$ of subsegments of $[a, b]$ such that:

(1) *the interiors of the segments Δ_i ($i = 1, 2, \ldots, n$) are pairwise disjoint;*
(2) *the set X is contained in the interior of $\cup \{\Delta_i : i = 1, 2, \ldots, n\}$;*
(3) $\sup(f|\Delta_i) - \inf(f|\Delta_i) < \varepsilon$ *for each $i \in \{1, 2, \ldots, n\}$.*

Proof By the assumption, X is a closed set in $[a, b]$. Therefore, X is compact. If x is an arbitrary point of X, then there exists an open interval $U(x) \subset [a, b]$ containing x and satisfying

$$\operatorname{diam}(f(\operatorname{cl}(U(x)))) < \varepsilon.$$

So we come to the open covering $\{U(x) : x \in X\}$ of X. In view of the compactness of X, there exists a finite subcovering

$$\{U(x_j) : j = 1, 2, \ldots, m\}$$

of X. Now, it is easy to see how to obtain from $\{\operatorname{cl}(U(x_j)) : j = 1, 2, \ldots, m\}$ the required family $\{\Delta_i : i = 1, 2, \ldots, n\}$ of subsegments of $[a, b]$. □

Lemma C.2 *Let $f : [a, b] \to \mathbb{R}$ be a bounded function and let Y be a subset of $[a, b]$. Suppose that:*

(1) $\lambda^*(Y) > 0$, *where λ^* is the outer measure produced by λ;*
(2) f *is discontinuous at each point of Y.*

Then f is not integrable in the Riemann sense.

Proof By virtue of Theorem C.1, it suffices to show that there exists a real $r > 0$ such that

$$s_1(f, \mathcal{D}) - s_0(f, \mathcal{D}) > r$$

for all decompositions $\mathcal{D}$ of $[a,b]$. Let us define

$$Y_m = \{y \in Y : O_f(y) > 1/m\} \quad (m = 1, 2, \ldots).$$

Clearly, we may write

$$Y = \cup\{Y_m : m = 1, 2, \ldots\}.$$

Since $\lambda^*(Y) > 0$, there exists some natural number $m > 0$ for which $\lambda^*(Y_m) > 0$. Denote $\lambda^*(Y_m)$ by δ.

Now, consider any finite decomposition $\mathcal{D} = \{\Delta_i : i = 1, 2, \ldots, n\}$ of the segment $[a, b]$.

We shall say that a segment Δ_i belonging to this decomposition is admissible if $\lambda^*(\Delta_i \cap Y_m) > 0$.

Let $\{\Delta_i : i \in J\}$, where $J \subset \{1, 2, \ldots, n\}$, stand for the family of all admissible segments. Observe that

$$\lambda(\cup\{\Delta_i : i \in J\}) \geq \lambda^*(Y_m) = \delta.$$

Moreover, if $i \in J$, then there are points of Y_m which belong to the interior of Δ_i, so

$$\sup(f|\Delta_i) - \inf(f|\Delta_i) > 1/m.$$

Consequently,

$$\sum\{(\sup(f|\Delta_i) - \inf(f|\Delta_i))\lambda(\Delta_i) : i \in J\} > \delta/m.$$

The latter formula trivially implies that

$$s_1(f, \mathcal{D}) - s_0(f, \mathcal{D}) > \delta/m,$$

and we can take $r = \delta/m$. This ends the proof of Lemma C.2. $\square$

The following classical result was obtained by Lebesgue. It provides a very nice characterization of all real-valued Riemann integrable functions on $[a, b]$.

Theorem C.3 *A bounded function $f : [a, b] \to \mathbb{R}$ is integrable in the Riemann sense if and only if the set of all discontinuity points of f is of λ-measure zero.*

Proof Suppose first that f is Riemann integrable and, as usual, denote by $D(f)$ the set of all discontinuity points of f (see Chap. 6). By virtue of Lemma C.2, we must have

$$\lambda^*(D(f)) = 0$$

C Comparing the Riemann and Lebesgue Integrals

which is equivalent to $\lambda(D(f)) = 0$.

Conversely, suppose that f is bounded and $\lambda(D(f)) = 0$. Choose a real constant $\tau > 0$ satisfying

$$\sup\{|f(x)| : x \in [a, b]\} \leq \tau.$$

Further, take arbitrarily a real number $\varepsilon > 0$ and represent the set $D(f)$ in the form

$$D(f) = \cup\{Y_m : m = 1, 2, \ldots\},$$

where

$$Y_m = \{x \in [a, b] : O_f(x) \geq 1/m\}.$$

For a sufficiently large natural number m, we have $1/m < \varepsilon$ and, clearly, $\lambda(Y_m) = 0$. Denote by U some open subset of $[a, b]$ such that

$$Y_m \subset U, \qquad \lambda(U) < \varepsilon,$$

and put $X = [a, b] \setminus U$. Using Lemma C.1, we infer that there exists a finite family $\{\Delta_i : i = 1, 2, \ldots, n\}$ of subsegments of $[a, b]$ such that:

(1) the interiors of the segments Δ_i ($i = 1, 2, \ldots, n$) are pairwise disjoint;
(2) the set X is contained in the interior of $\cup \{\Delta_i : i = 1, 2, \ldots, n\}$;
(3) $\sup(f|\Delta_i) - \inf(f|\Delta_i) < \varepsilon$ for each $i \in \{1, 2, \ldots, n\}$.

It is not hard to see that the family $\{\Delta_i : i = 1, 2, \ldots, n\}$ can be extended to a finite decomposition $\mathcal{D}$ of $[a, b]$. Then a straightforward calculation gives us that, for this $\mathcal{D}$, we have

$$s_1(f, \mathcal{D}) - s_0(f, \mathcal{D}) \leq \varepsilon((b - a) + 2\tau).$$

According to Theorem C.1, the function f is integrable in the Riemann sense.

The proof of Theorem C.3 is thus complete. □

Remark C.1 It directly follows from Theorem C.3 that any real-valued continuous function on $[a, b]$ is integrable in the Riemann sense. Also, one of the consequences of Theorem C.3 is the fact that the family of all Riemann integrable functions on $[a, b]$ is an algebra of functions on $[a, b]$ and, simultaneously, is a lattice of functions on $[a, b]$. Actually, the Riemann integral is a positive linear functional on this family and is a restriction of the Lebesgue integral considered as a positive linear functional on the family of all Lebesgue integrable functions on $[a, b]$.

Remark C.2 It is easy to show that if a function $f : [a, b] \to \mathbb{R}$ (not necessarily bounded) is such that $\lambda(D(f)) = 0$, then f is Lebesgue measurable. Therefore, Theorem C.3 yields once again that any Riemann integrable function on $[a, b]$ is Lebesgue measurable.

Exercises

1. Let C denote the classical Cantor set on the unit segment $[0, 1]$ and let Z be an arbitrary subset of C.
   Verify that the characteristic function χ_Z is Riemann integrable and its Riemann integral equals zero.
   Conclude from this result that there are $2^{\mathbf{c}}$ many real-valued functions on $[0, 1]$ which are integrable in the Riemann sense (as usual, $\mathbf{c}$ stands for the cardinality of the continuum).

2. Show that on the segment $[0, 1]$ there exists a nowhere dense closed set F of strictly positive Lebesgue measure. Moreover, for any ε from the open interval $]0, 1[$, a nowhere dense closed subset F of $[0, 1]$ can be found whose λ-measure is equal to ε.
   Using Theorem C.3, infer from this fact that the upper semicontinuous characteristic function χ_F of F is not integrable in the Riemann sense.

   Remark C.3 Exercise 2 indicates that, even among real-valued bounded semicontinuous functions on $[0, 1]$, there are Riemann non-integrable ones. Obviously, this is a very weak feature of the Riemann integral, because semicontinuous functions are important in many topics of mathematical analysis (cf. Chaps. 9 and 10).

3. Prove that there exists a partition $\{X, Y\}$ of $[0, 1]$ such that:

   (a) both sets X and Y are Borel (hence Lebesgue measurable);
   (b) for every non-degenerate subsegment Δ of $[0, 1]$, the sets $X \cap \Delta$ and $Y \cap \Delta$ are of strictly positive Lebesgue measure.

   Argue as follows. Denote by $\{\Delta_n : n \in \mathbb{N}\}$ the sequence of all non-degenerate subsegments of $[0, 1]$ with rational endpoints and construct by recursion two countable families

   $$\{X_n : n \in \mathbb{N}\}, \qquad \{Y_n : n \in \mathbb{N}\}$$

   of nowhere dense closed sets in $[0, 1]$ satisfying the following two conditions:

   (i) $X_n \cap Y_m = \emptyset$ for any $n \in \mathbb{N}$ and $m \in \mathbb{N}$;
   (ii) $\lambda(X_n \cap \Delta_n) > 0$ and $\lambda(Y_n \cap \Delta_n) > 0$ for each $n \in \mathbb{N}$.

   Then put

   $$X = \cup\{X_n : n \in \mathbb{N}\}, \qquad Y = [0, 1] \setminus X.$$

   Finally, verify that the characteristic function χ_X of X does not belong to the first Baire class, i.e., there exists no sequence $\{f_n : n \in \mathbb{N}\}$ of real-valued continuous functions on $[0, 1]$ such that $\chi_X = \lim_{n \to \infty} f_n$.

For this purpose, take into account the fact that χ_X is everywhere discontinuous on [0, 1] (cf. Exercise 21 from Chap. 30).

4. Generalize the previous exercise and demonstrate that there exists a partition $\{Z_n : n \in \mathbb{N}\}$ of [0, 1] such that:

   (a) all Z_n ($n = 0, 1, 2, \ldots$) are Borel subsets of [0, 1];
   (b) for every non-degenerate subsegment Δ of [0, 1] and for any $n \in \mathbb{N}$, the set $\Delta \cap Z_n$ is of strictly positive Lebesgue measure.

5. Prove the assertions formulated in Remark C.1.

6. Show that if a function $f : [a, b] \to \mathbb{R}$ (not necessarily bounded) is such that $\lambda(D(f)) = 0$, then f is Lebesgue measurable.

   In addition, demonstrate that, for every strictly positive real ε not exceeding 1, there exists a non-Lebesgue measurable function

   $$g : [0, 1] \to [0, 1]$$

   such that $\lambda(D(g)) = \varepsilon$.

7. Give an example of an increasing sequence $\{f_n : n \in \mathbb{N}\}$ of real-valued functions such that:

   (a) every f_n ($n \in \mathbb{N}$) is a continuous function from [0, 1] into [0, 1];
   (b) the function $\lim_{n \to \infty} f_n$ is not integrable in the Riemann sense.

   For this purpose, keep in mind Remark C.3.

8. Let a sequence $\{g_n : n \in \mathbb{N}\}$ of Riemann integrable functions on a segment $[a, b]$ be uniformly convergent to a function $g : [a, b] \to \mathbb{R}$.
   Prove that g is also integrable in the Riemann sense.
   For this purpose, take into account the inclusion

   $$D(g) \subset \cup \{D(g_n) : n \in \mathbb{N}\}$$

   and then use Theorem C.3.

9. Verify that any real-valued bounded monotone function on a segment $[a, b]$ is integrable in the Riemann sense.
   Moreover, any real-valued bounded function f on $[a, b]$ satisfying the inequality $\operatorname{card}(D(f)) \leq \operatorname{card}(\mathbb{N})$ is Riemann integrable on $[a, b]$ (although $D(f)$ can be everywhere dense in $[a, b]$).

10. Let $f : [a, b] \to \mathbb{R}$ be a function and let X be a closed subset of $[a, b]$. Suppose that, for some real $\varepsilon > 0$, the inequality $O_f(x) < \varepsilon$ holds true whenever $x \in X$. Show that there exists a real $\delta > 0$ having the following property:
    For any segment $\Delta \subset [a, b]$ which intersects X and whose length is less than δ, the relation $\sup(f|\Delta) - \inf(f|\Delta) \leq \varepsilon$ is fulfilled.

11. Let $f : [a, b] \to \mathbb{R}$ be a function. Take an arbitrary decomposition $\mathcal{D}$ of the segment $[a, b]$ into finitely many subsegments, i.e., put

    $$\mathcal{D} = \{\Delta_i : i = 1, 2, \ldots, n\}.$$

Choose any points $x_i \in \Delta_i$, where $i = 1, 2, \ldots, n$, and consider the value

$$r(f, \mathcal{D}, x_1, x_2, \ldots, x_n) = \sum \{f(x_i)\lambda(\Delta_i) : i = 1, 2, \ldots, n\}.$$

Also, define

$$l(\mathcal{D}) = \max(\lambda(\Delta_1), \lambda(\Delta_2), \ldots, \lambda(\Delta_n)).$$

The traditional definition of the Riemann integral of f is as follows: The value $r \in \mathbb{R}$ is the Riemann integral of f if, for each real $\varepsilon > 0$, there exists a real $\delta > 0$ such that

$$|r - r(f, \mathcal{D}, x_1, x_2, \ldots, x_n)| < \varepsilon$$

whenever $l(\mathcal{D}) < \delta$ and $x_1 \in \Delta_1, x_2 \in \Delta_2, \ldots, x_n \in \Delta_n$.

Prove that the traditional definition of the Riemann integral is equivalent to the definition presented in this appendix.

Argue step by step. First, suppose that f is Riemann integrable in the sense of the traditional definition and denote by r the Riemann integral of f. Take arbitrarily a real $\varepsilon > 0$. There exists a real $\delta > 0$ such that

$$|r - r(f, \mathcal{D}, x_1, x_2, \ldots, x_n)| < \varepsilon$$

for every decomposition $\mathcal{D}$ of $[a, b]$ with $l(\mathcal{D}) < \delta$ and for any points $x_1 \in \Delta_1, x_2 \in \Delta_2, \ldots, x_n \in \Delta_n$. Infer that

$$|r(f, \mathcal{D}, x_1, x_2, \ldots, x_n) - r(f, \mathcal{D}, y_1, y_2, \ldots, y_n)| < 2\varepsilon$$

whenever $\{x_1, y_1\} \subset \Delta_1, \{x_2, y_2\} \subset \Delta_2, \ldots, \{x_n, y_n\} \subset \Delta_n$. Consequently,

$$s_1(f, \mathcal{D}) - s_0(f, \mathcal{D}) \le 2\varepsilon,$$

which shows that f is bounded and is Riemann integrable by the definition given in this appendix (see Theorem C.1).

Conversely, suppose that f is Riemann integrable according to the definition of this appendix, and let r be the value of the Riemann integral of f (in view of Theorem C.2, r is simultaneously the value of Lebesgue integral of f). Since f is bounded, one may write

$$\sup\{|f(x)| : x \in [a, b]\} \le \tau,$$

where $\tau > 0$ is some real constant. Further, take arbitrarily a real $\varepsilon > 0$. By Theorem C.3, the set $D(f)$ is of λ-measure zero. Define

C Comparing the Riemann and Lebesgue Integrals

$$Y = \{x \in [a,b] : O_f(x) \geq \varepsilon\}.$$

Since $Y \subset D(f)$, the equality $\lambda(Y) = 0$ holds. Consider an open subset U of $[a,b]$ such that

$$Y \subset U, \qquad \lambda(U) < \varepsilon,$$

and put $X = [a,b] \setminus U$. Observe that X is compact and

$$X \cap Y = \emptyset, \qquad (\forall x \in X)(O_f(x) < \varepsilon).$$

By virtue of Exercise 10, there exists a real $\delta > 0$ having the following property: For any segment $\Delta \subset [a,b]$ which intersects X and whose length is less than δ, the relation $\sup(f|\Delta) - \inf(f|\Delta) \leq \varepsilon$ holds true.
Let $\mathcal{D} = \{\Delta_i : i = 1, 2, \ldots, n\}$ be any decomposition of $[a,b]$ into subsegments with $l(\mathcal{D}) < \delta$. Check that

$$|r(f, \mathcal{D}, x_1, x_2, \ldots, x_n) - r(f, \mathcal{D}, y_1, y_2, \ldots, y_n)| \leq \varepsilon((b-a) + 2\tau)$$

whenever $\{x_1, y_1\} \subset \Delta_1, \{x_2, y_2\} \subset \Delta_2, \ldots, \{x_n, y_n\} \subset \Delta_n$.
Deduce from the latter formula that

$$s_1(f, \mathcal{D}) - s_0(f, \mathcal{D}) \leq \varepsilon((b-a) + 2\tau).$$

Keeping in mind the text preceding Theorem C.2, verify that

$$s_0(f, \mathcal{D}) - \varepsilon \leq r \leq s_1(f, \mathcal{D}) + \varepsilon.$$

Finally, take into account the trivial fact that

$$s_0(f, \mathcal{D}) \leq r(f, \mathcal{D}, x_1, x_2, \ldots, x_n) \leq s_1(f, \mathcal{D})$$

and conclude that

$$|r - r(f, \mathcal{D}, x_1, x_2, \ldots, x_n)| \leq \varepsilon + \varepsilon((b-a) + 2\tau),$$

which yields the Riemann integrability of f in the traditional sense.

12. Start with the traditional definition of the Riemann integral and check that no real-valued unbounded function on a segment $[a,b]$ can be integrable in the Riemann sense.

13. Show that:

    (a) if a function f is Riemann (Lebesgue) integrable, then the function $|f|$ is also Riemann (Lebesgue) integrable;

(b) there exists a non-Lebesgue measurable function g for which the function $|g|$ is Riemann integrable.

14. In the Euclidean space $\mathbb{R}^m$, where $m \geq 1$, consider a non-degenerate right parallelepiped

$$\Delta = [a_1, b_1] \times [a_2, b_2] \times \ldots \times [a_m, b_m]$$

and let $f : \Delta \to \mathbb{R}$ be a bounded function.

Similarly to the one-dimensional case $\mathbb{R}^1 = \mathbb{R}$, introduce the notion of m-dimensional Riemann integral of f on Δ and demonstrate that all results presented in this appendix have their natural analogs for the m-dimensional Riemann integral.

Appendix D
The Lax–Milgram Theorem

In this appendix we present the Lax–Milgram theorem concerning continuous bilinear elliptic forms on a real Hilbert space (see [116]). This important result is useful in many situations and, especially, is applicable to various problems in the general theory of partial differential equations. Actually, the Lax–Milgram theorem is a clever extension of the Riesz theorem on the representation of a continuous linear functional on a real Hilbert space (see Chap. 29). For this reason, we decided to include the Lax–Milgram theorem and its proof in our lecture course. However, we do not touch upon numerous applications of this celebrated result.

Let E and F be two real vector spaces and let $B : E \times F \to \mathbb{R}$ be a functional of two variables $x \in E$ and $y \in F$.

Recall that B is a *bilinear form* (or a *bilinear functional*) on $E \times F$ if the partial functionals

$$B(x, \cdot) : F \to \mathbb{R}, \qquad B(\cdot, y) : E \to \mathbb{R}$$

are linear for all $x \in E$ and for all $y \in F$.

It is clear that in this case one has

$$(\forall x \in E)(B(x, 0_F) = 0), \qquad (\forall y \in F)(B(0_E, y) = 0),$$

where 0_E and 0_F are the zero vectors in E and F, respectively.

If E and F are two topological vector spaces over $\mathbb{R}$, then it makes sense to consider continuous bilinear functionals on the product topological vector space $E \times F$.

The following theorem gives a useful characterization of continuous bilinear functionals on the topological product of two normed vector spaces.

Theorem D.1 *Let $(E, || \cdot ||)$ and $(F, || \cdot ||)$ be normed vector spaces over $\mathbb{R}$ and let $B : E \times F \to \mathbb{R}$ be a bilinear functional.*

Then the following three assertions are equivalent:

(1) B is continuous at the point $(0_E, 0_F)$;
(2) there exists a real constant $\alpha > 0$ such that $|B(x, y)| \leq \alpha \|x\| \|y\|$ for any vectors $x \in E$ and $y \in F$;
(3) B is continuous on $E \times F$.

Proof Let us first show that (1) $\Rightarrow$ (2).

Assume (1). Then there exists a real $\delta > 0$ for which $|B(x, y)| \leq 1$ whenever $\|x\| \leq \delta$ and $\|y\| \leq \delta$. Take arbitrarily two nonzero vectors $x \in E$ and $y \in F$. Clearly, there are two real constants $p > 0$ and $q > 0$ such that $\|px\| = \delta$ and $\|qy\| = \delta$. We may write

$$|B(px, qy)| \leq 1, \qquad |B(x, y)| \leq 1/pq = (1/\delta^2)(\|x\|\|y\|).$$

So, putting $\alpha = 1/\delta^2$, we get (2).

Now, let us demonstrate the validity of the implication (2) $\Rightarrow$ (3).

Suppose (2) and take arbitrarily a point $(x, y) \in E \times F$.

Let $\{(x_n, y_n) : n \in \mathbb{N}\}$ be any sequence of points in $E \times F$ converging to (x, y). Then

$$\lim_{n \to \infty} \|x_n - x\| = 0, \qquad \lim_{n \to \infty} \|y_n - y\| = 0.$$

We may write

$$|B(x, y) - B(x_n, y_n)| \leq |B(x, y) - B(x_n, y)| + |B(x_n, y) - B(x_n, y_n)|$$
$$\leq \alpha \|x - x_n\| \|y\| + \alpha \|x_n\| \|y - y_n\|,$$

whence (3) immediately follows (in view of the boundedness of $\{x_n : n \in \mathbb{N}\}$ in the space $(E, \|\cdot\|)$).

The implication (3) $\Rightarrow$ (1) is trivial.

Theorem D.1 has thus been proved. $\square$

Theorem D.2 *Let $(E, \|\cdot\|)$ be a Banach space over $\mathbb{R}$ and let $T : E \to E$ be a continuous linear operator. Suppose that there exists a real constant $\gamma > 0$ such that $\|T(x)\| \geq \gamma \|x\|$ for every $x \in E$.*

Then the range of T is a closed vector subspace of E.

Proof As usual, denote by $\mathrm{ran}(T)$ the range of T and consider in E any adherent point y of $\mathrm{ran}(T)$. We have to show that $y \in \mathrm{ran}(T)$.

Evidently, there exists a sequence $\{x_n : n \in \mathbb{N}\}$ of points in E such that the sequence $\{T(x_n) : n \in \mathbb{N}\}$ converges to y. In particular, we see that $\{T(x_n) : n \in \mathbb{N}\}$ is a Cauchy sequence in E. Further, the inequality

$$\|T(x_m) - T(x_n)\| \geq \gamma \|x_m - x_n\|,$$

D The Lax–Milgram Theorem

where m and n are any natural numbers, implies at once that $\{x_n : n \in \mathbb{N}\}$ is also a Cauchy sequence in E. Since E is complete, this sequence converges to some point $x \in E$. By virtue of the continuity of T we finally obtain

$$y = \lim_{n \to \infty} T(x_n) = T(x).$$

So $y \in \operatorname{ran}(T)$, and the last relation ends the proof. □

Let $(E, \|\cdot\|)$ be a normed vector space over $\mathbb{R}$ and let $B : E \times E \to \mathbb{R}$ be a bilinear functional.

This B is called an *elliptic bilinear functional* if there exists a real constant $\beta > 0$ such that $|B(x,x)| \geq \beta \|x\|^2$ for all $x \in E$.

Remark D.1 Let $(H, \langle \cdot, \cdot \rangle)$ be a pre-Hilbert space over $\mathbb{R}$ equipped with the norm $\|\cdot\|$ induced by the inner product in H, i.e.,

$$\|x\| = \langle x, x \rangle^{1/2} \qquad (x \in H).$$

It is clear that $\langle \cdot, \cdot \rangle$ is an elliptic bilinear functional for the above-mentioned canonical norm in H (here we obviously have $\beta = 1$).

Now, we are able to formulate and prove the important result of Lax and Milgram.

Theorem D.3 *Let $(H, \langle \cdot, \cdot \rangle)$ be a real Hilbert space and let*

$$B : H \times H \to \mathbb{R}$$

be a continuous bilinear elliptic functional.

Then, for every continuous linear functional f on H, there exists a unique vector $w \in H$ such that

$$(\forall x \in H)(B(x, w) = f(x))$$

or, equivalently, $B(\cdot, w) = f$.

Proof To begin the argument, consider an arbitrary element $u \in H$. Obviously, the partial mapping

$$x \to B(x, u) \qquad (x \in H)$$

is a continuous linear functional on H. According to the Riesz representation theorem (see Chap. 29), there is a unique element $v \in H$ such that

$$(\forall x \in H)(B(x, u) = \langle x, v \rangle).$$

Let us put $T(u) = v$. In this manner, we define the operator $T : H \to H$.

It is not difficult to check that this operator is linear (we leave the details to the reader).

Since B is a continuous bilinear form, we may write

$$|\langle x, T(u)\rangle| = |B(x, u)| \leq \alpha \|x\| \|u\|$$

for all $x \in H$ (where $\alpha > 0$ is a real constant associated with B by virtue of (2) of Theorem D.1). Taking $x = T(u)$, we get

$$\|T(u)\|^2 \leq \alpha \|T(u)\| \|u\|, \qquad \|T(u)\| \leq \alpha \|u\|,$$

which gives us the continuity of T. Further, since B is elliptic,

$$\beta \|u\|^2 \leq |B(u, u)| = |\langle u, T(u)\rangle| \leq \alpha \|u\| \|T(u)\|$$

for all $u \in H$ (where $\beta > 0$ is a second real constant associated with the bilinear elliptic form B). Consequently,

$$\beta \|u\| \leq \alpha \|T(u)\| \qquad (u \in H).$$

Now, if $T(x) = T(y)$, then $T(x - y) = 0_H$ and, in view of

$$\beta \|x - y\| \leq \alpha \|T(x - y)\| = 0,$$

we infer $\|x - y\| = 0$, which means that T is an injective operator.

Let us demonstrate that T is a surjective operator as well. For this purpose, suppose otherwise, i.e., $\operatorname{ran}(T) \neq H$. According to Theorem D.2, $\operatorname{ran}(T)$ is a closed vector subspace of H and differs from H. So there exists a nonzero vector $z \in H$ orthogonal to $\operatorname{ran}(T)$. In other words, we have $\langle z, T(u)\rangle = 0$ for all $u \in H$. Taking $u = z$, we come to the relation

$$0 = |\langle z, T(z)\rangle| = |B(z, z)| \geq \beta \|z\|^2,$$

which implies $z = 0_H$, contradicting the definition of z. Therefore, T is a bijective continuous linear operator from H onto H. The inverse linear operator

$$T^{-1} : H \to H$$

is continuous, too, in view of the inequality

$$\|T^{-1}(x)\| \leq (\alpha/\beta) \|x\| \qquad (x \in H).$$

Remembering the definition of T, we can write

$$B(x, T^{-1}(v)) = \langle x, v\rangle \qquad (x \in H, \ v \in H).$$

D The Lax–Milgram Theorem

Now, consider an arbitrary continuous linear functional f on H. By the Riesz representation theorem, there exists a unique vector $v \in H$ such that $f(x) = \langle x, v \rangle$ for all $x \in H$. Putting $w = T^{-1}(v)$, we obtain $B(\cdot, w) = f$.

It only remains to check that w is unique. Indeed, let $w' \in H$ have the same property, i.e., $B(\cdot, w') = f$. Then

$$0 = |f(x) - f(x)| = |B(x, w) - B(x, w')| = |B(x, w - w')|$$

for all $x \in H$. Taking $x = w - w'$ and keeping in mind the relation

$$0 = |B(w - w', w - w')| \geq \beta \|w - w'\|^2,$$

we conclude that $w = w'$.

This completes the proof of the Lax–Milgram theorem. □

Remark D.2 As mentioned in Remark D.1, the inner product $\langle \cdot, \cdot \rangle$ on $H \times H$ is a special case of a continuous bilinear elliptic functional. It is easy to see that if B coincides with $\langle \cdot, \cdot \rangle$, then the associated T is the identity operator on H, and in this case we once again come to the Riesz representation theorem.

Remark D.3 In the proof of Theorem D.3, the continuity of the inverse operator T^{-1} follows also from a much deeper theorem of Banach which states that if g is any bijective continuous linear operator acting from a Banach space $(E, \|\cdot\|)$ onto a Banach space $(F, \|\cdot\|)$, then the inverse linear operator g^{-1} is also continuous (see, e.g., [38, 104] or Exercise 9 of this appendix). It should be noticed that, for separable Banach spaces (and, more generally, for Polish topological groups), the above-mentioned Banach theorem is a direct consequence of some profound facts of classical descriptive set theory (see Exercise 6 below).

Exercises

1. Let $(H, \langle \cdot, \cdot \rangle)$ be a real Hilbert space, Z be a nonempty closed convex subset of H, and let $x \in H \setminus Z$.
   Show that, for every point $z \in Z$, the inequality $\langle z - x, \mathrm{pr}_Z(x) - x \rangle > 0$ holds true, which means that the angle between the vectors $z - x$ and $\mathrm{pr}_Z(x) - x$ is always acute.

2. Let $(H, \langle \cdot, \cdot \rangle)$ be a real pre-Hilbert space, let Z be a subset of H, and let $\{Z_1, Z_2\}$ be a partition of Z. Consider any point x belonging to the convex hull of Z (i.e., $x \in \mathrm{conv}(Z)$ which means that x is representable as a convex combination of some finite family of points from Z).
   Prove that there exist two points $z_1 \in Z_1$ and $z_2 \in Z_2$ such that

   $$\langle z_1 - x, z_2 - x \rangle \leq 0.$$

Compare this result with Exercise 1.

3. Let $(H, \langle \cdot, \cdot \rangle)$ be a real pre-Hilbert space, let Z be a subset of H, and let x be a point in H belonging to $\mathrm{conv}(Z) \setminus Z$. Suppose also that, for any two points $z \in Z$ and $z' \in Z$, among the three non-negative values

$$||z - z'||, \qquad ||x - z||, \qquad ||x - z'||$$

at least two are equal to each other.
Demonstrate that the point x is equidistant from all points of Z.
For this purpose, use the result of Exercise 2.

4. Let $(G, \cdot)$ be a Baire topological group, $(H, \cdot)$ be a topological group with a countable base, and let $f : G \to H$ be a group homomorphism (in the algebraic sense).
Supposing that this f has the Baire property (see Chap. 30), show that f is a continuous mapping.

5. Let $(G, \cdot)$ and $(H, \cdot)$ be two Polish groups and let $f : G \to H$ be a group isomorphism (in the algebraic sense).
Supposing that f has the Baire property, prove that the inverse isomorphism f^{-1} also has the Baire property.
For this purpose, take into account that f is a continuous mapping (in view of Exercise 4) and then apply the theorem of classical descriptive set theory, according to which the f-image of any Borel subset of G is an analytic subset of H (see Exercise 12 from Chap. 33). Finally, keep in mind the fact that all analytic sets in H have the Baire property (see, e.g., [82, 109]).

6. Let $(G, \cdot)$ and $(H, \cdot)$ be two Polish groups and let $f : G \to H$ be a group isomorphism (in the algebraic sense) having the Baire property.
Show that the inverse isomorphism f^{-1} is a continuous mapping, so f turns out to be an isomorphism of the topological group G onto the topological group H.
For this purpose, use the results of Exercises 4 and 5.

7. Let a sequence $(a_0, a_1, \ldots, a_n, \ldots) \in \mathbb{R}^\mathbb{N}$ satisfy the following condition:
For every sequence $x = (x_0, x_1, \ldots, x_n, \ldots)$ from the real Hilbert space ℓ_2, the partial sums

$$x_0 a_0 + x_1 a_1 + \cdots + x_n a_n \qquad (n \in \mathbb{N})$$

are uniformly bounded (but the common bound for these sums depends on x).
Demonstrate that $(a_0, a_1, \ldots, a_n, \ldots) \in \ell_2$.
For this purpose, suppose otherwise, i.e., $\sum \{a_n^2 : n \in \mathbb{N}\} = +\infty$.
Let $n(0) = 0$ and define recursively a strictly increasing sequence of nonzero natural numbers $n(1), n(2), \ldots, n(k), \ldots$ such that

$$a_{n(k)}^2 + a_{n(k)+1}^2 + \cdots + a_{n(k+1)-1}^2 \geq 1$$

for any $k \in \mathbb{N}$. Then define

$$b_k = (a_{n(k)}^2 + a_{n(k)+1}^2 + \cdots + a_{n(k+1)-1}^2)^{1/2} \qquad (k \in \mathbb{N})$$

and, for each natural index $i \in [n(k), n(k+1) - 1]$, put

$$y_i = (1/(k+1))(a_i/b_k).$$

Verify that $(y_0, y_1, \ldots, y_i, \ldots) \in \ell_2$, but

$$\lim_{n \to \infty} (y_0 a_0 + y_1 a_1 + \cdots + y_n a_n) = +\infty.$$

The obtained contradiction yields the required result.

8. Infer the Riesz theorem on linear continuous functionals defined on a real Hilbert space (see Chap. 29) from the Radon–Nikodym theorem (see Chap. 26). For this purpose, argue as follows. Let H be a real Hilbert space and let f be a linear continuous functional on H. According to Exercise 9 of Chap. 29, f is completely determined by its restriction to some closed separable vector subspace of H. In view of this, it suffices to consider only the case when H itself is separable. If H is simultaneously finite-dimensional, then the situation is more or less clear. So, one may assume that H is separable and infinite-dimensional. But then H is isomorphic to the standard space $L_2([-\pi, \pi[)$, where $[-\pi, \pi[$ is equipped with the Lebesgue measure λ. So, without loss of generality, H can be identified with $L_2([-\pi, \pi[)$. Thus,

$$|f(\phi)| \leq d\|\phi\| \qquad (\phi \in L_2([-\pi, \pi[)),$$

where $d = d_f$ is some real non-negative constant and $\|\phi\|$ denotes, as usual, the L_2-norm of ϕ.

Further, for each λ-measurable subset X of $[-\pi, \pi[$, put

$$\nu(X) = f(\chi_X)$$

and verify that ν is a signed measure absolutely continuous with respect to λ. Therefore, by virtue of the Radon–Nikodym theorem, there exists a λ-integrable function ψ on $[-\pi, \pi[$ such that

$$\nu(X) = \int_X \psi(x) \, d\lambda(x) \qquad (X \in \mathrm{dom}(\lambda)).$$

Starting with this formula, show that, for any step-function $s \in L_2([-\pi, \pi[)$, one has

$$f(s) = \int_{-\pi}^{\pi} s(x) \psi(x) \, d\lambda(x).$$

Then deduce from the above equality that

$$f(\phi) = \int_{-\pi}^{\pi} \phi(x)\psi(x) \, d\lambda(x)$$

for every bounded function $\phi \in L_2([-\pi, \pi[)$.

Keep in mind the fact that the space $L_2([-\pi, \pi[)$ admits an orthonormal basis consisting of uniformly bounded functions (e.g., the ordinary trigonometric basis). Denote this basis by $\{\psi_n : n \in \mathbb{N}\}$ and introduce the real numbers

$$a_n = \int_{-\pi}^{\pi} \psi_n(x)\psi(x) \, d\lambda(x) \qquad (n \in \mathbb{N}).$$

Let now ϕ be an arbitrary function from $L_2([-\pi, \pi[)$ and let

$$s_n = x_0\psi_0 + x_1\psi_1 + \cdots + x_n\psi_n \qquad (n \in \mathbb{N}),$$

where

$$x_i = \int_{-\pi}^{\pi} \psi_i(x)\phi(x) \, d\lambda(x) \qquad (i = 0, 1, \ldots, n).$$

Observe that

$$\lim_{n \to \infty} \|\phi - s_n\| = 0,$$

$$\lim_{n \to \infty} f(s_n) = f(\phi),$$

$$f(s_n) = x_0 a_0 + x_1 a_1 + \cdots + x_n a_n.$$

Verify that the sums

$$x_0 a_0 + x_1 a_1 + \cdots + x_n a_n \qquad (n \in \mathbb{N})$$

are uniformly bounded by some real positive constant (depending on ϕ). Using Exercise 7, infer the relation $\psi \in L_2([-\pi, \pi[)$, which readily implies

$$f(\phi) = \sum \{x_n a_n : n \in \mathbb{N}\} = \int_{-\pi}^{\pi} \phi(x)\psi(x) \, d\lambda(x).$$

Obviously, this yields the Riesz theorem for the given linear continuous functional f.

Remark D.4 It is useful to compare the above exercise with Exercise 12 of Chap. 29.

9. Prove the classical result of Banach stating that if E and F are two real Banach spaces and $g : E \to F$ is a bijective continuous linear operator, then its inverse $g^{-1} : F \to E$ is also continuous.

For this purpose, argue step by step. First, observe that the equality

$$\cup \{g(B(0, m+1)) : m \in \mathbb{N}\} = F$$

holds, where $B(0, m+1)$ denotes the ball in E with center 0 and radius $m+1$. Since F is complete, there exists a natural number m such that the set $\text{cl}(g(B(0, m+1)))$ has nonempty interior. Keeping in mind the convexity and symmetry of $\text{cl}(g(B(0, m+1)))$, infer that some ball in F with center 0 is entirely contained in $\text{cl}(g(B(0, m+1)))$. Moreover, applying an appropriate homothety, one may suppose that there exists a real $\delta > 0$ such that the ball $B(0, \delta)$ in F is a subset of $\text{cl}(g(B(0, 1)))$, where $B(0, 1)$ is the unit ball in E. Equivalently, one can write

$$B(0, r) \subset \text{cl}(g(B(0, r/\delta)))$$

for any real $r > 0$, where $B(0, r)$ denotes the ball in F with center 0 and radius r and, accordingly, $B(0, r/\delta)$ denotes the ball in E with center 0 and radius r/δ.

Now, take arbitrarily a nonzero vector $y \in B(0, \delta)$ and consider a series $\sum \{\varepsilon_n : n \in \mathbb{N}\}$ of strictly positive real numbers such that

$$\varepsilon_0 = ||y||, \qquad (1/\delta) \sum \{\varepsilon_n : n \in \mathbb{N}\} \leq 2.$$

Further, construct by recursion a sequence $\{x_n : n \in \mathbb{N} \setminus \{0\}\} \subset E$ satisfying the following two conditions:

(a) $||y - g(x_1) - g(x_2) - \cdots - g(x_n)|| \leq \varepsilon_n$ for each $n = 1, 2, \ldots$;
(b) $||x_{n+1}|| \leq \varepsilon_n/\delta$ for each $n \in \mathbb{N}$.

In view of the inclusion $B(0, ||y||) \subset \text{cl}(g(B(0, \varepsilon_0/\delta)))$, there exists a vector $x_1 \in E$ such that

$$||x_1|| \leq \varepsilon_0/\delta, \qquad ||y - g(x_1)|| \leq \varepsilon_1.$$

Supposing that the vectors $x_1, \ldots, x_n$ in E have already been determined and taking into account the inclusion

$$B(0, \varepsilon_n) \subset \text{cl}(g(B(0, \varepsilon_n/\delta))),$$

one can find a vector $x_{n+1} \in E$ such that $||x_{n+1}|| \leq \varepsilon_n/\delta$ and

$$||y - g(x_1) - g(x_2) - \cdots - g(x_n) - g(x_{n+1})|| \leq \varepsilon_{n+1}.$$

Proceeding in this manner, one comes to the sequence $\{x_n : n = 1, 2, \ldots\} \subset E$ for which the conditions (a) and (b) are fulfilled. Deduce that this sequence also satisfies the following condition:

(c) the series $x_1 + x_2 + \cdots + x_n + \cdots$ converges to some vector $x \in E$ such that $||x|| \leq 2$ and $x = g^{-1}(y)$.

Conclude from (c) that the linear operator g^{-1} is bounded on the ball $B(0, \delta) \subset F$, so g^{-1} is continuous.

10. For a real normed vector space $(E, ||\cdot||)$, demonstrate that the following two assertions are equivalent:

(a) E is complete with respect to $||\cdot||$ (i.e., E is a Banach space);
(b) every series $\sum \{x_n : n \in \mathbb{N}\}$ of elements of E such that

$$\sum \{||x_n|| : n \in \mathbb{N}\} < +\infty$$

is convergent in E.

The implication (a) $\Rightarrow$ (b) is trivial. To show the validity of the converse implication, suppose (b) and take any Cauchy sequence $\{y_n : n \in \mathbb{N}\}$ in E. Obviously, there exists a strictly increasing sequence of natural numbers $n(0), n(1), \ldots, n(k), \ldots$ such that

$$||y_{n(k)} - y_{n(k+1)}|| \leq 1/2^k \qquad (k \in \mathbb{N}).$$

Consider the series $\sum \{x_k : k \in \mathbb{N}\}$ in E, where

$$x_k = y_{n(k)} - y_{n(k+1)} \qquad (k \in \mathbb{N}).$$

According to (b), this series converges to some vector $x \in E$ or, equivalently,

$$\lim_{k \to \infty} (y_{n(0)} - y_{n(k+1)}) = x,$$

from which it follows that $\{y_{n(k+1)} : k \in \mathbb{N}\}$ converges to $y_{n(0)} - x$. Since the given sequence $\{y_n : n \in \mathbb{N}\}$ is fundamental in E, it also converges to the same vector $y_{n(0)} - x$.

11. Let $(E, ||\cdot||)$ be a real normed vector space and let H be a closed vector subspace of E. Consider the quotient vector space E/H and, for each $X \in E/H$, put

$$||X|| = \inf\{||x|| : x \in X\}.$$

Check that the above formula defines a norm on E/H and the canonical surjection from E onto E/H is a continuous linear operator having the property that the images of all open subsets of E are open subsets of E/H.

D The Lax–Milgram Theorem

12. Let $(E, ||\cdot||)$ be a real Banach space and let H be a closed vector subspace of E.

    Show that quotient vector space E/H equipped with the norm described in Exercise 11 is also a Banach space.

    For this purpose, argue as follows. In view of Exercise 10, it suffices to verify that if a series $\sum\{X_n : n \in \mathbb{N}\}$ in E/H is such that

    $$\sum\{||X_n|| : n \in \mathbb{N}\} < +\infty,$$

    then this series converges in E/H. Keeping in mind the definition of the norm in E/H, one can choose for every $n \in \mathbb{N}$ an element $x_n \in X_n$ satisfying the relation $||x_n|| \leq 2||X_n||$. So one has

    $$\sum\{||x_n|| : n \in \mathbb{N}\} < +\infty.$$

    Since E is a Banach space, the series $\sum\{x_n : n \in \mathbb{N}\}$ converges to some $x \in E$. It remains to check that the series $\sum\{X_n : n \in \mathbb{N}\}$ converges to $x + H$ in E/H.

13. Prove another classical result of Banach stating that if E and F are two real Banach spaces and $g : E \to F$ is a surjective continuous linear operator, then g is an open mapping (i.e., the g-images of all open subsets of E are open in F).

    For this purpose, denote $g^{-1}(0)$ by H and express g in the form

    $$g = f \circ h,$$

    where h is the canonical surjection from E onto E/H and f is a bijective continuous linear operator from E/H onto F. Taking into account Exercises 11 and 12, reduce the above-mentioned result of Banach to its special case formulated in Exercise 9.

Appendix E
Resolvable Topological Spaces

The material of the present appendix is of a purely topological flavor and is not necessary for most topics which are discussed in this lecture course.

Here we want to show that the class of all resolvable topological spaces (see Chap. 6) includes the class of all nonempty metric spaces without isolated points and the class of all nonempty locally compact topological spaces without isolated points.

Our further argument relies on some properties of infinite cardinal numbers and on the method of transfinite induction.

We begin with several auxiliary notions and lemmas.

Let E be an arbitrary topological space.

A family $\mathcal{B}$ of nonempty open subsets of E is called a π-*base* (or a *pseudo-base*) of E if, for every nonempty open set $U \subset E$, there exists a set $V \in \mathcal{B}$ such that $V \subset U$.

The following lemma is an immediate consequence of the above definition.

Lemma E.1 *If $\mathcal{B}$ is a base of E, then $\mathcal{B} \setminus \{\emptyset\}$ is a π-base of E.*

In general, the converse assertion is not true (see, e.g., Exercise 1).

The next lemma slightly generalizes the result presented in Exercise 7 of Chap. 6.

Lemma E.2 *Suppose that every member of some π-base of a nonempty topological space E is a resolvable subspace of E.*

Then E itself is a resolvable space.

Proof Denote by $\mathcal{B}$ a π-base of E, all members of which are resolvable subspaces of E, and consider a subfamily $\mathcal{B}'$ of $\mathcal{B}$ having the following two properties:

(i) $\mathcal{B}'$ is disjoint;
(ii) $\mathcal{B}'$ is maximal by inclusion.

Notice that the existence of such $\mathcal{B}'$ is a simple consequence of the Kuratowski–Zorn lemma (which is one of numerous equivalents of the Axiom of Choice within **ZF** set theory).

Now, denoting $\mathcal{B}'$ by $\{V_j : j \in J\}$, consider in E the open set

$$V = \cup\{V_j : j \in J\}.$$

We assert that $\mathrm{cl}(V) = E$, i.e., V is everywhere dense in E. Suppose for a moment otherwise. In this case, the open set $E \setminus \mathrm{cl}(V)$ is not empty, so there exists a $W \in \mathcal{B}$ such that

$$W \subset E \setminus \mathrm{cl}(V).$$

Then the disjoint family $\mathcal{B}' \cup \{W\} \subset \mathcal{B}$ indicates that $\mathcal{B}'$ is not maximal by inclusion. The obtained contradiction establishes the equality $\mathrm{cl}(V) = E$.

Further, according to our assumption, all spaces V_j ($j \in J$) are resolvable. This means that, for each $j \in J$, there exists a partition $\{X_j, Y_j\}$ of V_j such that both sets X_j and Y_j are everywhere dense in V_j. Let us put

$$X = \cup\{X_j : j \in J\}, \qquad Y = \cup\{Y_j : j \in J\}.$$

Then, taking into account the disjointness of $\mathcal{B}' = \{V_j : j \in J\}$, we infer that $X \cap Y = \emptyset$, and it is clear that the sets X and Y are everywhere dense in V. Since $\mathrm{cl}(V) = E$, both X and Y are everywhere dense in E as well.

Lemma E.2 has thus been proved. □

A topological space E is called *isodyne* if, for each nonempty open set $U \subset E$, one has the equality $\mathrm{card}(U) = \mathrm{card}(E)$.

Accordingly, an open subset U of a topological space E is called isodyne if, for each nonempty open set $V \subset U$, one has $\mathrm{card}(V) = \mathrm{card}(U)$.

Lemma E.3 *For any topological space E, the family of all nonempty open isodyne subsets of E is a π-base of E.*

Proof Take an arbitrary nonempty open set U in E. We must show that there exists a nonempty isodyne open set V entirely contained in U. If U is isodyne, then there is nothing to prove. Otherwise, there exists a nonempty open set $V_0 \subset U$ such that

$$\mathrm{card}(V_0) < \mathrm{card}(U).$$

Let us proceed by induction. Suppose that a nonempty open set $V_k \subset U$ has already been defined for a natural number k. If V_k is isodyne, then the required result follows. If V_k is not isodyne, then there exists a nonempty open subset V_{k+1} of V_k such that

$$\mathrm{card}(V_{k+1}) < \mathrm{card}(V_k).$$

E Resolvable Topological Spaces

Since the cardinal numbers are well-ordered by the standard ordering, the process of defining the sets V_k should be stopped at some $n \in \mathbb{N}$. Obviously, for this n, the nonempty open set V_n is isodyne, which completes the proof. □

The next lemma is purely set-theoretical and, in fact, copies the classical Bernstein construction (cf. [12, 100, 107, 109, 141]).

Lemma E.4 *Let E be an infinite ground set and let $\{E_i : i \in I\}$ be a family of subsets of E satisfying the following conditions:*

(1) $\mathrm{card}(I) \leq \mathrm{card}(E)$;
(2) $(\forall i \in I)(\mathrm{card}(E_i) = \mathrm{card}(E))$.

Then there exist two disjoint sets $X \subset E$ and $Y \subset E$ such that

$$X \cap E_i \neq \emptyset, \qquad Y \cap E_i \neq \emptyset$$

for all indices $i \in I$.

Proof Without loss of generality, we may suppose that $I \sim E$, i.e., I and E are equinumerous sets. Moreover, we may assume that the set I is well-ordered by some ordering $\preceq$ and this $\preceq$ is such that any proper initial subinterval of I has cardinality strictly less than $\mathrm{card}(I)$.

Now, we will construct the required disjoint sets X and Y by using the method of transfinite induction (transfinite recursion).

Suppose that, for an index $j \in I$, the two partial injective families of elements $\{x_i : i \prec j\} \subset E$ and $\{y_i : i \prec j\} \subset E$ have already been determined so that

$$\{x_i : i \prec j\} \cap \{y_i : i \prec j\} = \emptyset.$$

Consider the set $Z_j = \{x_i : i \prec j\} \cup \{y_i : i \prec j\}$ and observe that

$$\mathrm{card}(Z_j) < \mathrm{card}(E) = \mathrm{card}(E_j), \qquad \mathrm{card}(E_j \setminus Z_j) = \mathrm{card}(E_j).$$

Choose two distinct elements x and y from $E_j \setminus Z_j$ and put

$$x_j = x, \qquad y_j = y.$$

Proceeding in this manner, we obtain the two injective families

$$\{x_i : i \in I\}, \qquad \{y_i : i \in I\}$$

of elements of E. Finally, we define

$$X = \{x_i : i \in I\}, \qquad Y = \{y_i : i \in I\}.$$

By virtue of our construction, we have $X \cap Y = \emptyset$ and

$$x_i \in X \cap E_i, \qquad y_i \in Y \cap E_i$$

for each index $i \in I$. So $X \cap E_i \neq \emptyset$ and $Y \cap E_i \neq \emptyset$. This yields the desired result. □

Lemma E.5 *Let E be an infinite isodyne topological space for which there exists at least one π-base whose cardinality does not exceed* $\mathrm{card}(E)$.

Then E is a resolvable topological space.

Proof Actually, this is a direct consequence of Lemma E.4. Indeed, let $\{U_i : i \in I\}$ be a π-base of E such that $\mathrm{card}(I) \leq \mathrm{card}(E)$. Since E is isodyne, we have

$$(\forall i \in I)(\mathrm{card}(U_i) = \mathrm{card}(E)).$$

Applying Lemma E.4 to E and to $\{U_i : i \in I\}$, we get two disjoint subsets X and Y of E such that $X \cap U_i \neq \emptyset$ and $Y \cap U_i \neq \emptyset$ for each $i \in I$. The latter means that both sets X and Y are everywhere dense in E. Therefore, E is a resolvable space and the proof is complete. □

Recall that a topological space E satisfies the first countability axiom if any point x of E possesses a countable fundamental system of neighborhoods of x (or, in other words, any $x \in E$ possesses a countable local base).

Lemma E.6 *Every infinite topological space E satisfying the first countability axiom has a base (hence a π-base) whose cardinality is equal to* $\mathrm{card}(E)$.

Proof Indeed, it suffices to consider the family

$$\mathcal{B} = \{U_n(x) : x \in E, \, n \in \mathbb{N}\},$$

where $\{U_n(x) : n \in \mathbb{N}\}$ denotes a countable local base of open neighborhoods of a point $x \in E$. Clearly, this $\mathcal{B}$ is a base of E and the cardinality of $\mathcal{B}$ is exactly $\mathrm{card}(E)$. □

Lemma E.7 *Let E be a topological space and let $\mathbf{a}$ be a cardinal number.*

If E has a π-base with cardinality not exceeding $\mathbf{a}$, then E has a π-base with cardinality not exceeding $\mathbf{a}$, all members of which are isodyne open sets.

Consequently, if E has a base with cardinality not exceeding $\mathbf{a}$, then E has a π-base with cardinality not exceeding $\mathbf{a}$, all members of which are isodyne open sets.

Proof Denote by $\mathcal{B} = \{U_i : i \in I\}$ some π-base of E such that $\mathrm{card}(I) \leq \mathbf{a}$. According to Lemma E.3, for each index $i \in I$, there exists a nonempty open isodyne set $V_i \subset U_i$. Now, it is clear that the family

$$\mathcal{B}' = \{V_i : i \in I\}$$

turns out to be a π-base of E whose cardinality is less than or equal to **a**. By definition of $\mathcal{B}'$, all members of $\mathcal{B}'$ are isodyne. □

Lemma E.8 *Let $(E, \mathcal{T})$ be an isodyne topological space such that*

$$\mathrm{card}(\mathcal{T}) > 2$$

and let $(E, \mathcal{T})$ satisfy the first countability axiom.
Then $(E, \mathcal{T})$ is a resolvable space.

Proof The assumptions that E is isodyne and $\mathrm{card}(\mathcal{T}) > 2$ readily imply that E is infinite and does not contain isolated points. Further, by virtue of Lemmas E.6 and E.7, E possesses a π-base with cardinality not exceeding $\mathrm{card}(E)$. Therefore, the desired result is an immediate consequence of Lemma E.5. □

Summarizing the above, we are able to prove the following.

Theorem E.1 *Suppose that a nonempty topological space E is such that:*

(1) *all singletons in E are closed;*
(2) *E satisfies the first countability axiom;*
(3) *E does not have isolated points.*

Then E is a resolvable space.
In particular, every nonempty metric space without isolated points is a resolvable space.

Proof It follows from (1) and (3) of this theorem that any nonempty open subset of E is infinite. According to Lemmas E.3 and E.8, there exists a π-base of E, all members of which are resolvable subspaces of E. So, applying Lemma E.2, we come to the required result. □

By using a similar argument, one can obtain analogous results for certain nonmetrizable topological spaces. For instance, let us formulate a result due to Hewitt [67].

Theorem E.2 *Every nonempty locally compact topological space without isolated points is resolvable.*

The proof of this theorem is sketched in some exercises below.

Exercises

1. On the real line $\mathbb{R}$ consider the family of all half-open intervals of the form

$$[a, b[\quad (a \in \mathbb{R},\ b \in \mathbb{R},\ a \le b).$$

Since the intersection of any two such intervals is also of the same form, this family can be considered as a base of the topology S, which is called the *Sorgenfrey topology* of $\mathbb{R}$ (see [40, 84]). Accordingly, the space $(\mathbb{R}, S)$ is called the *Sorgenfrey line*.

Verify that $(\mathbb{R}, S)$ possesses the following properties:

(a) S contains the standard order topology of $\mathbb{R}$;
(b) $(\mathbb{R}, S)$ is a separable resolvable space (e.g., $\mathbb{Q}$ is everywhere dense in $(\mathbb{R}, S)$);
(c) $(\mathbb{R}, S)$ has a countable π-base;
(d) $(\mathbb{R}, S)$ is a regular space;
(e) $(\mathbb{R}, S)$ is a hereditarily Lindelöf space;
(f) the cardinality of any base of $(\mathbb{R}, S)$ is equal to $\mathbf{c}$ (consequently, there exists a π-base of $(\mathbb{R}, S)$ which is not a base of $(\mathbb{R}, S)$).

Taking into account the above properties, conclude that $(\mathbb{R}, S)$ is a nonmetrizable perfectly normal space.

2. Let E be a topological space satisfying the following conditions:

   (i) E is separable (i.e., E contains a countable everywhere dense subset);
   (ii) there exists a closed discrete set D in E of cardinality $\mathbf{c}$.

Show that E cannot be a normal space.

For this purpose, suppose on the contrary that E is normal. Observe that, in view of (i), the cardinality of the family of all real-valued continuous functions on E does not exceed $\mathbf{c}$. On the other hand, every subset X of D is also closed and discrete in E, and the function $\chi_X | D$ is continuous on D. Let f_X be a continuous mapping from E to $[0, 1]$ extending $\chi_X | D$. Then the cardinality of the family $\{f_X : X \subset D\}$ is equal to $2^\mathbf{c}$, so one obtains a contradiction which yields that E is not normal.

Let now F denote the topological product $(\mathbb{R}, S) \times (\mathbb{R}, S)$, where $(\mathbb{R}, S)$ is the Sorgenfrey line (see Exercise 1). This F is usually called the *Sorgenfrey plane*. Check that:

(a) F is separable (e.g., the set $\mathbb{Q} \times \mathbb{Q}$ is everywhere dense in F);
(b) the set $D = \{(x, y) \in \mathbb{R}^2 : x + y = 0\}$ is a closed discrete subset of F and $\mathrm{card}(D) = \mathbf{c}$.

Conclude that F is not normal. Therefore, in general, the topological product of two perfectly normal spaces is not a normal space.

3. Recall that a Hausdorff topological space E is *locally compact* if, for every point $x \in E$, there exists a compact neighborhood $U(x)$ of x.

Verify that:

(a) any closed subset of a locally compact space is also locally compact;
(b) any open subset of a locally compact space is also locally compact.

E Resolvable Topological Spaces 435

4. Let E be a locally compact topological space and let $e \notin E$.
   Prove that there exists a topology $\mathcal{T}$ on the set $E^* = E \cup \{e\}$ satisfying the following conditions:

   (a) the space $(E^*, \mathcal{T})$ is compact;
   (b) the topology induced by $\mathcal{T}$ on E coincides with the original topology of E;
   (c) e is an isolated point in E^* if and only if E is compact.

   For this purpose, consider in E^* the family of all complements of compact subsets of E as the family of all open neighborhoods of e with respect to the required topology $\mathcal{T}$.

   Remark E.1 Recall that the space $(E^*, \mathcal{T})$ is Alexandrov's one-point compactification of E (cf. [19, 40, 84]).

5. Let E be an arbitrary infinite compact topological space.
   Demonstrate that there exists a base of E whose cardinality does not exceed card(E).
   For this purpose, argue as follows. Take any two distinct points x and y from E and choose their open neighborhoods $U(x)$ and $V(y)$ such that

   $$U(x) \cap V(y) = \emptyset.$$

   Then consider the family

   $$\mathcal{F} = \{(U(x), V(y)) : (x, y) \in E \times E, \ x \neq y\}.$$

   Clearly, the cardinality of this family does not exceed card(E).
   Verify that the family of all intersections of the form

   $$U_1(x) \cap U_2(x) \cap \cdots \cap U_n(x) \qquad (n \in \mathbb{N}, \ x \in E),$$

   where

   $$\{U_1(x), U_2(x), \ldots, U_n(x)\} \subset \mathrm{pr}_1(\mathcal{F}),$$

   is a base of E whose cardinality also does not exceed card(E).

6. Using the results of Exercises 4 and 5, show that if E is an infinite locally compact topological space, then there exists a base $\mathcal{B}$ of E with card($\mathcal{B}$) $\leq$ card(E).

7. Prove Hewitt's theorem stating that if E is a nonempty locally compact topological space without isolated points, then E is a resolvable space.
   For this purpose, use Lemma E.2 and the preceding exercises.

8. Let E be a ground set and let Φ be a filter of subsets of E. Define a topology $\mathcal{T}_\Phi$ on E by putting $\mathcal{T}_\Phi = \{\emptyset\} \cup \Phi$.
   This $\mathcal{T}_\Phi$ is usually called the topology canonically associated with the given filter Φ.

Check that:

(a) any nonempty open set in $(E, \mathcal{T}_\Phi)$ is everywhere dense in E;
(b) if Φ is an ultrafilter, then $(E, \mathcal{T}_\Phi)$ is an irresolvable topological space.

9. An ultrafilter Φ of subsets of an infinite ground set E is called *countably complete* if, for every family $\{X_n : n \in \mathbb{N}\}$ of subsets of E, the relation

$$E = \cup\{X_n : n \in \mathbb{N}\}$$

implies that there exists an $n \in \mathbb{N}$ such that $X_n \in \Phi$.
Observe that if, for some element $x \in E$, one has

$$\Phi = \{X \subset E : x \in X\},$$

then Φ is countably complete.
On the other hand, the existence of countably complete ultrafilters not containing singletons cannot be established within the framework of **ZFC** set theory, because such ultrafilters are closely connected with very large cardinal numbers (see, for instance, [76, 80, 108, 111]).
Let Φ be a countably complete ultrafilter on a ground set E, which contains no singletons in E, and let $\mathcal{T}_\Phi$ denote the topology on E canonically associated with this Φ.
Show that:

(a) there are no isolated points in the space $(E, \mathcal{T}_\Phi)$;
(b) if $f : E \to \mathbb{R}$ is an arbitrary function, then there exists a nonempty set $U \in \mathcal{T}_\Phi$ such that the restriction $f|U$ is constant (so $f|U$ trivially is continuous on U);
(c) for any function $f : E \to \mathbb{R}$, the closed set E is not of the form $D(f)$ (where $D(f)$ denotes the set of all discontinuity points of f).

In order to prove (b), use the countable completeness of Φ and define by induction a sequence $\{\Delta_n : n \in \mathbb{N}\}$ of segments in $\mathbb{R}$ such that

$$\Delta_{n+1} \subset \Delta_n \quad (n \in \mathbb{N}),$$

$$\lim_{n \to \infty} \lambda(\Delta_n) = 0,$$

$$f^{-1}(\Delta_n) \in \Phi \quad (n \in \mathbb{N}).$$

Then infer that, for the point $t \in \cap\{\Delta_n : n \in \mathbb{N}\}$, the relation $f^{-1}(t) \in \Phi$ holds true, and obtain the required result.

10. Check that if a topological space F has an everywhere dense resolvable subspace, then F is resolvable, too.
On the other hand, give an example of a resolvable topological space $(E, \mathcal{T})$ which contains an everywhere dense irresolvable subspace.

E Resolvable Topological Spaces

For this purpose, pick any two distinct elements x and y, take an infinite set A such that $A \cap \{x, y\} = \emptyset$ and equip A with a nontrivial ultrafilter Φ of subsets of A. Then put

$$E = \{x, y\} \cup A, \qquad \mathcal{T} = \{\emptyset\} \cup \{\{x, y\} \cup Z : Z \in \Phi\}.$$

Verify that the topological space $(E, \mathcal{T})$ is resolvable (more precisely, both singletons $\{x\}$ and $\{y\}$ are everywhere dense in E), but the subspace A of E is everywhere dense in E and irresolvable (in view of (b) of Exercise 8).

11. Let $\{E_i : i \in I\}$ be a family of nonempty topological spaces. Prove the following assertions:

    (a) if at least one of the spaces E_i ($i \in I$) is resolvable, then the topological product $\prod \{E_i : i \in I\}$ is resolvable;
    (b) if all spaces E_i ($i \in I$) are resolvable, then their topological sum $\sum \{E_i : i \in I\}$ is resolvable;
    (c) any bijective continuous image of a resolvable space is resolvable.

12. Let E be a Hausdorff topological space and let Z be an everywhere dense subset of E.
    Prove that the inequality $\mathrm{card}(E) \leq \mathrm{card}(\mathcal{P}(\mathcal{P}(Z)))$ holds true.
    For this purpose, consider a mapping $g : E \to \mathcal{P}(\mathcal{P}(Z))$ defined by the formula

    $$g(x) = \{U \cap Z : U \in \mathcal{U}(x)\} \qquad (x \in E),$$

    where $\mathcal{U}(x)$ denotes the filter of all neighborhoods of x.
    Check that g is an injection and obtain the required result.

13. Define by recursion $\mathbf{b}_0 = \mathrm{card}(\mathbb{N})$ and $\mathbf{b}_{n+1} = 2^{\mathbf{b}_n}$ for any natural number n. Then put $\mathbf{b} = \sum \{\mathbf{b}_n : n \in \mathbb{N}\}$.
    Show that:

    (i) $\mathbf{b}^\omega = 2^{\mathbf{b}} > \mathbf{b}$;
    (ii) if E is a set with $\mathrm{card}(E) = \mathbf{b}$, then it is impossible to equip E with a topology $\mathcal{T}$ so that $(E, \mathcal{T})$ would be a Hausdorff isodyne space of second Baire category.

    For (ii), use the result of Exercise 12.

Remark E.2 In connection with Exercise 13, it can be proved assuming **GCH** that if $\mathbf{a}$ is an infinite cardinal, then the following two assertions are equivalent:

(*) $\mathbf{a}$ is an ω-power, i.e., $\mathbf{a}^\omega = \mathbf{a}$;
(**) any set E with $\mathrm{card}(E) = \mathbf{a}$ admits a topology $\mathcal{T}$ such that $(E, \mathcal{T})$ is a Hausdorff isodyne space of second Baire category.

For a proof of this equivalence, see [88].

14. Let $(E, \leq)$ be a linearly ordered dense set containing at least two distinct elements.

Show that E, equipped with its order topology, is a resolvable topological space.

15. Denote by $\mathcal{D}$ the family of all those subsets X of $\mathbb{R}$ which satisfy the relation $\lambda_*(\mathbb{R} \setminus X) = 0$, where λ_* is the inner measure produced by the Lebesgue measure λ on $\mathbb{R}$.

Let $\mathcal{F}$ be a family of functions acting from $\mathbb{R}$ into $\mathbb{R}$.

A function $g : \mathbb{R} \to \mathbb{R}$ is called *universal* for $\mathcal{F}$ (with respect to the family $\mathcal{D}$) if, for any $f \in \mathcal{F}$, one has $\{x \in \mathbb{R} : g(x) = f(x)\} \in \mathcal{D}$.

Demonstrate that if $\mathcal{F}$ is the family of all real-valued λ-measurable functions on $\mathbb{R}$, then there exists a universal function g for $\mathcal{F}$ (with respect to $\mathcal{D}$) and check that such a g cannot be λ-measurable.

For this purpose, use the method of transfinite recursion and construct a partition of $\mathbb{R}$ into continuum many Bernstein subsets of $\mathbb{R}$ (cf. Exercise 13 from Chap. 4).

16. Show that there exists a continuous function $f : \mathbb{R} \to \mathbb{R}$ such that, for every Bernstein subset B of $\mathbb{R}$, one has $f(B) = \mathbb{R}$.

For this purpose, consider a Peano type continuous surjective function $g : \mathbb{R} \to \mathbb{R}^2$, take $f = \mathrm{pr}_2 \circ g$ and verify that this f is as required.

17. Check that there exists a σ-algebra $\mathcal{A}$ of subsets of $\mathbb{R}$ such that:

(a) $\mathrm{card}(\mathcal{A}) = 2^{\mathbf{c}}$;
(b) $\{x\} \in \mathcal{A}$ for each point $x \in \mathbb{R}$;
(c) for any Borel subset B of $\mathbb{R}$, if $B \in \mathcal{A}$, then either B is countable or B is co-countable.

For this purpose, use a partition of $\mathbb{R}$ into continuum many Bernstein subsets.

Bibliography

1. S.I. Adian, P.S. Novikov, On a semicontinuous function. Uchen. Zap. Moskov. Gos. Ped. Inst. **138**(3), 3–10 (1958) (in Russian)
2. T. Apostol, *Mathematical Analysis*, 2nd edn. (Addison–Wesley, New York, 1974)
3. A. Ascherl, J. Lehn, Two principles of extending probability measures. Manuscr. Math. **21**, 43–50 (1977)
4. R. Baire, *Leçons sur les Fonctions Discontinues* (Gauthier-Villars, Paris, 1905)
5. S. Banach, Sur l'équation fonctionnelle $f(x+y) = f(x) + f(y)$. Fund. Math. **1**, 123 (1920)
6. S. Banach, Sur le probleme de la mesure. Fund. Math. **4**, 7–33 (1923)
7. S. Banach, Über die Baire'sche Kategorie gewisser Funktionenmengen. Stud. Math. **3**, 174–179 (1931)
8. S. Banach, *Théorie des Opérations Linéaires*, 2nd edn. (Chelsea Publishing Company, New York, 1978)
9. S. Banach, C. Kuratowski, Sur une généralisation du probléme de la mesure. Fundam. Math. **14**, 127–131 (1929)
10. S. Banach, H. Steinhaus, Sur le principe de la condensation de singularités. Fundam. Math. **9**, 50–61 (1927)
11. J.L. Bell, D.H. Fremlin, A geometric form of the Axiom of Choice. Fundam. Math. **77**, 167–170 (1972)
12. F. Bernstein, Zur Theorie der trigonometrischen Reihen. Sitzungsber. Sächs. Akad. Wiss. Leipzig. Math.-Natur. **60**, 325–338 (1908)
13. C.E. Blair, The Baire category theorem implies the principle of dependent choices. Bull. Acad. Polon. Sci. Ser. Math. **25**, 933–934 (1977)
14. A. Blass, Existence of bases implies the Axiom of Choice. Contemp. Math. **31**, 31–33 (1984)
15. J. Blazek, E. Borák, J. Malý, On Köpcke and Pompeiu functions. Čas. pro pest. mat. **103**, 53–61 (1978)
16. H. Blumberg, New properties of all real functions. Trans. Amer. Math. Soc. **24**, 113–128 (1922)
17. V. Bogachev, *Measure Theory* (Springer, Berlin-Heidelberg, 2007)
18. N. Bourbaki, *Theory of Sets* (Hermann et Cie, Paris, 1968)
19. N. Bourbaki, *General Topology* (Springer, Berlin, 1989)
20. N. Bourbaki, *Integration*, vols. I, II (Springer, Berlin, 2004)
21. J.B. Brown, Restriction theorems in real analysis. Real Anal. Exchange **20**(2), 510–526 (1994–1995)
22. J.B. Brown, K. Prikry, Variations on Lusin's theorem. Trans. Amer. Math. Soc. **302**(1), 77–86 (1987)

23. A. Bruckner, *Differentiation of Real Functions* (Springer, Berlin, 1978)
24. A. Bruckner, J.L. Leonard, Derivatives. Amer. Math. Mon. **73**(4), 24–56 (1966)
25. J. Brzuchowski, J. Cichon, E. Grzegorek, C. Ryll-Nardzewski, On the existence of nonmeasurable unions. Bull. Acad. Pol. Sci. Ser. Sci. Math. **27**, 447–448 (1979)
26. L. Bukovský, *The Structure of the Real Line* (Birkhäuser, Basel, 2011)
27. V.V. Buldygin, A.B. Kharazishvili, *Geometric Aspects of Probability Theory and Mathematical Statistics* (Kluwer Academic Publishers, Dordrecht, 2000)
28. L. Carleson, On convergence and growth of partial sums of Fourier series. Acta Math. **116**(1), 135–157 (1966)
29. J. Cichon, M. Morayne, R. Ralowski, Cz. Ryll-Nardzewski, Sz. Zeberski, On nonmeasurable unions. Topol. Appl. **154**(4), 884–893 (2007)
30. K. Ciesielski, How good is Lebesgue measure? Math. Intell. **11**(2), 54–58 (1989)
31. K. Ciesielski, Set-theoretic real analysis. J. Appl. Anal. **3**(2), 143–190 (1997)
32. K.C. Ciesielski, J.B. Seoane-Sepúlveda, Differentiability versus continuity: restriction and extension theorems and monstrous examples. Bull. Amer. Math. Soc. **56**(2), 211–260 (2019)
33. C. Dellacherie, *Capacités et Processus Stochastiques* (Springer, Berlin, 1972)
34. A. Denjoy, Sur les fonctions dérivées sommables. Bull. Soc. Math. France **43**, 161–248 (1915)
35. J. Dieudonne, *The Foundations of Modern Analysis* (Academic Press, New York, 1960)
36. G. Drago, P.D. Lamberti, P. Toni, A "bouquet" of discontinuous functions for beginners in mathematical analysis. Amer. Math. Mon. **118**(9), 799–811 (2011)
37. N. Dunford, Integration and linear operations. Trans. Amer. Math. Soc. **40**, 474–494 (1936)
38. N. Dunford, J.T. Schwartz, *Linear Operators: General Theory*, Part 1 (Wiley-Interscience, New York, 1988)
39. D. Egoroff (D. Egorov), Sur les suites de fonctions mesurables. C. R. Acad. Sci. Paris *CLII*, 244–246 (1911)
40. R. Engelking, *General Topology* (PWN, Warszawa, 1985)
41. P. Erdös, R.D. Mauldin, The nonexistence of certain invariant measures. Proc. Amer. Math. Soc. **59**, 321–322 (1976)
42. M.P. Ershov, Extension of measures and stochastic equations. Probab. Theory Appl. **19**(3), 457–471 (1974) (in Russian)
43. M. Evans, On continuous functions and the approximate symmetric derivatives. Colloq. Math. **31**, 129–136 (1974)
44. G.M. Fichtenholz, *Course in Differential and Integral Calculus*, vols. I–III (Fizmatlit, Moscow, 2001–2003) (in Russian)
45. G. Folland, *Real Analysis: Modern Techniques and their Applications*, 2nd edn. (Wiley, New York, 1999)
46. D.H. Fremlin, *Consequences of Martin's Axiom* (Cambridge University Press, Cambridge, 1984)
47. D.H. Fremlin, Measure-additive coverings and measurable selectors, Diss. Math. **260**, 1–120 (1987)
48. D.H. Fremlin, *Measure Theory, vol. 4: Topological Measure Spaces* (University of Essex, Torres Fremlin, 2006)
49. H. Friedman, On definability of nonmeasurable sets. Canad. J. Math. **32**(3), 653–656 (1980)
50. L. Fuchs, *Infinite Abelian Groups*, vol. 1 (Academic Press, New York, 1970)
51. J.L. Gamez-Merino, G.A. Munoz-Fernandez, V.M. Sanchez, J.B. Seoane-Sepulveda, Sierpiński–Zygmund functions and other problems of lineability. Proc. Amer. Math. Soc. **138**(11), 3863–3876 (2010)
52. R.J. Gardner, W.F. Pfeffer, Borel measures, in *Handbook of Set-Theoretic Topology*, ed. by K. Kunen, J.E. Vaughan (North-Holland, Amsterdam, 1984), pp. 961–1043
53. B.R. Gelbaum, J.M.H. Olmsted, *Counterexamples in Analysis* (Holden-Day, San Francisco, 1964)
54. I.M. Gelfand, A.N. Kolmogorov, On rings of continuous functions on topological spaces. Dokl. Akad. Nauk SSSR **22**(1), 11–15 (1939) (in Russian)
55. L. Gillman, M. Jerison, *Rings of Continuous Functions* (Van Nostrand, Princeton, 1960)

Bibliography

56. B.V. Gnedenko, A.N. Kolmogorov, *Limit Distributions for Sums of Independent Random Variables* (Gosudarstv. Izdat. Tehn.-Teor. Lit., Moscow-Leningrad, 1949) (in Russian)
57. C. Goffman, Everywhere differentiable functions and the density topology. Proc. Amer. Math. Soc. **51**, 250–251 (1975)
58. C. Goffman, C.J. Neugebauer, T. Nishiura, Density topology and approximate continuity. Duke Math. J. **28**, 497–505 (1961)
59. T. Gowers (ed.), *The Princeton Companion to Mathematics* (Princeton University Press, Princeton, 2008)
60. G.L.M. Groenewegen, A.C.M. van Rooij, *Spaces of Continuous Functions* (Atlantis Press/Springer, Amsterdam/Paris, 2016)
61. V.I. Gurarii, Subspaces and bases in spaces of continuous functions. Dokl. Akad. Nauk SSSR **167**, 971–973 (1966) (in Russian)
62. V.I. Gurarii, Linear spaces composed of everywhere nondifferentiable functions. C. R. Acad. Bulgare Sci. **44**, 13–16 (1991) (in Russian)
63. P.R. Halmos, *Measure Theory* (D. Van Nostrand, New York, 1950)
64. P.R. Halmos, *Naive Set Theory* (Van Nostrand Reinhold, New York, 1960)
65. G. Hamel, Eine Basis aller Zahlen und die unstetigen Lösungen der Funktionalgleichung $f(x+y) = f(x) + f(y)$. Math. Ann. **60**, 459–462 (1905)
66. H. Herrlich, *Axiom of Choice* (Springer, Berlin, 2006)
67. E. Hewitt, A problem of set theoretic topology. Duke Math. J. **10**(2), 309–333 (1943)
68. E. Hewitt, Rings of real-valued continuous functions. Trans. Amer. Math. Soc. **64**, 45–99 (1948)
69. E. Hewitt, K. Ross, *Abstract Harmonic Analysis*, vol. 1 (Springer, Berlin, 1963)
70. E. Hewitt, K. Stromberg, *Real and Abstract Analysis* (Springer, New York, 1965)
71. E.W. Hobson, *Theory of Functions of a Real Variable*, vol. II (Dover, New York, 1957)
72. D.H. Hyers, S.M. Ulam, Approximately convex functions. Proc. Amer. Math. Soc. **3**, 821–828 (1952)
73. V. Jarník, Sur la dérivabilité des fonctions continues. Spisy Privodov Fak. Univ. Karlovy **129**, 3–9 (1934)
74. J. Jasiński, I. Reclaw, Restrictions to continuous and pointwise discontinuous functions. Real Anal. Exchange **23**(1), 161–174 (1997–1998)
75. T.J. Jech, *The Axiom of Choice* (North-Holland, Amsterdam, 1973)
76. T.J. Jech, *Set Theory* (Springer, Berlin, 2003)
77. R.B. Jensen, Independence of the axiom of dependent choices from the countable axiom of choice. J. Symbol. Logic **31**, 294 (1966)
78. S. Kaczmarz, H. Steinhaus, *Theorie der Orthogonalreihen*, Monogr. Matem. (Warszawa–Lwow, Warsaw, 1935)
79. S. Kakutani, Two fixed-point theorems concerning bicompact convex sets. Proc. Imp. Acad. Tokyo **14**, 242–245 (1938)
80. A. Kanamori, *The Higher Infinite* (Springer, Berlin-Heidelberg, 2003)
81. Y. Katznelson, K. Stromberg, Everywhere differentiable, nowhere monotone functions. Amer. Math. Mon. **81**, 349–354 (1974)
82. A.S. Kechris, *Classical Descriptive Set Theory* (Springer, New York, 1995)
83. J.L. Kelley, The Tychonoff product theorem implies the Axiom of Choice. Fundam. Math. **37**, 75–76 (1950)
84. J.L. Kelley, *General Topology* (Springer, New York, 1991)
85. A.B. Kharazishvili, On certain types of invariant measures. Dokl. Akad. Nauk SSSR **222**(3), 538–540 (1975) (in Russian)
86. A.B. Kharazishvili, The absolute nonmeasurability of the unit ball in an infinite-dimensional separable Hilbert space. Bull. Acad. Sci. GSSR **107**(1), 17–20 (1982) (in Russian)
87. A.B. Kharazishvili, *Invariant Extensions of the Lebesgue Measure* (Tbilisi University Press, Tbilisi, 1983) (in Russian)
88. A.B. Kharazishvili, Some properties of isodyne topological spaces. Bull. Acad. Sci. GSSR **127**(2), 261–264 (1987) (in Russian)

89. A.B. Kharazishvili, Some remarks on the property (N) of Luzin. Ann. Math. Silesianae **9**, 33–42 (1995)
90. A.B. Kharazishvili, Sup-measurable and weakly sup-measurable mappings in the theory of ordinary differential equations. J. Appl. Anal. **3**(2), 211–224 (1997)
91. A.B. Kharazishvili, *Applications of Point Set Theory in Real Analysis* (Kluwer Academic Publishers, Dordrecht, 1998)
92. A.B. Kharazishvili, *Nonmeasurable Sets and Functions* (Elsevier, Amsterdam, 2004)
93. A.B. Kharazishvili, On additive absolutely nonmeasurable Sierpiński–Zygmund functions. Real Anal. Exchange **31**(2), 553–560 (2005–2006)
94. A.B. Kharazishvili, On measurable Sierpiński–Zygmund functions. J. Appl. Anal. **12**(2), 283–292 (2006)
95. A.B. Kharazishvili, *Topics in Measure Theory and Real Analysis* (Atlantis Press/World Scientific, Amsterdam/Paris, 2009)
96. A.B. Kharazishvili, On a relationship between the measurability and continuity of real-valued functions. Georgian Math. J. **17**(4), 649–661 (2010)
97. A.B. Kharazishvili, Measurability properties of Vitali sets. Amer. Math. Mon. **118**(10), 693–703 (2011)
98. A.B. Kharazishvili, Some remarks concerning monotone and continuous restrictions of real-valued functions. Proc. A. Razmadze Math. Inst. **157**, 11–21 (2011)
99. A.B. Kharazishvili, *Set Theoretical Aspects of Real Analysis* (Chapman and Hall/CRC, Boca Raton, 2015)
100. A.B. Kharazishvili, *Strange Functions in Real Analysis* (3rd edn.) (Chapman and Hall/CRC, Boca Raton, 2017)
101. A.B. Kharazishvili, On the Steinhaus property and ergodicity via the measure-theoretic density of sets. Real Anal. Exchange **44**(1), 217–228 (2019)
102. A.B. Kharazishvili, *Notes on Real Analysis and Measure Theory* (Springer, Berlin-Heidelberg, 2022)
103. A.B. Kharazishvili, A.P. Kirtadze, On the measurability of functions with respect to certain classes of measures. Georgian Math. J. **11**(3), 489–494 (2004)
104. A.N. Kolmogorov, S.V. Fomin, *Introductory Real Analysis* (Dover Publications, New York, 1975)
105. P. Komjáth, V. Totik, *Problems and Theorems in Classical Set Theory* (Springer, New York, 2006)
106. N. Kryloff, N.O. Bogoliouboff, La theorie generale de la measure dans son application a l'etude des systemes dynamiques de la mecanique non lineaire. Ann. Math. **38**, 65–113 (1937)
107. M. Kuczma, *An Introduction to the Theory of Functional Equations and Inequalities: Cauchy's Equation and Jensen's Inequality* (PWN, Katowice, 1985)
108. K. Kunen, *Set Theory* (North-Holland, Amsterdam, 1980)
109. K. Kuratowski, *Topology*, vol. 1 (Academic Press, New York, 1966)
110. K. Kuratowski, On the concept of strongly transitive systems in topology. Ann. Mat. Pura Appl. **98**(4), 357–363 (1974)
111. K. Kuratowski, A. Mostowski, *Set Theory* (North-Holland, Amsterdam, 1967)
112. K. Kuratowski, Cz. Ryll-Nardzewski, A general theorem on selectors. Bull. Acad. Polon. Sci. Ser. Math. **13**(6), 397–402 (1965)
113. K. Kuratowski, S. Ulam, *Quelques propriétés topologiques du produit combinatoire*. Fundam. Math. **19**, 248–251 (1932)
114. A.G. Kurosh, *The Theory of Groups* (Izd. Nauka, Moscow, 1967) (in Russian)
115. Y. Kuznetsova, On continuity of measurable group representations and homomorphisms. Stud. Math. **210**, 197–208 (2012)
116. P.D. Lax, A.N. Milgram, Parabolic equations, in *Contributions to the Theory of Partial Differential Equations*, Annals of Mathematics Studies, vol. 33 (Princeton University Press, 1964), pp. 167–190
117. H. Lebesgue, Sur les fonctions représentables analytiquement. J. Math. Pures et Appl. Ser. 6 **1**(fasc. 2), 139–216 (1905)

118. N. Lusin (N. Luzin), Sur une probléme de M. Baire. C. R. Acad. Sci. Paris **158**, 1259 (1914)
119. N. Lusin (N. Luzin), *Leçons sur les Ensembles Analytiques et leurs Applications* (Gauthier-Villars, Paris, 1930)
120. N.N. Lusin (N.N. Luzin), *The Integral and Trigonometric Series* (Izd. GITTL, Moscow, 1956) (in Russian)
121. N.N. Luzin, *Collected Works*, vols. 1–3 (Izd. Acad. Nauk SSSR, Moscow, 1953–1959) (in Russian)
122. G.W. Mackey, Borel structures in groups and their duals. Trans. Amer. Math. Soc. **85**, 134–169 (1957)
123. J. Marcinkiewicz, Sur les nombres dérivés. Fundam. Math. **24**, 305–308 (1935)
124. A.A. Markov, Some theorems on Abelian sets. Doklady Akad. Nauk SSSR **1**, 299–302 (1936) (in Russian)
125. G. Matusik, On the lattice generated by Hamel functions. Real Anal. Exchange **36**(1), 65–78 (2010)
126. R.D. Mauldin, The existence of nonmeasurable sets. Amer. Math. Mon. **86**, 45–46 (1979)
127. R.D. Mauldin, The set of continuous nowhere differentiable functions. Pacific J. Math. **83**(1), 199–205 (1979); **121**(1), 119–120 (1986)
128. S. Mazurkiewicz, Sur les fonctions non-derivables. Stud. Math. **3**, 92–94 (1931)
129. S. Mazurkiewicz, Sur les suites de fonctions continues. Fundam. Math. **18**, 114–117 (1932)
130. E. Michael, Continuous selections, I. Ann. Math. **63**(2), 361–382 (1956)
131. E. Michael, Selected selection theorems. Amer. Math. Mon. **63**, 233–238 (1956)
132. A.W. Miller, Special subsets of the real line, in *Handbook of Set-Theoretic Topology*, ed. by K. Kunen, J.E. Vaughan (North-Holland, Amsterdam, 1984), pp. 201–233
133. J. Mycielski, On the paradox of the sphere. Fundam. Math. **42**, 348–355 (1955)
134. I.P. Natanson, *Theory of Functions of a Real Variable*, 3rd edn. (Izdat. Nauka, Moscow, 1974) (in Russian)
135. T. Natkaniec, *Almost Continuity* (WSP, Bydgoszcz, 1992)
136. T. Natkaniec, H. Rosen, An example of an additive almost continuous Sierpiński–Zygmund function. Real Anal. Exchange **30**, 261–266 (2004–2005)
137. H. Okamoto, M. Wunsch, A geometric construction of continuous strictly increasing singular functions. Proc. Japan Acad. Ser. A Math. Sci. **83**(7), 114–118 (2007)
138. W. Orlicz, Zur Theorie der Differentialgleichung $y' = f(x, y)$. Bull. Acad. Polon. Sci. Sér. A, 221–228 (1932)
139. W. Orlicz, *Collected Papers*, vols. 1, 2 (PWN, Warszawa, 1988)
140. A.J. Ostaszewski, Beyond Lebesgue and Baire III: Steinhaus' theorem and its descendants. Topol. Appl. **160**(10), 1144–1154 (2013)
141. J.C. Oxtoby, *Measure and Category* (Springer, Berlin, 1971)
142. J. Pawlikowski, The Hahn–Banach theorem implies the Banach–Tarski paradox. Fundam. Math. **138**, 21–22 (1991)
143. I.G. Petrovskii, *Lectures on the Theory of Ordinary Differential Equations* (GITTL, Moscow, 1949) (in Russian)
144. Z. Piotrowski, Separate and joint continuity. Real Anal. Exchange **11**, 293–322 (1985–1986)
145. Sh.S. Pkhakadze, On the iterated integrals. Proc. A. Razmadze Math. Inst. **20**, 167–209 (1954) (in Russian)
146. Sh.S. Pkhakadze, The Theory of Lebesgue measure. Proc. A. Razmadze Math. Inst. **25**, 3–272 (1958) (in Russian)
147. K. Plotka, Sum of Sierpiński–Zygmund and Darboux like functions. Topol. Appl. **122**(3), 547–564 (2002)
148. D. Repovs, P.V. Semenov, E. Michael's theory of continuous selections; development and applications. Uspekhi Mat. Nauk. **49**(6 (300)), 151–190 (1994) (in Russian)
149. M. Riesz, Sur le probleme des moments. III. Arkiv Mat. Astron. Fysik **17**(16), 1–52 (1923)
150. L. Rodriguez-Piazza, Every separable Banach space is isometric to a space of continuous nowhere differentiable functions. Proc. Amer. Math. Soc. **123**(12), 3649–3654 (1995)
151. K.A. Ross, *Elementary Analysis: The Theory of Calculus* (Springer, New York, 1980)

152. W. Rudin, *Principles of Mathematical Analysis* (McGraw-Hill, New York, 1976)
153. W. Rudin, *Real and Complex Analysis* (McGraw-Hill Book Company, New York, 1987)
154. S. Saks, *Theory of the Integral* (Warszawa-Lwów, Warsaw, 1937)
155. E. Schechter, *Handbook of Analysis and its Foundations* (Academic Press, New York, 1997)
156. W. Sierpiński, L'axiome de M. Zermelo et son role dans la Théorie des Ensembles et l'Analyse. Bull. Intern. Acad. Sci. Cracovie, Ser. A, 97–152 (1918)
157. W. Sierpiński, Sur un théoréme équivalent á l'hypothése du continu. Bull. Int. Acad. Sci. Cracovie Ser. A, 1–3 (1919)
158. W. Sierpiński, Sur la question de la mesurabilité de la base de M. Hamel. Fundam. Math. **1**, 105–111 (1920)
159. W. Sierpiński, Sur un probléme concernant les ensembles mesurables superficiellement. Fundam. Math. **1**, 112–115 (1920)
160. W. Sierpiński, Sur l'équation fonctionnelle $f(x+y) = f(x) + f(y)$. Fundam. Math. **1**, 116–122 (1920)
161. W. Sierpiński, Sur les fonctions convexes mesurables. Fundam. Math. **1**, 125–128 (1920)
162. W. Sierpiński, Les exemples effectifs et l'axiome du choix. Fundam. Math. **2**, 112–118 (1921)
163. W. Sierpiński, Sur un probleme concernant les fonctions semi-continues. Fundam. Math. **28**, 1–6 (1937)
164. W. Sierpiński, Fonctions additives non complètement additives et fonctions nonmesurables. Fundam. Math. **30**, 96–99 (1938)
165. W. Sierpiński, L'hypothèse généralisée du continu et l'axiome du choix. Fundam. Math. **34**, 1–5 (1947)
166. W. Sierpiński, *Hypothése du Continu*, 2nd edn. (Chelsea Publishing, New York, 1956)
167. W. Sierpiński, *Cardinal and Ordinal Numbers* (PWN, Warszawa, 1958)
168. W. Sierpiński, A. Zygmund, Sur une fonction qui est discontinue sur tout ensemble de puissance du continu. Fundam. Math. **4**, 316–318 (1923)
169. S. Solecki, On sets nonmeasurable with respect to invariant measures. Proc. Amer. Math. Soc. **119**(1), 115–124 (1993)
170. S. Solecki, Measurability properties of sets of Vitali's type. Proc. Amer. Math. Soc. **119**(3), 897–902 (1993)
171. H. Steinhaus, Sur les distances des points dans les ensembles de mesure positive. Fundam. Math. **1**, 93–104 (1920)
172. M. Stone, The generalized Weierstrass approximation theorem. Math. Mag. **21**(4), 167–184 (1948)
173. E. Szpilrajn (E. Marczewski), Sur l'extension de la mesure lebesguienne. Fundam. Math. **25**, 551–558 (1935)
174. E. Szpilrajn (E. Marczewski), The characteristic function of a sequence of sets and some of its applications. Fundam. Math. **31**, 207–223 (1938)
175. E. Szpilrajn (E. Marczewski), On problems of the theory of measure. Uspekhi Mat. Nauk **1**(2 (12)), 179–188 (1946) (in Russian)
176. F. Tall, The density topology. Pacific J. Math. **62**, 275–284 (1976)
177. T. Tao, *An Introduction to Measure Theory* (Amer. Math. Soc., Providence, 2011)
178. G.P. Tolstov, A note on D.F. Egorov's theorem. Dokl. Akad. Nauk SSSR **22**, 309–311 (1939) (in Russian)
179. G.P. Tolstov, On partial derivatives. Izv. Akad. Nauk SSSR Ser. Mat. **13**(5), 425–446 (1949) (in Russian)
180. S. Ulam, Zur Masstheorie in der allgemeinen Mengenlehre. Fundam. Math. **16**, 140–150 (1930)
181. G. Vitali, *Sul problema della misura dei gruppi di punti di una retta* (Tip. Gamberini e Parmeggiani, Bologna, 1905)
182. M.L. Wage, Products of Radon spaces. Uspekhi Mat. Nauk **35**(3), 151–153 (1980) (in Russian)
183. S. Wagon, *The Banach–Tarski Paradox* (Cambridge University Press, Cambridge, 1985)
184. C. Weil, On nowhere monotone functions. Proc. Amer. Math. Soc. **56**, 388–389 (1976)

185. H.E. White Jr., Topological spaces in which Blumberg's theorem holds. Proc. Amer. Math. Soc. **44**, 454–462 (1974)
186. E.T. Whittaker, G.N. Watson, *A Course of Modern Analysis*, vols. 1, 2 (Cambridge University Press, Cambridge, 1990)
187. Z. Zahorski, Sur la premiére dérivée. Trans. Amer. Math. Soc. **69**, 1–54 (1950)
188. P. Zakrzewski, Measures on algebraic-topological structures, in: *Handbook of Measure Theory*, ed. by E. Pap (North-Holland, Amsterdam, 2002), pp. 1091–1130
189. P. Zakrzewski, On a construction of universally small sets. Real Anal. Exchange **28**(1), 215–221 (2002–2003)
190. E. Zermelo, Beweis dass jede Menge wohlgeordnet werden kann. Math. Ann. **59**, 514–516 (1904)
191. A. Zygmund, *Trigonometric Series*, vol. 1 (Cambridge University Press, Cambridge, 2002)

Index

A
Absolutely continuous signed measure, 273
Absolutely measurable function, 171, 249
Absolutely measurable set, 171, 249
Absolutely nonmeasurable function, 171, 358
Absolutely nonmeasurable set, 171
Absolute null topological space, 357
Accumulation point of a set, 101
Adherent point of a set, 92
Admissible algebra of functions, 335
Affine function, 15, 78
Affine linear manifold, 309
Alexandrov's theorem, 199
Algebraic basis of a vector space, 184
Algebra of continuous functions, 333
Algebra of sets, 133
Almost G-invariant function, 233
Almost G-invariant set, 211
Almost transitive action, 323
Analytic set, 364
Analytic space, 250
Antisymmetric binary relation, 8
A-operation, 364
Atom of a measure, 257, 262
Axiom of Archimedes, 37
Axiom of Choice (AC), 1
Axiom of Dependent Choice (DC), 109
Axiom of separation, 1

B
Baire property, 321, 322
Baire property of a function, 107
Baire theorem on category, 109
Banach–Kuratowski–Pettis theorem, 324
Banach–Mazur theorem, 343
Banach–Steinhaus theorem, 105, 107
Base set, 1
Bernstein set, 259
Bijection, 12
Bilinear form, 417
Binary operation, 12
Binary relation, 3
Boolean, 2
Borel–Cantelli lemma, 140
Borel mapping, 322
Borel measurable function, 107

C
Cantor–Bernstein theorem, 26
Cantor's Axiom, 37
Cantor set, 45
Cantor's inequality, 28
Capacity, 366
Carathéodory measure induced by an outer measure, 166
Carathéodory's theorem, 142
Cardinality, 23
Cardinality of the continuum (c), 31
Cardinal measurable in Ulam's sense, 195
Cardinal number, 23
Cardinal-valued invariant, 123
Cartesian product of sets, 3
Characteristic function, 13
Charge, 271
Choquet's theorem, 366
Co-meager set, 105

448   Index

Commutative divisible group, 245
Commutative ring, 18
Commutative topological group, 105
Compact space, 47
Compact support of a function, 289
Completely regular space, 122
Complete measure, 169
Complete metric space, 17
Completion of a measure, 169
Composition of binary relations, 4
Concave function, 77
Condensation point of a set, 115
Conditional completeness, 90
Cone in a real vector space, 89
Continuous measure, 357
Continuum Hypothesis (CH), 190
Convergent family of sets, 139
Convergent in measure sequence of functions, 386
Convergent sequence of sets, 132
Converse binary relation, 3
Convex cone in a real vector space, 90
Convex function, 77
Convex set in a real vector space, 90
Countable (anti)chain condition, 197
Countable set, 30
Countable subadditivity of a function, 53
Countably additive ideal of sets, 134
Countably additive measure, 142
Countably additive measure on a ring of sets, 150
Countably additive measure on a semiring of sets, 151
Countably complete filter of sets, 134
Countably complete ultrafilter, 436
Countably quasi-compact space, 101

D
Darboux property, 229
Darboux property of a function, 74
Decreasing function, 65
Dedekind complete linearly ordered field, 37
Dedekind completeness of $\mathbb{R}$, 37
δ-filter of sets, 134
Diagonal, 3
Dieudonné measure, 205
Diffuse measure, 357
Dini's theorem, 99
Direct sum of measures, 402
Dirichlet function, 16
Domain of a binary relation, 3
Dominated convergence theorem, 382
Dual space, 215

E
Egorov's theorem, 376
Elliptic bilinear functional, 419
Empty binary relation, 3
Empty set, 2
Empty unary relation, 2
Equinumerous sets, 23
Equivalence classes, 8
Equivalence of functions with respect to a measure, 274
Equivalence relation, 8
Equivalent measures, 210
Ergodic measure, 213
Extension of a partial function, 11

F
Factor set, 8
Filter in a partially ordered set, 196
Filter of sets, 134
Finitely additive measure, 141
    on a ring of sets, 150
    on a semiring of sets, 151
Finite set, 24
Finite set in the standard sense, 36
Finite subadditivity of a function, 53
First Baire category, 105
First Baire class, 108
First category, 105
First projection of a binary relation, 3
Fixed point of a mapping, 19
Floor function, 100
Fractional part, 100
Freely acting transformation group, 209
Function, 11
Functional relation, 11

G
G-absolutely nonmeasurable set, 235
G-congruence of sets, 209
Gelfand–Kolmogorov theorem, 351
Generalized Cantor space, 266
Generalized Continuum Hypothesis (GCH), 191
Generalized Luzin set, 363
Generalized sequence (net), 350
G-invariant algebra of sets, 172
G-invariant measure, 172
G-negligible set, 236
G-orbit, 209
Graph, 14
Ground set, 1
G-thick set, 236

Index

H
Hahn–Banach theorem, 105, 178
Hahn–Jordan decomposition theorem, 271
Hahn–Jordan representation of a signed measure, 273
Hamel basis of $\mathbb{R}$, 227
Henry's theorem, 208
Hereditarily Lindelöf topological space, 117
Hewitt–Savage theorem, 403
Hewitt's theorem, 433
Hilbert dimension, 315
Hilbert's logical operator, 23
Hilbert's theorem, 354
Horizontal section, 6

I
Ideal, 347
Ideal of sets, 133
Idempotent element of a ring, 346
Identity transformation, 12
Image of a set with respect to a binary relation, 4
Incompatible elements of a partially ordered set, 197
Inconsistent elements of a partially ordered set, 197
Increasing function, 65
Independent family of sets, 227
Indicator, 13
Infinite set, 26
Infinite set in the standard sense, 36
Injection, 12
Inner measure produced by a finite countably additive measure, 158
Inner regular measure, 201
Integrable function, 377
Invariant measure, 210
Involution, 13
Irresolvable topological space, 60, 327
Isodyne topological space, 118, 430

J
Jensen's inequality, 78

K
König's inequality, 40
Kuratowski–Zorn lemma, 9

L
Lattice of functions, 90
Lax–Milgram theorem, 419

Lebesgue p-space associated with μ, 386
Left derivative, 80
Limit of a family of sets, 139
Limit of a sequence of sets, 132
Lindelöf topological space, 117
Linear basis of a vector space, 184
Linear functional with finite support, 287
Linear ordering, 9
Local compactness of $\mathbb{R}$, 37
Locally bounded function, 47
Locally compact Hausdorff space, 434
Lomnicki–Ulam theorem, 399
Lower convex function, 77
Lower Darboux sum, 406
Lower limit of a family of sets, 139
Lower limit of a sequence of sets, 131
Lower Riemann integral, 406
Lower semicontinuous function, 95
 at a point, 87
Luzin set, 191
Luzin space, 194
Luzin's theorem, 192

M
Marczewski's theorem, 142
Martin's Axiom (MA), 197
Meager set, 105
Measurable envelope of a set, 157
Measurable function, 375
Measurable function in Luzin's sense, 254
Measurable function with respect to a countably additive measure, 171
Measurable hull of a set, 157
Measurable in the Borel sense, 322
Measurable in the sense of Carathéodory, 163
Measurable kernel of a set, 158
Measurable partial function, 375
Measurable set, 153
Metrically transitive measure, 213
Metric space, 17
Mid-point convex function, 77
Modular function, 146
Monotone function, 65
μ-approximable, 206
 from above, 206, 295
 from below, 206, 295
 set, 295
μ-equivalent partial functions, 376
μ-integrable function, 380, 381, 384
μ-integral, 381, 384
μ^*-measure, 134
Multi-function, 14

N

n-ary relation, 10
Natural number in von Neumann's sense, 25
Noether ring, 353
Non-atomic measure, 257, 262
Non-meager set, 105
Nontrivial idempotent, 346
Normal topological space, 128

O

Open mapping, 427
Ordered field of rational numbers ($\mathbb{Q}$), 37
Ordered pair, 2
Ordinal in von Neumann's sense, 23, 35
Ordinal type, 33
Orthonormal basis, 267
Orthonormal Hilbert basis, 315
Oscillation function, 54
Oscillation of a function at a point, 49
Outer measure, 134
    associated with a finitely additive measure, 142
Outer regular measure, 201

P

Parallelogram formula, 307, 313
Parseval's equality, 315
Partial function, 11
Partially ordered vector space, 176
Partially pre-ordered vector space, 175
Partial ordering, 8
Partial quasi-ordering, 8
Peano curve, 120
Perfectly normal space, 128
Perfect probability space, 261
Periodic function, 223
π-base of a topological space, 328, 429
Point of continuity of a function, 57
Point of discontinuity of a function, 57
Polish topological vector space, 184
Positive linear functional, 281, 287
Power set, 2
Pre-Hilbert space, 307
Pre-image of a set with respect to a binary relation, 4
Prime ideal, 352
Principle of condensation of singularities, 105, 107
Principle of inclusion and exclusion, 147
Probability measure, 211
Probability space, 261
Product measure, 256, 392, 401
Product of a family of sets, 18
Product σ-algebra, 392
Product topology, 121
Projection, 18, 309
Proper subset, 2
Pseudo-base of a topological space, 328, 429
Pseudo-metric, 135

Q

Quasi-compact space, 47
Quasi-invariant measure, 210
Quasi-metric, 135
Quotient set, 8

R

Radon charge, 300
Radon measure, 199, 250
    associated with a positive linear functional, 297
Radon–Nikodym derivative, 277
Radon–Nikodym theorem, 275
Radon signed measure, 300
Radon space, 199, 250
Random variable, 261
Range of a binary relation, 4
Rasiowa–Sikorski lemma, 196
Real Hilbert space, 310
Real line ($\mathbb{R}$), 2
Real-valued measurable cardinal, 195
Real-valued measurable function, 171
Reflexive binary relation, 3
Regular measure, 201
Regular outer measure, 136
Regular topological space, 128
Relatively measurable function, 171, 358
Relatively measurable set, 171
Residual set, 105
Resolvable topological space, 60, 327
Restriction of a partial function, 11
Reverse binary relation, 3
Riemann integrable, 406
Riemann integral, 297
Riesz extension theorem, 176
Riesz representation theorem, 277
Right derivative, 80
Ring of sets, 149

S

Second Baire category, 105
Second category, 105
Second projection of a binary relation, 4

Index 451

Section, 6
    horizontal, 6
    vertical, 6
Selector of a multi-function, 14
Selectors, 18
Semicontinuous function at a point, 87
Seminorm, 179
Semiring of sets, 150
Separability of $\mathbb{R}$, 37
Separable metric space, 17
Separates points, 333
Separation theorem, 126
Set of all natural numbers ($\mathbb{N}$), 26
Set-valued function, 14
Sierpiński–Zygmund type function, 44
σ-algebra of sets, 133
σ-finite G-quasi-invariant measure, 211
σ-finite measure, 170
σ-ideal of sets, 134
Signed measure, 271
Simple discontinuity point of a function, 73
Singleton, 2
Solvable group, 244
Sorgenfrey line, 434
Sorgenfrey plane, 434
Sorgenfrey topology, 268, 434
Space equipped with a transformation group, 209
Step-function, 377
Stone–Weierstrass theorem, 337
Strictly concave function, 78
Strictly convex function, 78
Strictly decreasing function, 65
Strictly increasing function, 65
Strictly monotone function, 65
Strong Darboux property, 229
Strongly inaccessible cardinal, 195
Strongly paradoxical group, 244
Subadditive mapping, 106
Sublinear functional, 178
Successor ordinal, 35
Super-strong Darboux property, 229
Surjection, 12
Suslin's condition for a measure, 212
Suslin space, 250
Symmetric binary relation, 7
Symmetric closure of a binary relation, 7
Symmetric difference of sets, 18
Symmetric group, 13

T
Tarski's theorem, 142
τ-additivity, 252

τ-smoothness, 252
Ternary relation, 10
Topological weight of a space, 122
Topology canonically associated with a filter, 435
Total binary relation, 3
Totally imperfect set, 259
Totally incompatible subset of a partially ordered set, 197
Totally inconsistent subset of a partially ordered set, 197
Totally irresolvable space, 327
Total unary relation, 2
Transitive binary relation, 7
Transitive closure of a binary relation, 8
Transitive group, 209
Transitively acting transformation group, 209
Transitive set, 34
Trigonometric Fourier series, 111
Trivial idempotents, 346
Two-valued measurable cardinal, 265

U
Ulam's transfinite matrix, 189
Ultrafilter, 181
Ultraideal, 181
Unary operation, 12
Unary relation, 1
Uncountable set, 30
Uniformization of a multi-function, 14
Uniform relation, 11
Uniqueness property, 170
Universal binary relation, 6
Universal function, 44, 438
Universally measurable function, 171, 249
Universally measurable set, 171, 249
Universal measure zero, 193, 357
Upper convex function, 77
Upper Darboux sum, 406
Upper limit of a family of sets, 139
Upper limit of a sequence of sets, 131
Upper Riemann integral, 406
Upper semicontinuity of a measure, 142
Upper semicontinuous function, 95
    at a point, 87

V
Variation of a measure, 273
Vertical section, 6
Vitali's equivalence relation, 8
Vitali set, 244
Vitali's partition of $\mathbb{R}$, 8

W
Weak form of AC, 36
Weakly inaccessible cardinal, 195
Well-ordered set, 25
Well-ordering, 28

Z
Zero-one law, 403
ZFC set theory, 1
ZF set theory, 1

The manufacturer's authorised representative in the EU is Springer Nature Customer Service Centre GmbH, Europaplatz 3, 69115 Heidelberg, Germany. If you have any concerns regarding our products, please contact ProductSafety@springernature.com

Printed and bound by CPI Group (UK) Ltd, Croydon, CR0 4YY

26/03/2026

02078916-0009